VOLUME ONE HUNDRED AND FOURTEEN

ADVANCES IN
PROTEIN CHEMISTRY AND STRUCTURAL BIOLOGY

Molecular Chaperones in Human Disorders

VOLUME ONE HUNDRED AND FOURTEEN

ADVANCES IN PROTEIN CHEMISTRY AND STRUCTURAL BIOLOGY

Molecular Chaperones in Human Disorders

Edited by

ROSSEN DONEV

MicroPharm Ltd,
United Kingdom

ELSEVIER

ACADEMIC PRESS

An imprint of Elsevier

Academic Press is an imprint of Elsevier
50 Hampshire Street, 5th Floor, Cambridge, MA 02139, United States
525 B Street, Suite 1650, San Diego, CA 92101, United States
The Boulevard, Langford Lane, Kidlington, Oxford OX5 1GB, United Kingdom
125 London Wall, London, EC2Y 5AS, United Kingdom

First edition 2019

ISBN: 978-0-12-815557-8
ISSN: 1876-1623

For information on all Academic Press publications
visit our website at https://www.elsevier.com/books-and-journals

Publisher: Zoe Kruze
Acquisition Editor: Ashlie M. Jackman
Editorial Project Manager: Teresa Pons-Ferrer
Production Project Manager: James Selvam
Cover Designer: Vicky Pearson Esser

Typeset by SPi Global, India

Contents

Contributors

Amanda B. Abildgaard
The Linderstrøm-Lang Centre for Protein Science, Department of Biology, University of Copenhagen, Copenhagen, Denmark

Fahmeda Ahmed
Department of Chemistry, King's College London, London, United Kingdom

Nikolai O. Artemyev
Department of Molecular Physiology and Biophysics; Department of Ophthalmology and Visual Sciences, The University of Iowa Carver College of Medicine, Iowa City, IA, United States

Jennifer M. Austin
Department of Chemistry, King's College London, London, United Kingdom

Rashi Benarroch
Department of Chemistry, King's College London, London, United Kingdom

Isabel Betancor-Fernández
Centre for Biomedical Research on Rare Diseases (CIBERER), Hospital Universitario de Canarias, ITB, University of La Laguna, Tenerife, Spain

Celine Brooks
Department of Ophthalmology, West Virginia University, Morgantown, WV, United States

Johannes Buchner
Center for Integrated Protein Science Munich CIPSM at the Department Chemie, Technische Universität München, Garching, Germany

Chiranjib Chakraborty
Department of Biotechnology, Adamas University, Kolkata, West Bengal, India

J. Priyadharshini Christy
Department of Integrative Biology, School of Bio Sciences and Technology, VIT, Vellore, Tamil Nadu, India

Lene Clausen
The Linderstrøm-Lang Centre for Protein Science, Department of Biology, University of Copenhagen, Copenhagen, Denmark

Vinay Dahiya
Center for Integrated Protein Science Munich CIPSM at the Department Chemie, Technische Universität München, Garching, Germany

Elif Esvap
Department of Chemical Engineering, Boğaziçi University, Istanbul, Turkey

José Ángel Fernández-Higuero
Biofisika Institute (CSIC, UPV/EHU) and Department of Biochemistry and Molecular Biology, Faculty of Science and Technology, University of the Basque Country (UPV/EHU), Bilbao, Spain

C. George Priya Doss
Department of Integrative Biology, School of Bio Sciences and Technology, VIT, Vellore, Tamil Nadu, India

Sarah K. Gersing
The Linderstrøm-Lang Centre for Protein Science, Department of Biology, University of Copenhagen, Copenhagen, Denmark

Rasmus Hartmann-Petersen
The Linderstrøm-Lang Centre for Protein Science, Department of Biology, University of Copenhagen, Copenhagen, Denmark

Rivka L. Isaacson
Department of Chemistry, King's College London, London, United Kingdom

Sharada Iyer
Department of Integrative Biology, School of Bio Sciences and Technology, VIT, Vellore, Tamil Nadu, India

E. Judith
Department of Integrative Biology, School of Bio Sciences and Technology, VIT, Vellore, Tamil Nadu, India

Kresten Lindorff-Larsen
The Linderstrøm-Lang Centre for Protein Science, Department of Biology, University of Copenhagen, Copenhagen, Denmark

Noel Mesa-Torres
Department of Physical Chemistry, University of Granada, Granada, Spain

Ravit Mesika
Department of Biological Chemistry, The Alexander Silberman Institute of Life Sciences, The Hebrew University of Jerusalem, Jerusalem, Israel

Arturo Muga
Biofisika Institute (CSIC, UPV/EHU) and Department of Biochemistry and Molecular Biology, Faculty of Science and Technology, University of the Basque Country (UPV/EHU), Bilbao, Spain

Elif Ozkirimli Olmez
Department of Chemical Engineering, Boğaziçi University, Istanbul, Turkey

Hatice Asuman Ozkara
Department of Medical Biochemistry, Hacettepe University, Faculty of Medicine, Ankara, Turkey

Angel L. Pey
Department of Physical Chemistry, University of Granada, Granada, Spain

Dana Reichmann
Department of Biological Chemistry, The Alexander Silberman Institute of Life Sciences, The Hebrew University of Jerusalem, Jerusalem, Israel

Eduardo Salido
Centre for Biomedical Research on Rare Diseases (CIBERER), Hospital Universitario de Canarias, ITB, University of La Laguna, Tenerife, Spain

R. Siva
Department of Integrative Biology, School of Bio Sciences and Technology, VIT, Vellore, Tamil Nadu, India

Maxim Sokolov
Department of Ophthalmology, West Virginia University, Morgantown, WV, United States

Amelie Stein
The Linderstrøm-Lang Centre for Protein Science, Department of Biology, University of Copenhagen, Copenhagen, Denmark

Iftikhar Aslam Tayubi
Faculty of Computing and Information Technology, King Abdulaziz University, Rabigh, Saudi Arabia

D. Thirumal Kumar
Department of Integrative Biology, School of Bio Sciences and Technology, VIT, Vellore, Tamil Nadu, India

Kutlu O. Ulgen
Department of Chemical Engineering, Boğaziçi University, Istanbul, Turkey

Ravi P. Yadav
Department of Molecular Physiology and Biophysics, The University of Iowa Carver College of Medicine, Iowa City, IA, United States

Z. Begum Yagci
Department of Chemical Engineering, Boğaziçi University, Istanbul, Turkey

Hatem Zayed
Department of Biomedical Sciences, College of Health and Sciences, Qatar University, Doha, Qatar

Preface

Proteins consist of one or more polypeptide chains that need to fold into a specific three-dimensional shape in order to be able to perform their biological function. However, not all proteins can spontaneously fold correctly under in vitro and/or in vivo conditions. It is known that although thermodynamics determines protein folding, kinetics and unspecific intermolecular interactions need also to be considered. Correct folding in vivo of a number of proteins requires assistance from other proteins, named molecular chaperones, that play major role in dedicated machinery that prevents aggregation and assists the process of protein folding in the cell in an ATP-dependent manner. The term "molecular chaperone" defines a set of structurally unrelated proteins which main function is to promote efficient folding of newly synthesized polypeptide chains and association of assembled subunits, prevent their aggregation, and ensure protein homeostasis in cells.

This thematic volume of the *Advances in Protein Chemistry and Structural Biology* is focused on the role of molecular chaperones in a number of conformational or protein misfolding human diseases such as neurodegenerative diseases, certain types of hereditary cancer, retinal disorders, metabolic disorders, amyloidosis, and type II diabetes. The aim of this volume is to promote further basic and translational research in this very exciting and central field in understanding mechanisms underlying a number of human diseases and identifying new strategies for their efficient treatment. Therefore, this volume will be particularly useful to researchers working on molecular mechanisms regulating protein folding, assembly of macromolecular structures, and design of therapeutics for treatment of different disorders. This volume is also intended as a scholarly material for use by students interested and/or working on projects in the field.

Dr. Rossen Donev
MicroPharm Ltd
United Kingdom

CHAPTER ONE

Functional principles and regulation of molecular chaperones

Vinay Dahiya, Johannes Buchner*
Center for Integrated Protein Science Munich CIPSM at the Department Chemie, Technische Universität München, Garching, Germany
*Corresponding author: e-mail address: johannes.buchner@tum.de

Contents

Abstract

To be able to perform their biological function, a protein needs to be correctly folded into its three dimensional structure. The protein folding process is spontaneous and does not require the input of energy. However, in the crowded cellular environment where there is high risk of inter-molecular interactions that may lead to protein molecules sticking to each other, hence forming aggregates, protein folding is assisted. Cells have evolved robust machinery called molecular chaperones to deal with the protein folding problem and to maintain proteins in their functional state. Molecular chaperones promote efficient folding of newly synthesized proteins, prevent their aggregation and ensure protein homeostasis in cells. There are different classes of molecular chaperones functioning in a complex interplay. In this review, we discuss the principal characteristics of different classes of molecular chaperones, their structure-function relationships, their mode of regulation and their involvement in human disorders.

Advances in Protein Chemistry and Structural Biology, Volume 114
ISSN 1876-1623
https://doi.org/10.1016/bs.apcsb.2018.10.001

1. Protein folding

Proteins are polymers of amino acids and the final step of information transfer for the genome. A polypeptide chain needs to adopt a unique three dimensional structure in order to be functional in a process known as protein folding. In late 1950s, Anfinsen and coworkers showed that all the information required to properly fold polypeptide chains is contained in the amino acid sequence of a protein (Anfinsen, 1973; Anfinsen, Haber, Sela, & White, 1961). However, it became clear from in vitro refolding experiments that not all proteins can spontaneously refold in a test tube, but are subjected to unspecific aggregation of polypeptide chains (Zettlmeissl, Rudolph, & Jaenicke, 1979). Also, under cellular conditions upon overexpression of mammalian proteins in *E. coli*, insoluble cytosolic aggregates termed inclusion bodies were observed (Marston, 1986). Though Anfinsen's principle still holds true, it is now well known that although thermodynamics determines protein folding, kinetics and unspecific inter-molecular interactions need to be considered. Therefore, assistance from "accessory proteins" popularly called "molecular chaperones" is required in vivo. The view that Anfinsen put forward has changed considerably over the last decades with the discovery of the dedicated machinery that prevents aggregation and assists the process of protein folding in the cell in an ATP-dependent manner.

1.1 In vitro versus in vivo protein folding

The linear chain of amino acid select the most stable conformation spontaneously and usually quite rapidly from a myriad of possibilities. Decades of work has established an excessive body of knowledge of how proteins fold; however, the prediction of the three dimensional fold from a linear amino acid sequence still remains a major challenge. Protein folding in the cell is still inaccessible to a detailed analysis of the folding pathways and kinetic mechanisms. In vitro refolding studies starting with unfolding the protein in strong denaturants like urea or guanidinium chloride (GdnCl) followed by its dilution in refolding buffer allowed deducing folding pathways and kinetics. In pioneering experiments, Anfinsen (1973) and Anfinsen et al. (1961) used native ribonuclease (RNase). The enzyme was denatured in urea in the presence of a reductant and upon removal of the denaturant and reducing agent by dialysis, RNase could be recovered with native enzymatic activity. Hence, given the conditions are favorable, proteins can fold spontaneously from their unfolded state without the assistance of additional factors in an exothermic reaction. This concept has been confirmed in many subsequent studies ranging from small, single domain proteins to oligomeric proteins (Jaenicke, 1987). These studies revealed that the process of protein folding is not a simple two step N ↔ U process but is characterized by the presence of various partially folded intermediates (Arai & Kuwajima, 1996; Batey, Scott, & Clarke, 2006; Jamin & Baldwin, 1996; Malhotra & Udgaonkar, 2014). Such intermediates make the folding energy landscape rugged by acting as kinetic traps (Fig. 1) which slow down the overall rate of the folding reaction. For some proteins it was found that they are not amenable to in vitro refolding especially large or oligomeric proteins. Their folding results in the formation of kinetically trapped, slow folding, hydrophobic intermediates that tend to self-associate and aggregate (Dahiya & Chaudhuri, 2013). It is possible to manipulate and control the in vitro refolding conditions to create an ideal situation for a protein to fold. However, in contrast, every protein has to fold under identical conditions in vivo. As aggregation is a higher order reaction (Walter & Buchner, 2002), it becomes more pronounced under cellular conditions where the environment is extremely crowded due to high concentrations of proteins and other macromolecules present at 300–400 g/L (Zimmerman & Trach, 1991). The problem is further potentiated by the presence of partially folded or misfolded states which are at high risk of aggregation primarily because hydrophobic residues are still exposed in them favoring inter-molecular interactions that often result in the formation of amorphous aggregates (Fig. 1). These may give rise to fibrillar aggregates called amyloids (Fig. 1) which are characterized by cross β structures and are the cause of several neurodegenerative diseases (Chiti & Dobson, 2017). Therefore, the protein folding process in cells requires molecular chaperones which can bind to the exposed hydrophobic regions of the incompletely folded polypeptide chains and hence prevent their aggregation (Fig. 1).

2. The concept of molecular chaperones

The term "molecular chaperone" is used to define a functional set of structurally unrelated proteins that are required for the correct folding of newly synthesized polypeptide chains and unstable protein conformers. They act as "accessory proteins" and do not themselves form part of the structure of the proteins they assist in folding and neither do they provide conformational information to the process of protein folding. Instead they

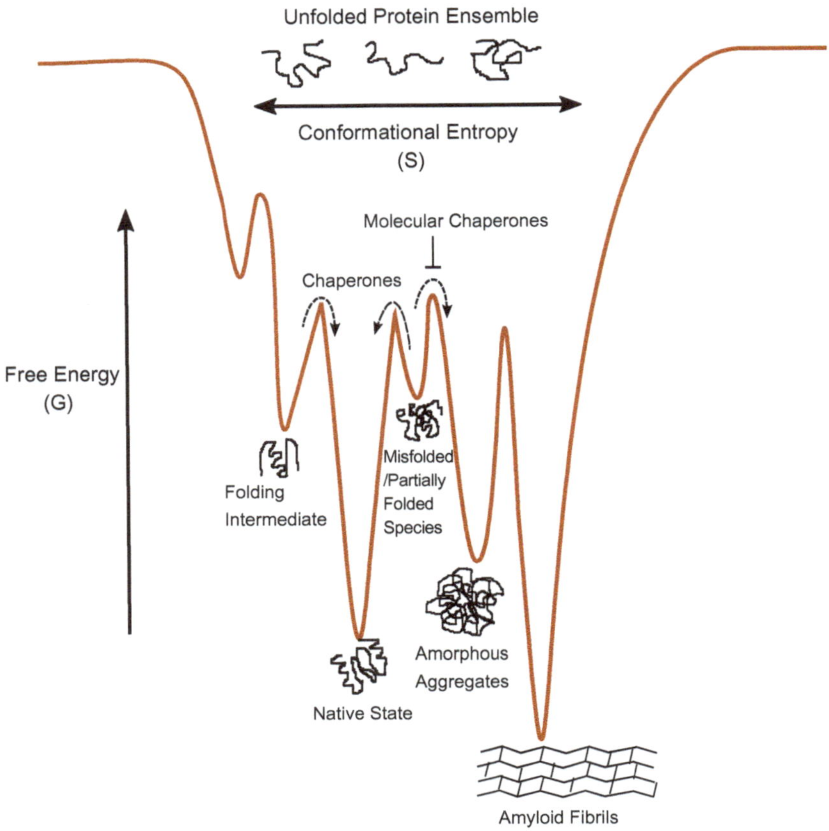

Fig. 1 Protein folding energy landscape. The vertical axis shows the free energy of each conformation that a protein can adopt on its folding route while the width represents the conformational entropy. As we move down the "funnel-like" landscape, the free energy and entropy of a folding protein decreases and a unique native structure is generated from an ensemble of denatured states. The energy landscape of proteins is often rugged due to the formation of intermediates located in local minima which act as "kinetic traps." Such traps slow down the folding reaction and are often responsible for protein aggregation due to inter-molecular interactions. The cellular environment favors inter-molecular interactions among hydrophobic regions of partially folded protein states and thus the formation of amorphous aggregates or ordered amyloid fibrils. Interaction with molecular chaperones enables proteins to prevent these interactions or even kinetic traps by inducing certain conformational rearrangements.

bind to the "sticky" surfaces of the proteins thereby preventing incorrect interactions within and between polypeptide chains which can result in wrong conformers populated on the folding pathway. In consequence, molecular chaperones are not against Anfinsen's principle of self-assembly of the linear polypeptide chain; they are important to render the protein folding process efficient in cells in general and allow specific regulation of activity on the level of protein folding. Nature has invented the concept of molecular chaperones several times independently. The different representations are quite different in structure and specific mechanism. Accordingly, molecular chaperones are categorized into different chaperone families with the members of the same family sharing common structural and functional features. These families are classified by the molecular weight of its members and are historically referred to as the heat shock protein (Hsp) families though they are not only required under stress but also are necessary under normal cellular conditions (Solis et al., 2016). In this review, we focus on the functional principles and mode of regulation of selected classes of chaperones including recent advancements.

2.1 The GroE molecular chaperone system

Bacterial GroE, one of the best characterized molecular chaperone machines, is essential at all temperatures in *E. coli* (Fayet, Ziegelhoffer, & Georgopoulos, 1989). GroE consists of two components: GroEL and GroES. GroEL is a double ring complex consisting of two times seven subunits. It was first identified in the context of T4 phage infection. Genetic defects in the GroE operon affected the ability of phages to assemble and compromised their infectivity (Georgopoulos, Hendrix, Kaiser, & Wood, 1972; Takano & Kakefuda, 1972). Later it was shown that RuBisCo, an enzyme responsible for the carbon fixation in cells, required a homolog of GroEL for the correct assembly of its large and small subunits in chloroplasts (Barraclough & Ellis, 1980; Goloubinoff, Christeller, Gatenby, & Lorimer, 1989; Hemmingsen et al., 1988). Soon it was found that proteins imported to mitochondria failed to reach to their native state when Hsp60, a GroEL homolog in mitochondria, was mutated (Cheng et al., 1989). A monomeric iron sulfur protein was among the proteins to be affected. This observation made clear that the role of GroEL is not restricted to protein assembly but it is also needed for the correct folding of monomers. These double ring shaped, ATP-hydrolyzing chaperones were later termed "chaperonins" (Hemmingsen et al., 1988).

2.1.1 Architecture of GroEL and GroES

The seven subunit rings of a GroEL molecule are stacked back to back to each other (Fig. 2A). It forms a cylindrical structure with a height of 145 Å and width of 135 Å. The striking feature of the GroEL cylinder is

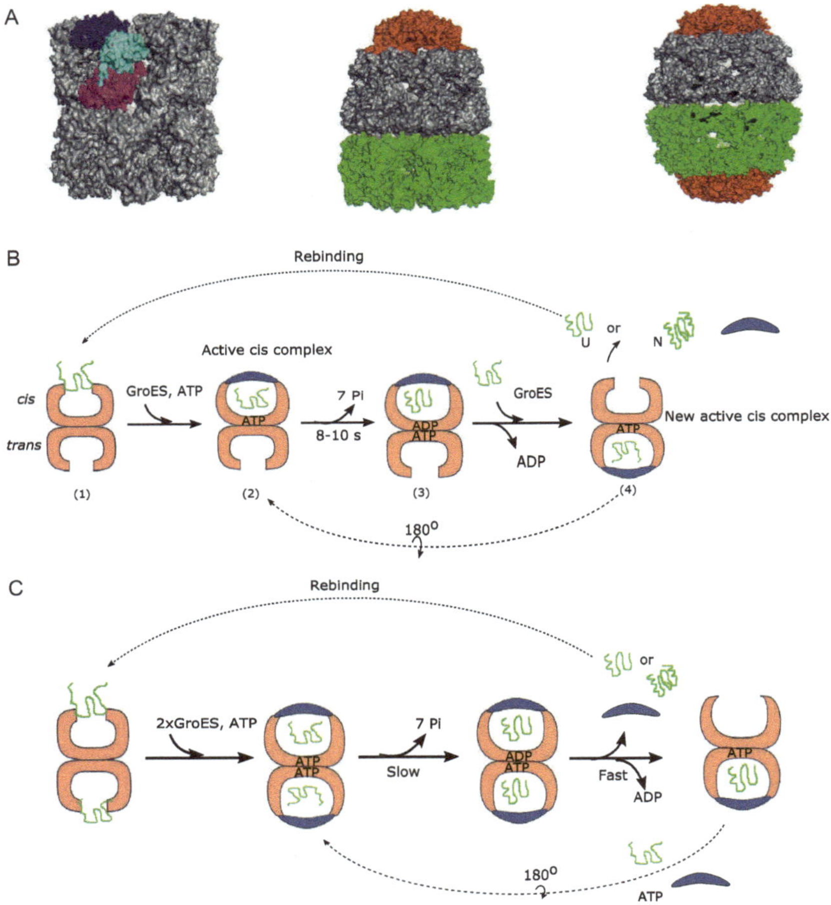

Fig. 2 GroEL structure and reaction cycle. (A) Crystal structure of GroEL from *E. coli* (PDB 1GRL) (Braig et al., 1994). The apical domain (blue), intermediate domain (cyan) and equatorial domain (pink) of a single subunit are shown in the left panel. The middle panel depicts the crystal structure of the asymmetric chaperonin complex GroEL-GroES-$(ADP)_7$ (PDB 1AON) (Xu, Horwich, & Sigler, 1997). The two GroEL rings stacked back to back are shown in gray and green. GroES (red) binds to only one ring giving a bullet-shaped appearance. Binding of GroES causes upward and outward movements of the apical domains increasing the volume of the central cavity to $\sim$80 Å^3. The right panel shows the crystal structure of the "football"-shaped GroEL-$(GroES)_2$-$(ADP\text{-}BeFx)_{14}$ complex (PDB 4PKN) (Fei et al., 2014). In the presence of the substrate, the GroEL-GroES reaction cycle operates via an intermediate in which two GroES subunits (shown in red) are bound to opposite ends of both the rings. (B) Asymmetric GroEL-GroES reaction cycle. Non-native substrate polypeptide binds to the open ring of GroEL via hydrophobic interactions (1). GroES in the presence of ATP binds to the same ring as the substrate forming cis ring (2). Large conformational changes in the apical domain upon GroES

(Continued)

the central cavity in either ring with a diameter of 45 Å (Braig, Adams, & Brünger, 1995; Braig et al., 1994). Each subunit comprises an apical, intermediate and equatorial domain (Fig. 2A, first panel). The apical domain participates in both substrate and GroES binding (Braig et al., 1993; Langer et al., 1992). The equatorial domain forms the central part of the GroEL cylinder and contains the ATP binding site. Furthermore, the C-terminal tails emerge from the equatorial domain into the central cavity and close it at the equator (Thiyagarajan, Henderson, & Joachimiak, 1996). The equatorial domain is the major site of contact both within and between the rings. The apical and equatorial domains of GroEL are connected by the flexible intermediate domain whose motions are important for the structural rearrangements during the functional cycle of GroEL. It regulates allosteric communication between the apical and equatorial domains (Roseman et al., 1996; Saibil, Fenton, Clare, & Horwich, 2013)

GroEL requires the single ring co-chaperonin GroES for function. It also exhibits a ring structure and consists of seven subunits of 10 kDa each (Hunt et al., 1996). It binds to GroEL in the presence of nucleotide (Langer et al., 1992; Saibil, Dong, Wood, & auf der Mauer, 1991) producing an upward and outward opening of GroEL apical domains which doubles the volume of the central channel giving a dome-shaped appearance to the GroEL/GroES complex (Chen et al., 1994) (Fig. 2A, second panel). A notable feature of GroES is the presence of mobile loops that mediates its interaction with the apical domains of GroEL (Landry et al., 1993; Richardson, Schwager, Landry, & Georgopoulos, 2001). GroES competes with the substrate proteins for the same apical binding site in GroEL (Chen et al., 1994). In the presence of ATP, this together with a rotation of the GroEL rings obscuring the substrate binding sites leads to the release of polypeptide into the central

Fig 2—cont'd binding lead to the release of substrate inside the GroEL cavity and switche the cavity wall from hydrophobic to hydrophilic allowing the folding of the substrate during the time of ATP hydrolysis (3). Hydrolysis of ATP weakens the cis complex and permits binding of ATP to the *trans* ring. At this stage, a transient separation of two GroEL rings has been suggested (Yan et al., 2018). Subsequent binding of GroES and the substrate to the *trans* ring results in the discharge of ADP, GroES and the substrate from the *cis* ring (4). The released substrate could be in a folded state (N) or is still unfolded (U) in which case it re-binds to GroEL and a new cycle is started with the previous *trans* ring now acting as the new cis ring. (C) Symmetric reaction cycle. In the presence of the substrate, discharge of ADP is fast and ATP hydrolysis becomes the rate limiting step of the cycle resulting in the accumulation of the symmetric "football" intermediate. In the symmetric cycle, protein can fold simultaneously in both the rings. The bullet complex is also formed transiently after ATP hydrolysis.

cavity enabling its folding (Mayhew et al., 1996; Weissman et al., 1995). Different shaped GroEL-GroES complexes have been observed depending on the stoichiometry of GroEL:GroES. GroES can either bind at one end of the GroEL to form a bullet-like structure (Fig. 2A, second panel) (Haldar et al., 2015; Hayer-Hartl et al., 2016; Horwich, Farr, & Fenton, 2006; Inobe et al., 2008; Langer et al., 1992; Xu et al., 1997) or at both ends to form a football-like structure (Fig. 2A, third panel) (Azem, Kessel, & Goloubinoff, 1994; Fei et al., 2014; Harris, Pluckthun, & Zahn, 1994; Koike-Takeshita, Arakawa, Taguchi, & Shimamura, 2014; Llorca, Marco, Carrascosa, & Valpuesta, 1994; Schmidt et al., 1994a, 1994b). Though the importance of symmetric "football" complexes has been debated, several studies have shown that they too are functional forms of the GroEL catalytic cycle (Grallert & Buchner, 2001; Sameshima, Iizuka, Ueno, & Funatsu, 2010; Yang, Ye, & Lorimer, 2013; Ye & Lorimer, 2013).

GroEL belongs to the family of Hsp60 chaperones which are further divided into two groups based on their architecture. Group I chaperonins include GroEL in eubacteria and Hsp60 in mitochondria or chloroplasts. Group II chaperonins are found in archaebacteria and eukaryotic cells and are referred to as the TRiC/CCT complex (Cowan & Lewis, 2001; Frydman, 2001; Hartl, Bracher, & Hayer-Hartl, 2011). Group II chaperonins consist of eight subunits per ring and do not require GroES like co-chaperone but have evolved a built-in lid in their structure that replaced the need for GroES (Jaime et al., 2007; Leitner et al., 2012; Lopez, Dalton, & Frydman, 2015). Compared to Group I chaperonins our knowledge on Group II chaperonins is still limited.

2.1.2 The functional cycle of GroEL/GroES

Non-native states of proteins first bind to GroEL followed by a time-limited attempt at folding in the central GroEL cavity upon simultaneous binding of ATP and GroES. After ATP hydrolysis, the polypeptide is released into the bulk solution whether or not it has become committed to reaching the native state. The protein molecules which have not yet attained the native state again bind to GroEL and in this way undergoes multiple cycles of GroEL binding and release unless they are completely folded (Lin, Puchalla, Shoup, & Rye, 2013; Todd, Viitanen, & Lorimer, 1994; Weissman, Kashi, Fenton, & Horwich, 1994).

The GroEL/GroES system provides proteins an isolated environment inside the GroE cage in which they can fold without being affected by aggregation (Buchner et al., 1991). The reaction cycle of GroEL/GroES can be divided into four distinct but interdependent phases (Fig. 2B): (a) binding of the substrate protein by GroEL; (b) release of the bound substrate into the central cavity by the simultaneous binding of GroES and ATP to form the *cis* active complex which allows folding of the substrate during the time of ATP hydrolysis; (c) hydrolysis of ATP resulting in the destabilization of the high-energy *cis* complex and priming release of the bound substrate, GroES, and ADP; (d) binding of ATP to the *trans* ring, finally causing discharge of GroES, ADP and entrapped substrate protein from the *cis* side. Below we describe our current understanding of each phase and aspects which are still under debate/investigation.

2.1.2.1 Substrate binding to GroEL

GroEL preferentially binds to unfolding intermediates through their exposed hydrophobic regions thereby preventing incorrect intra- and inter-molecular interactions. The substrate binding sites involve residues with non-polar side chains on the cavity-facing side of the apical domain (Buckle, Zahn, & Fersht, 1997; Fenton, Kashi, Furtak, & Norwich, 1994). The structural plasticity of the apical domain allows GroEL to bind to diverse substrates (Ashcroft et al., 2002; Fei, Yang, LaRonde-LeBlanc, & Lorimer, 2013). The substrate protein can adopt several structural states when bound to more than one apical domain. This has vast implications on substrate structural remodeling induced by the GroEL chaperonin (Ashcroft et al., 2002; Elad et al., 2007; Farr et al., 2000; Fei et al., 2013; Horst et al., 2005; Sharma et al. 2008). GroEL is capable of interacting with different substrate conformations ranging from unfolded (Horst et al., 2005) to compact intermediates (Dahiya & Chaudhuri, 2014; Sharma et al., 2008) to structured forms (de Beus, Doyle, & Teschke, 2000; Lilie & Buchner, 1995). This immense binding heterogeneity enables GroEL also to unfold misfolded conformers thus influencing the folding pathway of substrate proteins (Chakraborty et al., 2010; Georgescauld et al., 2014; Sharma et al., 2008; Tang et al., 2006). An iterative annealing model has been proposed according to which the substrate protein undergoes repeated cycles of GroEL binding and release unfolding the substrate every time it binds to GroEL (Todd, Lorimer, & Thirumalai, 1996; Todd et al., 1994). Highly frustrated systems or kinetically trapped intermediates are thus directed to the productive folding pathway (Lin, Madan, & Rye, 2008; Lin et al., 2013; Sharma et al., 2008). GroEL has also been found to accelerate the folding rates by directly altering or modifying the folding energy landscape of its bound substrates (Beissinger, Rutkat, & Buchner, 1999; Chakraborty et al., 2010; Chan & Dill, 1996; Sparrer, Lilie, & Buchner, 1996). It achieves this either by channelizing the folding preferentially through one of the folding routes out of many other routes available (Bhutani & Udgaonkar, 2001; Patra & Udgaonkar, 2007) or by populating different on-pathway intermediates (Dahiya & Chaudhuri, 2014; Georgescauld et al., 2014; Weaver et al., 2017). These conformational changes in the substrate are a product of GroEL binding but whether these concomitant changes are really required for accelerated folding by GroEL needs further investigation.

2.1.2.2 GroES binding and substrate encapsulation

Protein folding by GroEL requires the co-chaperone GroES which upon binding to GroEL releases the bound substrate into the GroEL folding chamber (Martin, Mayhew, Langer, & Hartl, 1993; Sparrer & Buchner, 1997; Weissman et al. 1995) (Fig. 2B). A prerequisite for GroES binding are the structural changes induced in GroEL domains by ATP. Binding of ATP to the *cis* ring occupied by polypeptide generates allosterically coupled movements between the intermediate and apical domains causing an elevation and counter-clockwise twist of the apical domains making them competent for GroES binding (Horwich et al., 2006). Association of GroES further promotes much larger *en bloc* domain movements which cause a greater elevation and twisting of the apical domains (Horwich et al., 2006). This moves the hydrophobic residues involved in polypeptide binding away from the cavity-facing side which leads to the release of the bound substrate into the central hydrophilic chamber. These large domain changes increase the volume of the central cavity to ~175,000 Å^3 enough for accommodating substrate proteins of ~60 kDa size (Hayer-Hartl et al. 2016; Xu et al., 1997).

Obligate GroE substrates often comprise of the complex TIM barrel fold and depend on active folding by GroEL and GroES/ATP (Fujiwara et al., 2010; Kerner et al. 2005). For them, binding of GroES switches the GroEL machinery from the binding to the folding mode by changing the lining of the cavity wall from hydrophobic to hydrophilic residues allowing the folding of substrates in an isolated environment. Hence GroE can be viewed as an Anfinsen cage (Saibil et al., 1993). In this context, requirement of GroES is found to be obligatory only under non-permissive folding conditions (Schmidt et al., 1994a, 1994b), arguing in favor of a passive role of GroEL cage in folding. Further, some proteins are able to fold by transient interaction with GroEL/ATP and do not depend on encapsulation within the central cavity (Chaudhuri, Verma, & Maheshwari, 2009; Kerner et al. 2005), although a physiological requirement of the chaperonin for such substrates seems unlikely. A different mechanism of GroEL/GroES assisted folding has been described for large proteins with molecular mass >60 kDa. For these proteins which cannot be accommodated inside the cavity, a *trans* mechanism of folding has been proposed where GroES binds to the *trans* side of the substrate protein (Chaudhuri et al., 2001; Dahiya & Chaudhuri, 2014; Paul, Singh, Mishra, & Chaudhuri, 2007). This promotes the release of the substrate owing to the GroES-induced transition of the distal ring from a high to a low affinity state for bound substrate (Inbar & Horovitz, 1997).

2.1.2.3 Substrate release

The residence time of the substrate inside the GroEL cage is governed by ATP hydrolysis which is the longest phase of the GroEL reaction cycle amounting to ~8–10 s (Rye et al., 1999). When the seven ATPs in the *cis* ring are hydrolyzed to ADP, the association of GroES to GroEL weakens. However, without the removal of GroES, the substrate protein cannot be released into the solution. Discharge of GroES is achieved by the inter-ring negative cooperativity within the GroEL rings (Burston, Ranson, & Clarke, 1995). Upon ATP hydrolysis in the *cis* ring, structural changes occur in the *trans* ring which increases the affinity of ATP in the *trans* ring. Binding of ATP to the *trans* ring then causes the destabilization of the *cis* complex and discharge of GroES, ADP and substrate (Fig. 2B). The released substrate can be either in the native conformation or fold spontaneously in solution. If the released substrate fails to reach the native state, then it re-binds to the open GroEL ring (Fenton & Horwich, 2003; Horwich et al., 2006) (Fig. 2B). GroES binding to the same ring as ATP and substrate forms the new active *cis* ring. In this way, GroEL exchanges its rings back and forth between active and dissociating chambers resulting in the productive folding of the substrates. This mechanism for substrate release operates for proteins which are folded within the central cavity of GroEL. However, how does release occur for large substrate proteins which are not encapsulated within the GroE cage? These proteins are not capped underneath GroES and hence GroES does not impose any restriction on them for the release. Still for the larger substrates GroES and ATP binding to the *trans* ring is necessary for their release and productive folding (Chaudhuri et al., 2001). Hence, even for large proteins which cannot benefit from encapsulation within GroEL cage, structural rearrangement by GroEL/GroES has been found to stimulate attainment of their native structure supporting an active role of GroEL in folding its substrates (Dahiya & Chaudhuri, 2014).

2.1.3 Symmetric versus asymmetric complexes

During the functional cycle, GroEL undergoes a series of conformational transitions which are coupled to GroES and ATP binding (Skjærven, Cuellar, Martinez, & Valpuesta, 2015). These conformational changes in GroEL are regulated by positive and negative cooperativity present within and in between the rings (Saibil et al., 2013; Yifrach & Horovitz, 1995). There has been lot of debate over the years about the shape and composition of this active complex. GroES binding at 1:1 M ratio to GroEL results in the formation of asymmetric or bullet-shaped complexes which are considered to be the dominant species formed during the GroEL cycle (Gruber & Horovitz, 2016; Hayer-Hartl et al., 2016; Horwich et al., 2006; Yan et al., 2018) (Fig. 2B). However, since GroEL consists of two rings it can bind simultaneously to both the rings at 1:2 M ratio forming (GroEL):$(GroES)_2$ symmetric football-shaped complexes (Fig. 2C) (Azem et al., 1994; Fei et al., 2014; Grallert & Buchner, 2001; Schmidt et al., 1994a, 1994b; Sparrer, Rutkat, & Buchner, 1997; Taguchi, 2015; Weiss, Jebara, Nisemblat, & Azem, 2016; Yang et al., 2013; Ye & Lorimer, 2013). Inter-ring negative cooperativity has been implicated in the formation of asymmetric complexes (Gruber & Horovitz, 2016; Horovitz & Willison, 2005).

An alternative model suggests that the reaction cycle proceeds via formation of functional symmetric complexes. (Fei et al. 2014; Koike-Takeshita, Yoshida, & Taguchi, 2008; Sameshima et al., 2010; Sparrer et al., 1997; Yamamoto & Ando, 2016; Yang et al., 2013; Ye & Lorimer, 2013). Formation of asymmetric and symmetric complexes is not mutually exclusive and the presence of substrate proteins has been suggested to switch the GroEL:GroES cycle from asymmetric to symmetric (Sameshima et al., 2010; Yang et al., 2013; Ye & Lorimer, 2013). Symmetric complexes are found to be populated under physiological conditions (37 °C, in the presence of non-folded substrates and 0.1–0.2 M K^+ ions) (Gruber & Horovitz, 2016); however, their formation is argued at physiological ratios of [ATP]/[ADP] and in the presence of foldable substrates (Haldar et al., 2015). It has been recently shown that defects in inter-ring negative cooperativity result in the population of symmetric complexes suggesting that they are not the functional intermediate in the GroEL reaction pathway

(Yan et al., 2018). Still if the symmetric intermediate had not been a catalytic form and just an off-pathway species then it should not lead to active refolding of GroEL model substrate proteins (Beissinger et al., 1999; Ben-Zvi, Chatellier, Fersht, & Goloubinoff, 1998; Sparrer & Buchner, 1997; Sparrer et al., 1997). Dissociation of ADP from the *cis* ring is the rate limiting step of the GroEL cycle which will prevent formation of symmetric complexes and allow asymmetric complex to be populated in a steady state scenario. However, in the presence of unfolded substrate, symmetric complexes can be detected as shown by recent FRET studies (Fei et al., 2014; Ye & Lorimer, 2013). The presence of substrate catalyzes ADP/ATP exchange thereby populating symmetric complexes while in the absence of substrate asymmetric complexes should dominate (Iizuka & Funatsu, 2016; Ye & Lorimer, 2013). Compared to the asymmetric complexes, the symmetric complexes differ at the GroEL/GroES interface and also at the interface between the two rings that possibly allows parallel processing of substrates by both the rings rather than alternating the folding rings (Fei et al., 2014). Hence, there are clear evidences that symmetric football species can be the functional form in vivo. However, investigations of GroEL cycling carried out in the presence of obligate substrate protein are still ongoing. Analyzing the importance of asymmetric versus symmetric complexes in vivo could add to resolve this outstanding issue.

2.1.4 Active versus passive folding

Another matter of debate is the mechanism by which GroEL/GroES chaperonin system facilitates the folding of its substrate proteins. In one view, GroE is simply a passive aggregation inhibiting box while in another view it participates actively in the folding process: the passive cage model suggests that GroEL cavity acts as an infinite dilution container or the "Anfinsen" cage which facilitates the folding of proteins by isolating them from the crowded cellular environment and allowing them to fold in the absence of aggregation (Ellis, 2001; Horwich, Apetri, & Fenton, 2009). In contrast, the active cage model stresses that the central cavity is involved in actively promoting the folding of proteins (Hayer-Hartl et al., 2016). The passive cage model is based upon the observations that the rate of spontaneous folding of proteins was similar to their GroEL-assisted folding rates at temperatures where aggregation does not limit the spontaneous folding of proteins (Horwich et al., 2009). At physiological temperatures where spontaneous folding of proteins is hampered by aggregation, GroEL could accelerate their folding by preventing the formation of off-pathway kinetic intermediates that diverges the proteins from their correct folding pathway hence, slowing down the overall folding rates. In this context, the GroEL chamber facilitates or accelerates folding by preventing aggregation and increasing the overall yield of the folding reaction. The proponents of passive cage model argue that if the cavity wall of the GroEL is involved in active folding then one expects to see differences in the folding rates and recovery of native substrate proteins when the wall residues are mutated. However, substantial differences were not observed even with multiple mutations in the cavity wall (Tang et al., 2008). These observations were further substantiated by the findings that the folding trajectory of DHFR inside the *cis* cavity and free in solution does not differ in the absence of reversible aggregation allowing direct comparison of the folding routes (Horst et al., 2007).

On the other side, there is large body of evidence suggesting active role of the GroEL cavity in protein folding (Hayer-Hartl et al., 2016). It has been widely shown using several substrate proteins that the GroEL cavity directly accelerates the folding of the substrate proteins by altering their folding routes (Chakraborty et al., 2010; Weaver & Rye, 2014; Weaver et al., 2017) even when no aggregation was detected in the absence of GroE (Georgescauld et al., 2014). The active cage model attributes the acceleration by the cavity to its highly charged nature and the destabilization of the kinetically trapped species by their steric confinement that supports the attainment of the native state of the protein (Hayer-Hartl et al., 2016). Support for an active role of GroEL in protein folding also comes from the iterative annealing model which states that substrates can undergo repeated cycles of GroEL binding and release accompanied with their unfolding (Todd et al., 1996, 1994). Such GroEL-induced unfolding of substrates enables breaking of non-native contacts giving the substrate multiple chances of folding.

2.2 The Hsp70 system

The Hsp70 chaperone family members are ~70 kDa large ATP-hydrolyzing enzymes found in archaea, bacteria, the eukaryotic cytosol and organelles. They form a central hub of the molecular chaperone network and are involved in diverse functions ranging from unfolding of substrates (Bukau, Weissman, & Horwich, 2006; De Los Rios et al., 2006; Rodriguez et al., 2008; Sharma et al., 2010; Sousa et al., 2016) to refolding (Groemping, Seidel, & Reinstein, 2005; Hu, Mayer, & Tomita, 2006;

Mayer, Kies, Kammermeier, & Buchner, 2000; Mayer, Rudiger, & Bukau, 2000; Mayer et al., 2000), disaggregation (Nillegoda et al., 2015; Schlee, Beinker, Akhrymuk, & Reinstein, 2004), degradation (Fernandez-Fernandez et al., 2017) to substrate handover (Morgner et al., 2015; Wegele et al., 2006) and protein transport across membranes (Chacinska et al., 2009; Craig, 2018). This broad functionality of Hsp70 is exercised with the help of specialized Hsp70 members that are assisted by different co-chaperones to fulfill the specific needs of the cell (Craig & Marszalek, 2017; Feige et al., 2009; Hendershot, 2004; Mayer, Kies, et al., 2000; Mayer, Rudiger, et al., 2000; Mayer et al., 2000; Nillegoda et al., 2015; Sousa & Lafer, 2015). Hsp70 accounts for 0.5–2% of the total protein content in unstressed cells which increases around twofold under heat shock conditions (Goloubinoff, 2017). The most extensively studied Hsp70 family member is the bacterial homolog DnaK whose function is regulated by its co-chaperone DnaJ and the nucleotide exchange factor GrpE. In the eukaryotic cytosol, there are two forms of Hsp70. Hsc70 which is the constitutively expressed form and the heat inducible form is Hsp70. Apart from these, there are two compartment-specific Hsp70s. BiP or Grp78 or HSPA5 is an essential ER Hsp70 chaperone involved in protein folding and quality control in the ER (Behnke, Feige, & Hendershot, 2015) while the mitochondrial Hsp70 known as HSPA9 or mortalin is important in Fe-S clusters biogenesis and protein translocation in mitochondria (Dutkiewicz, Nowak, Craig, & Marszalek, 2017). In general, all Hsp70 family members bind to hydrophobic peptide regions of unfolded, misfolded or aggregated proteins in an ATP-dependent manner (Mayer & Bukau, 2005). However, native states of some regulatory proteins also interact with Hsp70; here Hsp70 works in close cooperation with other chaperones like Hsp90 to control their activation (Kirschke et al., 2014; Pratt & Toft, 2003). Further, Hsp70 also binds to well-folded clathrin coats where it acts as an "uncoating ATPase" (Sousa & Lafer, 2015). Together with its J co-chaperone, Hsp70 mediates the disassembly of clathrin coats thereby assisting the process of vesicular transport in the cells (Krantz et al., 2013; Sousa et al., 2016; Ungewickell et al., 1995)

2.2.1 Structural framework of Hsp70

All members of Hsp70 family are comprised of two highly conserved and functional domains: the N-terminal ATP binding domain (NBD) and the C-terminal substrate binding domain (SBD) (Fig. 3A). The N-terminal domain (1–383, 44 kDa) consists of four subdomains—IA (1–39,

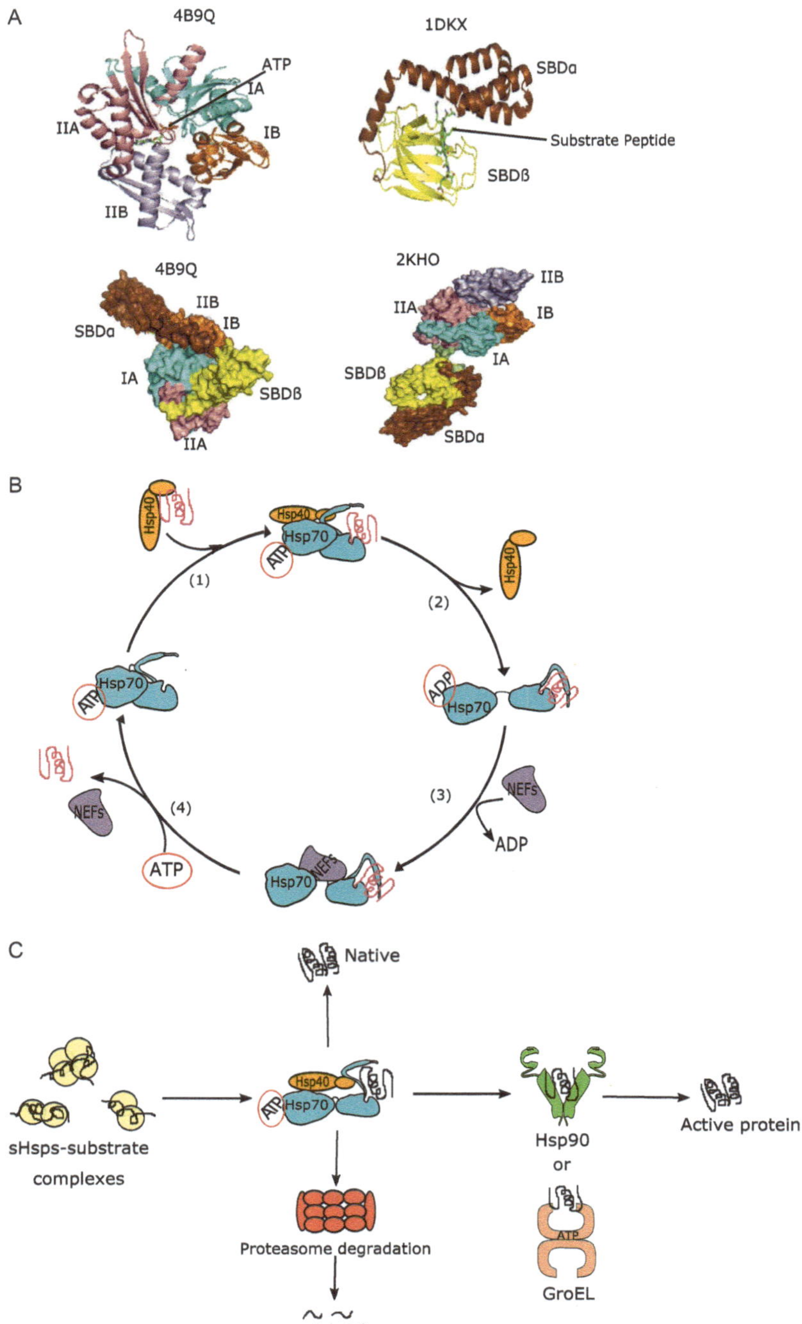

Fig. 3 See figure legend on next page.

116–188); IB (40–115); IIA (189–228, 307–385); IIB (229–306) all of which have α/β topology (Flaherty, DeLuca-Flaherty, & McKay, 1990). The subdomains IA and IB form the right lobe while IIA and IIB form the left lobe of the NBD (Fig. 3A, first panel) (Flaherty et al., 1990). The striking feature of the NBD architecture is the deep cleft present in between the right and the left lobe (Fig. 3A, first panel). The nucleotide binds in the center of this cleft which contains residues contributed by all the subdomains (Fig. 3A). The NBD is connected to the SBD by a 10–12 residue linker which has a conserved hydrophobic sequence (Vogel, Mayer, & Bukau, 2006). The linker mediates inter-domain communication and is responsible for coupling ATP binding and hydrolysis to the conformational changes in the SBD (English, Sherman, Meng, & Gierasch, 2017; Swain et al., 2007; Vogel et al., 2006). The substrate binding domain [SBD] (396–638, ~27 kDa) contains a

Fig. 3 Structure and functional cycle of Hsp70. (A) The upper left panel shows the crystal structure of the nucleotide binding domain (NBD) of *E. coli* DnaK (PDB 4B9Q) (Kityk, Kopp, Sinning, & Mayer, 2012). The NBD consists of four subdomains: IA (cyan), IB (orange), IIA (pink), IIB (purple). Subdomains IA and IB form the right lobe while subdomains IIA and IIB form the left lobe. There is a cleft in between in which ATP binds (shown as sticks). The upper right panel shows the crystal structure of the closed state of the substrate binding domain (SBD) of DnaK in complex with peptide substrate (PDB 1DKX) (Zhu et al., 1996). The SBDβ subdomain is shown in yellow and SBDα in brown. The bound peptide represented by green sticks directly contacts the SBDβ and is enclosed by SBDα. The lower left panel depicts the structure of DnaK in the ATP bound open conformation when the NBD and the SBD are docked onto each other. Colors are the same as in the upper panels. The lower right panel represents the structure of DnaK in the ADP bound closed conformation where the NBD and the SBD are undocked (PDB 2KHO) (Bertelsen, Chang, Gestwicki, & Zuiderweg, 2009). The linker connecting the two domains is shown in green. As can be seen, the SBDα packs against the SBDβ in the closed conformation thereby locking the bound substrate. (B) Hsp70 cycle. Unfolded or non-native substrate (shown in pink) is first captured by Hsp40 and handed over to Hsp70 in the ATP bound state (1). Hsp40, Hsp70 and substrate forms a stable complex in the presence of ATP. Hsp40 stimulates ATP hydrolysis reaction by Hsp70 leading to the closed Hsp70 state where the ATP and substrate binding domains are separated and the bound substrate is locked by the lid subdomain of SBD (2). Hsp40 then dissociates from the complex. Dissociation of ADP is accelerated by nucleotide exchange factors (NEFs) (3). Subsequent binding of ATP introduces a conformational change in Hsp70 opening up the substrate binding cleft and releasing the bound substrate (4). NEFs will also dissociate. (C) Hsp70 is the central component of protein homeostasis network of molecular chaperones. Hsp70 accepts non-native proteins from sHsps. These can be either refolded by Hsp70 to the native state or if the damage is too much then they can be passed on the proteasomal degradation machinery for proteolysis. Hsp70 can also transfer the substrate to downstream folding chaperones like Hsp90 or GroEL which enable final maturation of the client proteins.

β-sandwich subdomain, SBDβ, that directly contacts the substrates and a α-helical segment, SBDα, which forms the substrate enclosing lid subdomain (Zhu et al. 1996; Zuiderweg et al., 2013) (Fig. 3A, second panel). The lid acts as a latch that locks the substrate in the Hsp70-bound state thereby increasing its lifetime in the complex (Fig. 3A, second panel) (Moro, Fernández-Sáiz, & Muga, 2004; Zhu et al. 1996).

2.2.2 Functional cycle of Hsp70

Hsp70 exhibits a high degree of conformational dynamicity and adopts two discreet structural states depending on the nucleotide bound. In the ATP bound state, the SBD has low affinity for substrates resulting in their fast association and dissociation rates (Mayer, 2013; Mayer & Bukau, 2005; Zuiderweg et al., 2013). ATP hydrolysis results in the ADP bound state in which the peptide dissociation rate is low and the affinity of the SBD for bound substrates increases by one to two orders of magnitude (Schmid, Baici, Gehring, & Christen, 1994; Takeda & McKay, 1996). Hence, ATP hydrolysis acts as a molecular switch controlling the conformational states and substrate affinity of Hsp70. The allostery in between the NBD and SBD drives the reaction cycle of Hsp70.

In the nucleotide-free state, Hsp70 exists in equilibrium between the open and closed states with the frequency of transition between the states determined by the nucleotides (Kityk et al., 2012; Mayer, Kies, et al., 2000; Mayer, Rudiger, et al., 2000; Mayer et al., 2000). The presence of ATP dramatically increases the population of the open state in which SBDβ and SBDα are docked on different parts of the NBD making SBDβ free to bind substrates (Fig. 3A, third panel). Substrate binding often increases the ATP hydrolysis rate of Hsp70 (Flynn, Pohl, Flocco, & Rothman, 1991; McCarty, Buchberger, Reinstein, & Bukau, 1995) hence driving the formation of the closed state where the α-helical lid packs against the SBDβ thereby locking the bound substrate (Fig. 3A, fourth panel) and markedly reducing the substrate dissociation rates (Moro et al., 2004). ATP binding to the NBD induces the conformational change in the SBD necessary for releasing the substrate from the locked state by removing the α-helical lid from the SBDβ. Hsp70 undergoes repeated ATP-dependent cycles of substrate binding and release during which the misfolded/unfolded substrate can undergo folding. However, it has been shown recently that the conformational transition of Hsp70 does not involve just two discreet states but consists of ensemble of conformations which differ in inter-domain distances and undergo rapid fluctuations between them (Mapa et al., 2010; Marcinowski et al., 2011; Rosam et al., 2018; Schlecht, Erbse, Bukau, & Mayer, 2011; Sikor et al., 2013). NMR experiments have directly captured an allosteric state of Hsp70 bound to substrate that is intermediate between the open and closed states (Zhuravleva, Clerico, & Gierasch, 2012). How this conformational heterogeneity relates to substrate folding remains to be seen.

2.2.3 Substrate binding by Hsp70

The β-sandwich subdomain of the SBD comprises the hydrophobic groove in which the peptide substrate can bind in an extended conformation (Fig. 3A, second panel). The peptide binding cleft of the SBD interacts with the substrate mainly by hydrophobic interactions but also forms hydrogen bonds with the peptide backbone thereby stabilizing the bound substrate. The substrate readily associates with Hsp70 in the presence of ATP due to faster k_{on} rates however due to equally fast k_{off} rates, it can also rapidly dissociate. This is avoided by ATP hydrolysis to form the ADP bound state in which the substrate is trapped and the complex is stable due to very low off rates.

There is no single binding motif of Hsp70; however, studies of peptide libraries have shown a preference for short stretches of hydrophobic amino acids (Fourie, Sambrook, & Gething, 1994; Rüdiger, Germeroth, Schneider-Mergener, & Bukau, 1997; Schneider et al., 2016; Van Durme et al., 2009). This binding nature explains why Hsp70 can interact with the plethora of substrate proteins in cells. Since, hydrophobic amino acids are predominantly found in the core of the folded protein and are exposed upon unfolding; Hsp70 preferentially interacts with unfolded or partially unfolded substrate states. Taking into account the characteristic hydrophobic signature of its clients, a predictive algorithm has been established by several labs to classify Hsp70 binding peptides (Behnke et al., 2016; Blond-Elguindi et al., 1993; Pandya et al., 2009; Rüdiger et al., 1997; Schneider et al., 2016; Srinivasan et al., 2012; Van Durme et al., 2009). Recently, a combined sequence- and structure-based predictive model has been developed for BiP, the ER specific Hsp70 (Schneider et al., 2016). It predicts peptide binding sequences for BiP with high accuracy and allows prediction of the correct binding mode of the peptide. Despite these advancements, our current understanding of the characteristics of the binding sequence of Hsp70 is still limited. Differences have been shown to exist in the binding specificity between different Hsp70 members (Behnke et al., 2016; Fourie et al., 1994; Gragerov & Gottesman, 1994). It is still unclear to what extent, the binding specifics of Hsp70s are influenced by their functional roles (Behnke et al., 2016). Besides, J-proteins can impart substrate specificity to Hsp70 (Kampinga & Craig, 2010; Vembar et al., 2010).

The conformation of the peptide substrate also plays an important role in determining Hsp70 binding. Hsp70 prefers extended peptide conformations to form stable complexes and this behavior is conserved among its different members (Marcinowski et al., 2013). This is compatible with its binding to unfolded conformations but in some cases also folded or near-native states provide accessible hydrophobic binding sites (Clerico, Tilitsky, Meng, & Gierasch, 2015). In this context, it is important to mention that besides favoring an unstructured state of the substrate, Hsp70 has been shown to further induce unfolding or expansion of the substrate (Kellner et al., 2014; Sharma et al., 2010), even when it binds the native state of a protein (Kirschke et al. 2014; Sousa et al., 2016). Substrates too act as allosteric effectors of Hsp70 conformation. Substrate binding induces conformational transitions in Hsp70, the efficiency of which are governed by substrate affinities (Clerico et al., 2015; Mayer, 2013). High affinity substrate peptides are more efficient in increasing the ATP hydrolysis rates of Hsp70 thereby accelerating the attainment of undocked or closed conformation of Hsp70 (Mayer, Kies, et al., 2000; Mayer, Rudiger, et al., 2000; Mayer et al., 2000). This may have impact on the response of Hsp70 to substrates which bind with varying affinities. Hence, the thermodynamics of substrate binding is directly coupled to Hsp70 function. This is reflected in different behaviors observed for non-native substrate proteins and peptides with respect to Hsp70 ATPase rate (Marcinowski et al., 2011; Mayer, Reinstein, & Buchner, 2003). Substrate proteins were found to both accelerate and decelerate the ATP hydrolysis rate of BiP (Laufen et al., 1999; Mayer et al., 2003), whereas substrate peptides are almost always found to stimulate ATPase activity of Hsp70 (Mayer et al., 2003). Further, peptide compared to protein substrates stabilizes different conformational states of Hsp70: peptides favor a more closed lid conformation (Kityk et al., 2012; Marcinowski et al., 2011) while binding protein substrates support a more open conformation of the lid (Marcinowski et al., 2011). Different responses of Hsp70 toward peptide and protein substrates may impart functional plasticity to Hsp70. Along with the substrates, Hsp70 partner proteins such as Hsp40s and members of the nucleotide exchange factors (NEFs) influence and regulate the conformational transitions and hence functional cycle of Hsp70.

2.2.4 Hsp70 co-chaperones: Hsp40 and nucleotide exchange factors

Hsp70 does not work in isolation but is assisted by two major sets of accessory factors: J-proteins and NEFs which regulate nucleotide binding to Hsp70 and hence its interactions with the substrates. J-proteins also known as Hsp40s are a diverse class of proteins which impart much of the functional diversity to Hsp70. The same Hsp70 can interact with many different J-proteins. There are around 41 J-proteins in eukaryotes thus far outnumbering Hsp70 isoforms (Kampinga & Craig, 2010) thereby making Hsp70 functionally versatile. J-proteins are multi-domain proteins defined by the structurally conserved amino terminal J-domain which is responsible for functional interaction with Hsp70 and stimulating its ATPase activity severalfold (Wall, Zylicz, & Georgopoulos, 1994). Hsp40 itself binds to hydrophobic residues of substrate proteins and acts as a scanning factor for the Hsp70 chaperone (Behnke et al., 2016; Jin, Zhuang, & Hendershot, 2009; Rüdiger, Schneider-Mergener, & Bukau, 2001; Shen & Hendershot, 2005). Synergistically with the substrate proteins, bacterial Hsp40 can stimulate the ATPase activity of Hsp70 severalfold at physiological concentrations of Hsp40 (Laufen et al., 1999). A recent crystal structure of DnaJ in complex with DnaK shows that DnaJ makes direct contacts with the DnaK SBD (Kityk, Kopp, & Mayer, 2018). This suggests that there is a direct handover of substrates from the J-protein to Hsp70 supported by accelerated ATPase cycle. (Kityk et al., 2018).

In the current model of the Hsp70 reaction cycle (Kityk et al., 2018; Laufen et al., 1999; Marcinowski et al., 2011; Mayer & Bukau, 2005; Rosam et al., 2018; Szabo et al., 1994), Hsp40 first binds rapidly and transiently to the unfolded client protein through its C-terminal substrate binding domain. The J-domain protein then interacts with Hsp70-ATP and presents the clients to the open substrate binding pocket of Hsp70 (Fig. 3B). The ATPase cycle of Hsp70 is then stimulated by Hsp40. ATP hydrolysis locks the substrate in the Hsp70 binding pocket (Fig. 3B). Hence, Hsp40 drives the reaction cycle of Hsp70 forward and ATP hydrolysis is no longer the rate limiting step of the cycle in the presence of Hsp40. The next step in the reaction cycle is the release of ADP and consequently the trapped substrate which allows the reversal to the initial structural state of Hsp70 to which ATP can bind and a fresh ATPase cycle could start. ADP dissociation which is the second slow reaction of Hsp70 cycle is stimulated by another class of Hsp70 co-chaperones, the nucleotide exchange factors or NEFs (Bracher & Verghese, 2015; Schuermann et al., 2008) (Fig. 3B). The best studied NEF is the bacterial GrpE whose human counterpart is the Bag family of proteins characterized by the conserved Bag domain (Bracher & Verghese, 2015). The mechanism by which NEFs exchange ADP for ATP can be explained by their relative affinities for the two states. NEFs bind to the ADP or the nucleotide-free state with high affinities (Mayer & Bukau, 2005). Upon binding to the ADP state, GrpE opens the nucleotide binding cleft of the NBD thereby accelerating the discharge of ADP as evident form the crystal structure of GrpE in complex with DnaK-NBD (Harrison et al., 1997). Since GrpE has low affinity for the ATP bound state, it is released once ATP is bound to Hsp70 and a new round of substrate binding starts (Fig. 3B). A new role of NEFs in eukaryotic cells has been uncovered recently. It has been shown that NEFs not only stimulate the dissociation of ADP but also the release of substrates from the SBD of Hsp70 by directly competing with the bound substrate (Gowda et al., 2018; Rosam et al., 2018). The largely unstructured N-terminal domain of Bap, a co-chaperone of the ER-resident Hsp70, BiP has been shown to affect conformation of BiP SBD promoting substrate release (Rosam et al., 2018). This functional role of NEFs seems to be conserved as also yeast NEF Fes1 and its human counterpart HspBP1 shows similar effects on substrate release (Gowda et al., 2018).

2.2.5 Effect of Hsp70 on its substrates

Hsp70s assist protein folding in various ways: they prevent aggregation by binding to unfolded conformers, refold misfolded substrates or solubilize aggregates (Goloubinoff & De Los Rios, 2007) or dissociate protein complexes such as clathrin cages (Krantz et al., 2013; Sousa et al., 2016). "Entropic pulling" has been proposed as a unifying mechanism for these disparate functions of Hsp70 (Goloubinoff & De Los Rios, 2007). It means that Hsp70 uses energy from ATP hydrolysis to generate an entropic force to unfold its substrates. However, whether Hsp70 participates actively in the folding process of its clients is still unclear. Many results can be explained by a passive role where Hsp70 binds to unfolded substrates and prevents their aggregation while giving them another chance to fold. Evidence has been presented regarding an unfoldase action of Hsp70 in vitro (De Los Rios et al., 2006; Palleros, Shi, Reid, & Fink, 1994; Rodriguez et al., 2008; Sharma et al., 2010; Sousa et al., 2016). The unfolding action of Hsp70 has recently been shown for a native substrate, the glucocorticoid receptor, where Hsp70 induces local unfolding in the hormone binding pocket of the receptor transforming it into a conformation more suited for its downstream function (Kirschke et al., 2014). Another action of Hsp70 on a well-folded substrate is clathrin cages where Hsc70 together with a J-protein is required for disassembly (Braell, Schlossman, Schmid, & Rothman, 1984; Krantz et al., 2013; Rothnie et al., 2011; Sousa et al., 2016). This suggests a mechanism of force generation by Hsp70 binding (Sousa et al., 2016). All these pieces of evidence suggest an active role of Hsp70 in the chaperoning of some substrates. An active role of Hsp70 in the protein folding process is also supported by its cooperation with the small heat shock proteins (sHsps) in the refolding of misfolded or aggregated proteins (Fig. 3C). Here, the misfolded proteins are first bound to an sHsp and then handed over to Hsp70 which, assisted by a J-protein, ensures efficient refolding in an ATP-dependent manner when conditions become favorable (Ehrnsperger, Gräber, Gaestel, & Buchner, 1997; Lee & Vierling 2000; Zwirowski et al., 2017). Given their diverse functions and cooperation with other chaperones, Hsp70s lie at the crossroad of protein homeostasis. Either the substrates released from Hsp70 can spontaneously refold or depending on the cellular conditions and the protein itself, Hsp70 can hand over the substrates to the degradation machinery or to the downstream chaperones like GroEL or Hsp90 for their final maturation (Fig. 3C).

2.3 The Hsp90 chaperone system

Hsp90 is a highly conserved ATP-dependent heat shock protein present from bacteria to man that acts as the central hub of protein homeostasis (Taipale, Jarosz, & Lindquist, 2010). Early studies in yeast proved that Hsp90 is essential at all growth temperatures (Borkovich et al., 1989). Hsp90 is the most complex of all molecular chaperones as its functional cycle is assisted by a large cohort of co-chaperones and furthermore it often collaborates with the Hsp70 system in its chaperone cycle. The peculiar thing about Hsp90 that sets it apart from other chaperones is that it is abundant also under unstressed conditions and present in amounts seemingly in excess to normal needs. This speaks of a prominent role of Hsp90 in the normal functioning of cells. Hsp90 has very specific set of client proteins that it chaperones. These include kinases, hormone receptors, E3 ligases and transcription factors among others (Taipale et al., 2012) which are dependent on Hsp90 for their final maturation or activation in cells.

A single copy of Hsp90 exists in bacteria where it is called HtpG (Bardwell & Craig, 1988; Honore, Mejean, & Genest, 2017). In contrast, Hsp90 exists in two isoforms in yeast and humans which are known as Hsc82 and Hsp82 or Hsp90β and Hsp90α, respectively. Hsc82 and Hsp90β are constitutively expressed while Hsp82 and Hsp90α are stress-inducible (Chen, Zhong, & Monteiro, 2006; Shen et al., 1997; Zhang et al., 1999). Susan Lindquist first put forward the idea that the presence of excess of Hsp90 acts as potentiator or buffer for genetic variation (Rutherford & Lindquist, 1998). Upon stress when there is an increased cellular demand for Hsp90, the normal buffering function for client proteins is decreased, phenotypic variation become effective causing the evolution of new traits (Geller et al., 2018; Hummel et al., 2017; Rohner et al., 2013; Rutherford & Lindquist, 1998). The first protein discovered that underscored a functional role of Hsp90 was the oncogenic kinase v-Src. Its activity was reduced drastically with decreased Hsp90 expression while its normal cellular counterpart c-Src remained less affected (Xu, Singer, & Lindquist, 1999). Further it was found that geldanamycin, earlier suggested to be Src inhibitor, actually binds to and inactivates Hsp90 hence disrupting the function of Src kinase (Whitesell et al., 1994). Many of the client proteins of Hsp90 are ligand binding proteins and one may speculate that interaction with Hsp90 stabilizes them or increases their folding competence. Whether Hsp90 actively promotes a conformational change in its clients is still unknown. The overall picture is further complicated by the vast number of Hsp90 co-chaperones which seem to adjust the Hsp90 machinery to clients while their exact roles for client processing are still enigmatic.

2.3.1 Hsp90 structure and function

Hsp90 is a homo-dimer with each subunit consisting of three highly conserved domains: the N-terminal ATP binding domain (NTD), the middle domain (MD) and the C-terminal dimerization domain (CTD) (Fig. 4A). The dimerization of Hsp90 is of moderate affinity (~60 nM) mediated through the CTD (Chadli et al., 2000; Richter, Muschler, Hainzl, & Buchner, 2001). The extension of the CTD, a negatively charged tail ending with the MEEVD motif, is the main binding site of TPR domain co-chaperones (Schopf, Biebl, & Buchner, 2017). The NTD contains the ATP binding pocket (Prodromou et al., 1997) which is also the binding site for natural Hsp90 inhibitors such as geldanamycin and radicicol, and synthetic purine scaffold inhibitors (Chiosis et al., 2002; Roe et al., 1999). In eukaryotes, the NTD of Hsp90 is connected to the MD by a long, charged and flexible linker that affects Hsp90 function by modulating NTD-MD contacts (Hainzl, Lapina, Buchner, & Richter, 2009; Jahn et al., 2014; Tsutsumi et al., 2012). The MD constitutes the major binding site for Hsp90 clients and co-chaperones (Genest et al., 2013; Lorenz et al.,

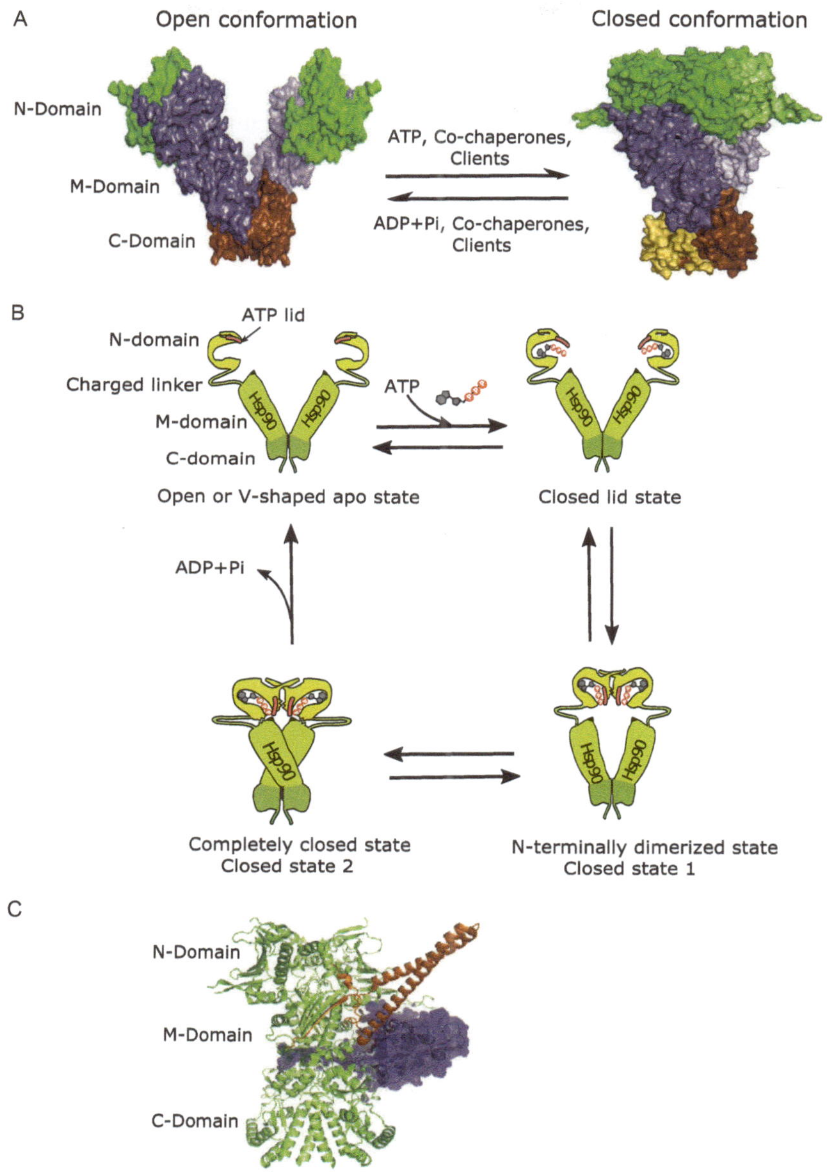

Fig. 4 Hsp90 structure and conformational cycle. (A) Open and closed conformations of Hsp90. Crystal structure of bacterial Hsp90, HtpG in the open conformation on the left (PDB 2IOQ) (Shiau, Harris, Southworth, & Agard, 2006). The closed conformation of yeast Hsp90 in complex with AMP-PNP is shown on the right (PDB 2CG9) (Ali et al., 2006). N-terminal domain (NTD) in green, middle domain (MD) in purple and C-terminal domain (CTD) in brown. ATP, client and co-chaperone binding can alter the equilibrium

(Continued)

2014; Retzlaff et al., 2010; Schmid et al. 2012). Recently, an allosteric switch point reporting on client interaction has been mapped to the MD of Hsp90 (Rutz et al., 2018). However, there is also evidence that all three Hsp90 domains contribute to client binding (Hagn et al., 2011; Karagoz et al., 2014; Park, Borin, Martinez-Yamout, & Dyson, 2011; Park, Kostic, & Dyson, 2011).

In the absence of ATP, Hsp90 has a V-shaped conformation (Fig. 4A). ATP binding induces a conformational rearrangement of the N-terminal lid segment which closes upon the ATP binding pocket. Concomitant structural changes follow (Fig. 4B) which first lead to the formation of the closed state 1 in which the N-terminal domains are dimerized and finally the closed state 2, the completely closed state of Hsp90 in which the MD and the NTD contact each other (Fig. 4A) (Ali et al., 2006; Shiau et al., 2006). This completes the ATPase site as a residue of the MD is important for catalysis. After ATP hydrolysis, Hsp90 returns to the apo state with the release of ADP and Pi (Fig. 4B). All the above conformational transitions are isoenergetic as revealed by single molecule experiments (Hellenkamp et al., 2017; Mickler et al., 2009; Southworth & Agard, 2008). The rate of ATP hydrolysis by Hsp90 is rather low with ATPase rates of 0.1 min^{-1} in man and 1 min^{-1} in yeast (McLaughlin, Smith, & Jackson, 2002; Panaretou et al., 1998; Scheibel, Weikl, & Buchner, 1998; Scheibel et al. 1997). Also, the affinity of human Hsp90 for ATP is very low with a K_D of 400 μM (Scheibel et al., 1997). The conformational changes in Hsp90 are subject to allosteric regulation with residues far away from the ATP binding site affecting its ATPase activity (Rehn et al., 2016; Retzlaff et al., 2009; Weikl et al., 2000). How these conformational transitions in Hsp90 and its ATPase cycle contribute mechanistically to client activation is not yet understood. The conformational dynamics in Hsp90 affects its interactions with both clients and

Fig 4—cont'd between the two states. (B) Conformational cycle of Hsp90. Hsp90 exists in a conformational equilibrium of different states. In the apo state, Hsp90 exists in a "V"-shaped conformation. ATP binding leads to the closure of the lid (shown in red) in the NTD forming the first intermediate. Then, the NTDs dimerize forming the first closed state of Hsp90 in which the NTD and MD interact. Further change in the relative orientations of the N and M domain leads to the formation of a completely closed state of Hsp90 which is competent for ATP hydrolysis. Upon ATP hydrolysis, the NTDs dissociate resulting in the regeneration of the open state of Hsp90 and a new cycle can start. (C) Cryo-EM structure of the Hsp90-Cdc37-Cdk4 complex (PDB 5FWK) (Verba et al., 2016). The kinase is in an open conformation with an unfolded β4-β5 strand (shown in blue) cutting across the MD of Hsp90 (green). Cdc37 bound to both kinase and Hsp90 is depicted in brown.

co-chaperones (Li & Buchner, 2013; Lorenz et al., 2014) both of which target specific Hsp90 states (Hessling, Richter, & Buchner, 2009; Lee et al., 2012; Lorenz et al., 2014; Sullivan et al., 1997).

2.3.2 Hsp90 functional regulation by co-chaperones

More than 20 Hsp90 co-chaperones have been identified to date which function in different ways (Li & Buchner, 2013; Li et al., 2012; Taipale et al., 2010). Some Hsp90 co-chaperones stabilize different conformational states of Hsp90, thereby either accelerating or slowing down the ATPase activity of Hsp90 and affecting Hsp90 function. Why is the cytosolic Hsp90 in eukaryotes targeted by a large number of co-chaperones? It induces directionality and makes Hsp90 cycle more deterministic (Ratzke, Nguyen, Mayer, & Hugel, 2012). A comprehensive in vivo study on different Hsp90 co-chaperones in yeast revealed that different Hsp90 clients require specific sets of co-chaperones for activation (Sahasrabudhe et al., 2017). This suggests that Hsp90 can adjust itself to the specific client need through the cohort of co-chaperones.

The largest subset of Hsp90 co-chaperones comprises a tetratricopeptide or TPR domain which binds to the MEEVD sequence in the C-terminal tail of Hsp90 (Li et al., 2012; Schopf et al., 2017; Young, Obermann, & Hartl, 1998). The TPR domain is made up of helix-turn-helix motifs which form a groove into which the C-terminal tail of Hsp90 can bind (Chen & Smith, 1998; Das, Cohen, & Barford, 1998; Scheufler et al., 2000; Schmid et al., 2012). The best characterized of the TPR co-chaperones is the "Hsp70 and Hsp90 organizing protein" Hop whose homolog in yeast is Sti1. Hop/Sti1 physically links the Hsp70 and Hsp90 chaperone systems thereby assisting transfer of client proteins from Hsp70 to Hsp90 (Chen & Smith, 1998; Johnson, Schumacher, Ross, & Toft, 1998; Kirschke et al., 2014; Wegele et al., 2006). Hop consists of three functional TPR domains: TPR2A binds to Hsp90 while there are two TPR binding sites for Hsp70 (TPR1 and TPR2B) (Scheufler et al., 2000; Schmid et al., 2012). Hsp90 binding to Sti1 changes the binding sites for Hsp70 enabling efficient client transfer (Röhl et al., 2015). Though the C-terminal tail of Hsp90 is sufficient for Hop/Sti1 binding, additional interaction with the Hsp90 MD has also been reported which are required for ATPase inhibition (Schmid et al., 2012). Sti1 acts as a non-competitive inhibitor of the Hsp90 ATPase activity by stabilizing the open conformation of Hsp90 (Li, Richter, & Buchner, 2011; Prodromou et al., 1999; Richter et al., 2003). There is no homolog of eukaryotic Sti1/Hop in bacteria where

Hsp70 and Hsp90 interact directly (Genest et al., 2015). Interestingly, there has been recent evidence that yeast Hsp70 and Hsp90 can also associate without Sti1 (Kravats et al., 2018).

Among the other TPR-domain-containing co-chaperones are the peptidyl-prolyl isomerases (PPIases) which catalyze *cis-trans* isomerization events involving proline peptide bonds in proteins (Fischer, 1994; Schiene-Fischer, 2015). Three members of the PPIase family in humans are important Hsp90 co-chaperones: cyclophilin 40 or Cyp40 and two FK506 binding proteins, FKBP51 and FKBP52, while in yeast the closely related Cpr6 and Cpr7 are Hsp90 associated PPIases (Li et al., 2012; Mayr, Richter, Lilie, & Buchner, 2000; Schopf et al., 2017). Interestingly, the PPIase activity has been found to be not essential for the activation of the glucocorticoid receptor (GR) and the androgen receptor (AR) by Hsp90 (Riggs et al., 2007), but whether this holds true for the other clients as well needs to be addressed. Nevertheless, PPIases are important for the progression of Hsp90 chaperone cycle (Li et al., 2011; Li, Richter, Reinstein, & Buchner, 2013). Since, PPIases exhibit chaperone activity and the binding sites on PPIases for non-native substrates and Hsp90 are distinct (Mayr et al., 2000; Pirkl, Fischer, Modrow, & Buchner, 2001), this suggests that PPIases could be involved in client handover to Hsp90 or coordinate with Hsp90 in client activation; however, direct evidence is missing. At least for one of the stringent Hsp90 clients GR, PPIases have been shown to regulate its activation where opposing effects were found for FKBP51 and FKBP52 on GR activation by Hsp90 (Davies, Ning, & Sánchez, 2005; Riggs et al., 2007; Wochnik et al., 2005). Since, an increasing body of evidence now suggests that increased FKBP51 levels and dysregulation of GR are associated with mental illness (Fries, Gassen, & Rein, 2017), an in depth mechanistic understanding of the mode of regulation of GR by the two FKBPs will have implications for the prognosis of mental disorders.

In contrast to above co-chaperones, Cdc37 is a dedicated co-chaperone for kinases which is essential for recruitment of kinase clients to Hsp90 (Keramisanou et al., 2016). Cdc37 binds to kinases as well as the NTD of Hsp90 thereby inhibiting its ATPase activity. In a cryo-EM structure of Hsp90-Cdc37-Cdk4 kinase complex, Cdc37 wraps around MD of Hsp90 making extensive contacts with both Hsp90 and kinase (Verba et al., 2016). Due to its specificity for kinases, the Hsp90-Cdc37 interaction could be a promising target for designing drugs specifically against Hsp90-dependent kinases.

Aha1 (activator of the Hsp90 ATPase) is a potent stimulator of the Hsp90 ATPase activity (Panaretou et al., 2002). Aha1 binds stably to the MD and accelerates the formation of an N-terminally dimerized state of Hsp90 (Hessling et al., 2009; Meyer et al., 2004; Retzlaff et al., 2010). Aha1 drives the Hsp90 reaction cycle forward by displacing Hop/Sti1 (Li et al., 2013). Furthermore, the client interaction and their activation have been shown to be influenced by Aha1 (Harst, Lin, & Obermann, 2005; Holmes, Sharp, Hobbs, & Workman, 2008; Sahasrabudhe et al., 2017). Disruption of the Hsp90-Aha1 complex results in increased stabilization of mutant cystic fibrosis transmembrane conductance regulator (CFTR). The GR and the Aha1 binding site in the MD overlap which results in competition for binding (Lorenz et al., 2014; Wang et al., 2006). In this context, a chemical compound inhibiting the Hsp90-Aha1 interaction has been recently identified (Stiegler et al., 2017).

An Hsp90 co-chaperone which acts toward the end of the Hsp90 cycle is p23 (in humans or Sba1 in yeast). p23 stabilizes the completely closed state (closed state 2) of Hsp90 by binding tightly in the junction formed by the two N-terminally dimerized domains (Ali et al., 2006). p23 binding slows down the ATPase activity of Hsp90 which is important for client association and maturation (Johnson & Toft, 1994; Lorenz et al., 2014; McLaughlin et al., 2006; Richter, Walter, & Buchner, 2004). p23 can directly interact with non-native substrates and possesses chaperone activity which is mainly attributed to its disordered C-terminal tail (Bose, Weikl, Bugl, & Buchner, 1996; Freeman, Toft, & Morimoto, 1996; Weaver et al., 2000; Weikl et al., 2000). Recently, a direct interaction has been demonstrated between p53 and p23 (Wu et al., 2018). This chaperone function of p23 seems to be important for its involvement in a broad range of processes like chromatin remodeling, Golgi function and ribosome biogenesis (Echtenkamp et al., 2011, 2016; Zelin et al., 2012). Taken together, Hsp90 is an immensely dynamic molecular machine that adjusts its parts, the co-chaperones, to suit the needs of diverse clients.

2.3.3 Hsp90-client interactions

A protein can be classified as an Hsp90 client if it physically interacts with Hsp90 and if inhibition of Hsp90 leads to a decrease in protein activity. This decrease upon Hsp90 inhibition can be due to client degradation or aggregation (Theodoraki & Caplan, 2012; Wang, Luo, Qin, & Xia, 2018; Xu et al., 2001). A diverse repertoire of clients ranging from signal transducers via hormone receptors to transcription factors depends on Hsp90 for their maintenance. Hence, Hsp90 affects almost all kinds of cellular processes and acts as a central hub of protein homeostasis. A comprehensive list of Hsp90 clients is maintained by the Picard laboratory. No clear binding motif or sequence has been identified to date for Hsp90 clients unlike for Hsp70. Generally Hsp90 interacts at late stages of folding with partially folded structures. However, our understanding of the molecular determinants that makes a protein an Hsp90 client or the principles that govern this interaction are still limited. Since many Hsp90 clients require a binding partner such as a ligand or DNA to achieve their final native structure, it seems safe to assume that Hsp90 targets metastable and energetically unstable states. Hence client instability is one of the driving forces that determine Hsp90 binding which was directly shown for kinases (Boczek et al., 2015; Taipale et al., 2012). The oncogenic kinase v-src which is the less stable variant of c-src is highly dependent on Hsp90 for its activity while c-src interacts very weakly with Hsp90 (Boczek et al., 2015; Falsone et al., 2004). The differences in Hsp90 interaction were found to be correlated with the thermal instability of kinase domains and the cooperativity of the folding transition (Taipale et al., 2012). For other Hsp90 clients such as GR and p53, a similar picture emerges. Hsp90 interacts with the ligand binding domain (LBD) in the case of GR (Howard, Holley, Yamamoto, & Distelhorst, 1990; Giannoukos, Silverstein, Pratt, & Simons, 1999; Kaul et al., 2002) and with the DNA binding domain (DBD) of p53 (Hagn et al., 2011; Müller et al., 2004). However, instability might not be the sole determinant of client recognition by Hsp90. In vitro analysis of the interaction of the tumor suppressor protein p53 and Hsp90 revealed that an unstable cancer mutant of p53 displayed less Hsp90 binding compared to wild type p53 (King, Wawrzynow, Höhfeld, & Zylicz, 2001); however, in vivo p53 cancer mutants interact more strongly with Hsp90 than wild type p53 (Blagosklonny, Toretsky, Bohen, & Neckers, 1996; Whitesell et al., 1998). This indicates that the cellular environment and interactions with other factors may affect Hsp90-client binding. Additionally, surface features like charge and hydrophobicity of client rather than contiguous amino acid sequence seem important determinants for Hsp90 interaction (Citri et al., 2006; Xu et al., 2005). All these findings suggest that Hsp90 binding determinants are diverse and the Hsp90 client range can change in response to environmental signals (McClellan et al., 2007).

Another unanswered question is what Hsp90 does do its clients? Do clients undergo conformational changes as they pass through the Hsp90

ATPase cycle? The client binding region may span across all three domains of Hsp90 but a consensus seems to develop that the major binding site is in the MD (Hagn et al., 2011; Karagoz et al., 2014; Lorenz et al., 2014; Verba et al., 2016). It is reasonable to assume that since Hsp90 undergoes large changes at domain interfaces during the ATPase cycle, these changes may be injected into the bound client. In this context, the conformational states of clients bound to Hsp90 were investigated for p53. An unfolded or molten globule state of the p53 DNA binding domain (p53-DBD) has been suggested to bind to Hsp90 (Park, Borin, et al., 2011; Park, Kostic, et al., 2011; Rüdiger, Freund, Veprintsev, & Fersht, 2002). However, this is challenged by the finding that Hsp90 binding does not cause major structural change in the folded p53-DBD (Hagn et al., 2011). Recently the cryo-EM structure of the Hsp90-Cdc37-Cdk4 complex showed that Cdk4 is bound to Hsp90 in a conformation where the domains are separated resulting in an open, less compact state (Fig. 4C) (Verba et al., 2016). Further, analysis of client conformation by single molecule FRET revealed that Hsp90 stabilizes an open state out of an ensemble of conformations of the RISC Argonaute protein, AGO2 (Tsuboyama, Tadakuma, & Tomari, 2018). Together, these findings suggest that rather than acting as an active foldase or unfoldase, Hsp90 stabilizes or traps a certain client conformation that is folding or ligand binding competent. However, more work is needed on the active or passive role of Hsp90 in client folding.

2.3.4 Chaperoning cycle of Hsp90

For its maturation, client proteins pass through various Hsp90 complexes which include different Hsp90 co-chaperones and potentially Hsp70 with its cofactors (Fig. 5). The components of the complexes change during the cycle and dictate the association and dissociation of client to and from Hsp90. For some of the stringent Hsp90 clients, like SHRs and kinases, the identity of these complexes seems well established. However still, already in this group of clients differences exist and for other clients the maturation steps are still elusive. In early in vitro reconstitution experiments with the GR, five chaperone components were found to be important for GR maturation: Hsp70, Hsp40, Hop, Hsp90, p23 (Dittmar, Hutchison, Owens-Grillo, & Pratt, 1996; Pratt, Morishima, Murphy, & Harrell, 2006). Hsp70 binding to GR occurs before it is handed over to Hsp90 involving the adaptor protein Hop (Pratt & Toft, 1997) (Fig. 5). Recently, it has been shown that Hsp70 locally unfolds the GR-LBD and Hsp90 together with Hop and p23 reverses this effect (Kirschke et al., 2014).

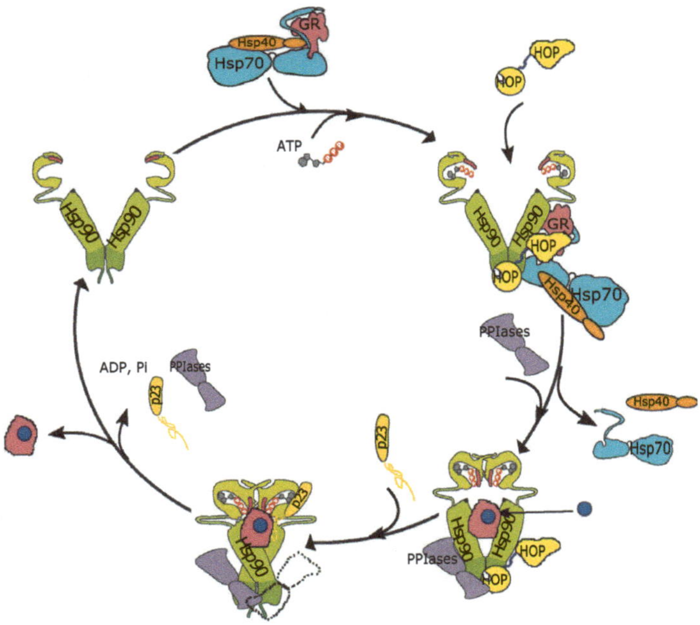

Fig. 5 Hsp90 chaperone cycle. The maturation cycle of glucocorticoid receptor (GR) by Hsp90 is shown. GR is transferred in a partially unfolded state to Hsp90 by Hsp70 and Hsp40 system via the adaptor co-chaperone Hop (Sti1 in yeast). Hop stabilizes the open state of Hsp90 inhibiting its ATPase activity and a stable complex is formed with the GR. Peptidyl-prolyl *cis-trans* isomerase (PPIases) can bind replacing Hop together with the binding of p23. A second molecule of PPIases may bind to the second TPR site (shown in dashed lines). These events lead to the formation of completely closed state of Hsp90. After ATP hydrolysis, GR and p23 are released. Since hormone-bound GR can also interact with Hsp90, additional trigger rendering GR into a "non-client" can be envisioned.

However, clients, including GR and p53, can also interact with Hsp90 directly (Lorenz et al., 2014; Müller et al., 2004; Walerych et al., 2010). Whether prior interaction with Hsp70 is necessary for all Hsp90 clients and how Hsp70 interaction enables clients in their downstream Hsp90 pathway remain enigmatic. Once, GR forms a complex with Hsp90, PPIases are recruited which enable dissociation of Hop/Sti1 from the complex (Fig. 5) and allow the Hsp90 ATPase cycle to proceed (Lorenz et al., 2014). Finally, p23 associates to the Hsp90-GR-PPIase complex in an ATP-dependent manner stabilizing the GR complex (Fig. 5) (Lorenz et al., 2014; Pratt & Toft, 1997). It is still unclear how the release of GR from Hsp90 takes place. Hormone binding can be speculated to trigger GR release from Hsp90; however, hormone-bound GR has also been shown to bind to Hsp90

(Kirschke et al., 2014; Lorenz et al., 2014). This suggests that there is another level of regulation that mediates release of GR from Hsp90.

For kinases, the decision whether a kinase will interact with Hsp90 or not is taken by the co-chaperone Cdc37 which senses instability in the kinase conformation (Keramisanou et al., 2016). Further, Cdc37 stabilizes an open conformation of Hsp90 (Roe et al., 2004). What the subsequent steps are and how the closed conformation of Hsp90 observed in the cryo-EM structure (Verba et al., 2016) is achieved are not known.

2.3.5 Hsp90 in cancer

It has been found that Hsp90 expression increases in cancer cells as a result of the proteo-toxic environment (Ferrarini, Heltai, Zocchi, & Rugarli, 1992). As Hsp90 is involved in the stabilization and activation of key regulatory proteins, Hsp90 overexpression in cancerous cells promotes their maturation. In this context, it has been shown that many cancer-causing proteins like p53 cancer mutants or oncogenic kinases are highly dependent on Hsp90 for their maintenance (Whitesell & Lindquist, 2005). In the case of p53, association with Hsp90 prevents the degradation of mutant protein (Blagosklonny et al., 1996; Sepehrnia, Paz, Dasgupta, & Momand, 1996; Whitesell et al., 1998). This destroys the normal cellular function of the wild type protein by forming inactive tetramers leading to cancer progression (Oren & Rotter, 2010). For the reasons given, Hsp90 has become an important target for anti-cancer drugs. Several Hsp90 inhibitors have been developed over recent years making use of the unique binding site for ATP. Some of them displayed promising anti-tumor effects and have entered clinical trials. Cancer cells are generally more sensitive to Hsp90 inhibitors compared to normal cells (Chiosis & Neckers, 2006). Among the cancer cells two types were identified differing in sensitivity to Hsp90 inhibition (Rodina et al., 2016). In the more sensitive cells, Hsp90 is present in large multi-molecular complexes containing Hsp70 and co-chaperones and clients which changes the microenvironment of the tumor (Rodina et al., 2016).

The first Hsp90 inhibitors discovered were geldanamycin and radicicol. Both of them act by competing with ATP for the ATP binding site in the N-terminal domain of Hsp90 (Schulte et al., 1998; Stebbins et al., 1997; Whitesell et al., 1994). However, both geldanamycin and radicicol are not useful as anti-tumor drugs because of high toxicity or stability issues, respectively (Sreedhar, Soti, & Csermely, 2004; Supko, Hickman, Grever, & Malspeis, 1995). There are other inhibitors that bind to the

CTD, disrupting interactions of Hsp90 with TPR domain co-chaperones. Novobiocin is one such inhibitor that does not trigger a heat shock response which is found to be associated with the NTD inhibitors (Whitesell & Lindquist, 2005). Further, inhibitors that target specific Hsp90-cochaperone complexes are currently being developed like derrubone, withaferin A and celastrol which disrupt the Cdc37-Hsp90 interaction and gedunin, an inhibitor of the Hsp90-p23 complex (Garg, Khandelwal, & Blagg, 2016). Targeting Hsp90 and oncogenic client proteins simultaneously seems to reduce oncogene switching in cancer cells (Pashtan et al., 2008) and could serve a promising anti-cancer therapy. PU-H71 is an Hsp90 inhibitor in clinical trials that interestingly binds more strongly to Hsp90 when it is present in complexes with co-chaperones and client proteins (Rodina et al., 2016). Hence, there is still room for targeting Hsp90 with mechanistically different compounds with better specificity and less toxic effects.

2.4 Small heat shock proteins (sHsps)

Small heat shock proteins or sHsps are a diverse class of molecular chaperones which are conserved across species and found in all kingdoms of life (Kriehuber et al., 2010; Vierling, 1997). They are small proteins of 15–40 kDa that often accumulate in cells under stress (Tissiéres, Mitchell, & Tracy, 1974). sHsps differ in their sequences and sizes, suggesting a large functional diversity within this class of molecular chaperones (Franck et al., 2004; Haslbeck & Buchner, 2002a, 2002b). The human genome codes for 10 different sHsps (HspB1–HspB10). Some are ubiquitously present like HspB1 (or Hsp27) while others are tissue-specific like the αA-crystallin (or HspB5) which is localized to the eye lens (Garrido, Paul, Seigneuric, & Kampinga, 2012). sHsps act as a first line of defense against stress, protecting proteins in non-native states and assisting their refolding in cooperation with Hsp70 (Fig. 3C). Being active players in preventing irreversible protein aggregation and channeling the misfolded or damaged proteins back to the folding route, sHsps play a critical role in maintaining protein homeostasis (Strauch & Haslbeck, 2016). A unique property of sHsps is their ability to form oligomers of various sizes which are found to be important for their function.

2.4.1 Structure of sHsps

Despite their sequence and oligomer size diversity, sHsps share a common structural framework consisting of an α-crystallin domain (ACD) which derives its name from the most prominent sHsp family member, the α-crystallin protein of the eye lens. The ACD is flanked by a flexible and non-conserved N-terminal (NTR) and a C-terminal region (CTR) (Haslbeck & Buchner, 2002a, 2002b; Haslbeck, Peschek, Buchner, & Weinkauf, 2016; Kappe, Leunissen, & de Jong, 2002; Kriehuber et al., 2010). Structurally, the ACD consists predominantly of β-strands exhibiting an immunoglobulin like β-sandwich fold (Hanazono et al., 2013; Jehle et al., 2010; Kim, Kim, & Kim, 1998; Laganowsky et al., 2010; Van Montfort, Slingsby, & Vierling, 2001; van Montfort et al., 2001). The NTR exhibits conformational flexibility. It seems to be largely unfolded with some helical elements (Fleckenstein et al., 2015; Hanazono et al., 2013; Jehle et al., 2011; Kim et al., 1998; van Montfort et al., 2001). The NTR is also a major site for post-translational modifications in sHsps and hence plays an important role in regulating their activity in vivo (Arrigo, 2013; Gaestel, 2002; Ito et al., 1997; van den IJssel, Overkamp, Bloemendal, & de Jong, 1998). Another flexible region of the sHsps is the CTR which is essential for the assembly of oligomers (Delbecq & Klevit, 2013). The CTR is believed to contribute to the solubility of oligomers of sHsps owing to its hydrophilic and polar nature (Smulders et al., 1996)

The oligomer size of sHsps ranges from 2-mer to 50-mer with the larger oligomer assemblies forming a hollow ball-like or barrel structure with openings toward the interior (Fig. 6A, left panel) (Basha et al., 2013; Bepperling et al., 2012; Fleckenstein et al., 2015; Hanazono et al., 2013; Kim et al., 1998; Peschek et al., 2009, 2013; van Montfort et al., 2001). The building block of the larger oligomers is the dimer which undergoes a hierarchical arrangement of subunits to form higher order oligomers (Basha, O'Neill, & Vierling, 2012; Delbecq & Klevit, 2013; Haslbeck, Franzmann, Weinfurtner, & Buchner, 2005). A dimer is characterized by two subunits of β6 strands which are reciprocally swapped to form a β-sandwich between monomers (Fig. 6A, middle panel) (Bepperling et al., 2012; Hanazono et al., 2013; Kim et al., 1998; van Montfort et al., 2001) This is commonly referred to as "bacterial" type of dimer or "β6-swapped dimer." Another type of dimer exists in metazoans, which is found in the α-crystallins. In this type, dimer interface is formed by interaction of antiparallel β6+7 strands from each protomer (Fig. 6A, right panel). Though the ACD domain mediates stable dimer formation, both the N-terminal and C-terminal regions are essential for oligomer assembly (Bagneris et al., 2009; Baranova et al., 2011; Basha et al., 2012; Chen et al., 2010; Garrido et al., 2012; Haslbeck et al., 2005; Laganowsky et al. 2010). The oligomerization process of sHsps is highly dynamic with

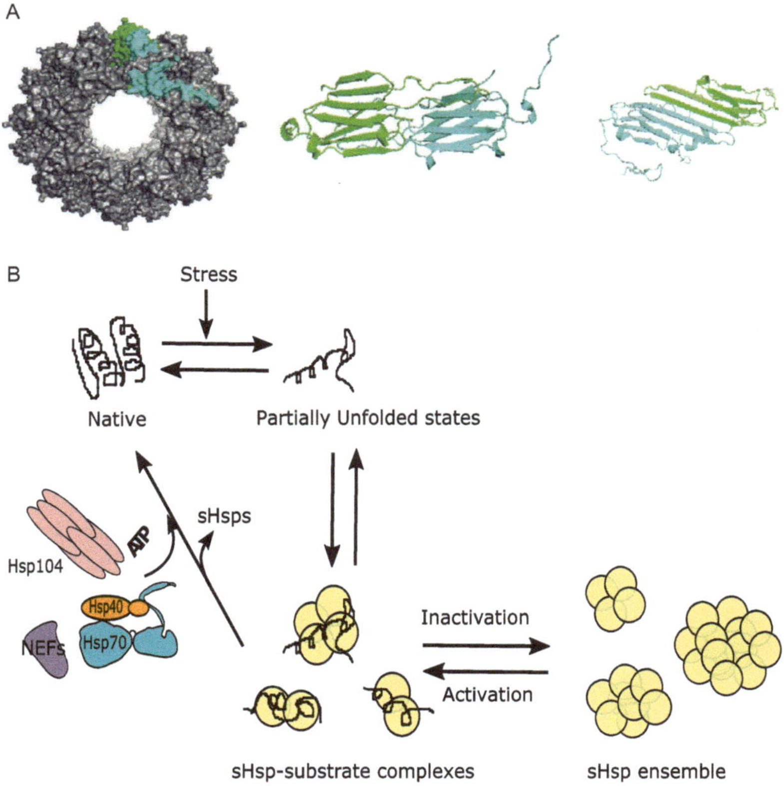

Fig. 6 Structure-Function of sHsp. (A) Surface representation of crystal structure of oligomer (32-mer) of sHsp Sip1 from *C. elegans* (PDB 4YDZ) (Fleckenstein et al., 2015) where two protomers of a dimer are in green and cyan (left). Dimers give rise to higher ordered oligomers that form a ball-like structure with openings to the inside of the sphere. The middle panel is the β6-swapped dimer arrangement of *M. jannaschii* Hsp 16.5 (PDB 1SHS) (Kim et al., 1998) and the right panel is the "metazoan" type β6+7 anti-parallel dimer (PDB 4YDZ). (B) Model for the chaperone function of sHsps. sHsps exist in an ensemble of higher order oligomers which dissociate into smaller species upon stresses like heat shock. The activated sHsp dimers bind to unfolded substrates in an ATP-independent manner which sequesters the denatured or misfolded substrates and prevent them from irreversible aggregation. Bound substrates in complexes with sHsp are released and reactivated by cooperation with ATP-dependent chaperone families like Hsp70-Hsp40 and/or Hsp100 (ClpB in bacteria). Efficient refolding by Hsp70 requires Hsp40 in eukaryotes (Ehrnsperger et al., 1997) and both DnaJ and GrpE in bacteria (Veinger, Diamant, Buchner, & Goloubinoff, 1998).

oligomers of different sizes existing in equilibrium with each other (Basha et al., 2012; Braun et al., 2011; Haslbeck, Weinkauf, & Buchner, 2015; Stengel et al., 2010). Rapid subunit association and dissociation can occur within the same sHsp or hetero-oligomers are generated when different sHsps interact (Bova, McHaourab, Han, & Fung, 2000; Haslbeck et al., 2016; Lentze et al., 2004; Sobott, Benesch, Vierling, & Robinson, 2002). This has functional implications for sHsps as the change in oligomer size and/or composition influences the exposed hydrophobic regions in sHsps, hence affecting substrate binding.

2.4.2 Binding of substrates by sHsps

Despite extensive research on sHsps, it still remains enigmatic how they recognize their substrates. sHsps probably bind to surface exposed hydrophobic regions of non-native or misfolded states; however, no binding motif on substrates is known so far. sHsps can bind to a large number of non-native proteins in one complex. Proteomic approaches have advanced our knowledge of the substrates that are associated with sHsps revealing significant diversity in the substrates bound (Basha et al., 2004; Bepperling et al., 2012; Fan et al., 2014; Fu et al., 2013; Haslbeck et al., 2004; Peschek et al., 2013). Though sHsps display promiscuous binding to unfolded proteins, substrate preferences begin to be identified (Fleckenstein et al., 2015; Fu et al., 2013; Mymrikov et al., 2016). In this context, differences in the substrate specificity have also been identified for different human sHsp members (Mymrikov et al., 2016) suggesting that members of sHsp family may have overlapping yet distinct substrate spectra.

The substrate binding site on sHsps is still enigmatic as all three regions, the NTR, the ACD and the CTR, have been shown to contribute to the substrate binding by sHsps (Hochberg et al., 2014; Lee, Roseman, Saibil, & Vierling, 1997; Lentze & Narberhaus, 2004; Treweek, Rekas, Walker, & Carver, 2010; Treweek et al., 2007). Since, the NTR and CTR sequences are non-conserved and variable, their potential involvement in substrate binding can explain how different sHsps can bind to different substrates. Furthermore, potentially different binding sites may be used by sHsps for different kinds of substrates (Mainz et al., 2015). One of the important questions is which oligomeric state of sHsps preferentially interacts with the substrate and whether this is substrate-dependent. In recent years evidence has been reported that small oligomeric states of sHsps bind to substrates better than higher oligomeric states where the hydrophobic residues are buried unless required upon need (Basha et al., 2012; Haslbeck et al., 2008; Peschek et al., 2013; van Montfort et al., 2001). However, some Hsps do not dissociate at temperatures of substrate denaturation and still effectively protect the substrate from aggregation indicating that different sHsps may utilize different mechanisms of substrate binding and chaperoning (Basha, Jones, Wysocki, & Vierling, 2010). This is further complicated by the post-translational modifications in sHsps which regulate their oligmerization and hence their binding affinities to the substrate (McHaourab, Godar, & Stewart, 2009; Shashidharamurthy, Koteiche, Dong, & Mchaourab, 2005). Substrate binding has consequences for sHsps as well. Substrates can shift the equilibrium toward lower order oligomers (Franzmann, Menhorn, Walter, & Buchner, 2008) or are capable of directly affecting the sHsp structure as recently shown by the crystal structure of HspB6 in complex with 14-3-3 regulatory protein where the NTR was found to be significantly ordered (Sluchanko et al., 2017).

2.4.3 Chaperone activity of sHsps

sHsps have been suggested to act as chaperones by preventing irreversible aggregation of heat or stress denatured proteins in an ATP-independent manner in vitro (Horwitz, 1992; Jakob, Gaestel, Engel, & Buchner, 1993; Lee, Pokala, & Vierling, 1995; Torok et al., 2001). sHsps bind to partially unfolded intermediates of proteins generated by various stress conditions keeping them in a folding-competent state and preventing their irreversible loss until they are taken care of by the downstream processing machinery (Fig. 6B). The chaperone activity of sHsps is usually assessed by in vitro aggregation assays where model substrate proteins are subjected to heat-induced aggregation and the activity of sHsps in preventing substrate aggregation is monitored by light scattering experiments (Buchner, Grallert, & Jakob, 1998; Jakob et al., 1993). The activity of sHsps toward cellular lysates can be quantified by their ability to maintain the proteins in the soluble state. In this context, sHsps are generally less efficient in preventing aggregation of large proteins suggesting that the mass ratio of substrate to sHsps is important than the molar ratio for sHsp chaperoning (Haslbeck & Vierling, 2015).

Generally, sHsps exist in a conformational ensemble of different oligomeric states where higher order oligomers predominate. Stress conditions like high temperature shift the equilibrium to the activated state of sHsps in which smaller oligomers are the dominant species (Fig. 6B). Non-native proteins can then bind to the activated sHsps often forming very stable complexes in which sHsps are found to be present again in larger oligomers (Haslbeck et al., 1999; Jakob et al., 1993; McHaourab et al., 2009; Stromer, Ehrnsperger, Gaestel, & Buchner, 2003). Since they do not possess ATPase activity of their own, sHsps are incapable of refolding the substrates. The substrates are refolded either spontaneously (Haslbeck et al., 2015; McHaourab et al., 2009) or are handed over to the Hsp70/Hsp40 or

Hsp100 system for their further reactivation (Ehrnsperger et al., 1997; Lee & Vierling 2000; Strozecka et al., 2012; Ungelenk et al., 2016) (Fig. 6B). Recently, Bag-3 has been shown to physically link sHsps and Hsp70 and promote refolding of client proteins by accelerating the de-oligomerization of sHsps (Rauch et al., 2017)

Oligomerization of sHsps is regulated by a change in the environmental conditions like temperature, pH or by phosphorylation of sHsps (Fleckenstein et al., 2015; Haslbeck et al., 1999; Peschek et al., 2013; Rajagopal et al., 2015). All the above mentioned conditions tend to regulate the process of subunit exchange and structural transitions in sHsps oligomers thereby changing the binding affinities for the substrates and chaperone activity. Interestingly, for the developmental-specific sHsp Sip1 in *Caenorhabditis elegans*, substrate specificity is imparted by a pH-regulated switch that activates Sip1 at the acidic conditions found in embryos highlighting how changes in the oligomer structure contributes to specific functions of sHsps. Phosphorylation acts as another mode of regulation for the chaperone activity of sHsps. For αB crystallin, mimicking phosphorylation of the NTR by glutamate mutations results in the destabilization of the oligomers leading to increased chaperone activity (Peschek et al., 2013). Hence, in the absence of ATPase activity or co-chaperones, sHsps have evolved complex ways of regulating their function mainly by controlling their oligomer size and composition.

During evolution, sHsps seem to have become functionally diversified yet share working principle of keeping the misfolded states of proteins in a folding-competent conformation to improve cellular fitness (Ungelenk et al., 2016). In vivo they may not prevent aggregation but deposit proteins in aggregates that can be recycled (Kampinga et al., 1995; Mogk et al., 2003; Specht, Miller, Mogk, & Bukau, 2011). Yeast Hsp42 has been shown to promote organized aggregation which can be cyto-protective under repeated stress while another sHsp, Hsp26, facilitates Hsp70-mediated refolding post stress (Ungelenk et al., 2016). Hence, sHsps do not only serve the role of "holdases" but together function to combat protein misfolding and aggregation through their functional diversity. Not surprisingly, nature has evolved a large number of small heat shock proteins to perform diverse functions suiting cellular needs.

2.4.4 Diverse functions of sHsps

sHsps are not only required under stress conditions but are essential for housekeeping functions in the cell (Mymrikov, Seit-Nebi, & Gusev,

2011). For example, yeast Hsp42 is involved in maintaining protein homeostasis under physiological conditions (Haslbeck et al., 2004; Specht et al., 2011). Different functions have been described for different members of sHsp family like protein folding, maintenance of cytoskeleton, combating oxidative stress, prevention of neuropathies, cancer and protein misfolding diseases (Garrido et al., 2012; Mymrikov et al., 2011). Of special mention is the essential requirement of sHsps in the maintenance of the refractive index of the eye in vertebrates (Horwitz, 2003). The α-crystallins which comprise the two homologous proteins, αA and αB crystallin, are the most prominent proteins of the eye lens. They are believed to prevent protein aggregation in the highly crowded environment of the eye lens and maintain its transparency (Horwitz, 2003). Mutants in α-crystallins have altered chaperone activity and are associated with cataract (Andley, Hamilton, & Ravi, 2008; Brady et al., 1997; Cheng et al., 2010; Huang et al., 2009; Xi et al., 2008).

sHsps are also found to be upregulated in neurodegenerative diseases (Carra et al., 2013). HspB5 and αB crystallin were shown to possess neuro-protective capacity in brain (Carra et al., 2017). Moreover, HspB1 and HspB5 can stably bind to and prevent α-synuclein amyloid fibril formation (Cox et al., 2016; Hochberg et al., 2014). The mechanism underlying their neuroprotective effects is still unclear. They may either balance the cellular redox state or prevent the activation of the apoptotic cascade. Understanding these mechanisms can help in targeting a specific member in a specific cell or tissue type for therapeutic purposes. Interestingly, upregulation of sHsps has also been found to be associated with counteracting proteostasis decline and increased longevity during aging (Hsu, Murphy, & Kenyon, 2003; Walther et al., 2015). In this context, sequestering proteins into aggregates by sHsps is a protective strategy to manage the burden of the proteostasis system during aging and could prove critical in regulating the lifespan of an organism.

2.5 Conclusions and perspectives

Enormous progress has been made in past years in understanding the process of protein folding and how chaperones are involved in it. Structural and functional aspects of different chaperone systems have been elucidated and our knowledge on specific roles of each chaperone class has extended. From both in vitro and in vivo studies, it is now well established that molecular chaperones are not only required to enable a protein to reach its native state but some of them, like Hsp90, are needed by selected "client" proteins to maintain their active state also under physiological conditions. Besides, with the development of advanced biophysical methods, we now know that chaperones are capable of inducing/supporting conformational changes in proteins and, in doing so, they do not work in isolation but there is interplay of different chaperone systems in the processing of client proteins. However, we are still far from achieving a comprehensive understanding of the chaperone networks operating in the cell. In this context, how different chaperone systems cooperate in maintaining proteostasis during the lifespan of an organism is still an important issue that may contribute to the treatment of protein folding diseases. A number of pathological conditions are now known to be caused by defects in protein homeostasis. Understanding the complex regulatory circuit of the proteostasis network and why this circuit fails during diseases is a major challenge for the chaperone field. Key questions include investigating if and how the proteome changes during aging or in pathological conditions, what the characteristic of a healthy proteome versus a diseased proteome is, and why only some proteins form aggregates or amyloids while others are successfully degraded. Further, can we tweak the proteostasis network in an aberrant condition to regain the normal state? Addressing these questions will be important for therapeutic intervention.

References

Ali, M. M., et al. (2006). Crystal structure of an Hsp90-nucleotide-p23/Sba1 closed chaperone complex. *Nature, 440*(7087), 1013–1017.

Andley, U. P., Hamilton, P. D., & Ravi, N. (2008). Mechanism of insolubilization by a single-point mutation in αA-crystallin linked with hereditary human cataracts. *Biochemistry, 47*(36), 9697–9706.

Anfinsen, C. B. (1973). Principles that govern the folding of protein chains. *Science, 181*(4096), 223–230.

Anfinsen, C. B., Haber, E., Sela, M., & White, F. H., Jr. (1961). The kinetics of formation of native ribonuclease during oxidation of the reduced polypeptide chain. *Proceedings of the National Academy of Sciences of the United States of America, 47*, 1309–1314.

Arai, M., & Kuwajima, K. (1996). Rapid formation of a molten globule intermediate in refolding of alpha-lactalbumin. *Folding & Design, 1*(4), 275–287.

Arrigo, A. P. (2013). Human small heat shock proteins: Protein interactomes of homo- and hetero-oligomeric complexes: An update. *FEBS Letters, 587*(13), 1959–1969.

Ashcroft, A. E., et al. (2002). Structural plasticity and noncovalent substrate binding in the GroEL apical domain: A study using electrospray ionization mass spectrometry and fluorescence binding studies. *Journal of Biological Chemistry, 277*(36), 33115–33126.

Azem, A., Kessel, M., & Goloubinoff, P. (1994). Characterization of a functional GroEL14 (GroES7)2 chaperonin hetero-oligomer. *Science, 265*(5172), 653–656.

Bagneris, C., et al. (2009). Crystal structures of alpha-crystallin domain dimers of alphaB-crystallin and Hsp20. *Journal of Molecular Biology, 392*(5), 1242–1252.

Baranova, E. V., et al. (2011). Three-dimensional structure of alpha-crystallin domain dimers of human small heat shock proteins HSPB1 and HSPB6. *Journal of Molecular Biology, 411*(1), 110–122.

Bardwell, J. C., & Craig, E. A. (1988). Ancient heat shock gene is dispensable. *Journal of Bacteriology, 170*(7), 2977–2983.
Barraclough, R., & Ellis, R. J. (1980). Protein synthesis in chloroplasts. IX. Assembly of newly-synthesized large subunits into ribulose bisphosphate carboxylase in isolated intact pea chloroplasts. *Biochimica et Biophysica Acta, 608*(1), 19–31.
Basha, E., Jones, C., Wysocki, V., & Vierling, E. (2010). Mechanistic differences between two conserved classes of small heat shock proteins found in the plant cytosol. *Journal of Biological Chemistry, 285*(15), 11489–11497.
Basha, E., O'Neill, H., & Vierling, E. (2012). Small heat shock proteins and alpha-crystallins: Dynamic proteins with flexible functions. *Trends in Biochemical Sciences, 37*(3), 106–117.
Basha, E., et al. (2004). The identity of proteins associated with a small heat shock protein during heat stress in vivo indicates that these chaperones protect a wide range of cellular functions. *Journal of Biological Chemistry, 279*(9), 7566–7575.
Basha, E., et al. (2013). An unusual dimeric small heat shock protein provides insight into the mechanism of this class of chaperones. *Journal of Molecular Biology, 425*(10), 1683–1696.
Batey, S., Scott, K. A., & Clarke, J. (2006). Complex folding kinetics of a multidomain protein. *Biophysical Journal, 90*(6), 2120–2130.
Behnke, J., Feige, M. J., & Hendershot, L. M. (2015). BiP and its nucleotide exchange factors Grp170 and Sil1: Mechanisms of action and biological functions. *Journal of Molecular Biology, 427*(7), 1589–1608.
Behnke, J., et al. (2016). Members of the Hsp70 family recognize distinct types of sequences to execute ER quality control. *Molecular Cell, 63*(5), 739–752.
Beissinger, M., Rutkat, K., & Buchner, J. (1999). Catalysis, commitment and encapsulation during GroE-mediated folding. *Journal of Molecular Biology, 289*(4), 1075–1092.
Ben-Zvi, A. P., Chatellier, J., Fersht, A. R., & Goloubinoff, P. (1998). Minimal and optimal mechanisms for GroE-mediated protein folding. *Proceedings of the National Academy of Sciences of the United States of America, 95*(26), 15275–15280.
Bepperling, A., et al. (2012). Alternative bacterial two-component small heat shock protein systems. *Proceedings of the National Academy of Sciences of the United States of America, 109*(50), 20407–20412.
Bertelsen, E. B., Chang, L., Gestwicki, J. E., & Zuiderweg, E. R. (2009). Solution conformation of wild-type E. coli Hsp70 (DnaK) chaperone complexed with ADP and substrate. *Proceedings of the National Academy of Sciences of the United States of America, 106*(21), 8471–8476.
Bhutani, N., & Udgaonkar, J. B. (2001). GroEL channels the folding of thioredoxin along one kinetic route. *Journal of Molecular Biology, 314*(5), 1167–1179.
Blagosklonny, M. V., Toretsky, J., Bohen, S., & Neckers, L. (1996). Mutant conformation of p53 translated in vitro or in vivo requires functional HSP90. *Proceedings of the National Academy of Sciences of the United States of America, 93*(16), 8379–8383.
Blond-Elguindi, S., et al. (1993). Affinity panning of a library of peptides displayed on bacteriophages reveals the binding specificity of BiP. *Cell, 75*(4), 717–728.
Boczek, E. E., et al. (2015). Conformational processing of oncogenic v-Src kinase by the molecular chaperone Hsp90. *Proceedings of the National Academy of Sciences of the United States of America, 112*(25), E3189–E3198.
Borkovich, K. A., et al. (1989). hsp82 is an essential protein that is required in higher concentrations for growth of cells at higher temperatures. *Molecular and Cellular Biology, 9*(9), 3919–3930.
Bose, S., Weikl, T., Bugl, H., & Buchner, J. (1996). Chaperone function of Hsp90-associated proteins. *Science, 274*(5293), 1715–1717.
Bova, M. P., McHaourab, H. S., Han, Y., & Fung, B. K. (2000). Subunit exchange of small heat shock proteins. Analysis of oligomer formation of alphaA-crystallin and Hsp27 by fluorescence resonance energy transfer and site-directed truncations. *Journal of Biological Chemistry, 275*(2), 1035–1042.

Bracher, A., & Verghese, J. (2015). The nucleotide exchange factors of Hsp70 molecular chaperones. *Frontiers in Molecular Biosciences*, *2*, 10.
Brady, J. P., et al. (1997). Targeted disruption of the mouse αA-crystallin gene induces cataract and cytoplasmic inclusion bodies containing the small heat shock protein αB-crystallin. *Proceedings of the National Academy of Sciences of the United States of America*, *94*(3), 884–889.
Braell, W. A., Schlossman, D. M., Schmid, S. L., & Rothman, J. E. (1984). Dissociation of clathrin coats coupled to the hydrolysis of ATP: Role of an uncoating ATPase. *Journal of Cell Biology*, *99*(2), 734–741.
Braig, K., Adams, P. D., & Brünger, A. T. (1995). Conformational variability in the refined structure of the chaperonin GroEL at 2.8 Å resolution. *Nature Structural Biology*, *2*, 1083.
Braig, K., et al. (1993). A polypeptide bound by the chaperonin groEL is localized within a central cavity. *Proceedings of the National Academy of Sciences of the United States of America*, *90*(9), 3978.
Braig, K., et al. (1994). The crystal structure of the bacterial chaperonin GroEL at 2.8 Å. *Nature*, *371*(6498), 578–586.
Braun, N., et al. (2011). Multiple molecular architectures of the eye lens chaperone alphaB-crystallin elucidated by a triple hybrid approach. *Proceedings of the National Academy of Sciences of the United States of America*, *108*(51), 20491–20496.
Buchner, J., Grallert, H., & Jakob, U. (1998). Analysis of chaperone function using citrate synthase as nonnative substrate protein. *Methods in Enzymology*, *290*, 323–338.
Buchner, J., et al. (1991). GroE facilitates refolding of citrate synthase by suppressing aggregation. *Biochemistry*, *30*(6), 1586–1591.
Buckle, A. M., Zahn, R., & Fersht, A. R. (1997). A structural model for GroEL–polypeptide recognition. *Proceedings of the National Academy of Sciences of the United States of America*, *94*(8), 3571–3575.
Bukau, B., Weissman, J., & Horwich, A. (2006). Molecular chaperones and protein quality control. *Cell*, *125*(3), 443–451.
Burston, S. G., Ranson, N. A., & Clarke, A. R. (1995). The origins and consequences of asymmetry in the chaperonin reaction cycle. *Journal of Molecular Biology*, *249*(1), 138–152.
Carra, S., et al. (2013). Different anti-aggregation and pro-degradative functions of the members of the mammalian sHSP family in neurological disorders. *Philosophical Transactions of the Royal Society of London. Series B, Biological Sciences*, *368*(1617).
Carra, S., et al. (2017). The growing world of small heat shock proteins: From structure to functions. *Cell Stress & Chaperones*, *22*(4), 601–611.
Chacinska, A., et al. (2009). Importing mitochondrial proteins: Machineries and mechanisms. *Cell*, *138*(4), 628–644.
Chadli, A., et al. (2000). Dimerization and N-terminal domain proximity underlie the function of the molecular chaperone heat shock protein 90. *Proceedings of the National Academy of Sciences of the United States of America*, *97*(23), 12524–12529.
Chakraborty, K., et al. (2010). Chaperonin-catalyzed rescue of kinetically trapped states in protein folding. *Cell*, *142*(1), 112–122.
Chan, H. S., & Dill, K. A. (1996). A simple model of chaperonin-mediated protein folding. *Proteins*, *24*(3), 345–351.
Chaudhuri, T. K., Verma, V. K., & Maheshwari, A. (2009). GroEL assisted folding of large polypeptide substrates in Escherichia coli: Present scenario and assignments for the future. *Progress in Biophysics and Molecular Biology*, *99*(1), 42–50.
Chaudhuri, T. K., et al. (2001). GroEL/GroES-mediated folding of a protein too large to be encapsulated. *Cell*, *107*(2), 235–246.
Chen, S., & Smith, D. F. (1998). Hop as an adaptor in the heat shock protein 70 (Hsp70) and hsp90 chaperone machinery. *Journal of Biological Chemistry*, *273*(52), 35194–35200.

Chen, B., Zhong, D., & Monteiro, A. (2006). Comparative genomics and evolution of the HSP90 family of genes across all kingdoms of organisms. *BMC Genomics*, 7, 156.
Chen, S., et al. (1994). Location of a folding protein and shape changes in GroEL-GroES complexes imaged by cryo-electron microscopy. *Nature*, *371*(6494), 261–264.
Chen, J., et al. (2010). Regions outside the alpha-crystallin domain of the small heat shock protein Hsp26 are required for its dimerization. *Journal of Molecular Biology*, *398*(1), 122–131.
Cheng, M. Y., et al. (1989). Mitochondrial heat-shock protein hsp60 is essential for assembly of proteins imported into yeast mitochondria. *Nature*, *337*(6208), 620–625.
Cheng, C., et al. (2010). Altered chaperone-like activity of α-crystallins promotes cataractogenesis. *Journal of Biological Chemistry*, *285*(52), 41187–41193.
Chiosis, G., & Neckers, L. (2006). Tumor selectivity of Hsp90 inhibitors: The explanation remains elusive. *ACS Chemical Biology*, *1*(5), 279–284.
Chiosis, G., et al. (2002). Development of a purine-scaffold novel class of Hsp90 binders that inhibit the proliferation of cancer cells and induce the degradation of Her2 tyrosine kinase. *Bioorganic and Medicinal Chemistry*, *10*(11), 3555–3564.
Chiti, F., & Dobson, C. M. (2017). Protein misfolding, amyloid formation, and human disease: A summary of progress over the last decade. *Annual Review of Biochemistry*, *86*, 27–68.
Citri, A., et al. (2006). Hsp90 recognizes a common surface on client kinases. *Journal of Biological Chemistry*, *281*(20), 14361–14369.
Clerico, E. M., Tilitsky, J. M., Meng, W., & Gierasch, L. M. (2015). How hsp70 molecular machines interact with their substrates to mediate diverse physiological functions. *Journal of Molecular Biology*, *427*(7), 1575–1588.
Cowan, N. J., & Lewis, S. A. (2001). Type II chaperonins, prefoldin, and the tubulin-specific chaperones. *Advances in Protein Chemistry*, *59*, 73–104.
Cox, D., et al. (2016). Small heat-shock proteins prevent alpha-synuclein aggregation via transient interactions and their efficacy is affected by the rate of aggregation. *Journal of Biological Chemistry*, *291*(43), 22618–22629.
Craig, E. A. (2018). Hsp70 at the membrane: Driving protein translocation. *BMC Biology*, *16*, 11.
Craig, E. A., & Marszalek, J. (2017). How do J-proteins get Hsp70 to do so many different things? *Trends in Biochemical Sciences*, *42*(5), 355–368.
Dahiya, V., & Chaudhuri, T. K. (2013). Functional intermediate in the refolding pathway of a large and multidomain protein malate synthase G. *Biochemistry*, *52*(26), 4517–4530.
Dahiya, V., & Chaudhuri, T. K. (2014). Chaperones GroEL/GroES accelerate the refolding of a multidomain protein through modulating on-pathway intermediates. *Journal of Biological Chemistry*, *289*(1), 286–298.
Das, A. K., Cohen, P. W., & Barford, D. (1998). The structure of the tetratricopeptide repeats of protein phosphatase 5: Implications for TPR-mediated protein-protein interactions. *EMBO Journal*, *17*(5), 1192–1199.
Davies, T. H., Ning, Y.-M., & Sánchez, E. R. (2005). Differential control of glucocorticoid receptor hormone-binding function by tetratricopeptide repeat (TPR) proteins and the immunosuppressive ligand FK506. *Biochemistry*, *44*(6), 2030–2038.
de Beus, M. D., Doyle, S. M., & Teschke, C. M. (2000). GroEL binds a late folding intermediate of phage P22 coat protein. *Cell Stress & Chaperones*, *5*(3), 163–172.
De Los Rios, P., et al. (2006). Hsp70 chaperones accelerate protein translocation and the unfolding of stable protein aggregates by entropic pulling. *Proceedings of the National Academy of Sciences of the United States of America*, *103*(16), 6166–6171.
Delbecq, S. P., & Klevit, R. E. (2013). One size does not fit all: The oligomeric states of alphaB crystallin. *FEBS Letters*, *587*(8), 1073–1080.

Dittmar, K. D., Hutchison, K. A., Owens-Grillo, J. K., & Pratt, W. B. (1996). Reconstitution of the steroid receptor hsp90 heterocomplex assembly system of rabbit reticulocyte lysate. *Journal of Biological Chemistry*, *271*(22), 12833–12839.

Dutkiewicz, R., Nowak, M., Craig, E. A., & Marszalek, J. (2017). Fe-S cluster Hsp70 chaperones: The ATPase cycle and protein interactions. *Methods in Enzymology*, *595*, 161–184.

Echtenkamp, F. J., et al. (2011). Global functional map of the p23 molecular chaperone reveals an extensive cellular network. *Molecular Cell*, *43*(2), 229–241.

Echtenkamp, F. J., et al. (2016). Hsp90 and p23 molecular chaperones control chromatin architecture by maintaining the functional pool of the RSC chromatin remodeler. *Molecular Cell*, *64*(5), 888–899.

Ehrnsperger, M., Gräber, S., Gaestel, M., & Buchner, J. (1997). Binding of non-native protein to Hsp25 during heat shock creates a reservoir of folding intermediates for reactivation. *EMBO Journal*, *16*(2), 221–229.

Elad, N., et al. (2007). Topologies of a substrate protein bound to the chaperonin GroEL. *Molecular Cell*, *26*(3), 415–426.

Ellis, R. J. (2001). Molecular chaperones: Inside and outside the anfinsen cage. *Current Biology*, *11*(24), R1038–R1040.

English, C. A., Sherman, W., Meng, W., & Gierasch, L. M. (2017). The Hsp70 interdomain linker is a dynamic switch that enables allosteric communication between two structured domains. *Journal of Biological Chemistry*, *292*, 14765–14774.

Falsone, S. F., et al. (2004). Oncogenic mutations reduce the stability of SRC kinase. *Journal of Molecular Biology*, *344*(1), 281–291.

Fan, Q., et al. (2014). Identification of proteins that interact with alpha A-crystallin using a human proteome microarray. *Molecular Vision*, *20*, 117–124.

Farr, G. W., et al. (2000). Multivalent binding of nonnative substrate proteins by the chaperonin GroEL. *Cell*, *100*(5), 561–573.

Fayet, O., Ziegelhoffer, T., & Georgopoulos, C. (1989). The groES and groEL heat shock gene products of *Escherichia coli* are essential for bacterial growth at all temperatures. *Journal of Bacteriology*, *171*(3), 1379–1385.

Fei, X., Yang, D., LaRonde-LeBlanc, N., & Lorimer, G. H. (2013). Crystal structure of a GroEL-ADP complex in the relaxed allosteric state at 2.7 Å resolution. *Proceedings of the National Academy of Sciences of the United States of America*, *110*(32), E2958–E2966.

Fei, X., Ye, X., LaRonde, N. A., & Lorimer, G. H. (2014). Formation and structures of GroEL:GroES2 chaperonin footballs, the protein-folding functional form. *Proceedings of the National Academy of Sciences of the United States of America*, *111*(35), 12775–12780.

Feige, M. J., et al. (2009). An unfolded C(H)1 domain controls the assembly and secretion of IgG antibodies. *Molecular Cell*, *34*(5), 569–579.

Fenton, W. A., & Horwich, A. L. (2003). Chaperonin-mediated protein folding: Fate of substrate polypeptide. *Quarterly Reviews of Biophysics*, *36*(2), 229–256.

Fenton, W. A., Kashi, Y., Furtak, K., & Norwich, A. L. (1994). Residues in chaperonin GroEL required for polypeptide binding and release. *Nature*, *371*, 614.

Fernandez-Fernandez, M. R., et al. (2017). Hsp70—A master regulator in protein degradation. *FEBS Letters*, *591*(17), 2648–2660.

Ferrarini, M., Heltai, S., Zocchi, M. R., & Rugarli, C. (1992). Unusual expression and localization of heat-shock proteins in human tumor cells. *International Journal of Cancer*, *51*(4), 613–619.

Fischer, G. (1994). Peptidyl-prolyl cis/trans isomerases and their effectors. *Angewandte Chemie International Edition*, *33*(14), 1415–1436.

Flaherty, K. M., DeLuca-Flaherty, C., & McKay, D. B. (1990). Three-dimensional structure of the ATPase fragment of a 70K heat-shock cognate protein. *Nature*, *346*, 623.

Fleckenstein, T., et al. (2015). The chaperone activity of the developmental small heat shock protein Sip1 is regulated by pH-dependent conformational changes. *Molecular Cell, 58*(6), 1067–1078.

Flynn, G. C., Pohl, J., Flocco, M. T., & Rothman, J. E. (1991). Peptide-binding specificity of the molecular chaperone BiP. *Nature, 353*, 726.

Fourie, A. M., Sambrook, J. F., & Gething, M. J. (1994). Common and divergent peptide binding specificities of hsp70 molecular chaperones. *Journal of Biological Chemistry, 269*(48), 30470–30478.

Franck, E., et al. (2004). Evolutionary diversity of vertebrate small heat shock proteins. *Journal of Molecular Evolution, 59*(6), 792–805.

Franzmann, T. M., Menhorn, P., Walter, S., & Buchner, J. (2008). Activation of the chaperone Hsp26 is controlled by the rearrangement of its thermosensor domain. *Molecular Cell, 29*(2), 207–216.

Freeman, B. C., Toft, D. O., & Morimoto, R. I. (1996). Molecular chaperone machines: Chaperone activities of the cyclophilin Cyp-40 and the steroid aporeceptor-associated protein p23. *Science, 274*(5293), 1718–1720.

Fries, G. R., Gassen, N. C., & Rein, T. (2017). The FKBP51 glucocorticoid receptor co-chaperone: Regulation, function, and implications in health and disease. *International Journal of Molecular Sciences, 18*(12), 2614–2644.

Frydman, J. (2001). Folding of newly translated proteins in vivo: The role of molecular chaperones. *Annual Review of Biochemistry, 70*, 603–647.

Fu, X., et al. (2013). In vivo substrate diversity and preference of small heat shock protein IbpB as revealed by using a genetically incorporated photo-cross-linker. *Journal of Biological Chemistry, 288*(44), 31646–31654.

Fujiwara, K., et al. (2010). A systematic survey of in vivo obligate chaperonin-dependent substrates. *The EMBO Journal, 29*(9), 1552–1564.

Gaestel, M. (2002). sHsp-phosphorylation: Enzymes, signaling pathways and functional implications. *Progress in Molecular and Subcellular Biology, 28*, 151–169.

Garg, G., Khandelwal, A., & Blagg, B. S. J. (2016). Anticancer inhibitors of Hsp90 function: Beyond the usual suspects. In J. Isaacs & L. Whitesell (Eds.), *Advances in cancer research: Vol. 129. Hsp90 in cancer: Beyond the usual suspects* (pp. 51–88). Cambridge, MA: Academic Press. (Chapter Three).

Garrido, C., Paul, C., Seigneuric, R., & Kampinga, H. H. (2012). The small heat shock proteins family: The long forgotten chaperones. *The International Journal of Biochemistry & Cell Biology, 44*(10), 1588–1592.

Geller, R., et al. (2018). Hsp90 shapes protein and RNA evolution to balance trade-offs between protein stability and aggregation. *Nature Communications, 9*(1), 1781.

Genest, O., et al. (2013). Uncovering a region of heat shock protein 90 important for client binding in E. coli and chaperone function in yeast. *Molecular Cell, 49*(3), 464–473.

Genest, O., et al. (2015). Hsp70 and Hsp90 of E. coli directly interact for collaboration in protein remodeling. *Journal of Molecular Biology, 427*(24), 3877–3889.

Georgescauld, F., et al. (2014). GroEL/ES chaperonin modulates the mechanism and accelerates the rate of TIM-barrel domain folding. *Cell, 157*(4), 922–934.

Georgopoulos, C. P., Hendrix, R. W., Kaiser, A. D., & Wood, W. B. (1972). Role of the host cell in bacteriophage morphogenesis: Effects of a bacterial mutation on T4 head assembly. *Nature: New Biology, 239*(89), 38–41.

Giannoukos, G., Silverstein, A. M., Pratt, W. B., & Simons, S. S. (1999). The seven amino acids (547–553) of rat glucocorticoid receptor required for steroid and Hsp90 binding contain a functionally independent LXXLL motif that is critical for steroid binding. *Journal of Biological Chemistry, 274*(51), 36527–36536.

Goloubinoff, P. (2017). Editorial: The HSP70 molecular chaperone machines. *Frontiers in Molecular Biosciences, 4*, 1.

Goloubinoff, P., Christeller, J. T., Gatenby, A. A., & Lorimer, G. H. (1989). Reconstitution of active dimeric ribulose bisphosphate carboxylase from an unfoleded state depends on two chaperonin proteins and Mg-ATP. *Nature*, *342*(6252), 884–889.

Goloubinoff, P., & De Los Rios, P. (2007). The mechanism of Hsp70 chaperones: (entropic) pulling the models together. *Trends in Biochemical Sciences*, *32*(8), 372–380.

Gowda, N. K. C., et al. (2018). Nucleotide exchange factors Fes1 and HspBP1 mimic substrate to release misfolded proteins from Hsp70. *Nature Structural and Molecular Biology*, *25*(1), 83–89.

Gragerov, A., & Gottesman, M. E. (1994). Different peptide binding specificities of hsp70 family members. *Journal of Molecular Biology*, *241*(2), 133–135.

Grallert, H., & Buchner, J. (2001). Review: A structural view of the GroE chaperone cycle. *Journal of Structural Biology*, *135*(2), 95–103.

Groemping, Y., Seidel, R., & Reinstein, J. (2005). Balance of ATPase stimulation and nucleotide exchange is not required for efficient refolding activity of the DnaK chaperone. *FEBS Letters*, *579*(25), 5713–5717.

Gruber, R., & Horovitz, A. (2016). Allosteric mechanisms in chaperonin machines. *Chemical Reviews*, *116*(11), 6588–6606.

Hagn, F., et al. (2011). Structural analysis of the interaction between Hsp90 and the tumor suppressor protein p53. *Nature Structural and Molecular Biology*, *18*(10), 1086–1093.

Hainzl, O., Lapina, M. C., Buchner, J., & Richter, K. (2009). The charged linker region is an important regulator of Hsp90 function. *Journal of Biological Chemistry*, *284*(34), 22559–22567.

Haldar, S., et al. (2015). Chaperonin-assisted protein folding: Relative population of asymmetric and symmetric GroEL:GroES complexes. *Journal of Molecular Biology*, *427*(12), 2244–2255.

Hanazono, Y., et al. (2013). Nonequivalence observed for the 16-meric structure of a small heat shock protein, SpHsp16.0, from Schizosaccharomyces pombe. *Structure*, *21*(2), 220–228.

Harris, J. R., Pluckthun, A., & Zahn, R. (1994). Transmission electron microscopy of GroEL, GroES, and the symmetrical GroEL/ES complex. *Journal of Structural Biology*, *112*(3), 216–230.

Harrison, C. J., et al. (1997). Crystal structure of the nucleotide exchange factor GrpE bound to the ATPase domain of the molecular chaperone DnaK. *Science*, *276*(5311), 431–435.

Harst, A., Lin, H., & Obermann, W. M. (2005). Aha1 competes with Hop, p50 and p23 for binding to the molecular chaperone Hsp90 and contributes to kinase and hormone receptor activation. *Biochemical Journal*, *387*(Pt. 3), 789–796.

Hartl, F. U., Bracher, A., & Hayer-Hartl, M. (2011). Molecular chaperones in protein folding and proteostasis. *Nature*, *475*, 324.

Haslbeck, M., & Buchner, J. (2002a). Chaperone function of sHsps. *Progress in Molecular and Subcellular Biology*, *28*, 37–59.

Haslbeck, M., & Buchner, J. (2002b). Chaperone function of sHsps. In A.-P. Arrigo & W. E. G. Müller (Eds.), *Small stress proteins* (pp. 37–59). Berlin, Heidelberg: Springer Berlin Heidelberg.

Haslbeck, M., Franzmann, T., Weinfurtner, D., & Buchner, J. (2005). Some like it hot: The structure and function of small heat-shock proteins. *Nature Structural and Molecular Biology*, *12*(10), 842–846.

Haslbeck, M., Peschek, J., Buchner, J., & Weinkauf, S. (2016). Structure and function of alpha-crystallins: Traversing from in vitro to in vivo. *Biochimica et Biophysica Acta*, *1860*(1 Pt. B), 149–166.

Haslbeck, M., & Vierling, E. (2015). A first line of stress defense: Small heat shock proteins and their function in protein homeostasis. *Journal of Molecular Biology*, *427*(7), 1537–1548.

Haslbeck, M., Weinkauf, S., & Buchner, J. (2015). Regulation of the chaperone function of small Hsps. In R. M. Tanguay & L. E. Hightower (Eds.), *The big book on small heat shock proteins* (pp. 155–178). Cham: Springer International Publishing.

Haslbeck, M., et al. (1999). Hsp26: A temperature-regulated chaperone. *The EMBO Journal*, *18*(23), 6744–6751.

Haslbeck, M., et al. (2004). Hsp42 is the general small heat shock protein in the cytosol of Saccharomyces cerevisiae. *The EMBO Journal*, *23*(3), 638–649.

Haslbeck, M., et al. (2008). Structural dynamics of archaeal small heat shock proteins. *Journal of Molecular Biology*, *378*(2), 362–374.

Hayer-Hartl, M., Bracher, A., & Hartl, F. U. (2016). The GroEL-GroES chaperonin machine: A nano-cage for protein folding. *Trends in Biochemical Sciences*, *41*(1), 62–76.

Hellenkamp, B., et al. (2017). Multidomain structure and correlated dynamics determined by self-consistent FRET networks. *Nature Methods*, *14*(2), 174–180.

Hemmingsen, S. M., et al. (1988). Homologous plant and bacterial proteins chaperone oligomeric protein assembly. *Nature*, *333*(6171), 330.

Hendershot, L. M. (2004). The ER function BiP is a master regulator of ER function. *The Mount Sinai Journal of Medicine, New York*, *71*(5), 289–297.

Hessling, M., Richter, K., & Buchner, J. (2009). Dissection of the ATP-induced conformational cycle of the molecular chaperone Hsp90. *Nature Structural and Molecular Biology*, *16*, 287.

Hochberg, G. K., et al. (2014). The structured core domain of alphaB-crystallin can prevent amyloid fibrillation and associated toxicity. *Proceedings of the National Academy of Sciences of the United States of America*, *111*(16), E1562–E1570.

Holmes, J. L., Sharp, S. Y., Hobbs, S., & Workman, P. (2008). Silencing of HSP90 cochaperone AHA1 expression decreases client protein activation and increases cellular sensitivity to the HSP90 inhibitor 17-allylamino-17-demethoxygeldanamycin. *Cancer Research*, *68*(4), 1188–1197.

Honore, F. A., Mejean, V., & Genest, O. (2017). Hsp90 is essential under heat stress in the bacterium *Shewanella oneidensis*. *Cell Reports*, *19*(4), 680–687.

Horovitz, A., & Willison, K. R. (2005). Allosteric regulation of chaperonins. *Current Opinion in Structural Biology*, *15*(6), 646–651.

Horst, R., et al. (2005). Direct NMR observation of a substrate protein bound to the chaperonin GroEL. *Proceedings of the National Academy of Sciences of the United States of America*, *102*(36), 12748–12753.

Horst, R., et al. (2007). Folding trajectories of human dihydrofolate reductase inside the GroEL GroES chaperonin cavity and free in solution. *Proceedings of the National Academy of Sciences of the United States of America*, *104*(52), 20788–20792.

Horwich, A. L., Apetri, A. C., & Fenton, W. A. (2009). The GroEL/GroES cis cavity as a passive anti-aggregation device. *FEBS Letters*, *583*(16), 2654–2662.

Horwich, A. L., Farr, G. W., & Fenton, W. A. (2006). GroEL – GroES-mediated protein folding. *Chemical Reviews*, *106*(5), 1917–1930.

Horwitz, J. (1992). Alpha-crystallin can function as a molecular chaperone. *Proceedings of the National Academy of Sciences of the United States of America*, *89*(21), 10449–10453.

Horwitz, J. (2003). Alpha-crystallin. *Experimental Eye Research*, *76*(2), 145–153.

Howard, K. J., Holley, S. J., Yamamoto, K. R., & Distelhorst, C. W. (1990). Mapping the HSP90 binding region of the glucocorticoid receptor. *Journal of Biological Chemistry*, *265*(20), 11928–11935.

Hsu, A.-L., Murphy, C. T., & Kenyon, C. (2003). Regulation of aging and age-related disease by DAF-16 and heat-shock factor. *Science*, *300*(5622), 1142–1145.

Hu, B., Mayer, M. P., & Tomita, M. (2006). Modeling Hsp70-mediated protein folding. *Biophysical Journal*, *91*(2), 496–507.

Huang, Q., et al. (2009). Mechanism of cataract formation in αA-crystallin Y118D mutation. *Investigative Ophthalmology & Visual Science, 50*(6), 2919–2926.
Hummel, B., et al. (2017). The evolutionary capacitor HSP90 buffers the regulatory effects of mammalian endogenous retroviruses. *Nature Structural and Molecular Biology, 24*, 234.
Hunt, J. F., et al. (1996). The crystal structure of the GroES co-chaperonin at 2.8 Å resolution. *Nature, 379*(6560), 37–45.
Iizuka, R., & Funatsu, T. (2016). Chaperonin GroEL uses asymmetric and symmetric reaction cycles in response to the concentration of non-native substrate proteins. *Biophysics and Physicobiology, 13*, 63–69.
Inbar, E., & Horovitz, A. (1997). GroES promotes the T to R transition of the GroEL ring distal to GroES in the GroEL-GroES complex. *Biochemistry, 36*(40), 12276–12281.
Inobe, T., et al. (2008). Asymmetry of the GroEL-GroES complex under physiological conditions as revealed by small-angle X-ray scattering. *Biophysical Journal, 94*(4), 1392–1402.
Ito, H., et al. (1997). Phosphorylation of alphaB-crystallin in response to various types of stress. *Journal of Biological Chemistry, 272*(47), 29934–29941.
Jaenicke, R. (1987). Folding and association of proteins. *Progress in Biophysics and Molecular Biology, 49*(2), 117–237.
Jahn, M., et al. (2014). The charged linker of the molecular chaperone Hsp90 modulates domain contacts and biological function. *Proceedings of the National Academy of Sciences of the United States of America, 111*(50), 17881–17886.
Jaime, M. B., et al. (2007). The inter-ring arrangement of the cytosolic chaperonin CCT. *EMBO Reports, 8*(3), 252–257.
Jakob, U., Gaestel, M., Engel, K., & Buchner, J. (1993). Small heat shock proteins are molecular chaperones. *Journal of Biological Chemistry, 268*(3), 1517–1520.
Jamin, M., & Baldwin, R. L. (1996). Refolding and unfolding kinetics of the equilibrium folding intermediate of apomyoglobin. *Nature Structural Biology, 3*, 613.
Jehle, S., et al. (2010). Solid-state NMR and SAXS studies provide a structural basis for the activation of αB-crystallin oligomers. *Nature Structural and Molecular Biology, 17*(9), 1037–1042.
Jehle, S., et al. (2011). N-terminal domain of αB-crystallin provides a conformational switch for multimerization and structural heterogeneity. *Proceedings of the National Academy of Sciences of the United States of America, 108*(16), 6409–6414.
Jin, Y., Zhuang, M., & Hendershot, L. M. (2009). ERdj3, a luminal ER DnaJ homologue, binds directly to unfolded proteins in the mammalian ER: Identification of critical residues. *Biochemistry, 48*(1), 41–49.
Johnson, B. D., Schumacher, R. J., Ross, E. D., & Toft, D. O. (1998). Hop modulates hsp70/hsp90 interactions in protein folding. *Journal of Biological Chemistry, 273*(6), 3679–3686.
Johnson, J. L., & Toft, D. O. (1994). A novel chaperone complex for steroid receptors involving heat shock proteins, immunophilins, and p23. *Journal of Biological Chemistry, 269*(40), 24989–24993.
Kampinga, H. H., & Craig, E. A. (2010). The HSP70 chaperone machinery: J proteins as drivers of functional specificity. *Nature Reviews. Molecular Cell Biology, 11*(8), 579–592.
Kampinga, H. H., et al. (1995). Thermal protein denaturation and protein aggregation in cells made thermotolerant by various chemicals: Role of heat shock proteins. *Experimental Cell Research, 219*(2), 536–546.
Kappe, G., Leunissen, J. A., & de Jong, W. W. (2002). Evolution and diversity of prokaryotic small heat shock proteins. *Progress in Molecular and Subcellular Biology, 28*, 1–17.
Karagoz, G. E., et al. (2014). Hsp90-tau complex reveals molecular basis for specificity in chaperone action. *Cell, 156*(5), 963–974.
Kaul, S., et al. (2002). Mutations at positions 547–553 of rat glucocorticoid receptors reveal that hsp90 binding requires the presence, but not defined composition, of a seven-amino

acid sequence at the amino terminus of the ligand binding domain. *Journal of Biological Chemistry*, *277*(39), 36223–36232.

Kellner, R., et al. (2014). Single-molecule spectroscopy reveals chaperone-mediated expansion of substrate protein. *Proceedings of the National Academy of Sciences of the United States of America*, *111*(37), 13355–13360.

Keramisanou, D., et al. (2016). Molecular mechanism of protein kinase recognition and sorting by the Hsp90 kinome-specific cochaperone Cdc37. *Molecular Cell*, *62*(2), 260–271.

Kerner, M. J., et al. (2005). Proteome-wide analysis of chaperonin-dependent protein folding in *Escherichia coli*. *Cell*, *122*(2), 209–220.

Kim, K. K., Kim, R., & Kim, S. H. (1998). Crystal structure of a small heat-shock protein. *Nature*, *394*(6693), 595–599.

King, F. W., Wawrzynow, A., Höhfeld, J., & Zylicz, M. (2001). Co-chaperones Bag-1, Hop and Hsp40 regulate Hsc70 and Hsp90 interactions with wild-type or mutant p53. *The EMBO Journal*, *20*(22), 6297–6305.

Kirschke, E., et al. (2014). Glucocorticoid receptor function regulated by coordinated action of the Hsp90 and Hsp70 chaperone cycles. *Cell*, *157*(7), 1685–1697.

Kityk, R., Kopp, J., & Mayer, M. P. (2018). Molecular mechanism of J-domain-triggered ATP hydrolysis by Hsp70 chaperones. *Molecular Cell*, *69*(2), 227–237, (e224).

Kityk, R., Kopp, J., Sinning, I., & Mayer, M. P. (2012). Structure and dynamics of the ATP-bound open conformation of Hsp70 chaperones. *Molecular Cell*, *48*(6), 863–874.

Koike-Takeshita, A., Arakawa, T., Taguchi, H., & Shimamura, T. (2014). Crystal structure of a symmetric football-shaped GroEL:GroES2-ATP14 complex determined at 3.8 Å reveals rearrangement between two GroEL rings. *Journal of Molecular Biology*, *426*(21), 3634–3641.

Koike-Takeshita, A., Yoshida, M., & Taguchi, H. (2008). Revisiting the GroEL-GroES reaction cycle via the symmetric intermediate implied by novel aspects of the GroEL(D398A) mutant. *Journal of Biological Chemistry*, *283*(35), 23774–23781.

Krantz, K. C., et al. (2013). Clathrin coat disassembly by the yeast Hsc70/Ssa1p and Auxilin/Swa2p proteins observed by single-particle burst analysis spectroscopy. *Journal of Biological Chemistry*, *288*(37), 26721–26730.

Kravats, A. N., et al. (2018). Functional and physical interaction between yeast Hsp90 and Hsp70. *Proceedings of the National Academy of Sciences of the United States of America*, *115*(10), E2210–e2219.

Kriehuber, T., et al. (2010). Independent evolution of the core domain and its flanking sequences in small heat shock proteins. *FASEB Journal*, *24*(10), 3633–3642.

Laganowsky, A., et al. (2010). Crystal structures of truncated alphaA and alphaB crystallins reveal structural mechanisms of polydispersity important for eye lens function. *Protein Science*, *19*(5), 1031–1043.

Landry, S. J., et al. (1993). Characterization of a functionally important mobile domain of GroES. *Nature*, *364*, 255.

Langer, T., et al. (1992). Chaperonin-mediated protein folding: GroES binds to one end of the GroEL cylinder, which accommodates the protein substrate within its central cavity. *EMBO Journal*, *11*(13), 4757–4765.

Laufen, T., et al. (1999). Mechanism of regulation of Hsp70 chaperones by DnaJ cochaperones. *Proceedings of the National Academy of Sciences of the United States of America*, *96*(10), 5452–5457.

Lee, G. J., Pokala, N., & Vierling, E. (1995). Structure and in vitro molecular chaperone activity of cytosolic small heat shock proteins from pea. *Journal of Biological Chemistry*, *270*(18), 10432–10438.

Lee, G. J., Roseman, A. M., Saibil, H. R., & Vierling, E. (1997). A small heat shock protein stably binds heat-denatured model substrates and can maintain a substrate in a folding-competent state. *EMBO Journal*, *16*(3), 659–671.

Lee, G. J., & Vierling, E. (2000). A small heat shock protein cooperates with heat shock protein 70 systems to reactivate a heat-denatured protein. *Plant Physiology*, *122*(1), 189–198.

Lee, C.-T., et al. (2012). Dynamics of the regulation of Hsp90 by the co-chaperone Sti1. *The EMBO Journal*, *31*(6), 1518–1528.

Leitner, A., et al. (2012). The molecular architecture of the eukaryotic chaperonin TRiC/CCT. *Structure*, *20*(5), 814–825.

Lentze, N., & Narberhaus, F. (2004). Detection of oligomerisation and substrate recognition sites of small heat shock proteins by peptide arrays. *Biochemical and Biophysical Research Communications*, *325*(2), 401–407.

Lentze, N., et al. (2004). Temperature and concentration-controlled dynamics of rhizobial small heat shock proteins. *European Journal of Biochemistry*, *271*(12), 2494–2503.

Li, J., & Buchner, J. (2013). Structure, function and regulation of the hsp90 machinery. *Biomedical Journal*, *36*(3), 106–117.

Li, J., Richter, K., & Buchner, J. (2011). Mixed Hsp90-cochaperone complexes are important for the progression of the reaction cycle. *Nature Structural & Molecular Biology*, *18*(1), 61–66.

Li, J., Richter, K., Reinstein, J., & Buchner, J. (2013). Integration of the accelerator Aha1 in the Hsp90 co-chaperone cycle. *Nature Structural and Molecular Biology*, *20*(3), 326–331.

Li, J., Soroka, J., & Buchner, J. (2012). The Hsp90 chaperone machinery: Conformational dynamics and regulation by co-chaperones. *Biochimica et Biophysica Acta, Molecular Cell Research*, *1823*(3), 624–635.

Lilie, H., & Buchner, J. (1995). Interaction of GroEL with a highly structured folding intermediate: Iterative binding cycles do not involve unfolding. *Proceedings of the National Academy of Sciences of the United States of America*, *92*(18), 8100–8104.

Lin, Z., Madan, D., & Rye, H. S. (2008). GroEL stimulates protein folding through forced unfolding. *Nature Structural & Molecular Biology*, *15*, 303.

Lin, Z., Puchalla, J., Shoup, D., & Rye, H. S. (2013). Repetitive protein unfolding by the trans ring of the GroEL-GroES chaperonin complex stimulates folding. *Journal of Biological Chemistry*, *288*(43), 30944–30955.

Llorca, O., Marco, S., Carrascosa, J. L., & Valpuesta, J. M. (1994). The formation of symmetrical GroEL-GroES complexes in the presence of ATP. *FEBS Letters*, *345*(2–3), 181–186.

Lopez, T., Dalton, K., & Frydman, J. (2015). The mechanism and function of group II chaperonins. *Journal of Molecular Biology*, *427*(18), 2919–2930.

Lorenz, O. R., et al. (2014). Modulation of the Hsp90 chaperone cycle by a stringent client protein. *Molecular Cell*, *53*(6), 941–953.

Mainz, A., et al. (2015). The chaperone αB-crystallin uses different interfaces to capture an amorphous and an amyloid client. *Nature Structural and Molecular Biology*, *22*, 898.

Malhotra, P., & Udgaonkar, J. B. (2014). High-energy intermediates in protein unfolding characterized by thiol labeling under nativelike conditions. *Biochemistry*, *53*(22), 3608–3620.

Mapa, K., et al. (2010). The conformational dynamics of the mitochondrial Hsp70 chaperone. *Molecular Cell*, *38*(1), 89–100.

Marcinowski, M., et al. (2011). Substrate discrimination of the chaperone BiP by autonomous and cochaperone-regulated conformational transitions. *Nature Structural & Molecular Biology*, *18*(2), 150–158.

Marcinowski, M., et al. (2013). Conformational selection in substrate recognition by Hsp70 chaperones. *Journal of Molecular Biology*, *425*(3), 466–474.

Marston, F. A. (1986). The purification of eukaryotic polypeptides synthesized in *Escherichia coli*. *Biochemical Journal*, *240*(1), 1–12.

Martin, J., Mayhew, M., Langer, T., & Hartl, U. (1993). The reaction cycle of GroEL and GroES in chaperonin-assisted protein folding. *Nature*, *366*, 228.

Mayer, M. P. (2013). Hsp70 chaperone dynamics and molecular mechanism. *Trends in Biochemical Sciences*, *38*(10), 507–514.

Mayer, M. P., & Bukau, B. (2005). Hsp70 chaperones: Cellular functions and molecular mechanism. *Cellular and Molecular Life Sciences*, *62*(6), 670–684.

Mayer, M., Kies, U., Kammermeier, R., & Buchner, J. (2000). BiP and PDI cooperate in the oxidative folding of antibodies in vitro. *Journal of Biological Chemistry*, *275*(38), 29421–29425.

Mayer, M., Reinstein, J., & Buchner, J. (2003). Modulation of the ATPase cycle of BiP by peptides and proteins. *Journal of Molecular Biology*, *330*(1), 137–144.

Mayer, M. P., Rudiger, S., & Bukau, B. (2000). Molecular basis for interactions of the DnaK chaperone with substrates. *Biological Chemistry*, *381*(9–10), 877–885.

Mayer, M. P., et al. (2000). Multistep mechanism of substrate binding determines chaperone activity of Hsp70. *Nature Structural and Molecular Biology*, 7(7), 586–593.

Mayhew, M., et al. (1996). Protein folding in the central cavity of the GroEL-GroES chaperonin complex. *Nature*, *379*(6564), 420–426.

Mayr, C., Richter, K., Lilie, H., & Buchner, J. (2000). Cpr6 and Cpr7, two closely related Hsp90-associated immunophilins from *Saccharomyces cerevisiae*, differ in their functional properties. *Journal of Biological Chemistry*, *275*(44), 34140–34146.

McCarty, J. S., Buchberger, A., Reinstein, J., & Bukau, B. (1995). The role of ATP in the functional cycle of the DnaK chaperone system. *Journal of Molecular Biology*, *249*(1), 126–137.

McClellan, A. J., et al. (2007). Diverse cellular functions of the Hsp90 molecular chaperone uncovered using systems approaches. *Cell*, *131*(1), 121–135.

McHaourab, H. S., Godar, J. A., & Stewart, P. L. (2009). Structure and mechanism of protein stability sensors: Chaperone activity of small heat shock proteins. *Biochemistry*, *48*(18), 3828–3837.

McLaughlin, S. H., Smith, H. W., & Jackson, S. E. (2002). Stimulation of the weak ATPase activity of human hsp90 by a client protein. *Journal of Molecular Biology*, *315*(4), 787–798.

McLaughlin, S. H., et al. (2006). The co-chaperone p23 arrests the Hsp90 ATPase cycle to trap client proteins. *Journal of Molecular Biology*, *356*(3), 746–758.

Meyer, P., et al. (2004). Structural basis for recruitment of the ATPase activator Aha1 to the Hsp90 chaperone machinery. *The EMBO Journal*, *23*(6), 1402–1410.

Mickler, M., et al. (2009). The large conformational changes of Hsp90 are only weakly coupled to ATP hydrolysis. *Nature Structural and Molecular Biology*, *16*(3), 281–286.

Mogk, A., et al. (2003). Small heat shock proteins, ClpB and the DnaK system form a functional triade in reversing protein aggregation. *Molecular Microbiology*, *50*(2), 585–595.

Morgner, N., et al. (2015). Hsp70 forms antiparallel dimers stabilized by post-translational modifications to position clients for transfer to Hsp90. *Cell Reports*, *11*(5), 759–769.

Moro, F., Fernández-Sáiz, V., & Muga, A. (2004). The lid subdomain of DnaK is required for the stabilization of the substrate-binding site. *Journal of Biological Chemistry*, *279*(19), 19600–19606.

Müller, L., et al. (2004). hsp90 regulates the activity of wild type p53 under physiological and elevated temperatures. *Journal of Biological Chemistry*, *279*(47), 48846–48854.

Mymrikov, E. V., Seit-Nebi, A. S., & Gusev, N. B. (2011). Large potentials of small heat shock proteins. *Physiological Reviews*, *91*(4), 1123–1159.

Mymrikov, E. V., et al. (2016). The chaperone activity and substrate spectrum of human small heat shock proteins. *Journal of Biological Chemistry*.

Nillegoda, N. B., et al. (2015). Crucial HSP70 co-chaperone complex unlocks metazoan protein disaggregation. *Nature*, *524*(7564), 247–251.

Oren, M., & Rotter, V. (2010). Mutant p53 gain-of-function in cancer. *Cold Spring Harbor Perspectives in Biology*, *2*(2), a001107.

Palleros, D. R., Shi, L., Reid, K. L., & Fink, A. L. (1994). hsp70-protein complexes. Complex stability and conformation of bound substrate protein. *Journal of Biological Chemistry, 269*(18), 13107–13114.

Panaretou, B., et al. (1998). ATP binding and hydrolysis are essential to the function of the Hsp90 molecular chaperone in vivo. *The EMBO Journal, 17*(16), 4829–4836.

Panaretou, B., et al. (2002). Activation of the ATPase activity of Hsp90 by the stress-regulated cochaperone Aha1. *Molecular Cell, 10*(6), 1307–1318.

Pandya, M. J., et al. (2009). Interaction of human heat shock protein 70 with tumor-associated peptides. *Biological Chemistry, 390*(4), 305–312.

Park, S. J., Borin, B. N., Martinez-Yamout, M. A., & Dyson, H. J. (2011). The client protein p53 adopts a molten globule-like state in the presence of Hsp90. *Nature Structural and Molecular Biology, 18*, 537.

Park, S. J., Kostic, M., & Dyson, H. J. (2011). Dynamic interaction of Hsp90 with its client protein p53. *Journal of Molecular Biology, 411*(1), 158–173.

Pashtan, I., et al. (2008). Targeting Hsp90 prevents escape of breast cancer cells from tyrosine kinase inhibition. *Cell Cycle*, 7(18), 2936–2941.

Patra, A. K., & Udgaonkar, J. B. (2007). Characterization of the folding and unfolding reactions of single-chain monellin: Evidence for multiple intermediates and competing pathways. *Biochemistry, 46*(42), 11727–11743.

Paul, S., Singh, C., Mishra, S., & Chaudhuri, T. K. (2007). The 69 kDa *Escherichia coli* maltodextrin glucosidase does not get encapsulated underneath GroES and folds through trans mechanism during GroEL/GroES-assisted folding. *FASEB Journal, 21*(11), 2874–2885.

Peschek, J., et al. (2009). The eye lens chaperone α-crystallin forms defined globular assemblies. *Proceedings of the National Academy of Sciences of the United States of America, 106*(32), 13272–13277.

Peschek, J., et al. (2013). Regulated structural transitions unleash the chaperone activity of alphaB-crystallin. *Proceedings of the National Academy of Sciences of the United States of America, 110*(40), E3780–E3789.

Pirkl, F., Fischer, E., Modrow, S., & Buchner, J. (2001). Localization of the chaperone domain of FKBP52. *Journal of Biological Chemistry, 276*(40), 37034–37041.

Pratt, W. B., Morishima, Y., Murphy, M., & Harrell, M. (2006). Chaperoning of glucocorticoid receptors. *Handbook of Experimental Pharmacology, 172*, 111–138.

Pratt, W. B., & Toft, D. O. (1997). Steroid receptor interactions with heat shock protein and immunophilin chaperones. *Endocrine Reviews, 18*(3), 306–360.

Pratt, W. B., & Toft, D. O. (2003). Regulation of signaling protein function and trafficking by the hsp90/hsp70-based chaperone machinery. *Experimental Biology and Medicine, 228*(2), 111–133.

Prodromou, C., et al. (1997). Identification and structural characterization of the ATP/ADP-binding site in the Hsp90 molecular chaperone. *Cell, 90*(1), 65–75.

Prodromou, C., et al. (1999). Regulation of Hsp90 ATPase activity by tetratricopeptide repeat (TPR)-domain co-chaperones. *The EMBO Journal, 18*(3), 754–762.

Rajagopal, P., et al. (2015). A conserved histidine modulates HSPB5 structure to trigger chaperone activity in response to stress-related acidosis. *eLife, 4*, e07304.

Ratzke, C., Nguyen, M. N. T., Mayer, M. P., & Hugel, T. (2012). From a ratchet mechanism to random fluctuations evolution of Hsp90's mechanochemical cycle. *Journal of Molecular Biology, 423*(3), 462–471.

Rauch, J. N., et al. (2017). BAG3 is a modular, scaffolding protein that physically links heat shock protein 70 (Hsp70) to the small heat shock proteins. *Journal of Molecular Biology, 429*(1), 128–141.

Rehn, A., et al. (2016). Allosteric regulation points control the conformational dynamics of the molecular chaperone Hsp90. *Journal of Molecular Biology, 428*(22), 4559–4571.

Retzlaff, M., et al. (2009). Hsp90 is regulated by a switch point in the C-terminal domain. *EMBO Reports*, *10*(10), 1147–1153.
Retzlaff, M., et al. (2010). Asymmetric activation of the hsp90 dimer by its cochaperone aha1. *Molecular Cell*, *37*(3), 344–354.
Richardson, A., Schwager, F., Landry, S. J., & Georgopoulos, C. (2001). The importance of a mobile loop in regulating chaperonin/co-chaperonin interaction: Humans versus *Escherichia coli*. *Journal of Biological Chemistry*, *276*(7), 4981–4987.
Richter, K., Muschler, P., Hainzl, O., & Buchner, J. (2001). Coordinated ATP hydrolysis by the Hsp90 dimer. *Journal of Biological Chemistry*, *276*(36), 33689–33696.
Richter, K., Walter, S., & Buchner, J. (2004). The Co-chaperone Sba1 connects the ATPase reaction of Hsp90 to the progression of the chaperone cycle. *Journal of Molecular Biology*, *342*(5), 1403–1413.
Richter, K., et al. (2003). Sti1 is a non-competitive inhibitor of the Hsp90 ATPase: Binding prevents the n-terminal dimerization reaction during the atpase cycle. *Journal of Biological Chemistry*, *278*(12), 10328–10333.
Riggs, D. L., et al. (2007). Noncatalytic role of the FKBP52 peptidyl-prolyl isomerase domain in the regulation of steroid hormone signaling. *Molecular and Cellular Biology*, *27*(24), 8658–8669.
Rodina, A., et al. (2016). The epichaperome is an integrated chaperome network that facilitates tumour survival. *Nature*, *538*, 397.
Rodriguez, F., et al. (2008). Molecular basis for regulation of the heat shock transcription factor σ32 by the DnaK and DnaJ chaperones. *Molecular Cell*, *32*(3), 347–358.
Roe, S. M., et al. (1999). Structural basis for inhibition of the Hsp90 molecular chaperone by the antitumor antibiotics radicicol and geldanamycin. *Journal of Medicinal Chemistry*, *42*(2), 260–266.
Roe, S. M., et al. (2004). The mechanism of Hsp90 regulation by the protein kinase-specific cochaperone p50(cdc37). *Cell*, *116*(1), 87–98.
Röhl, A., et al. (2015). Hsp90 regulates the dynamics of its cochaperone Sti1 and the transfer of Hsp70 between modules. *Nature Communications*, *6*, 6655.
Rohner, N., et al. (2013). Cryptic variation in morphological evolution: HSP90 as a capacitor for loss of eyes in cavefish. *Science*, *342*(6164), 1372–1375.
Rosam, M., et al. (2018). Bap (Sil1) regulates the molecular chaperone BiP by coupling release of nucleotide and substrate. *Nature Structural & Molecular Biology*, *25*(1), 90–100.
Roseman, A. M., et al. (1996). The chaperonin ATPase cycle: Mechanism of allosteric switching and movements of substrate-binding domains in GroEL. *Cell*, *87*(2), 241–251.
Rothnie, A., et al. (2011). A sequential mechanism for clathrin cage disassembly by 70-kDa heat-shock cognate protein (Hsc70) and auxilin. *Proceedings of the National Academy of Sciences of the United States of America*, *108*(17), 6927–6932.
Rüdiger, S., Freund, S. M. V., Veprintsev, D. B., & Fersht, A. R. (2002). CRINEPT-TROSY NMR reveals p53 core domain bound in an unfolded form to the chaperone Hsp90. *Proceedings of the National Academy of Sciences of the United States of America*, *99*(17), 11085–11090.
Rüdiger, S., Germeroth, L., Schneider-Mergener, J., & Bukau, B. (1997). Substrate specificity of the DnaK chaperone determined by screening cellulose-bound peptide libraries. *EMBO Journal*, *16*(7), 1501–1507.
Rüdiger, S., Schneider-Mergener, J., & Bukau, B. (2001). Its substrate specificity characterizes the DnaJ co-chaperone as a scanning factor for the DnaK chaperone. *EMBO Journal*, *20*(5), 1042–1050.
Rutherford, S. L., & Lindquist, S. (1998). Hsp90 as a capacitor for morphological evolution. *Nature*, *396*(6709), 336–342.
Rutz, D. A., et al. (2018). A switch point in the molecular chaperone Hsp90 responding to client interaction. *Nature Communications*, *9*(1), 1472.

Rye, H. S., et al. (1999). GroEL-GroES cycling. *Cell, 97*(3), 325–338.
Sahasrabudhe, P., et al. (2017). The plasticity of the Hsp90 co-chaperone system. *Molecular Cell, 67*(6), 947–961.e5.
Saibil, H., Dong, Z., Wood, S., & auf der Mauer, A. (1991). Binding of chaperonins. *Nature, 353*(6339), 25–26.
Saibil, H. R., Fenton, W. A., Clare, D. K., & Horwich, A. L. (2013). Structure and allostery of the chaperonin GroEL. *Journal of Molecular Biology, 425*(9), 1476–1487.
Saibil, H. R., et al. (1993). ATP induces large quaternary rearrangements in a cage-like chaperonin structure. *Current Biology, 3*(5), 265–273.
Sameshima, T., Iizuka, R., Ueno, T., & Funatsu, T. (2010). Denatured proteins facilitate the formation of the football-shaped GroEL–(GroES)2 complex. *Biochemical Journal, 427*(2), 247–254.
Scheibel, T., Weikl, T., & Buchner, J. (1998). Two chaperone sites in Hsp90 differing in substrate specificity and ATP dependence. *Proceedings of the National Academy of Sciences of the United States of America, 95*(4), 1495–1499.
Scheibel, T., et al. (1997). ATP-binding properties of human Hsp90. *Journal of Biological Chemistry, 272*(30), 18608–18613.
Scheufler, C., et al. (2000). Structure of TPR domain–peptide complexes: Critical elements in the assembly of the Hsp70–Hsp90 multichaperone machine. *Cell, 101*(2), 199–210.
Schiene-Fischer, C. (2015). Multidomain peptidyl prolyl cis/trans isomerases. *Biochimica et Biophysica Acta, 1850*(10), 2005–2016.
Schlecht, R., Erbse, A. H., Bukau, B., & Mayer, M. P. (2011). Mechanics of Hsp70 chaperones enables differential interaction with client proteins. *Nature Structural and Molecular Biology, 18*(3), 345–351.
Schlee, S., Beinker, P., Akhrymuk, A., & Reinstein, J. (2004). A chaperone network for the resolubilization of protein aggregates: Direct interaction of ClpB and DnaK. *Journal of Molecular Biology, 336*(1), 275–285.
Schmid, D., Baici, A., Gehring, H., & Christen, P. (1994). Kinetics of molecular chaperone action. *Science, 263*(5149), 971–973.
Schmid, A. B., et al. (2012). The architecture of functional modules in the Hsp90 co-chaperone Sti1/Hop. *EMBO Journal, 31*(6), 1506–1517.
Schmidt, M., et al. (1994a). On the role of groES in the chaperonin-assisted folding reaction. Three case studies. *Journal of Biological Chemistry, 269*(14), 10304–10311.
Schmidt, M., et al. (1994b). Symmetric complexes of GroE chaperonins as part of the functional cycle. *Science, 265*(5172), 656–659.
Schneider, M., et al. (2016). BiPPred: Combined sequence- and structure-based prediction of peptide binding to the Hsp70 chaperone BiP. *Proteins, 84*(10), 1390–1407.
Schopf, F. H., Biebl, M. M., & Buchner, J. (2017). The HSP90 chaperone machinery. *Nature Reviews. Molecular Cell Biology, 18*, 345.
Schuermann, J. P., et al. (2008). Structure of the Hsp110:Hsc70 nucleotide exchange machine. *Molecular Cell, 31*(2), 232–243.
Schulte, T. W., et al. (1998). Antibiotic radicicol binds to the N-terminal domain of Hsp90 and shares important biologic activities with geldanamycin. *Cell Stress & Chaperones, 3*(2), 100–108.
Sepehrnia, B., Paz, I. B., Dasgupta, G., & Momand, J. (1996). Heat shock protein 84 forms a complex with mutant p53 protein predominantly within a cytoplasmic compartment of the cell. *Journal of Biological Chemistry, 271*(25), 15084–15090.
Sharma, S., et al. (2008). Monitoring protein conformation along the pathway of chaperonin-assisted folding. *Cell, 133*(1), 142–153.
Sharma, S. K., et al. (2010). The kinetic parameters and energy cost of the Hsp70 chaperone as a polypeptide unfoldase. *Nature Chemical Biology, 6*(12), 914–920.

Shashidharamurthy, R., Koteiche, H. A., Dong, J., & Mchaourab, H. S. (2005). Mechanism of chaperone function in small heat shock proteins: Dissociation of the HSP27 oligomer IS required for recognition and binding of destabilized T4 lysozyme. *Journal of Biological Chemistry, 280*(7), 5281–5289.

Shen, Y., & Hendershot, L. M. (2005). ERdj3, a stress-inducible endoplasmic reticulum dnaj homologue, serves as a CoFactor for BiP's interactions with unfolded substrates. *Molecular Biology of the Cell, 16*(1), 40–50.

Shen, Y., et al. (1997). Essential role of the first intron in the transcription of hsp90beta gene. *FEBS Letters, 413*(1), 92–98.

Shiau, A. K., Harris, S. F., Southworth, D. R., & Agard, D. A. (2006). Structural analysis of *E. coli* hsp90 reveals dramatic nucleotide-dependent conformational rearrangements. *Cell, 127*(2), 329–340.

Sikor, M., et al. (2013). Real-time observation of the conformational dynamics of mitochondrial Hsp70 by spFRET. *EMBO Journal, 32*(11), 1639–1649.

Skjærven, L., Cuellar, J., Martinez, A., & Valpuesta, J. M. (2015). Dynamics, flexibility, and allostery in molecular chaperonins. *FEBS Letters, 589*(19 Pt. A), 2522–2532.

Sluchanko, N. N., et al. (2017). Structural basis for the interaction of a human small heat shock protein with the 14-3-3 universal signaling regulator. *Structure, 25*(2), 305–316.

Smulders, R. H. P. H., et al. (1996). Immobilization of the C-terminal extension of bovine αA-crystallin reduces chaperone-like activity. *Journal of Biological Chemistry, 271*(46), 29060–29066.

Sobott, F., Benesch, J. L., Vierling, E., & Robinson, C. V. (2002). Subunit exchange of multimeric protein complexes. Real-time monitoring of subunit exchange between small heat shock proteins by using electrospray mass spectrometry. *Journal of Biological Chemistry, 277*(41), 38921–38929.

Solis, E. J., et al. (2016). Defining the essential function of yeast Hsf1 reveals a compact transcriptional program for maintaining eukaryotic proteostasis. *Molecular Cell, 63*(1), 60–71.

Sousa, R., & Lafer, E. M. (2015). The role of molecular chaperones in clathrin mediated vesicular trafficking. *Frontiers in Molecular Biosciences, 2*, 26.

Sousa, R., et al. (2016). Clathrin-coat disassembly illuminates the mechanisms of Hsp70 force generation. *Nature Structural and Molecular Biology, 23*, 821.

Southworth, D. R., & Agard, D. A. (2008). Species-dependent ensembles of conserved conformational states define the Hsp90 chaperone ATPase cycle. *Molecular Cell, 32*(5), 631–640.

Sparrer, H., & Buchner, J. (1997). How GroES regulates binding of nonnative protein to GroEL. *Journal of Biological Chemistry, 272*(22), 14080–14086.

Sparrer, H., Lilie, H., & Buchner, J. (1996). Dynamics of the GroEL-protein complex: Effects of nucleotides and folding mutants. *Journal of Molecular Biology, 258*(1), 74–87.

Sparrer, H., Rutkat, K., & Buchner, J. (1997). Catalysis of protein folding by symmetric chaperone complexes. *Proceedings of the National Academy of Sciences of the United States of America, 94*(4), 1096–1100.

Specht, S., Miller, S. B., Mogk, A., & Bukau, B. (2011). Hsp42 is required for sequestration of protein aggregates into deposition sites in Saccharomyces cerevisiae. *Journal of Cell Biology, 195*(4), 617–629.

Sreedhar, A. S., Soti, C., & Csermely, P. (2004). Inhibition of Hsp90: A new strategy for inhibiting protein kinases. *Biochimica et Biophysica Acta, 1697*(1–2), 233–242.

Srinivasan, S. R., et al. (2012). Molecular chaperones DnaK and DnaJ share predicted binding sites on most proteins in the *E. coli* proteome. *Molecular BioSystems, 8*(9), 2323–2333.

Stebbins, C. E., et al. (1997). Crystal structure of an Hsp90-geldanamycin complex: Targeting of a protein chaperone by an antitumor agent. *Cell, 89*(2), 239–250.

Stengel, F., et al. (2010). Quaternary dynamics and plasticity underlie small heat shock protein chaperone function. *Proceedings of the National Academy of Sciences of the United States of America, 107*(5), 2007–2012.

Stiegler, S. C., et al. (2017). A chemical compound inhibiting the Aha1-Hsp90 chaperone complex. *Journal of Biological Chemistry*.

Strauch, A., & Haslbeck, M. (2016). The function of small heat-shock proteins and their implication in proteostasis. *Essays in Biochemistry*, *60*(2), 163–172.

Stromer, T., Ehrnsperger, M., Gaestel, M., & Buchner, J. (2003). Analysis of the interaction of small heat shock proteins with unfolding proteins. *Journal of Biological Chemistry*, *278*(20), 18015–18021.

Strozecka, J., et al. (2012). Importance of N- and C-terminal regions of IbpA, *Escherichia coli* small heat shock protein, for chaperone function and oligomerization. *Journal of Biological Chemistry*, *287*(4), 2843–2853.

Sullivan, W., et al. (1997). Nucleotides and two functional states of hsp90. *Journal of Biological Chemistry*, *272*(12), 8007–8012.

Supko, J. G., Hickman, R. L., Grever, M. R., & Malspeis, L. (1995). Preclinical pharmacologic evaluation of geldanamycin as an antitumor agent. *Cancer Chemotherapy and Pharmacology*, *36*(4), 305–315.

Swain, J. F., et al. (2007). Hsp70 chaperone ligands control domain association via an allosteric mechanism mediated by the interdomain linker. *Molecular Cell*, *26*(1), 27–39.

Szabo, A., et al. (1994). The ATP hydrolysis-dependent reaction cycle of the *Escherichia coli* Hsp70 system DnaK, DnaJ, and GrpE. *Proceedings of the National Academy of Sciences of the United States of America*, *91*(22), 10345–10349.

Taguchi, H. (2015). Reaction cycle of chaperonin GroEL via symmetric "football" intermediate. *Journal of Molecular Biology*, *427*(18), 2912–2918.

Taipale, M., Jarosz, D. F., & Lindquist, S. (2010). HSP90 at the hub of protein homeostasis: Emerging mechanistic insights. *Nature Reviews. Molecular Cell Biology*, *11*, 515.

Taipale, M., et al. (2012). Quantitative analysis of HSP90-client interactions reveals principles of substrate recognition. *Cell*, *150*(5), 987–1001.

Takano, T., & Kakefuda, T. (1972). Involvement of a bacterial factor in morphogenesis of bacteriophage capsid. *Nature: New Biology*, *239*, 34.

Takeda, S., & McKay, D. B. (1996). Kinetics of peptide binding to the bovine 70 kDa heat shock cognate protein, a molecular chaperone. *Biochemistry*, *35*(14), 4636–4644.

Tang, Y.-C., et al. (2006). Structural features of the GroEL-GroES nano-cage required for rapid folding of encapsulated protein. *Cell*, *125*(5), 903–914.

Tang, Y. C., et al. (2008). Essential role of the chaperonin folding compartment in vivo. *The EMBO Journal*, *27*(10), 1458–1468.

Theodoraki, M. A., & Caplan, A. J. (2012). Quality control and fate determination of Hsp90 client proteins. *Biochimica et Biophysica Acta, Molecular Cell Research*, *1823*(3), 683–688.

Thiyagarajan, P., Henderson, S. J., & Joachimiak, A. (1996). Solution structures of GroEL and its complex with rhodanese from small-angle neutron scattering. *Structure*, *4*(1), 79–88.

Tissiéres, A., Mitchell, H. K., & Tracy, U. M. (1974). Protein synthesis in salivary glands of Drosophila melanogaster: Relation to chromosome puffs. *Journal of Molecular Biology*, *84*(3), 389–398.

Todd, M. J., Lorimer, G. H., & Thirumalai, D. (1996). Chaperonin-facilitated protein folding: Optimization of rate and yield by an iterative annealing mechanism. *Proceedings of the National Academy of Sciences of the United States of America*, *93*(9), 4030–4035.

Todd, M. J., Viitanen, P. V., & Lorimer, G. H. (1994). Dynamics of the chaperonin ATPase cycle: Implications for facilitated protein folding. *Science*, *265*(5172), 659–666.

Torok, Z., et al. (2001). Synechocystis HSP17 is an amphitropic protein that stabilizes heat-stressed membranes and binds denatured proteins for subsequent chaperone-mediated refolding. *Proceedings of the National Academy of Sciences of the United States of America*, *98*(6), 3098–3103.

Treweek, T. M., Rekas, A., Walker, M. J., & Carver, J. A. (2010). A quantitative NMR spectroscopic examination of the flexibility of the C-terminal extensions of the

molecular chaperones, alphaA- and alphaB-crystallin. *Experimental Eye Research*, *91*(5), 691–699.
Treweek, T. M., et al. (2007). Site-directed mutations in the C-terminal extension of human alphaB-crystallin affect chaperone function and block amyloid fibril formation. *PLoS One*, *2*(10), e1046.
Tsuboyama, K., Tadakuma, H., & Tomari, Y. (2018). Conformational activation of Argonaute by distinct yet coordinated actions of the Hsp70 and Hsp90 chaperone systems. *Molecular Cell*, *70*(4), 722–729.e4.
Tsutsumi, S., et al. (2012). Charged linker sequence modulates eukaryotic heat shock protein 90 (Hsp90) chaperone activity. *Proceedings of the National Academy of Sciences of the United States of America*, *109*(8), 2937–2942.
Ungelenk, S., et al. (2016). Small heat shock proteins sequester misfolding proteins in near-native conformation for cellular protection and efficient refolding. *Nature Communications*, 7, 13673.
Ungewickell, E., et al. (1995). Role of auxilin in uncoating clathrin-coated vesicles. *Nature*, *378*(6557), 632–635.
van den IJssel, P. R., Overkamp, P., Bloemendal, H., & de Jong, W. W. (1998). Phosphorylation of alphaB-crystallin and HSP27 is induced by similar stressors in HeLa cells. *Biochemical and Biophysical Research Communications*, *247*(2), 518–523.
Van Durme, J., et al. (2009). Accurate prediction of DnaK-peptide binding via homology modelling and experimental data. *PLoS Computational Biology*, *5*(8), e1000475.
Van Montfort, R., Slingsby, C., & Vierling, E. (2001). Structure and function of the small heat shock protein/alpha-crystallin family of molecular chaperones. *Advances in Protein Chemistry*, *59*, 105–156.
van Montfort, R. L., et al. (2001). Crystal structure and assembly of a eukaryotic small heat shock protein. *Nature Structural and Molecular Biology*, *8*(12), 1025–1030.
Veinger, L., Diamant, S., Buchner, J., & Goloubinoff, P. (1998). The small heat-shock protein IbpB from *Escherichia coli* stabilizes stress-denatured proteins for subsequent refolding by a multichaperone network. *Journal of Biological Chemistry*, *273*(18), 11032–11037.
Vembar, S. S., et al. (2010). J domain co-chaperone specificity defines the role of BiP during protein translocation. *Journal of Biological Chemistry*, *285*(29), 22484–22494.
Verba, K. A., et al. (2016). Atomic structure of Hsp90-Cdc37-Cdk4 reveals that Hsp90 traps and stabilizes an unfolded kinase. *Science*, *352*(6293), 1542–1547.
Vierling, E. (1997). The small heat shock proteins in plants are members of an ancient family of heat induced proteins. *Acta Physiologiae Plantarum*, *19*(4), 539–547.
Vogel, M., Mayer, M. P., & Bukau, B. (2006). Allosteric regulation of Hsp70 chaperones involves a conserved interdomain linker. *Journal of Biological Chemistry*, *281*(50), 38705–38711.
Walerych, D., et al. (2010). ATP binding to Hsp90 is sufficient for effective chaperoning of p53 protein. *Journal of Biological Chemistry*, *285*(42), 32020–32028.
Wall, D., Zylicz, M., & Georgopoulos, C. (1994). The NH2-terminal 108 amino acids of the *Escherichia coli* DnaJ protein stimulate the ATPase activity of DnaK and are sufficient for lambda replication. *Journal of Biological Chemistry*, *269*(7), 5446–5451.
Walter, S., & Buchner, J. (2002). Molecular chaperones—Cellular machines for protein folding. *Angewandte Chemie (International Ed. in English)*, *41*(7), 1098–1113.
Walther, D. M., et al. (2015). Widespread proteome remodeling and aggregation in aging C. elegans. *Cell*, *161*(4), 919–932.
Wang, T., Luo, S., Qin, H., & Xia, Y. (2018). Hsp90 inhibition renders iNOS aggregation and the clearance of iNOS aggregates by proteasomes requires SPSB2. *Free Radical Biology & Medicine*, *117*, 90–98.
Wang, X., et al. (2006). Hsp90 cochaperone Aha1 downregulation rescues misfolding of CFTR in cystic fibrosis. *Cell*, *127*(4), 803–815.

Weaver, J., & Rye, H. S. (2014). The C-terminal tails of the bacterial chaperonin GroEL stimulate protein folding by directly altering the conformation of a substrate protein. *Journal of Biological Chemistry, 289*(33), 23219–23232.

Weaver, A. J., et al. (2000). Crystal structure and activity of human p23, a heat shock protein 90 co-chaperone. *Journal of Biological Chemistry, 275*(30), 23045–23052.

Weaver, J., et al. (2017). GroEL actively stimulates folding of the endogenous substrate protein PepQ. *Nature Communications, 8*, 15934.

Wegele, H., et al. (2006). Substrate transfer from the chaperone Hsp70 to Hsp90. *Journal of Molecular Biology, 356*(3), 802–811.

Weikl, T., et al. (2000). C-terminal regions of Hsp90 are important for trapping the nucleotide during the ATPase cycle. *Journal of Molecular Biology, 303*(4), 583–592.

Weiss, C., Jebara, F., Nisemblat, S., & Azem, A. (2016). Dynamic complexes in the chaperonin-mediated protein folding cycle. *Frontiers in Molecular Biosciences, 3*, 80.

Weissman, J. S., Kashi, Y., Fenton, W. A., & Horwich, A. L. (1994). GroEL-mediated protein folding proceeds by multiple rounds of binding and release of nonnative forms. *Cell, 78*(4), 693–702.

Weissman, J. S., et al. (1995). Mechanism of GroEL action: Productive release of polypeptide from a sequestered position under GroES. *Cell, 83*(4), 577–587.

Whitesell, L., & Lindquist, S. L. (2005). HSP90 and the chaperoning of cancer. *Nature Reviews. Cancer, 5*(10), 761–772.

Whitesell, L., et al. (1994). Inhibition of heat shock protein HSP90-pp60v-src heteroprotein complex formation by benzoquinone ansamycins: Essential role for stress proteins in oncogenic transformation. *Proceedings of the National Academy of Sciences of the United States of America, 91*(18), 8324–8328.

Whitesell, L., et al. (1998). The physical association of multiple molecular chaperone proteins with mutant p53 is altered by geldanamycin, an hsp90-binding agent. *Molecular and Cellular Biology, 18*(3), 1517–1524.

Wochnik, G. M., et al. (2005). FK506-binding proteins 51 and 52 differentially regulate dynein interaction and nuclear translocation of the glucocorticoid receptor in mammalian cells. *Journal of Biological Chemistry, 280*(6), 4609–4616.

Wu, H., et al. (2018). Characterization of an Hsp90-independent interaction between co-chaperone p23 and transcription factor p53. *Biochemistry, 57*(6), 935–944.

Xi, J.-H., et al. (2008). Mechanism of small heat shock protein function in vivo: A knock-in mouse model demonstrates that the r49c mutation in αa-crystallin enhances protein insolubility and cell death. *Journal of Biological Chemistry, 283*(9), 5801–5814.

Xu, Z., Horwich, A. L., & Sigler, P. B. (1997). The crystal structure of the asymmetric GroEL-GroES-(ADP)7 chaperonin complex. *Nature, 388*(6644), 741–750.

Xu, Y., Singer, M. A., & Lindquist, S. (1999). Maturation of the tyrosine kinase c-src as a kinase and as a substrate depends on the molecular chaperone Hsp90. *Proceedings of the National Academy of Sciences of the United States of America, 96*(1), 109–114.

Xu, W., et al. (2001). Sensitivity of mature Erbb2 to geldanamycin is conferred by its kinase domain and is mediated by the chaperone protein Hsp90. *Journal of Biological Chemistry, 276*(5), 3702–3708.

Xu, W., et al. (2005). Surface charge and hydrophobicity determine ErbB2 binding to the Hsp90 chaperone complex. *Nature Structural & Molecular Biology, 12*, 120.

Yamamoto, D., & Ando, T. (2016). Chaperonin GroEL-GroES functions as both alternating and non-alternating engines. *Journal of Molecular Biology, 428*(15), 3090–3101.

Yan, X., et al. (2018). GroEL ring separation and exchange in the chaperonin reaction. *Cell, 172*(3), 605–617, (e611).

Yang, D., Ye, X., & Lorimer, G. H. (2013). Symmetric GroEL:GroES2 complexes are the protein-folding functional form of the chaperonin nanomachine. *Proceedings of the National Academy of Sciences of the United States of America, 110*(46), E4298–E4305.

Ye, X., & Lorimer, G. H. (2013). Substrate protein switches GroE chaperonins from asymmetric to symmetric cycling by catalyzing nucleotide exchange. *Proceedings of the National Academy of Sciences of the United States of America*, *110*(46), E4289–E4297.

Yifrach, O., & Horovitz, A. (1995). Nested cooperativity in the ATPase activity of the oligomeric chaperonin GroEL. *Biochemistry*, *34*(16), 5303–5308.

Young, J. C., Obermann, W. M. J., & Hartl, F. U. (1998). Specific binding of tetratricopeptide repeat proteins to the C-terminal 12-kDa domain of hsp90. *Journal of Biological Chemistry*, *273*(29), 18007–18010.

Zelin, E., et al. (2012). The p23 molecular chaperone and GCN5 acetylase jointly modulate protein-DNA dynamics and open chromatin status. *Molecular Cell*, *48*(3), 459–470.

Zettlmeissl, G., Rudolph, R., & Jaenicke, R. (1979). Reconstitution of lactic dehydrogenase. Noncovalent aggregation vs. reactivation. 1. Physical properties and kinetics of aggregation. *Biochemistry*, *18*(25), 5567–5571.

Zhang, S. L., et al. (1999). Regulation of human hsp90alpha gene expression. *FEBS Letters*, *444*(1), 130–135.

Zhu, X., et al. (1996). Structural analysis of substrate binding by the molecular chaperone DnaK. *Science*, *272*(5268), 1606–1614.

Zhuravleva, A., Clerico, E. M., & Gierasch, L. M. (2012). An interdomain energetic tug-of-war creates the allosterically active state in Hsp70 molecular chaperones. *Cell*, *151*(6), 1296–1307.

Zimmerman, S. B., & Trach, S. O. (1991). Estimation of macromolecule concentrations and excluded volume effects for the cytoplasm of *Escherichia coli*. *Journal of Molecular Biology*, *222*(3), 599–620.

Zuiderweg, E. R. P., et al. (2013). Allostery in the Hsp70 chaperone proteins. *Topics in Current Chemistry*, *328*, 99–153.

Zwirowski, S., et al. (2017). Hsp70 displaces small heat shock proteins from aggregates to initiate protein refolding. *EMBO Journal*, *36*(6), 783–796.

CHAPTER TWO

Protein stability and degradation in health and disease

Lene Clausen[†], Amanda B. Abildgaard[†], Sarah K. Gersing, Amelie Stein, Kresten Lindorff-Larsen*, Rasmus Hartmann-Petersen*
The Linderstrøm-Lang Centre for Protein Science, Department of Biology, University of Copenhagen, Copenhagen, Denmark
*Corresponding authors: e-mail address: lindorff@bio.ku.dk; rhpetersen@bio.ku.dk

Contents

Abstract

The cellular proteome performs highly varied functions to sustain life. Since most of these functions require proteins to fold properly, they can be impaired by mutations that affect protein structure, leading to diseases such as Alzheimer's disease, cystic fibrosis, and Lynch syndrome. The cell has evolved an intricate protein quality control (PQC) system that includes degradation pathways and a multitude of molecular chaperones and co-chaperones, all working together to catalyze the refolding or removal of aberrant proteins. Thus, the PQC system limits the harmful consequences of dysfunctional proteins, including those arising from disease-causing mutations. This complex system is still not fully understood. In particular the structural and sequence motifs that, when exposed, trigger degradation of misfolded proteins are currently under investigation.

[†] These authors contributed equally.

Advances in Protein Chemistry and Structural Biology, Volume 114
ISSN 1876-1623
https://doi.org/10.1016/bs.apcsb.2018.09.002

Moreover, several attempts are being made to activate or inhibit parts of the PQC system as a treatment for diseases. Here, we briefly review the present knowledge on the PQC system and list current strategies that are employed to exploit the system in disease treatment.

1. Introduction

Protein homeostasis (proteostasis) refers to an extensive network of components that acts to maintain proteins in the correct concentration, conformation, and subcellular location, to cooperatively achieve the stability and functional features of the proteome. During proteostasis, cells are in a state of dynamic equilibrium in which protein synthesis and protein folding are balanced with degradation. Because proteins participate in practically every process within the cell, it is essential to maintain and regulate proteostasis to preserve both cell function, organismal health, and to be able to respond and adapt to environmental challenges (Balch, Morimoto, Dillin, & Kelly, 2008; Hartl, Bracher, & Hayer-Hartl, 2011).

The majority of proteins need to adopt a relatively well-defined three-dimensional structure, their native conformation, in order to function. Anfinsen's *in vitro* studies on the folding of ribonuclease A demonstrated that it is the amino acid sequence that determines a protein's three-dimensional structure (Anfinsen, Haber, Sela, & White, 1961). Protein function is closely linked to the protein's ability to obtain various, stable conformational states; hence folding of a protein into its correct three-dimensional structure represents a compromise between its thermodynamic stability and flexibility. Accordingly, although the native state is thermodynamically favorable, it is often only slightly stable (5–15 kcal/mol) under physiological conditions (Bartlett & Radford, 2009; Dobson, Sali, & Karplus, 1998; Kim, Hipp, Bracher, Hayer-Hartl, & Hartl, 2013; Maxwell et al., 2005). Hence, failure in achieving or preserving the native conformation of proteins occurs regularly as a consequence of errors during translation, mutations, or chemical, environmental or physical stress conditions, giving rise to misfolded protein species. Such misfolded proteins typically expose hydrophobic regions that are susceptible to form toxic aggregates, which have been linked to several neurodegenerative diseases such as Alzheimer's and Parkinson's disease (Chiti & Dobson, 2006; Kim, Hipp, et al., 2013; Kim, Miller, Young, & Fields, 2013). In line with Anfinsen's studies, a number of newly translated proteins have the ability to fold spontaneously also in the cell. However, the majority of proteins require the assistance of molecular chaperones to achieve their native state and to avoid non-native interactions during the folding process (Hartl et al., 2011). In addition, as a means to avert such non-native interactions and to increase the efficiency of protein folding, ribosomes engaged in translation as polysomes are spatially arranged to maximize the distance between nascent polypeptide chains (Brandt et al., 2009; Kim, Hipp, et al., 2013; Kim, Miller, et al., 2013).

2. Protein quality control

With the purpose of preventing protein misfolding and aggregation, cells utilize very efficient and highly sensitive protein quality control mechanisms consisting of a repertoire of molecular chaperones and degradation systems. Molecular chaperones can assist in *de novo* protein folding by interacting with the nascent polypeptide chain during translation and recognize misfolded proteins to catalyze protein refolding (Hartl et al., 2011). The main chaperone systems that function in *de novo* folding and refolding, such as HSP70 and HSP90, are ATPases and have the ability to recognize and selectively associate with non-native proteins. Chaperone binding shields otherwise exposed segments of hydrophobic residues in non-native conformations of proteins, thus preventing protein aggregation. Then, protein folding is allowed to proceed during continuous cycles of ATP binding and hydrolysis, resulting in dynamic cycles of client binding and release until successful folding is achieved, or the client is targeted for degradation. Moreover, chaperones can collaborate with a cohort of co-chaperones that function in a number of ways. Co-chaperones can regulate the chaperone ATP cycle, assist in substrate loading or unloading, and recruit chaperones to various protein complexes and subcellular locations. How the chaperones engage in a specific cellular process thus depends on the distinct set of co-chaperones that participate in the chaperone/co-chaperone complex rather than the chaperone alone (Arndt, Rogon, & Hohfeld, 2007; Kim, Hipp, et al., 2013; Kim, Miller, et al., 2013; Rizzolo et al., 2017). Chaperones are not only involved in protein folding but also, when protein folding fails, in targeting misfolded proteins to the degradation machinery (Arndt et al., 2007; Kandasamy & Andreasson, 2018). The main intracellular protein degradation pathways include the autophagy-lysosome pathway and the ubiquitin-proteasome system (UPS). Autophagy is a regulated degradative route in which cytosolic components are engulfed and delivered to lysosomes for degradation. The autophagy system preferentially processes large cellular components, which exceed the spatial capacity of proteasomes, and is thus responsible for degradation of potentially dangerous large protein aggregates, whereas the UPS mainly targets small soluble proteins for degradation (Finley, 2009; Hershko & Ciechanover, 1998; Kettern, Dreiseidler, Tawo, & Hohfeld, 2010; Kriegenburg, Ellgaard, & Hartmann-Petersen, 2012). In the following we will focus primarily on the role of the UPS in PQC.

3. The UPS in protein quality control

Proteins destined for degradation through the UPS are first modified by conjugation to a chain of ubiquitin moieties and subsequently targeted to the 26S proteasome for degradation. This ubiquitin conjugation is catalyzed by three sequential enzymatic steps involving three distinct enzyme species (Deshaies & Joazeiro, 2009). First, at the expense of ATP hydrolysis, a ubiquitin-activating enzyme (E1) mediates the formation of a thioester bond between the C-terminal glycine residue in ubiquitin and the active site cysteine of the enzyme. Second, the activated ubiquitin moiety is transferred from the E1 to the active site cysteine of a ubiquitin-conjugating enzyme (E2). In the third step, a ubiquitin-protein ligase (E3) transfers the activated ubiquitin moiety from the E2 enzyme to an amino group in the target protein, usually the ε amino group in a lysine residue. Repeating this cascade adds other ubiquitin moieties to the first ubiquitin via internal ubiquitin lysine residues, thus creating a ubiquitin chain. The E3 enzymes are the most diverse class of enzymes in the ubiquitination process as they mediate substrate specificity. Accordingly, the human genome codes for more than 600 different E3 enzymes, while budding yeast has less than 100 different E3s (Deshaies & Joazeiro, 2009). As for how E3s recognize misfolded proteins, much still has to be learned. However, presumably they recognize some structural or sequence features, such as hydrophobicity, which are presented in misfolded proteins, but buried in native proteins. Adding to the complexity, it has been shown that while certain E3s interact directly with the misfolded protein, others rely on engaging substrates via chaperones or other adaptor proteins (Fig. 1).

Chaperone-associated E3 ubiquitin ligases may function as co-chaperones and can switch the chaperone activity from protein folding to protein degradation through the UPS. In this way, the UPS directly employs molecular chaperones as recognition factors for non-native proteins. The human protein CHIP (C-terminus of HSP70 interacting protein, also known as STUB1)

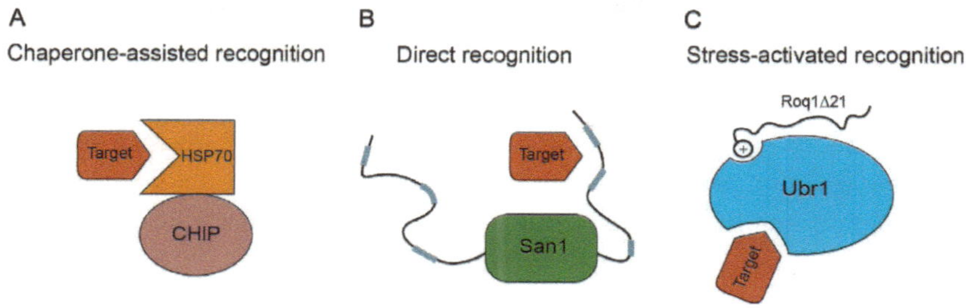

Fig. 1 E3 recognition of misfolded proteins. (A) A misfolded protein (target) can engage E3 ubiquitin-protein ligases indirectly through molecular chaperones such as HSP70 (orange) that associate with the E3, CHIP (pink). (B) Alternatively, a misfolded protein (target) may be recognized by E3 ubiquitin-protein ligases through a direct interaction, such as in the case of the yeast E3 San1 (green), that binds its targets through short hydrophobic regions (blue) embedded within its disordered N- and C-terminal arms (black threads). (C) During stress conditions, the budding yeast Roq1 protein is cleaved into the Roq1Δ21 fragment which binds Ubr1 (blue) and directs its E3 activity toward PQC targets.

is one of the best described chaperone-associated E3s, which was initially identified as an interaction partner of HSP70, but was later shown to bind HSP90 as well. CHIP engages HSP70 and HSP90 through its N-terminal tetratricopeptide repeat domain. Moreover, CHIP possesses a U-box, enabling it to interact with ubiquitin-conjugating enzymes and function as an E3 ubiquitin-protein ligase (Fig. 1A) (Arndt et al., 2007; Kettern et al., 2010; Kriegenburg et al., 2012). Although CHIP is only found in higher eukaryotes, yeast cells also rely on HSP70, HSP90, and other chaperones for the degradation of many misfolded proteins (Comyn, Young, Loewen, & Mayor, 2016; Fang et al., 2014; Guerriero, Weiberth, & Brodsky, 2013; Kandasamy & Andreasson, 2018; Kriegenburg et al., 2014; McClellan, Tam, Kaganovich, & Frydman, 2005; Park et al., 2007), suggesting that multiple PQC E3s collaborate directly or indirectly with the chaperone machinery.

An example of direct substrate recognition is provided by the yeast nuclear ubiquitin ligase San1 that has been demonstrated to bind directly to misfolded proteins via short hydrophobic regions scattered throughout its intrinsically disordered N- and C-terminal regions. These hydrophobic regions serve as substrate-recognition sites that identify misfolded proteins presumably by binding to exposed hydrophobic areas on the substrates. Thus, the intrinsically disordered regions provide San1 with great flexibility and Velcro-like arms that allow association with many, differently shaped, misfolded proteins independently of chaperones (Fig. 1B) (Enam, Geffen, Ravid, & Gardner, 2018; Fredrickson, Rosenbaum, Locke, Milac, & Gardner, 2011; Gardner, Nelson, & Gottschling, 2005; Rosenbaum et al., 2011). A human ortholog of San1 has yet to be identified, though bioinformatics analyses suggest that multiple human and yeast E3s are disordered and may operate in a San1-like manner (Boomsma, Nielsen, Lindorff-Larsen, Hartmann-Petersen, & Ellgaard, 2016).

Finally, since the presence of misfolded proteins may activate the cellular stress response, the PQC system is tightly connected with and regulated by stress signaling (Hohfeld & Hoppe, 2018; Jonikas et al., 2009; Lindholm, Korhonen, Eriksson, & Koks, 2017; Lyumkis et al., 2014; Poulsen et al., 2017). An example of this was recently uncovered by a series of elegant experiments in budding yeast (Szoradi et al., 2018), revealing mechanistic insight into the function of the PQC E3, Ubr1 (Eisele & Wolf, 2008). The data showed that upon stress, the stress response protein Roq1 is cleaved into the shorter Roq1Δ21 by the Ynm3 protease. Then, Roq1Δ21 interacts with the E3 Ubr1, transforming Ubr1's substrate specificity toward misfolded proteins (Fig. 1C) (Szoradi et al., 2018). In this way, by tuning the substrate preference of an otherwise non-PQC E3, the cells can rapidly adjust to stress conditions and increase their capacity for eliminating misfolded proteins.

4. Protein quality control in cellular compartments

Misfolding of proteins can occur in any cellular compartment. Hence, distinct PQC systems exist to maintain proteostasis in the ER, mitochondria, as well as the cytosol and nucleus. The quality control systems in the cytosol and nucleus are to some extent interlinked. In the cytosol, misfolded proteins are often bound by chaperones and ubiquitinated by the cytosolic E3 ubiquitin ligases Ubr1, Ubr2, CHIP, and E6-AP. Ribosomes are associated with specialized chaperones (Doring et al., 2017; Zhang, Sinning, & Rospert, 2017), and the specialized E3, Ltn1, also operates in the cytosol, where it associates with ribosomes to target nascent proteins upon ribosome stalling (Bengtson & Joazeiro, 2010). In yeast, most proteasomes are present in the nucleus along the inner face of the nuclear envelope (Wilkinson et al., 1998), while proteasomes in mammalian cells are uniformly present in the cytosol and nucleus (Brooks et al., 2000). This, combined with yeast cells having a closed mitosis where the nuclear membrane remains intact during cell division, may contribute to the nucleus being especially important for

PQC-mediated degradation in yeast cells (Enam et al., 2018; Nielsen, Poulsen, Rebula, & Hartmann-Petersen, 2014). Accordingly, it has been observed that misfolded cytosolic proteins are shuttled to the nucleus and then targeted for degradation by San1 (Fig. 2) (Amm & Wolf, 2016; Guerriero et al., 2013; Kriegenburg et al., 2012; Prasad, Xu, & Ng, 2018).

When proteins misfold in the ER lumen, the proteins are degraded through a process termed ER-associated degradation (ERAD). The ERAD pathway consists of multiple subdivisions with distinct specificities for different misfolded protein species (Christianson & Ye, 2014; Preston & Brodsky, 2017; Wangeline, Vashistha, & Hampton, 2017; Wu & Rapoport, 2018). For degradation, the substrate is first recognized by ER resident chaperones and transported back across the ER membrane into the cytosol through a

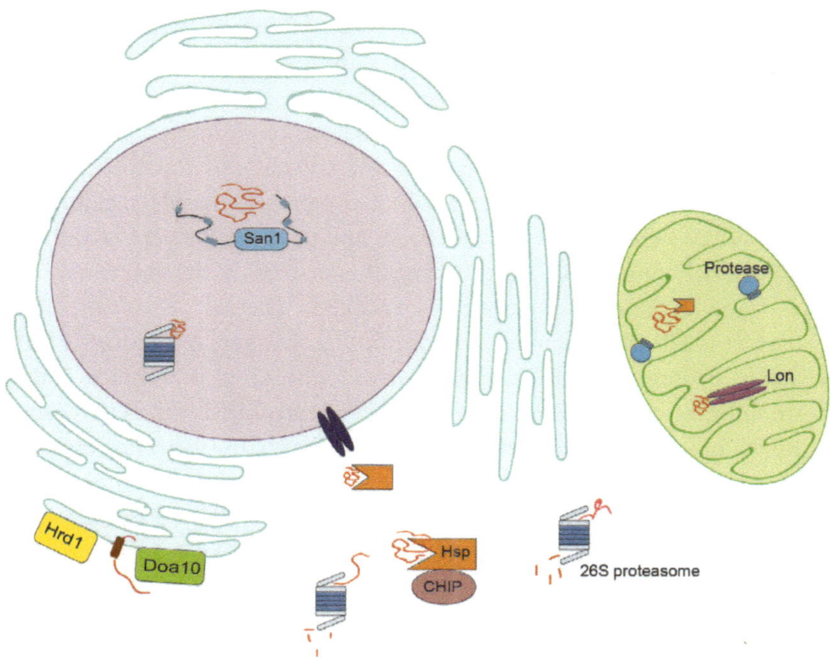

Fig. 2 Topography of PQC-linked degradation in the cell. In the cytosol, misfolded proteins (red) are often bound by chaperones (orange). If the native state is unattainable, the substrate protein is targeted for degradation by cytosolic E3 ligases such as CHIP (pink). Alternatively, chaperones mediate transport of the substrate to the nucleus, where it is recognized and ubiquitinated by a nuclear E3 such as San1 (blue). In the ERAD pathway, misfolded proteins are retrotranslocated over the ER membrane, and are ubiquitinated by membrane-bound E3 ligases such as Doa10 (green) and Hrd1 (yellow), and then degraded by the 26S proteasome. In the mitochondria, chaperones, Lon, and membrane-bound proteases maintain proteostasis.

mechanism known as retrotranslocation. Once a part of the substrate is exposed at the cytosolic side of the membrane, it is ubiquitinated by ubiquitin ligases associated with the membrane (Fig. 2). In both yeast and higher eukaryotes transmembrane E3 ligases, such as Hrd1 and Doa10, are involved in ERAD (Fig. 2). However, some ERAD E3s also participate in cytosolic/nuclear PQC (Bays, Gardner, Seelig, Joazeiro, & Hampton, 2001; Bordallo, Plemper, Finger, & Wolf, 1998; Maurer et al., 2016; Swanson, Locher, & Hochstrasser, 2001; Zattas & Hochstrasser, 2015). Following ubiquitination, the ERAD substrate is released into the cytosol where it is degraded by the proteasome.

In addition, PQC is particularly important in mitochondria, as cytotoxic reactive oxygen species produced through respiration can lead to oxidization and damage of nearby proteins. Furthermore, the complex structure of mitochondria gives rise to an intricate system of protein synthesis, import, and assembly. Hence, specific chaperones and proteases are found in every mitochondrial compartment. The accumulation of misfolded proteins triggers a mitochondrial stress response, which leads to an upregulation of proteases and chaperones, thus increasing the capacity of the PQC system. Degradation of misfolded mitochondrial proteins is mainly facilitated by mitochondrial HSP70s, the protease Lon, and proteases present in the inner mitochondrial membrane either facing the matrix or the intermembrane space (Fig. 2) (Bender, Lewrenz, Franken, Baitzel, & Voos, 2011; Savel'ev et al., 1998; Wagner, Arlt, van Dyck, Langer, & Neupert, 1994). However, proteins present in the outer mitochondrial membrane and the intermembrane-space are also degraded by the cytosolic UPS, in part relying on the E3 ubiquitin ligase Parkin. Finally, if the accumulation of misfolded proteins exceeds a certain threshold in mitochondria, these can either fuse with undamaged mitochondria, or undergo fission followed by mitophagy (mitochondrial autophagy), to eradicate those mitochondria that are beyond repair. Although different systems exist to maintain proteostasis in the cytosol, nucleus, ER, and mitochondria, there is a substantial overlap between the mechanisms and components in the different compartments. Hence PQC should be perceived as a coherent system.

5. Degradation signals

When it comes to selecting misfolded proteins for degradation, the UPS recognizes substrates through degradation signals, the so-called degrons, embedded within the protein sequence or structure. A degron is defined as the minimal element within a protein that is sufficient for targeting the protein for UPS-mediated degradation in any structural environment. This means that degrons can be synthetically transferred to other proteins and still function as degradation signals (Ravid & Hochstrasser, 2008). So far most characterized PQC degrons are short continuous peptide sequences, but similar to epitopes, some are likely to be non-continuous.

A classic example of a degron are the N-degrons, which are defined by certain amino acids that, when positioned in the N-terminus, lead to rapid proteasomal degradation of the protein (Bachmair & Varshavsky, 1989). The E3s and other components involved in this so-called N-end rule degradation pathway have been identified and characterized (Dissmeyer, Rivas, & Graciet, 2018; Tasaki et al., 2005; Varshavsky, 2017), and although the N-degrons are not directly linked to PQC, the PQC degrons are likely to operate by similar principles.

As for PQC degrons, i.e., structural and/or sequence regions that target misfolded proteins for degradation, these are, despite remarkable recent efforts (Geffen et al., 2016; Kim, Miller, et al., 2013; Maurer et al., 2016), currently not well defined. Presumably, they are hydrophobic in nature and found in internal parts of the protein structure that are not exposed in the native protein (Fig. 3).

A few PQC degrons have been characterized. For instance, the Deg1 degron in the yeast Mat2α transcription factor consists of 67 residues, of which 19 residues make out an essential segment that forms an amphipathic helix (Johnson, Swanson, Rakhilina, & Hochstrasser, 1998). Deg1 is recognized by the E3 Doa10, which is embedded in the ER-membrane and

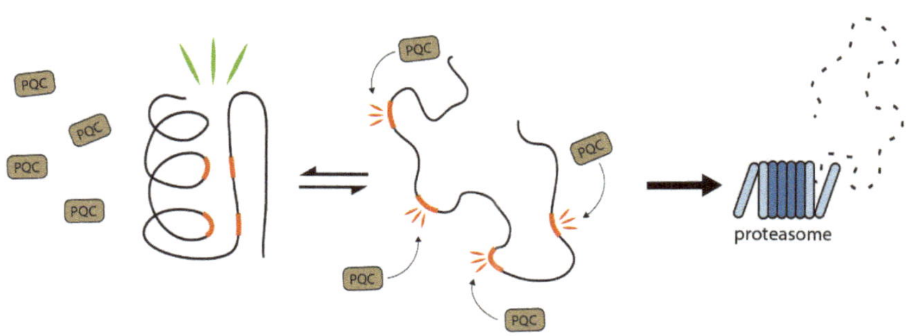

Fig. 3 Degradation signals (degrons) in misfolded proteins. Small structural or sequence elements buried within the native protein structure may act as degradation signals, degrons (orange patches), when exposed in misfolded proteins. The degrons are recognized by E3s and molecular chaperones of the protein quality control (PQC) system and target the misfolded protein for proteasomal degradation.

targets both soluble and membrane-anchored substrates for proteasomal degradation (Maurer et al., 2016; Ravid, Kreft, & Hochstrasser, 2006). The destabilizing effect of Deg1 is lost when hydrophobic residues of the 19 residue segment of Deg1 are mutated, indicating that hydrophobicity is needed for Doa10 recognition of Deg1 (Johnson et al., 1998; Ravid & Hochstrasser, 2008). This preference of Doa10 for hydrophobic degrons is in line with earlier studies on the artificial degron CL1 (Gilon, Chomsky, & Kulka, 1998, 2000; Metzger, Maurer, Dancy, & Michaelis, 2008).

In a recent proteome-wide approach to search for degrons within internal protein sequences, Ravid and colleagues carried out a screen of the yeast proteome (Geffen et al., 2016). They did this by fusing potential degron-sequences to the Ura3 protein, which catalyzes conversion of 5-fluoroorotic acid (5FOA) into the cytotoxic 5-fluorouracil. Thus, if the sequence is a degron, the cell would be viable in presence of 5FOA. This revealed multiple potential and physiologically relevant degrons, although no clear common sequence or structural elements were immediately evident.

Since both Doa10 and San1 recognition motifs involve hydrophobic interactions, it is likely that this feature would be seen in many PQC degrons. Interestingly, this overlaps with the chaperone-binding elements in misfolded proteins (Doring et al., 2017; Furth et al., 2011; Rudiger, Buchberger, & Bukau, 1997; Rudiger, Germeroth, Schneider-Mergener, & Bukau, 1997), and indicates that the various PQC components display shared substrate specificities. However, further systematic, large-scale studies, including linking degrons with their respective E3s, are needed before the PQC degrons can be accurately defined and perhaps, like the degrons involved in the turnover of cyclins and other cell cycle proteins (Chang & Barford, 2014), predicted by bioinformatics.

6. The consequence of missense mutations for protein stability

Since the amino acid sequence defines protein structure and function, a single mutation can alter a protein's features. Missense or nonsense mutations are the most prevalent types of genetic variation observed in coding regions (Lek et al., 2016). Nonsense mutations typically have obvious consequences, since introduction of a frame shift or a premature stop codon will result in a truncated, incomplete, misfolded, and often nonfunctional protein product. On the other hand, missense mutations or short in-frame deletions can cause disease through various mechanisms. For example, a missense mutation in or near the active site of an enzyme may abolish or reduce protein function by perturbing the characteristics of the catalytic groups. Missense mutations that affect a protein's surface may disrupt its ability to participate in various intermolecular interactions, hence making the protein unable to carry out its cellular function. However, many disease-causing missense mutations, in particular those affecting residues that are buried within the protein core, perturb the intramolecular interactions within the protein, which define the protein's structure, thermodynamic stability and/or folding rate (Dobson et al., 1998). Accordingly, such missense mutations may cause protein misfolding, degradation, and ultimately result in disease. This disease mechanism is seen in multiple genetic diseases, including cystic fibrosis, Lynch syndrome, phenylketonuria, and developmental disorders (Ahner, Nakatsukasa, Zhang, Frizzell, & Brodsky, 2007; Arlow, Scott, Wagenseller, & Gammie, 2013; Bross et al., 1999; Gersting et al., 2008; Nielsen et al., 2017; Pey, Desviat, Gamez, Ugarte, & Perez, 2003; Sahni et al., 2015; Suri et al., 2017; Underhaug, Aubi, & Martinez, 2012). Importantly, the structural consequences of missense mutations can vary from mild to severe, depending on their nature and position in the protein (Fig. 4).

Folding of a protein into its native state is a thermodynamically favorable process, but depending on the nature of the mutation, folding into the native state can be rendered unfavorable, causing the protein to unfold (Fig. 4). In such cases, the folding equilibrium of the protein will be skewed such that a larger fraction of missense proteins will exist in the unfolded state compared to the wild type. Since the PQC network only detects the misfolded conformations of the protein, structurally destabilizing mutations will cause a clearance of the protein while neutral mutations will not (Mathiassen et al., 2015; Nielsen et al., 2017). Accordingly, it was recently shown that the degree of structural destabilization correlates with the turnover (Nielsen et al., 2017), so the more dramatic a structural destabilization is, the more rapidly the protein is degraded (Fig. 4). Intriguingly, for a model protein, a global destabilization of only 3 kcal/mol was sufficient to trigger degradation (Nielsen et al., 2017). Although this value is likely dependent on how thermodynamically stable the wild type protein is, and whether the PQC recognizes locally or globally unfolded regions, a 3 kcal/mol destabilization is certainly not dramatic, and suggests that the PQC system is highly sensitive and may target proteins that are only slightly structurally perturbed and still functional (Enam et al., 2018; Gardner et al., 2005; Kampmeyer et al., 2017; Kriegenburg et al., 2014; Nielsen et al., 2017). The results also

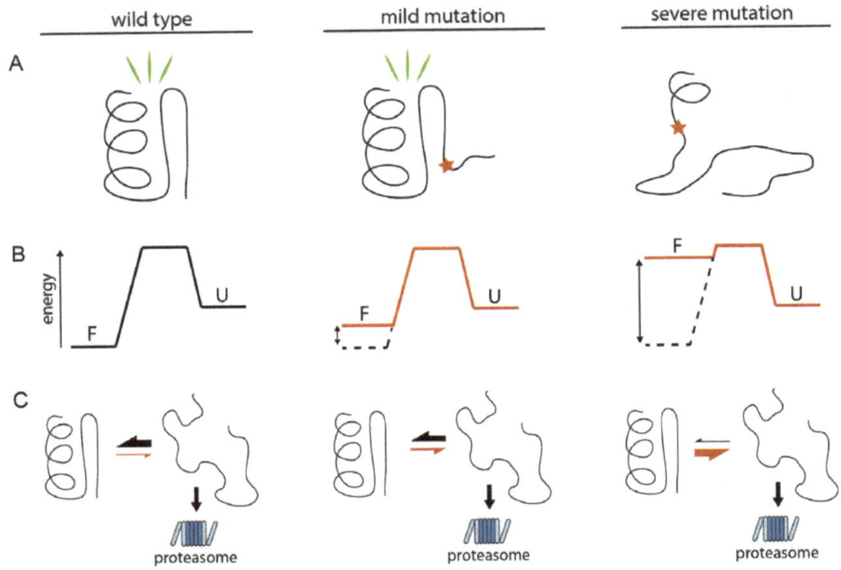

Fig. 4 Consequences of mutations on protein structures. (A) Schematic structures of a wild type protein, and variants carrying either mildly or severely structurally destabilizing mutations. A red star indicates the mutation. Green lines indicate that the protein is functional. (B) The folding energy landscapes of the protein variants in (A). The wild type protein is most stable in its native folded (F) state. The folded states of mildly and severely mutated variants are less stable, and they are more likely to unfold (U). (C) The folding equilibrium of the protein variants depicted in (A). Most of the wild type protein exists in its folded state. A larger fraction of the destabilized protein variants are found in their misfolded/unfolded states. The misfolded/unfolded proteins are targeted for degradation in the proteasome. In reality, many proteins likely display also local unfolding events and further research is needed to determine the structure, kinetics and thermodynamic properties of the states recognized by the PQC.

demonstrate that computational predictions of perturbed stability may be very useful both in assessing pathogenicity of mutations and in assigning molecular mechanisms of disease.

Finally, even synonymous mutations, which by definition do not cause changes to the amino acid sequence, and are therefore generally not expected to cause any changes to the function of the protein, have been shown to affect protein folding and stability. It has been demonstrated that synonymous single-nucleotide polymorphisms can alter the conformation of the corresponding protein by exchanging a common codon with a rare one (Kimchi-Sarfaty et al., 2007). Presumably, this effect is caused by rare codons delaying the rate of translation elongation, and thereby affecting the folding rate of a protein, which in turn may influence the structure of the nascent protein (Jacobs & Shakhnovich, 2017; Kirchner et al., 2017; Quax, Claassens, Soll, & van der Oost, 2015).

7. The role of chaperones in protein evolution and manifestation of genetic disorders

The general function of molecular chaperones in assisting protein folding also confers a central role in facilitating the evolution of new proteins. In the context of evolution, mutations offer the opportunity for proteins to acquire improved or novel activities. However, if such a beneficial mutation happens to result in a destabilization of the protein, it may lead to protein misfolding and degradation (Drummond, 2009; Geller, Pechmann, Acevedo, Andino, & Frydman, 2018). Therefore, evolution may require stabilizing mutations to precede destabilizing, but otherwise beneficial, mutations. In addition, it has been shown that evolution may make use of the HSP90 chaperone to buffer destabilizing mutations. HSP90 is expressed to a much higher level than required for growth (Borkovich, Farrelly, Finkelstein, Taulien, & Lindquist, 1989). This reservoir of excess HSP90 can serve as a capacitor, where HSP90 stabilizes unstable proteins, thereby concealing genetic variations and allowing mutations to accumulate, potentially leading to evolution of novel functions without the detrimental consequences of misfolding and degradation. Conversely, HSP90 can also serve as a potentiator, allowing genetic changes to have immediate phenotypic consequences by assisting in folding an otherwise unstable protein variant, thus allowing it to retain function instead of being targeted for degradation (Rutherford & Lindquist, 1998; Whitesell & Lindquist, 2005). One example of this has been observed for oncogenic mutations of the SRC tyrosine kinase. Normal c-Src interacts with HSP90 during folding and when the correct fold has been achieved, HSP90 dissociates (Xu, Singer, & Lindquist, 1999). By contrast, HSP90 displays a strong interaction with the viral c-Src homolog, v-Src. HSP90 interaction protects v-Src from degradation and increases its activity, leading to oncogenic transformation (Brugge, 1986; Whitesell, Mimnaugh, De Costa, Myers, & Neckers, 1994; Xu & Lindquist, 1993). This HSP90-dependency phenomenon has been recognized as a common feature for many oncogenic mutations, especially in a wide range of kinases. Thus, the protein-folding buffer has been shown to direct the manifestations of genetic variation in various model organisms. (Jarosz, Taipale, & Lindquist, 2010; Whitesell & Lindquist, 2005). However, HSP90 levels may become compromised under stress conditions and hence HSP90 will be unable to buffer the folding (and function) of destabilized mutant protein variants. One example of HSP90 functioning in this way to unmask cryptic variations has been suggested for the eye-size variation in the fresh water fish *Astyanax mexicanus*, which is normally found in creeks and rivers, but occasionally resides in caves, where they have been transformed eyeless. Studies indicate that the conditions of cave life disturb HSP90 function and unmask the same reduced eye size as that observed upon direct inhibition of HSP90 (Rohner et al., 2013). Importantly, these observations on wild populations support earlier studies of HSP90 capacitor function in *Drosophila* (Rutherford & Lindquist, 1998).

HSP90 can modify the effects of genetic mutations in humans by binding mutant proteins directly. There is a strong correlation between the severity of human genetic disorder and mutant proteins having greater physical interaction with HSP90 rather than HSP70. Indeed, these two chaperones have distinct biochemical functions with HSP70 preferentially binding to more disordered states and HSP90 to partially folded configurations. Thus, HSP90 binding to mutant proteins has the potential to both promote and inhibit the manifestation of human genetic diseases (Karras et al., 2017).

To further emphasize the importance of proper protein folding, protein misfolding and degradation have been associated with a number of devastating diseases, known as conformational or protein misfolding diseases (Carrell & Lomas, 1997), including several neurodegenerative diseases, certain types of hereditary cancer, metabolic disorders, amyloidosis and type II diabetes. Aberrant protein folding may lead to unusual binding interactions, mislocalization, degradation or aggregation causing either loss-of-function or gain-of-function toxicity. Loss-of-function conformational diseases, such as cystic fibrosis, are often caused by an inherited mutation, disturbing proper protein folding and causing the protein to have a shorter half-life and therefore an inadequate steady state level. Gain-of-function toxicity, as observed in several neurodegenerative diseases, is characterized by the accumulation of toxic intra- and/or extracellular protein aggregates, and may be caused by inherited mutations, perturbed PQC or environmental factors. The capacity of the cell to maintain proteostasis may gradually decline with age, which is why increasing age is recognized as a major risk factor for conformational diseases. Accordingly, several conformational diseases show delayed onset with symptoms usually taking place in late adulthood, indicating that the cellular capacity to buffer protein misfolding may become inadequate with age (Hipp, Park, & Hartl, 2014; Hohfeld & Hoppe, 2018; Walther et al., 2015).

8. The protein quality control pathway as a drug target for protein conformational diseases

Given the central importance of the PQC pathway in proteostasis, molecular chaperones, E3 enzymes and the proteasome may serve as rational targets for the design of therapeutics in genetic disorders associated with loss of protein activity due to aberrant protein folding and degradation (Kampmeyer et al., 2017; Muntau, Leandro, Staudigl, Mayer, & Gersting, 2014). As molecular chaperones cooperate with E3 enzymes in chaperone-assisted degradation, chaperone inhibitors also hold potential for treatment of protein misfolding diseases. Such a strategy has already been suggested in the form of histone deacetylase inhibitors (HDACis), a class of proteostasis regulators that has been demonstrated to have a beneficial effect in correcting the cellular phenotypes of a number of conformational diseases including cystic fibrosis (Hutt et al., 2010), Gaucher's disease (Lu et al., 2011), Niemann-Pick type C disease (Munkacsi et al., 2011; Pipalia et al., 2011), von Hippel-Lindau disease (Yang, Huntoon, Ksendzovsky, Zhuang, & Lonser, 2013) and neurodegenerative disorders (Coppede, 2014). HDACis induce HSP90 acetylation, thereby preventing HSP90 binding to client proteins, thus inhibiting chaperone function (Scroggins et al., 2007).

Importantly, just as chaperones can promote degradation through interaction with E3s, they can also limit degradation by refolding unstable proteins, thereby reducing the destabilizing effect of the mutation. Drugs that increase the cellular levels of chaperones have therefore been tested in an attempt to decrease the disease phenotype. Arimoclomol is a small-molecule drug that is currently undergoing clinical trials for a set of diseases, including the protein aggregation disease amyotrophic lateral sclerosis (ALS) and the lysosomal storage disorder Niemann-Pick disease (NPD). Arimoclomol functions by activating the transcription factor HSF-1, which amplifies transcription of HSP70 in stressed cells, and has been shown to relieve symptoms in mice models with NPD and related diseases (Kirkegaard et al., 2016, 2010).

Lastly, a complementary approach for targeting disease proteins has gained further traction in recent years. Proteolysis-targeting chimeras (PROTACs) are small molecules that can drive the interaction between a target protein and an E3 enzyme, leading to target ubiquitination and degradation by the proteasome (Buckley et al., 2012; Winter et al., 2015). This approach is used as an alternative to small molecules directly inhibiting those proteins. Progress in the area has recently been reviewed (Lai & Crews, 2017).

9. Concluding remarks

Early studies, performed even before the discovery of the ubiquitin-proteasome system, have shown that mutations may cause structural destabilization and degradation of proteins (Capecchi, Capecchi, Hughes, & Wahl, 1974). Since then we have learned much about the PQC system and many of the involved components, such as E3s and chaperones, have been identified and characterized in detail. However, several important issues remain. For instance, although the studies of CHIP have clearly demonstrated interplay between chaperones and the UPS, it is currently unknown to what extent other PQC E3s depend on chaperones for substrate recognition. Conversely, the direct target recognition through intrinsically disordered regions observed for San1 is likely employed by multiple E3s, but this remains to be shown. In addition, the nature of the degrons triggering PQC-mediated clearance is largely undefined, and although the PQC system is regulated by stress signaling it is currently unknown how and to what extent this affects the protein degradation and cell physiology as a whole. Finally, in recent years we have witnessed an explosion in studies of the formation and widespread importance of membrane-less organelles, often formed via the process of liquid-liquid phase-separation (LLPS). Many of the proteins involved in LLPS are aggregation prone, and a liquid-to-solid transition may be related to disease and accelerated by disease-causing mutations (Shin & Brangwynne, 2017). Thus, the cell likely has specialized, but so far incompletely understood, quality control mechanisms for membrane-less organelles (Alberti & Carra, 2018). As an example, recent studies have shown that SPOP, a substrate adaptor of the cullin3-RING ubiquitin ligase, can phase separate together with substrate proteins that undergo LLPS, thus providing a mechanism for UPS-mediated degradation of such proteins (Bouchard et al., 2018). Mutations in SPOP that interfere with phase separation perturb proteostasis, and thus provide a mechanism for mutant-driven oncogenesis.

As we continue to further our understanding of cellular PQC, we also increase our ability to understand, predict and modulate the role of PQC in disease. So far, many PQC-specific E3s have mostly been characterized in yeast cells, and it is therefore essential with further studies in higher eukaryotes, in particular to probe the therapeutic potential of PQC inhibition or activation for various genetic disorders and other protein misfolding diseases. Further, improved computational algorithms targeted at discovering which disease-causing mutations result from decreased cellular stability or otherwise perturbed proteostasis will aid in the long-term goal of diagnosis and treatment of human diseases.

Competing interests

No competing interests declared.

Funding

Our work, quoted here, was supported by the Lundbeck Foundation (to A.S., K.L.-L. and R. H.-P.), The Danish Cancer Society (to R.H.-P.), the Novo Nordisk Foundation (to K.L.-L. and R.H.-P.), the A.P. Møller Foundation (to R.H.-P.), Aase and Ejnar Danielsens Foundation (to R.H.-P.), the Danish Rheumatism Association (to R.H.-P.), Foundation Jochum (to R.H.-P.), and the Danish Council for Independent Research (Natural Sciences) (to R.H.-P.).

References

Ahner, A., Nakatsukasa, K., Zhang, H., Frizzell, R. A., & Brodsky, J. L. (2007). Small heat-shock proteins select deltaF508-CFTR for endoplasmic reticulum-associated degradation. *Molecular Biology of the Cell, 18*, 806–814.

Alberti, S., & Carra, S. (2018). Quality control of membraneless organelles. *Journal of Molecular Biology, 430*, 4711–4729.

Amm, I., & Wolf, D. H. (2016). Molecular mass as a determinant for nuclear San1-dependent targeting of misfolded cytosolic proteins to proteasomal degradation. *FEBS Letters, 590*, 1765–1775.

Anfinsen, C., Haber, E., Sela, M., & White, D. (1961). The kinetics of formation of native ribonuclease during oxidation of the reduced polypeptide chain. *Proceedings of the National Academy of Sciences of the United States of America, 47*, 1309–1314.

Arlow, T., Scott, K., Wagenseller, A., & Gammie, A. (2013). Proteasome inhibition rescues clinically significant unstable variants of the mismatch repair protein Msh2. *Proceedings of the National Academy of Sciences of the United States of America, 110*, 246–251.

Arndt, V., Rogon, C., & Hohfeld, J. (2007). To be, or not to be—Molecular chaperones in protein degradation. *Cellular and Molecular Life Sciences, 64*, 2525–2541.

Bachmair, A., & Varshavsky, A. (1989). The degradation signal in a short-lived protein. *Cell, 56*, 1019–1032.

Balch, W. E., Morimoto, R. I., Dillin, A., & Kelly, J. W. (2008). Adapting proteostasis for disease intervention. *Science, 319*, 916–919.

Bartlett, A. I., & Radford, S. E. (2009). An expanding arsenal of experimental methods yields an explosion of insights into protein folding mechanisms. *Nature Structural & Molecular Biology, 16*, 582–588.

Bays, N. W., Gardner, R. G., Seelig, L. P., Joazeiro, C. A., & Hampton, R. Y. (2001). Hrd1p/Der3p is a membrane-anchored ubiquitin ligase required for ER-associated degradation. *Nature Cell Biology, 3*, 24–29.

Bender, T., Lewrenz, I., Franken, S., Baitzel, C., & Voos, W. (2011). Mitochondrial enzymes are protected from stress-induced aggregation by mitochondrial chaperones and the Pim1/LON protease. *Molecular Biology of the Cell, 22*, 541–554.

Bengtson, M. H., & Joazeiro, C. A. (2010). Role of a ribosome-associated E3 ubiquitin ligase in protein quality control. *Nature, 467*, 470–473.

Boomsma, W., Nielsen, S. V., Lindorff-Larsen, K., Hartmann-Petersen, R., & Ellgaard, L. (2016). Bioinformatics analysis identifies several intrinsically disordered human E3 ubiquitin-protein ligases. *PeerJ*, *4*, e1725.

Bordallo, J., Plemper, R. K., Finger, A., & Wolf, D. H. (1998). Der3p/Hrd1p is required for endoplasmic reticulum-associated degradation of misfolded lumenal and integral membrane proteins. *Molecular Biology of the Cell*, *9*, 209–222.

Borkovich, K. A., Farrelly, F. W., Finkelstein, D. B., Taulien, J., & Lindquist, S. (1989). hsp82 is an essential protein that is required in higher concentrations for growth of cells at higher temperatures. *Molecular Cell Biology*, *9*, 3919–3930.

Bouchard, J. J., Otero, J. H., Scott, D. C., Szulc, E. M., Martin, E. W., Sabri, N., et al. (2018). Cancer mutations of the tumor suppressor SPOP disrupt the formation of active, phase-separated compartments. *Molecular Cell*, *72*, 19–36.

Brandt, F., Etchells, S. A., Ortiz, J. O., Elcock, A. H., Hartl, F. U., & Baumeister, W. (2009). The native 3D organization of bacterial polysomes. *Cell*, *136*, 261–271.

Brooks, P., Fuertes, G., Murray, R. Z., Bose, S., Knecht, E., Rechsteiner, M. C., et al. (2000). Subcellular localization of proteasomes and their regulatory complexes in mammalian cells. *The Biochemical Journal*, *346*(Pt. 1), 155–161.

Bross, P., Corydon, T. J., Andresen, B. S., Jorgensen, M. M., Bolund, L., & Gregersen, N. (1999). Protein misfolding and degradation in genetic diseases. *Human Mutation*, *14*, 186–198.

Brugge, J. S. (1986). Interaction of the Rous sarcoma virus protein pp60src with the cellular proteins pp50 and pp90. *Current Topics in Microbiology and Immunology*, *123*, 1–22.

Buckley, D. L., Van, M. I., Gareiss, P. C., Tae, H. S., Michel, J., Noblin, D. J., et al. (2012). Targeting the von Hippel-Lindau E3 ubiquitin ligase using small molecules to disrupt the VHL/HIF-1alpha interaction. *Journal of the American Chemical Society*, *134*, 4465–4468.

Capecchi, M. R., Capecchi, N. E., Hughes, S. H., & Wahl, G. M. (1974). Selective degradation of abnormal proteins in mammalian tissue culture cells. *Proceedings of the National Academy of Sciences of the United States of America*, *71*, 4732–4736.

Carrell, R. W., & Lomas, D. A. (1997). Conformational disease. *Lancet*, *350*, 134–138.

Chang, L., & Barford, D. (2014). Insights into the anaphase-promoting complex: A molecular machine that regulates mitosis. *Current Opinion in Structural Biology*, *29*, 1–9.

Chiti, F., & Dobson, C. M. (2006). Protein misfolding, functional amyloid, and human disease. *Annual Review of Biochemistry*, *75*, 333–366.

Christianson, J. C., & Ye, Y. (2014). Cleaning up in the endoplasmic reticulum: Ubiquitin in charge. *Nature Structural & Molecular Biology*, *21*, 325–335.

Comyn, S. A., Young, B. P., Loewen, C. J., & Mayor, T. (2016). Prefoldin promotes proteasomal degradation of cytosolic proteins with missense mutations by maintaining substrate solubility. *PLoS Genetics*, *12*, e1006184.

Coppede, F. (2014). The potential of epigenetic therapies in neurodegenerative diseases. *Frontiers in Genetics*, *5*, 220.

Deshaies, R. J., & Joazeiro, C. A. (2009). RING domain E3 ubiquitin ligases. *Annual Review of Biochemistry*, *78*, 399–434.

Dissmeyer, N., Rivas, S., & Graciet, E. (2018). Life and death of proteins after protease cleavage: Protein degradation by the N-end rule pathway. *The New Phytologist*, *218*, 929–935.

Dobson, C. M., Sali, A., & Karplus, M. (1998). Protein folding: A perspective from theory and experiment. *Angewandte Chemie (International Ed. in English)*, *37*, 868–893.

Doring, K., Ahmed, N., Riemer, T., Suresh, H. G., Vainshtein, Y., Habich, M., et al. (2017). Profiling Ssb-nascent chain interactions reveals principles of Hsp70-assisted folding. *Cell*, *170*, 298–311.

Drummond, D. A. (2009). Protein evolution: Innovative chaps. *Current Biology*, *19*, R740–R742.

Eisele, F., & Wolf, D. H. (2008). Degradation of misfolded protein in the cytoplasm is mediated by the ubiquitin ligase Ubr1. *FEBS Letters*, *582*, 4143–4146.
Enam, C., Geffen, Y., Ravid, T., & Gardner, R. G. (2018). Protein quality control degradation in the nucleus. *Annual Review of Biochemistry*, *87*, 725–749.
Fang, N. N., Chan, G. T., Zhu, M., Comyn, S. A., Persaud, A., Deshaies, R. J., et al. (2014). Rsp5/Nedd4 is the main ubiquitin ligase that targets cytosolic misfolded proteins following heat stress. *Nature Cell Biology*, *16*, 1227–1237.
Finley, D. (2009). Recognition and processing of ubiquitin-protein conjugates by the proteasome. *Annual Review of Biochemistry*, *78*, 477–513.
Fredrickson, E. K., Rosenbaum, J. C., Locke, M. N., Milac, T. I., & Gardner, R. G. (2011). Exposed hydrophobicity is a key determinant of nuclear quality control degradation. *Molecular Biology of the Cell*, *22*, 2384–2395.
Furth, N., Gertman, O., Shiber, A., Alfassy, O. S., Cohen, I., Rosenberg, M. M., et al. (2011). Exposure of bipartite hydrophobic signal triggers nuclear quality control of Ndc10 at the endoplasmic reticulum/nuclear envelope. *Molecular Biology of the Cell*, *22*, 4726–4739.
Gardner, R. G., Nelson, Z. W., & Gottschling, D. E. (2005). Degradation-mediated protein quality control in the nucleus. *Cell*, *120*, 803–815.
Geffen, Y., Appleboim, A., Gardner, R. G., Friedman, N., Sadeh, R., & Ravid, T. (2016). Mapping the landscape of a eukaryotic degronome. *Molecular Cell*, *63*, 1055–1065.
Geller, R., Pechmann, S., Acevedo, A., Andino, R., & Frydman, J. (2018). Hsp90 shapes protein and RNA evolution to balance trade-offs between protein stability and aggregation. *Nature Communications*, *9*, 1781.
Gersting, S. W., Kemter, K. F., Staudigl, M., Messing, D. D., Danecka, M. K., Lagler, F. B., et al. (2008). Loss of function in phenylketonuria is caused by impaired molecular motions and conformational instability. *American Journal of Human Genetics*, *83*, 5–17.
Gilon, T., Chomsky, O., & Kulka, R. G. (1998). Degradation signals for ubiquitin system proteolysis in Saccharomyces cerevisiae. *The EMBO Journal*, *17*, 2759–2766.
Gilon, T., Chomsky, O., & Kulka, R. G. (2000). Degradation signals recognized by the Ubc6p-Ubc7p ubiquitin-conjugating enzyme pair. *Molecular and Cellular Biology*, *20*, 7214–7219.
Guerriero, C. J., Weiberth, K. F., & Brodsky, J. L. (2013). Hsp70 targets a cytoplasmic quality control substrate to the San1p ubiquitin ligase. *The Journal of Biological Chemistry*, *288*, 18506–18520.
Hartl, F. U., Bracher, A., & Hayer-Hartl, M. (2011). Molecular chaperones in protein folding and proteostasis. *Nature*, *475*, 324–332.
Hershko, A., & Ciechanover, A. (1998). The ubiquitin system. *Annual Review of Biochemistry*, *67*, 425–479.
Hipp, M. S., Park, S. H., & Hartl, F. U. (2014). Proteostasis impairment in protein-misfolding and -aggregation diseases. *Trends in Cell Biology*, *24*, 506–514.
Hohfeld, J., & Hoppe, T. (2018). Ub and down: Ubiquitin exercise for the elderly. *Trends in Cell Biology*, *28*, 512–522.
Hutt, D. M., Herman, D., Rodrigues, A. P., Noel, S., Pilewski, J. M., Matteson, J., et al. (2010). Reduced histone deacetylase 7 activity restores function to misfolded CFTR in cystic fibrosis. *Nature Chemical Biology*, *6*, 25–33.
Jacobs, W. M., & Shakhnovich, E. I. (2017). Evidence of evolutionary selection for cotranslational folding. *Proceedings of the National Academy of Sciences of the United States of America*, *114*, 11434–11439.
Jarosz, D. F., Taipale, M., & Lindquist, S. (2010). Protein homeostasis and the phenotypic manifestation of genetic diversity: Principles and mechanisms. *Annual Review of Genetics*, *44*, 189–216.

Johnson, P. R., Swanson, R., Rakhilina, L., & Hochstrasser, M. (1998). Degradation signal masking by heterodimerization of MATalpha2 and MATa1 blocks their mutual destruction by the ubiquitin-proteasome pathway. *Cell*, *94*, 217–227.

Jonikas, M. C., Collins, S. R., Denic, V., Oh, E., Quan, E. M., Schmid, V., et al. (2009). Comprehensive characterization of genes required for protein folding in the endoplasmic reticulum. *Science*, *323*, 1693–1697.

Kampmeyer, C., Nielsen, S. V., Clausen, L., Stein, A., Gerdes, A. M., Lindorff-Larsen, K., et al. (2017). Blocking protein quality control to counter hereditary cancers. *Genes, Chromosomes and Cancer*, *56*, 823–831.

Kandasamy, G., & Andreasson, C. (2018). Hsp70-Hsp110 chaperones deliver ubiquitin-dependent and -independent substrates to the 26S proteasome for proteolysis in yeast. *Journal of Cell Science*, *131*, jcs210948.

Karras, G. I., Yi, S., Sahni, N., Fischer, M., Xie, J., Vidal, M., et al. (2017). HSP90 shapes the consequences of human genetic variation. *Cell*, *168*, 856–866.

Kettern, N., Dreiseidler, M., Tawo, R., & Hohfeld, J. (2010). Chaperone-assisted degradation: Multiple paths to destruction. *Biological Chemistry*, *391*, 481–489.

Kim, Y. E., Hipp, M. S., Bracher, A., Hayer-Hartl, M., & Hartl, F. U. (2013). Molecular chaperone functions in protein folding and proteostasis. *Annual Review of Biochemistry*, *82*, 323–355.

Kim, I., Miller, C. R., Young, D. L., & Fields, S. (2013). High-throughput analysis of in vivo protein stability. *Molecular & Cellular Proteomics*, *12*, 3370–3378.

Kimchi-Sarfaty, C., Oh, J. M., Kim, I. W., Sauna, Z. E., Calcagno, A. M., Ambudkar, S. V., et al. (2007). A "silent" polymorphism in the MDR1 gene changes substrate specificity. *Science*, *315*, 525–528.

Kirchner, S., Cai, Z., Rauscher, R., Kastelic, N., Anding, M., Czech, A., et al. (2017). Alteration of protein function by a silent polymorphism linked to tRNA abundance. *PLoS Biology*, *15*, e2000779.

Kirkegaard, T., Gray, J., Priestman, D. A., Wallom, K. L., Atkins, J., Olsen, O. D., et al. (2016). Heat shock protein-based therapy as a potential candidate for treating the sphingolipidoses. *Science Translational Medicine*, *8*, 355ra118.

Kirkegaard, T., Roth, A. G., Petersen, N. H., Mahalka, A. K., Olsen, O. D., Moilanen, I., et al. (2010). Hsp70 stabilizes lysosomes and reverts Niemann-Pick disease-associated lysosomal pathology. *Nature*, *463*, 549–553.

Kriegenburg, F., Ellgaard, L., & Hartmann-Petersen, R. (2012). Molecular chaperones in targeting misfolded proteins for ubiquitin-dependent degradation. *The FEBS Journal*, *279*, 532–542.

Kriegenburg, F., Jakopec, V., Poulsen, E. G., Nielsen, S. V., Roguev, A., Krogan, N., et al. (2014). A chaperone-assisted degradation pathway targets kinetochore proteins to ensure genome stability. *PLoS Genetics*, *10*, e1004140.

Lai, A. C., & Crews, C. M. (2017). Induced protein degradation: An emerging drug discovery paradigm. *Nature Reviews Drug Discovery*, *16*, 101–114.

Lek, M., Karczewski, K. J., Minikel, E. V., Samocha, K. E., Banks, E., Fennell, T., et al. (2016). Analysis of protein-coding genetic variation in 60,706 humans. *Nature*, *536*, 285–291.

Lindholm, D., Korhonen, L., Eriksson, O., & Koks, S. (2017). Recent insights into the role of unfolded protein response in ER stress in health and disease. *Frontiers in Cell and Development Biology*, *5*, 48.

Lu, J., Yang, C., Chen, M., Ye, D. Y., Lonser, R. R., Brady, R. O., et al. (2011). Histone deacetylase inhibitors prevent the degradation and restore the activity of glucocerebrosidase in Gaucher disease. *Proceedings of the National Academy of Sciences of the United States of America*, *108*, 21200–21205.

Lyumkis, D., Oliveira dos, P. D., Tahara, E. B., Webb, K., Bennett, E. J., Vinterbo, S., et al. (2014). Structural basis for translational surveillance by the large ribosomal subunit-associated protein quality control complex. *Proceedings of the National Academy of Sciences of the United States of America, 111*, 15981–15986.

Mathiassen, S. G., Larsen, I. B., Poulsen, E. G., Madsen, C. T., Papaleo, E., Lindorff-Larsen, K., et al. (2015). A two-step protein quality control pathway for a misfolded DJ-1 variant in fission yeast. *The Journal of Biological Chemistry, 290*, 21141–21153.

Maurer, M. J., Spear, E. D., Yu, A. T., Lee, E. J., Shahzad, S., & Michaelis, S. (2016). Degradation signals for ubiquitin-proteasome dependent cytosolic protein quality control (CytoQC) in yeast. *G3 (Bethesda, Md.), 6*, 1853–1866.

Maxwell, K. L., Wildes, D., Zarrine-Afsar, A., De Los Rios, M. A., Brown, A. G., Friel, C. T., et al. (2005). Protein folding: Defining a "standard" set of experimental conditions and a preliminary kinetic data set of two-state proteins. *Protein Science, 14*, 602–616.

McClellan, A. J., Tam, S., Kaganovich, D., & Frydman, J. (2005). Protein quality control: Chaperones culling corrupt conformations. *Nature Cell Biology, 7*, 736–741.

Metzger, M. B., Maurer, M. J., Dancy, B. M., & Michaelis, S. (2008). Degradation of a cytosolic protein requires endoplasmic reticulum-associated degradation machinery. *The Journal of Biological Chemistry, 283*, 32302–32316.

Munkacsi, A. B., Chen, F. W., Brinkman, M. A., Higaki, K., Gutierrez, G. D., Chaudhari, J., et al. (2011). An "exacerbate-reverse" strategy in yeast identifies histone deacetylase inhibition as a correction for cholesterol and sphingolipid transport defects in human Niemann-Pick type C disease. *The Journal of Biological Chemistry, 286*, 23842–23851.

Muntau, A. C., Leandro, J., Staudigl, M., Mayer, F., & Gersting, S. W. (2014). Innovative strategies to treat protein misfolding in inborn errors of metabolism: Pharmacological chaperones and proteostasis regulators. *Journal of Inherited Metabolic Disease, 37*, 505–523.

Nielsen, S. V., Poulsen, E. G., Rebula, C. A., & Hartmann-Petersen, R. (2014). Protein quality control in the nucleus. *Biomolecules, 4*, 646–661.

Nielsen, S. V., Stein, A., Dinitzen, A. B., Papaleo, E., Tatham, M. H., Poulsen, E. G., et al. (2017). Predicting the impact of Lynch syndrome-causing missense mutations from structural calculations. *PLoS Genetics, 13*, e1006739.

Park, S. H., Bolender, N., Eisele, F., Kostova, Z., Takeuchi, J., Coffino, P., et al. (2007). The cytoplasmic Hsp70 chaperone machinery subjects misfolded and endoplasmic reticulum import-incompetent proteins to degradation via the ubiquitin-proteasome system. *Molecular Biology of the Cell, 18*, 153–165.

Pey, A. L., Desviat, L. R., Gamez, A., Ugarte, M., & Perez, B. (2003). Phenylketonuria: Genotype-phenotype correlations based on expression analysis of structural and functional mutations in PAH. *Human Mutation, 21*, 370–378.

Pipalia, N. H., Cosner, C. C., Huang, A., Chatterjee, A., Bourbon, P., Farley, N., et al. (2011). Histone deacetylase inhibitor treatment dramatically reduces cholesterol accumulation in Niemann-Pick type C1 mutant human fibroblasts. *Proceedings of the National Academy of Sciences of the United States of America, 108*, 5620–5625.

Poulsen, E. G., Kampmeyer, C., Kriegenburg, F., Johansen, J. V., Hofmann, K., Holmberg, C., et al. (2017). UBL/BAG-domain co-chaperones cause cellular stress upon overexpression through constitutive activation of Hsf1. *Cell Stress & Chaperones, 22*, 143–154.

Prasad, R., Xu, C., & Ng, D. T. W. (2018). Hsp40/70/110 chaperones adapt nuclear protein quality control to serve cytosolic clients. *The Journal of Cell Biology, 217*, 2019–2032.

Preston, G. M., & Brodsky, J. L. (2017). The evolving role of ubiquitin modification in endoplasmic reticulum-associated degradation. *The Biochemical Journal, 474*, 445–469.

Quax, T. E., Claassens, N. J., Soll, D., & van der Oost, J. (2015). Codon bias as a means to fine-tune gene expression. *Molecular Cell, 59*, 149–161.

Ravid, T., & Hochstrasser, M. (2008). Diversity of degradation signals in the ubiquitin-proteasome system. *Nature Reviews Molecular Cell Biology*, *9*, 679–690.

Ravid, T., Kreft, S. G., & Hochstrasser, M. (2006). Membrane and soluble substrates of the Doa10 ubiquitin ligase are degraded by distinct pathways. *The EMBO Journal*, *25*, 533–543.

Rizzolo, K., Huen, J., Kumar, A., Phanse, S., Vlasblom, J., Kakihara, Y., et al. (2017). Features of the chaperone cellular network revealed through systematic interaction mapping. *Cell Reports*, *20*, 2735–2748.

Rohner, N., Jarosz, D. F., Kowalko, J. E., Yoshizawa, M., Jeffery, W. R., Borowsky, R. L., et al. (2013). Cryptic variation in morphological evolution: HSP90 as a capacitor for loss of eyes in cavefish. *Science*, *342*, 1372–1375.

Rosenbaum, J. C., Fredrickson, E. K., Oeser, M. L., Garrett-Engele, C. M., Locke, M. N., Richardson, L. A., et al. (2011). Disorder targets misorder in nuclear quality control degradation: A disordered ubiquitin ligase directly recognizes its misfolded substrates. *Molecular Cell*, *41*, 93–106.

Rudiger, S., Buchberger, A., & Bukau, B. (1997). Interaction of Hsp70 chaperones with substrates. *Nature Structural Biology*, *4*, 342–349.

Rudiger, S., Germeroth, L., Schneider-Mergener, J., & Bukau, B. (1997). Substrate specificity of the DnaK chaperone determined by screening cellulose-bound peptide libraries. *The EMBO Journal*, *16*, 1501–1507.

Rutherford, S. L., & Lindquist, S. (1998). Hsp90 as a capacitor for morphological evolution. *Nature*, *396*, 336–342.

Sahni, N., Yi, S., Taipale, M., Fuxman Bass, J. I., Coulombe-Huntington, J., Yang, F., et al. (2015). Widespread macromolecular interaction perturbations in human genetic disorders. *Cell*, *161*, 647–660.

Savel'ev, A. S., Novikova, L. A., Kovaleva, I. E., Luzikov, V. N., Neupert, W., & Langer, T. (1998). ATP-dependent proteolysis in mitochondria. m-AAA protease and PIM1 protease exert overlapping substrate specificities and cooperate with the mtHsp70 system. *The Journal of Biological Chemistry*, *273*, 20596–20602.

Scroggins, B. T., Robzyk, K., Wang, D., Marcu, M. G., Tsutsumi, S., Beebe, K., et al. (2007). An acetylation site in the middle domain of Hsp90 regulates chaperone function. *Molecular Cell*, *25*, 151–159.

Shin, Y., & Brangwynne, C. P. (2017). Liquid phase condensation in cell physiology and disease. *Science*, *22*, 357.

Suri, M., Evers, J. M. G., Laskowski, R. A., O'Brien, S., Baker, K., Clayton-Smith, J., et al. (2017). Protein structure and phenotypic analysis of pathogenic and population missense variants in STXBP1. *Molecular Genetics & Genomic Medicine*, *5*, 495–507.

Swanson, R., Locher, M., & Hochstrasser, M. (2001). A conserved ubiquitin ligase of the nuclear envelope/endoplasmic reticulum that functions in both ER-associated and Matalpha2 repressor degradation. *Genes & Development*, *15*, 2660–2674.

Szoradi, T., Schaeff, K., Garcia-Rivera, E. M., Itzhak, D. N., Schmidt, R. M., Bircham, P. W., et al. (2018). SHRED is a regulatory cascade that reprograms Ubr1 substrate specificity for enhanced protein quality control during stress. *Molecular Cell*, *70*, 1025–1037.

Tasaki, T., Mulder, L. C., Iwamatsu, A., Lee, M. J., Davydov, I. V., Varshavsky, A., et al. (2005). A family of mammalian E3 ubiquitin ligases that contain the UBR box motif and recognize N-degrons. *Molecular and Cellular Biology*, *25*, 7120–7136.

Underhaug, J., Aubi, O., & Martinez, A. (2012). Phenylalanine hydroxylase misfolding and pharmacological chaperones. *Current Topics in Medicinal Chemistry*, *12*, 2534–2545.

Varshavsky, A. (2017). The ubiquitin system, autophagy, and regulated protein degradation. *Annual Review of Biochemistry*, *86*, 123–128.

Wagner, I., Arlt, H., van Dyck, L., Langer, T., & Neupert, W. (1994). Molecular chaperones cooperate with PIM1 protease in the degradation of misfolded proteins in mitochondria. *The EMBO Journal, 13*, 5135–5145.

Walther, D. M., Kasturi, P., Zheng, M., Pinkert, S., Vecchi, G., Ciryam, P., et al. (2015). Widespread proteome remodeling and aggregation in aging C. elegans. *Cell, 161*, 919–932.

Wangeline, M. A., Vashistha, N., & Hampton, R. Y. (2017). Proteostatic tactics in the strategy of sterol regulation. *Annual Review of Cell and Developmental Biology, 33*, 467–489.

Whitesell, L., & Lindquist, S. L. (2005). HSP90 and the chaperoning of cancer. *Nature Reviews Cancer, 5*, 761–772.

Whitesell, L., Mimnaugh, E. G., De Costa, B., Myers, C. E., & Neckers, L. M. (1994). Inhibition of heat shock protein HSP90-pp60v-src heteroprotein complex formation by benzoquinone ansamycins: Essential role for stress proteins in oncogenic transformation. *Proceedings of the National Academy of Sciences of the United States of America, 91*, 8324–8328.

Wilkinson, C. R., Wallace, M., Morphew, M., Perry, P., Allshire, R., Javerzat, J. P., et al. (1998). Localization of the 26S proteasome during mitosis and meiosis in fission yeast. *The EMBO Journal, 17*, 6465–6476.

Winter, G. E., Buckley, D. L., Paulk, J., Roberts, J. M., Souza, A., Dhe-Paganon, S., et al. (2015). DRUG DEVELOPMENT. Phthalimide conjugation as a strategy for in vivo target protein degradation. *Science, 348*, 1376–1381.

Wu, X., & Rapoport, T. A. (2018). Mechanistic insights into ER-associated protein degradation. *Current Opinion in Cell Biology, 53*, 22–28.

Xu, Y., & Lindquist, S. (1993). Heat-shock protein hsp90 governs the activity of pp60v-src kinase. *Proceedings of the National Academy of Sciences of the United States of America, 90*, 7074–7078.

Xu, Y., Singer, M. A., & Lindquist, S. (1999). Maturation of the tyrosine kinase c-src as a kinase and as a substrate depends on the molecular chaperone Hsp90. *Proceedings of the National Academy of Sciences of the United States of America, 96*, 109–114.

Yang, C., Huntoon, K., Ksendzovsky, A., Zhuang, Z., & Lonser, R. R. (2013). Proteostasis modulators prolong missense VHL protein activity and halt tumor progression. *Cell Reports, 3*, 52–59.

Zattas, D., & Hochstrasser, M. (2015). Ubiquitin-dependent protein degradation at the yeast endoplasmic reticulum and nuclear envelope. *Critical Reviews in Biochemistry and Molecular Biology, 50*, 1–17.

Zhang, Y., Sinning, I., & Rospert, S. (2017). Two chaperones locked in an embrace: Structure and function of the ribosome-associated complex RAC. *Nature Structural & Molecular Biology, 24*, 611–619.

CHAPTER THREE

Chaperones and retinal disorders

Maxim Sokolov[a], Ravi P. Yadav[b], Celine Brooks[a], Nikolai O. Artemyev[b,c,*]

[a]Department of Ophthalmology, West Virginia University, Morgantown, WV, United States
[b]Department of Molecular Physiology and Biophysics, The University of Iowa Carver College of Medicine, Iowa City, IA, United States
[c]Department of Ophthalmology and Visual Sciences, The University of Iowa Carver College of Medicine, Iowa City, IA, United States
*Corresponding author: e-mail address: nikolai-artemyev@uiowa.edu

Contents

Abstract

Defects in protein folding and trafficking are a common cause of photoreceptor degeneration, causing blindness. Photoreceptor cells present an unusual challenge to the protein folding and transport machinery due to the high rate of protein synthesis, trafficking and the renewal of the outer segment, a primary cilium that has been modified into a specialized light-sensing compartment. Phototransduction components, such as rhodopsin and cGMP-phosphodiesterase, and multimeric ciliary transport

Advances in Protein Chemistry and Structural Biology, Volume 114
ISSN 1876-1623
https://doi.org/10.1016/bs.apcsb.2018.09.001

complexes, such as the BBSome, are hotspots for mutations that disrupt proteostasis and lead to the death of photoreceptors. In this chapter, we review recent studies that advance our understanding of the chaperone and transport machinery of phototransduction proteins.

1. Introduction

Vision is mediated by rod and cone photoreceptors located in the outermost layer of the retina. First, visual information is captured by the photoreceptors, then the electrical signal is transmitted to the inner retina neurons. From the inner neurons, the signal is relayed to the visual cortex in the brain through the optic nerve. Rod and cone photoreceptors are highly polarized neurons with a specialized sensory compartment called the outer segment. Outer segments contain the phototransduction machinery that transforms absorbed photons into an electrical signal by hyperpolarizing the plasma membrane, thus controlling photoreceptor synaptic output to downstream neurons. The outer segment (OS) is an unusual primary cilium due to the presence of stacked flattened membrane discs, which are separated from the plasma membranes in rods or the infoldings in cones (Pearring, Salinas, Baker, & Arshavsky, 2013). The disc membrane is populated by key phototransduction components including the light receptor from the GPCR family (rhodopsin, cone opsins), the visual G protein (transducin), and the visual effector protein cGMP-phosphodiesterase of the PDE6 family (PDE6). At the disk membrane, photoexcited rhodopsin (R*) stimulates GDP/GTP exchange on the α-subunit ($G\alpha_{t1}$) of heterotrimeric transducin, resulting in the dissociation of $G\alpha_{t1}$GTP from $G\beta_1\gamma_1$ and R*. $G\alpha_{t1}$ in the GTP-bound conformation activates the effector enzyme, PDE6, by displacing the inhibitory γ-subunits (Pγ) from the PDE6 catalytic core. cGMP hydrolysis by activated PDE6 leads to a closure of cGMP-gated channels in the plasma membrane. The lifetime of the activated complex between $G\alpha_t$GTP and PDE6 is controlled by the GTPase activity of $G\alpha_t$. Hydrolysis of GTP transforms the GTP-bound conformation of $G\alpha_t$ into the inactive GDP-bound state and allows for re-inhibition of PDE6AB by Pγ (Arshavsky, Lamb, & Pugh, 2002; Lamb & Pugh, 2006).

OS proteins are constantly turned over through the phagocytosis of the OS tips by the cells of the retinal pigment epithelium and the addition of the newly synthesized components at the base of the OS. Protein synthesis in photoreceptors commences in the inner segment (IS), a distinct compartment that contains cellular metabolism organelles. A slender connecting cilium joins the IS with the OS. All newly synthesized phototransduction proteins are trafficked across the cilium in order to reach the sensory compartment (Karan, Zhang, Li, Frederick, & Baehr, 2008). High turnover and protein trafficking rates in photoreceptors coupled with unique structural properties of phototransduction components place a high demand on protein folding, quality control, and transport machinery to maintain proteostasis (Kosmaoglou, Schwarz, Bett, & Cheetham, 2008). Defects in protein folding and trafficking are a common cause of photoreceptor degeneration leading to blindness (Athanasiou et al., 2013; Kosmaoglou et al., 2008; Tzekov, Stein, & Kaushal, 2011). Signaling components in rods and cones are either shared by both cell types, or more often, they are represented by highly homologous yet distinct isoforms. Accordingly, mutations may affect rod or cone specific isoforms, or affect a common rod/cone component. Depending on the origin of the primary insult (mutation), retinal degenerative diseases manifest as retinitis pigmentosa (RP) (Hamel, 2006; Hartong, Berson, & Dryja, 2006), cone-rod dystrophy (CORD) (Hamel, 2007), achromatopsia (ACHM) (Hirji, Aboshiha, Georgiou, Bainbridge, & Michaelides, 2018), and Leber congenital amaurosis (LCA) (Koenekoop, 2004; Stone, 2007). RP is characterized by the initial loss of rods, followed by the secondary death of cones. In CORD, cones degenerate prior to the death of rods. ACHM represents a loss of cone photoreceptor function leading to color blindness. LCA severely affects both rods and cones at early stages. Retinal degeneration is also a diagnostic feature of syndromic ciliopathies caused by defects in proteins involved in ciliary function and trafficking. Besides retinal dystrophy, ciliopathies such as Bardet-Biedl syndrome (BBS), nephronophthisis (NPHP), Senior-Loken syndrome (SLS), Joubert syndrome (JS), and Meckel-Gruber syndrome (MKS) are accompanied by obesity, polydactyly, neurological and renal abnormalities (Bujakowska, Liu, & Pierce, 2017; Rachel, Li, & Swaroop, 2012; Ronquillo, Bernstein, & Baehr, 2012). Growing evidence suggests that photoreceptor cells, in addition to classical and ubiquitous molecular chaperones from the heat shock protein superfamily, utilize photoreceptor-specific chaperones or co-chaperones to assist in protein folding, subunit assembly, and protein trafficking (Athanasiou et al., 2013; Yadav & Artemyev, 2017; Zhang et al., 2011).

The high-resolution crystal structure of rhodopsin paved the way for groundbreaking discoveries of its structure-function relationships (Palczewski et al., 2000). Rhodopsin is the most abundant protein in rods, and rhodopsin mutations are responsible for about 25% of autosomal dominant RP cases

(Malanson & Lem, 2009). The majority of more than 200 known autosomal-dominant RP (adRP) rhodopsin mutants are misfolded leading to protein retention in the ER. Current approaches to alleviate the consequences of rhodopsin misfolding through manipulation of the heat shock response, the unfolded protein response and ER chaperones, as well as the use of chemical chaperones to stabilize the protein are the subject of a recent comprehensive review (Athanasiou et al., 2018). Here, we will review advances in our understanding of chaperone-assisted folding/assembly of transducin and PDE6, as well as the role of trafficking chaperones in the ciliary transport of these signaling molecules.

2. Folding and assembly of transducin βγ: The role of chaperonin

Chaperonins are protein-folding ATPase complexes that are present in all living cells and share a ring-shape structure. The hollow center of the rings creates a cavity, which is accessible from the outside (Llorca et al., 2000; Zhang et al., 2010). Each chaperonin subunit is comprised of an equatorial or ATP-binding domain, an intermediate domain, and an apical domain (Dekker et al., 2011). Substrate binding occurs on the apical domain and the intermediate domain connects the equatorial and apical domains (Valpuesta, Martin-Benito, Gomez-Puertas, Carrascosa, & Willison, 2002). Chaperonins facilitate folding of their protein substrates by temporally encapsulating the nascent polypeptides in their central cavity. The central cavity provides a microenvironment, where other molecules that could disrupt folding are absent. When the cavity is open, substrate proteins can enter the cavity and bind to the apical domain of one or more subunits. Once substrate binds, ATP binding occurs, and hydrolysis of ATP causes the cavity to close, triggering folding of the encapsulated substrate. The original term "chaperonin" described ring-shaped cytosolic complexes called GroEL from eubacteria and heat-shock protein (hsp) 60s from chloroplasts and mitochondria (Hemmingsen et al., 1988). Their eukaryotic and archaeal homologs, named TCP-1 ring complex (TRiC) or CCT (chaperonin containing TCP-1) (Frydman et al., 1992; Gao, Thomas, Chow, Lee, & Cowan, 1992; Kubota, Hynes, Carne, Ashworth, & Willison, 1994) and the thermosome (Klumpp, Baumeister, & Essen, 1997), were discovered later and collectively designated as group II chaperonins, to distinguish them from the original group I, as reviewed in Gutsche, Essen, and Baumeister (1999).

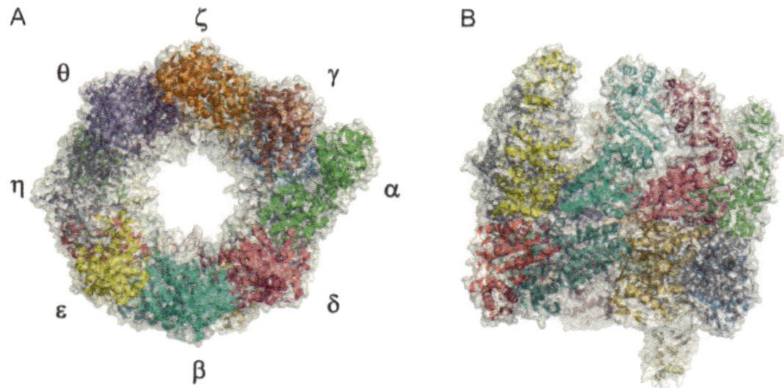

Fig. 1 Subunit arrangement of CCT. The CCT complex is comprised of two stacked rings, each made of eight homologous subunits (open form; cartoon representation; PDB ID: 4B2T) (A—top view, B—side view). Gβ associates with CCT *via* interaction of its Gγ-binding region with the γ-subunit of CCT. PhLP1 stabilizes Gβ and releases it from CCT prior to assembly of the Gβγ dimer.

CCT is ~1 MDa complex comprised of two stacked rings, each made of eight highly similar TCP-1 (tailless complex polypeptide 1) subunits, designated as α–θ in mammals, and as CCT1–8 in other organisms (Fig. 1A and B). Other chaperonins including the group I chaperonins usually consist of one repeating subunit, while the thermosome is made up of one to three subunits (Valpuesta et al., 2002). In contrast to group I chaperonins, CCT and the thermosome do not seem to require a co-chaperonin to serve as a lid to the central chamber. In GroEL, GroES caps the chamber and is necessary for folding, while Hsp10 acts as a lid for Hsp60. CCT creates a lid, which regulates access to the central cavity by combining the helical protrusions on the apical domain of each CCT subunit (Dekker et al., 2011). These helical protrusions are not found in group I chaperonins. Based on the central cavity volume, the theoretical size limit for a CCT substrate is 70 kDa; however, CCT apparently can fold significantly larger substrates by encapsulating individual domains rather than the entire protein (Russmann et al., 2012). CCT is constitutively active and essential for all eukaryotic cells, due to its role in the folding of actin and tubulin. In addition, CCT has a wide-ranging substrate specificity including a preference for proteins with the WD-40 repeat, which forms a beta-propeller structure (Willison, 2018). It would be safe to assume that a number of CCT substrates in specialized mammalian cells remain to be identified. Here, we will focus on the well-established role of CCT in the folding of the Gβ_1 subunit of the phototransduction G protein transducin.

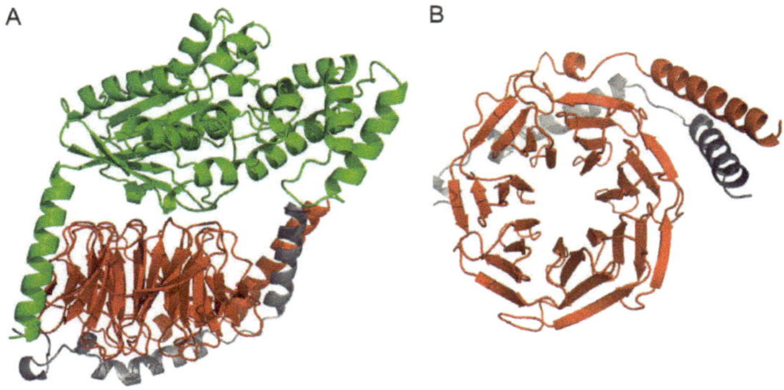

Fig. 2 Structural overview of heterotrimeric transducin. (A) Rod transducin is made of $G\alpha_{t1}$ (green), $G\beta_1$ (red), and $G\gamma_1$ subunits (gray) (cartoon representation PDB ID: 1GOT). In the cell, Gβ and Gγ are constitutively associated into $G\beta_1\gamma_1$ complex. (B) Gβ is the WD-40 repeat protein, which forms a seven-bladed β-propeller structure (top view).

Rods maintain an exceptionally high cellular level of transducin, a heterotrimeric G protein coupled to a light-activated receptor, rhodopsin, which is required for high light sensitivity of nocturnal vision. Rod-specific transducin is made of $G\alpha_{t1}$, $G\beta_1$, and $G\gamma_1$ subunits (Fig. 2A and B). In the cell, $G\beta_1$ and $G\gamma_1$ are constitutively associated into $G\beta_1\gamma_1$ subunits complex. $G\beta_1$ and four other G-protein β subunits that are expressed in eukaryotes were identified as *bona fide* CCT substrates (Garcia-Higuera, Gaitatzes, Smith, & Neer, 1998; Humrich et al., 2005; Valpuesta et al., 2002; Wells, Dingus, & Hildebrandt, 2006). The folding of $G\beta_1$ by CCT requires a co-chaperone, phosducin-like protein 1 (PhLP1) (Lukov, Hu, McLaughlin, Hamm, & Willardson, 2005). Analyses of $G\beta_1$ interactions with CCT utilizing chemical cross-linking and cryo-electron microscopy revealed that the characteristic 7-blade β-propeller structure of $G\beta_1$ is mostly present before the blades make contact with CCT, thus indicating that folding occurs before binding to CCT (Plimpton et al., 2015). However, the β-propeller structure of $G\beta_1$ remains partially unlocked and flexible at this stage. The folding of the β-propeller is finalized upon the binding of co-chaperone PhLP1 and the formation of a tertiary PhLP1-$G\beta_1$-CCT complex. PhLP1 intensively interacts with CCT, making contacts with at least three CCT subunits including TCP-1α, TCP-1γ, and TCP-1δ. These interactions appear to induce conformation changes of CCT that lock the β-propeller structure of $G\beta_1$ and cause $G\beta_1$ to rotate inside the folding cavity, which initiates its release (Plimpton et al., 2015). $G\beta_1$ is believed to be released in complex with PhLP1, which facilitates the assembly of the $G\beta_1\gamma_1$ dimer. $G\gamma_1$ cannot directly bind $G\beta_1$, unless it is in a

PhLP1-Gβ_1 complex. Accordingly, disrupting the PhLP1 function in rods either by expressing a dominant-negative mutant, $^{\Delta 1\text{-}48}$PhLP1 (Gao et al., 2013; Posokhova et al., 2011), or by knocking out *Pdcl* gene (Lai et al., 2013) decimated the formation of G$\beta_1\gamma_1$ dimer, which was accompanied by a virtual disappearance of Gγ_1. Binding of Gγ_1 leads to formation of apparently unstable tertiary PhLP1-G$\beta_1\gamma_1$ complex, which promptly dissociates into PhLP1 and G$\beta_1\gamma_1$ dimer. This step completes the folding and assembly of transducin G$\beta_1\gamma_1$ and recycles PhLP1. The assembled G$\beta_1\gamma_1$ dimer cannot rebind to CCT (Gao et al., 2013), because the α-helix of Gβ_1 that engages in coil-coil interaction with Gγ_1 (Gaudet, Bohm, & Sigler, 1996) binds to the TCP-1γ subunit of CCT (Plimpton et al., 2015). After assembly, another protein from the same family, phosducin assists in the trafficking of mature G$\beta_1\gamma_1$ into the rod sensory cilium.

The signaling function of mature transducin in rods is intertwined with another molecular chaperone, phosducin. In vertebrate photoreceptors, phosducin was first discovered as a major 33-kDa phosphoprotein that binds transducin G$\beta_1\gamma_1$ with a high affinity in the de-phosphorylated state (Gaudet et al., 1996; Lee, Brown, & Lolley, 1984; Lee, Lieberman, & Lolley, 1987). Initially phosducin was proposed to be a negative regulator of phototransduction, because binding of phosducin to G$\beta_1\gamma_1$ prevents the association of G$\beta_1\gamma_1$ with Gα_{t1} (Gaudet et al., 1996), and the phosphorylation status of phosducin *in vivo* is tightly controlled by the history of illumination (Song, Belcastro, Young, & Sokolov, 2007). Theoretically, phosducin would act as a negative regulator of phototransduction by modulating the amount of heterotrimeric transducin available for activation by rhodopsin (Lee et al., 1992; Wilkins, Bitensky, & Willardson, 1996; Yoshida et al., 1994). However, this function of phosducin could not be confirmed in phosducin knockout mice (Krispel et al., 2007). Instead, the analyses of this animal model indicated that phosducin acts as a trafficking chaperone during light-driven transducin translocation, an adaptive physiological response of rod photoreceptors to saturating light (Belcastro et al., 2012; Sokolov et al., 2002, 2004). Transducin withdrawal from the rod sensory cilium and its coupled receptor, rhodopsin, upon persistent overstimulation, was predicted based on a reversible redistribution of Gα_{t1} and G$\beta_1\gamma_1$ from the membrane to the aqueous phases (Sokolov et al., 2002). Phosducin facilitates transducin translocation by sequestering the hydrophobic farnesyl moiety of Gγ_1 inside the phosducin-G$\beta_1\gamma_1$ complex (Loew, Ho, Blundell, & Bax, 1998) (Fig. 3). The farnesyl moiety serves as a membrane anchor, which targets transducin to the cilium (Brooks et al., 2018). Some evidence suggests that phosducin

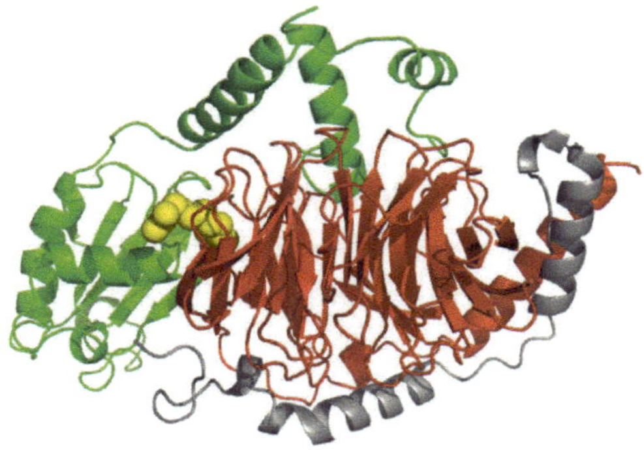

Fig. 3 Structural overview of the phosducin/G$\beta_1\gamma_1$ complex. The N-terminal (helical) domain and the C-terminal domain of phosducin (green) cover the top and side of the β-propeller of Gβ_1 (red). The binding of phosducin creates a small cavity between blades 6 and 7 of the β-propeller of Gβ_1, which sequesters the farnesyl moiety (yellow spheres) of Gγ_1 (gray) (PDB ID: 1A0R).

may also protect G$\beta_1\gamma_1$ from ubiquitination and degradation by the proteasome found in the rod cytosol (Obin et al., 2002). Correspondingly, phosducin-null rods fail to maintain normal G$\beta_1\gamma_1$ level possibly due to proteasomal degradation (Krispel et al., 2007; Sokolov et al., 2004). In summary, two closely related proteins from the phosducin family, PhLP1 and phosducin, serve as molecular chaperones for two distinct sub-populations of transducin G$\beta_1\gamma_1$. PhLP1 mediates the folding and assembly of nascent G$\beta_1\gamma_1$ in concert with the chaperonin CCT, whereas phosducin assists in the trafficking of mature G$\beta_1\gamma_1$ during light-induced transducin translocation.

3. CCT and the assembly of the BBSome

CCT also appears to be connected to the biogenesis of the BBSome, a protein complex that participates in the import and export of molecules to the cilium. BBS is an abbreviation for Bardet-Biedl syndrome, a genetic condition that is characterized by obesity, retinal degeneration, polydactyly, hypogonadism, renal failure, and learning difficulties (Forsythe & Beales, 2013). Based on the presence of mutations in BBS patients, 19 BBS genes were identified and linked to cilia malfunction (Zaghloul & Katsanis, 2009). Of the 19 BBS proteins, 7 proteins including BBS 1, 2, 4,5, 7, 8, and 9 were shown to form a complex coined the BBSome (Nachury et al., 2007). Later, BBS18 was shown to stabilize the BBSome, making the BBSome an octameric complex (Loktev et al., 2008). The BBSome is thought to form a planar membrane coat that moves along the cilium together with interflagellar transport (IFT) complex to enable docking and release of cargo molecules on both ends of the cilium; however, this process is not well understood (Klink et al., 2017). CCT was first linked to the BBSome, when BBS10 was found to share sequence homology with TCP-1 subunits of CCT (Stoetzel et al., 2006). Then it was demonstrated that BBS10 and two other BBS chaperonin-like proteins, BBS6 and BBS12, assemble with TCP-1α, TCP-1β, and TCP-1γ subunits of CCT into a CCT-like complex, which was essential for ciliary function (Seo et al., 2010). This BBS/CCT chaperonin is proposed to be involved in the folding of BBSome proteins and the assembly of the BBSome (Seo et al., 2010; Zhang, Yu, Seo, Stone, & Sheffield, 2012). The primary structure of BBS1, BBS2, BBS7, and BBS9 supports this notion that BBS/CCT chaperonin may fold BBS proteins. Based on their primary structure, these proteins are predicted to form the β-propeller fold (Jin et al., 2010), a prevalent tertiary structure of CCT substrates (Willison, 2018, p. 24). This prediction was confirmed for BBS1, which was shown to be a seven-bladed β-propeller, whose N-terminal domain interacts with the small GTPase ARL6 (BBS3) responsible for the recruitment of the BBSome to the ciliary membrane (Mourao, Nager, Nachury, & Lorentzen, 2014). The potential of CCT to fold BBS proteins or assemble the BBSome is further supported by the observation that the BBSome is responsive to the suppression of CCT; suppressing CCT in rod photoreceptors by a dominant-negative $^{\Delta 1-48}$PhLP1 transgene was shown to destabilize BBS2, BBS5, and BBS7 and to disrupt assembly of the BBSome (Sinha, Belcastro, Datta, Seo, & Sokolov, 2014). Confirming BBS proteins as CCT substrates and the assembly of the BBSome by the BBS/CCT chaperonin remains the subject of future investigation.

4. Does transducin α require a chaperone?

4.1 Ric8A as a candidate chaperone for transducin $G\alpha_t$-subunit

In general, Gα subunits are first folded in the cytosol and then transported to the endomembranes, where they form heterotrimers with Gβγ (Wedegaertner, 2012). The assembly of heterotrimeric rod transducin requires properly folded $G\alpha_{t1}$ and $G\beta_1\gamma_1$ (Willardson & Howlett, 2007). Assembly of heterotrimeric rod transducin occurs after $G\gamma_1$ associates with $G\beta_1$ in the cytoplasm and its C-terminal CAAX motif is modified with a farnesyl (C15) isoprenyl group (Lai, Perez-Sala, Canada, & Rando, 1990). The lipid modification targets

$G\beta_1\gamma_1$ to the ER membrane, where CAAX processing is completed *via* proteolytic removal of the AAX tripeptide and carboxymethylation of the terminal Cys by the ER-associated enzymes (Wedegaertner, 2012; Willardson & Howlett, 2007). The $G\alpha_{t1}$ subunit of transducin is co-translationally N-acylated at the N-terminal Gly2, since the initiating Met is removed. The acylation of $G\alpha_{t1}$ is heterogeneous involving myristate (C14:0), tetradecanoic (C14:1), tetradecadienoic acid (C14:2), and laurate (C12) moieties (Neubert, Johnson, Hurley, & Walsh, 1992). Little is known about the folding pathways of $G\alpha_{t1}$ prior to the formation of heterotrimeric transducin $G\alpha_{t1}\beta_1\gamma_1$ on the endomembranes of the IS. It is not even clear if a specialized chaperone is required for biogenesis of $G\alpha_{t1}$. However, several indicators support the idea of a chaperone assisted folding of $G\alpha_{t1}$. Compared to other $G\alpha$ subunits of the $G\alpha_{i/o}$ class, $G\alpha_{t1}$ is poorly expressed in heterologous cells, and thus chimeric $G\alpha_{t1}/G\alpha_i$ proteins are often expressed as a substitute (Skiba, Bae, & Hamm, 1996). *In vivo*, $G\alpha_{t1}$ is relatively unstable in the absence of $G\beta_1\gamma_1$. In rod photoreceptors depleted of $G\beta_1\gamma_1$ due to a lack of PhLP1 or $G\gamma_1$, levels of $G\alpha_{t1}$ are significantly decreased (Lai et al., 2013; Lobanova et al., 2008), whereas rods lacking $G\alpha_{t1}$ have normal levels of $G\beta_1\gamma_1$ (Calvert et al., 2000). Also, transducin is the second most abundant protein in the OS (Arshavsky et al., 2002), and its renewal requires that it is constantly produced at high levels with correctly folded tertiary structure. In $G\gamma_1$ KO mice, rods are proposed to degenerate, because the proteasome is overloaded with misfolded $G\beta_1$ (Lobanova, Finkelstein, Skiba, & Arshavsky, 2013). Moreover, insufficient capacity of the proteasome for large amounts of misfolded protein in photoreceptors appears to underlie retinal degeneration in other mouse models in which proteins are mistargeted and/or misfolded (Lobanova et al., 2013). Thus, correct and robust folding of $G\alpha_{t1}$ is imperative for the proteostasis and health of photoreceptors. Yet, in contrast to $G\beta\gamma$ the existence of a subunit-specific chaperone for $G\alpha_{t1}$ has not been discovered.

Growing evidence suggests that resistance to inhibitors of cholinesterase 8 (Ric8) proteins serves as ubiquitous chaperones for the $G\alpha$-subunits (Chan, Thomas, Sprang, & Tall, 2013; Gabay et al., 2011; Tall, Krumins, & Gilman, 2003). In *Drosophila* and *C. elegans*, a single isoform (Ric8) regulates the abundance and membrane localization of all $G\alpha$ proteins (Tall, 2013; Wilkie & Kinch, 2005). Vertebrates have two isoforms, Ric8A and Ric8B, with Ric8A regulating the $G\alpha_{i/o}$, $G\alpha_q$ and $G\alpha_{12/13}$ classes of G-protein α subunits, and Ric8B interacting with $G\alpha_s$ (Matsuzaki, 2005; Tall et al., 2003;

Von Dannecker, Mercadante, & Malnic, 2005). Ric8A was believed to positively regulate Gα subunits by acting as a non-GPCR guanine nucleotide exchange factor (GEFs). Ric8A preferentially binds GDP-bound Gα and causes GDP release. Like GPCRs, Ric8A forms a stable complex with Gα in an intermediate nucleotide-free state (Gα-empty), and the complex breaks apart when Gα binds GTP (Tall, 2013; Tall et al., 2003). It appears, however, that the effects of Ric8A (and Ric8B) on G-proteins can be explained largely by their chaperone activity (Papasergi, Patel, & Tall, 2015; Tall, 2013). In mouse ES cells lacking Ric8A, newly synthesized Gα subunits failed to associate with membranes and were rapidly degraded (Gabay et al., 2011). Furthermore, in cell-free translation systems, the expression of properly folded Gα subunits required Ric8A (Chan et al., 2013). Co-expression of Ric8A or Ric8B with Gα subunits in HEK293 cells or insect cells greatly elevated expression of Gα subunits (Chan et al., 2011). Therefore Ric8A behaves as *bona fide* chaperone for Gα. Although the role of Ric8A as a candidate chaperone for $G\alpha_{t1}$ has not been examined, Ric8A was shown to interact with and potentiate signaling of closely related gustducin Gα, a member of the transducin subfamily of the $G_{i/o}$ family (Fenech et al., 2009).

4.2 Progress in understanding the mechanism of the chaperone activity of Ric8

Strides have been made to understand the mechanism of Ric8A's chaperone activity. The classical structure of a Gα subunit features two key domains, the Ras-like (RD) domain and the α-helical domain (HD). GDP or GTP is bound to the RD domain, whereas the HD domain serves as a lid over the nucleotide binding site by interacting with the RD domain (Sprang, Chen, & Du, 2007). However, an agonist bound GPCR induces a Gα-empty state, in which the guanine-binding pocket is disrupted due to the reorganization of the purine ring binding P-loop and the phosphate binding β6-α5 loop. This reorganization abrogates the RD/HD interface and the domains are dynamically separated (Mahoney & Sunahara, 2016; Preininger, Meiler, & Hamm, 2013). Interestingly, comparable conformational changes in Gα have been identified after binding of Ric8A. A recent study using site-directed spin labeling and dipolar EPR spectroscopy demonstrated that Ric8a induced the separation of the RD and HD domains of Gα, which was accompanied by elevated structural plasticity of the nucleotide-binding site (Van Eps, Thomas, Hubbell, & Sprang, 2015). Further insights into the mechanism of Ric8A GEF activity were revealed with the use of hydrogen-deuterium exchange mass spectrometry. This study showed that Ric8A disrupts the secondary structure of RD surrounding the guanine-binding site, which perturbs the interface between RD and HD, causing domain separation and nucleotide release (Kant, Zeng, Thomas, Bothner, & Sprang, 2016). The intermediate complex of Ric8A with Gα-empty that underlies the GEF activity may in fact be similar to the folding intermediate during biosynthesis of Gα. A model has been proposed where the RD and HD domains are individually folded first with the assistance of HSP70/90, and possibly CCT, while Ric8A completes the folding by positioning the RH and HD domains and helping to organize the *de novo* nucleotide-binding pocket (Papasergi et al., 2015). Once the pocket is formed, Gα binds GTP and releases Ric8A. Subsequently, GTP is hydrolyzed, and GαGDP binds Gβγ yielding fully assembled heterotrimeric G-protein. Further progress in understanding the chaperone action of Ric8A is hampered by the lack of structures of either Ric8A or its complex with Gα. Ric8A has been predicted to have the armadillo-type fold that is characteristic of ARM repeat-containing proteins such as β-catenin and α-importin (Figueroa et al., 2009). Various models of Ric8A have been proposed, and they indicate a superhelical arrangement of 10 ARM repeats each comprised of three α-helices that are repeated in tandem. Strong evidence suggests that, similarly to GPCRs, Ric8A interacts with the C-terminus of Gα (Kant et al., 2016; Thomas et al., 2011). Unlike GPCRs, Ric8A does not activate heterotrimeric Gαβγ (Tall et al., 2003). Also, Ric8A appears to interact with the switch II region of Gα and possibly the RD/HD domain interface (Kant et al., 2016; Papasergi et al., 2015; Thomas et al., 2011). Thus, the GEF mechanisms of Ric8A and GPCRs, while sharing common features, are principally different.

5. AIPL1 is a specialized chaperone of PDE6

5.1 Relationships between PDE6 and AIPL1

PDE6s are central to photoreceptor physiology as the key effectors in the visual transduction cascade in rods and cones (Arshavsky et al., 2002). Furthermore, PDE6s are critical to the survival of rods and cones. Loss of function of PDE6 leads to elevation of intracellular cGMP levels, which causes photoreceptor cell death due to excessive opening of the cGMP-gated channels in the plasma membrane and unrestrained influx of Ca^{2+} (Bowes et al., 1990; Dryja, Rucinski, Chen, & Berson, 1999; Farber & Lolley, 1974; McLaughlin, Ehrhart, Berson, & Dryja, 1995; Pittler & Baehr, 1991; Wang, Tsang, & Chen, 2017). The rod PDE6 catalytic core is a heterodimer of PDE6A and PDE6B subunits, whereas cone PDE6C subunits form a catalytic homodimer (Cote, 2004). Mutations in PDE6AB cause RP (Dryja et al., 1999; McLaughlin et al., 1995), and mutations in PDE6C have been linked to ACHM (Chang et al., 2009; Grau et al., 2011; Thiadens et al., 2009). The second major reason for loss of function of PDE6 is malfunction of its specialized chaperone aryl hydrocarbon receptor (AhR)-interacting protein-like 1 (AIPL1). The *AIPL1* gene encoding AIPL1 was originally discovered due to its association with Leber congenital amaurosis (LCA) (Sohocki et al., 2000), and was named based on its high sequence identity (49%) and similar domain organization with the ubiquitously expressed aryl hydrocarbon receptor (AhR)-interacting protein (AIP) (Sohocki et al., 2000). The *AIPL1*-related LCA4 is one of the most clinically severe forms of LCA that causes blindness in early childhood (den Hollander, Roepman, Koenekoop, & Cremers, 2008; Kumaran, Moore, Weleber, & Michaelides, 2017). In contrast to AIP, AIPL1 is expressed exclusively in the retina and the pineal gland (Sohocki et al., 2000; van der Spuy et al., 2002). AIP and AIPL1 both contain an N-terminal FK506-binding protein (FKBP) domain and a C-terminal tetratricopeptide repeat (TPR)-domain with three tetratricopeptide repeats (Sohocki et al., 2000). The first strong indication that AIPL1 functions as a chaperone for PDE6 came from the analyses of the AIPL1 knockout and knockdown mouse models, which revealed dramatic reduction in protein levels and activity of PDE6 (Liu et al., 2004; Ramamurthy, Niemi, Reh, & Hurley, 2004). As a result, rod and cone photoreceptors in the mutant retina rapidly degenerate, thus recapitulating the LCA4 phenotype (Ramamurthy et al., 2004). The role of AIPL1 as the PDE6 chaperone was ultimately confirmed with successful heterologous expression of cone PDE6C in the presence of AIPL1 in HEK293 cells (Gopalakrishna, Boyd, & Artemyev, 2017; Gopalakrishna, Boyd, Yadav, & Artemyev, 2016).

5.2 Structural insights into the chaperone function of AIPL1

Until recently, structural information on AIPL1 had been lacking. Although AIPL1 has been shown to bind to PDE6A (Ramamurthy et al., 2003), the nature and the interface of this interaction were determined only recently. PDE6A and PDE6B subunits are farnesylated (C15) and geranylgeranylated (C20), respectively, at the C-terminal CAAX motifs (Anant et al., 1992), whereas cone PDE6C is geranylgeranylated (Christiansen & Ramamurthy, 2012). A direct binding of AIPL1 to the prenyl moieties of PDE6 was demonstrated with the use of a Fluorescence Resonance Energy Transfer (FRET) assay from AIPL1 Trp residues to the S-farnesyl-L-cysteine methyl ester (FC) probe or FC-labeled C-terminal peptide of PDE6 modified with a FRET acceptor group (Majumder, Gopalakrishna, Cheguru, Gakhar, & Artemyev, 2013; Yadav, Majumder, Gakhar, & Artemyev, 2015). The prenyl-binding site has been localized to the FKBP-domain of AIPL1 (Majumder et al., 2013). Interestingly, the highly homologous FKBP-domain of AIP does not bind to prenyl moieties, which is consistent with the inability of AIP to chaperone PDE6 in the HEK293T expression system (Gopalakrishna et al., 2016). These findings were puzzling since, despite a large body of structural data developed on FKBP-domain containing proteins, they are not known to bind prenyl lipid modifications. FKBPs play essential roles in fundamental cellular processes including protein folding and signaling (Hausch, 2015; MacMillan, 2013; Storer, Dickey, Galigniana, Rein, & Cox, 2011; Tong & Jiang, 2015). Often, FKBPs are active peptidylprolyl isomerases (PPIases) that catalyze the *cis-trans* conversion of peptidylprolyl bonds (Fanghanel & Fischer, 2004). FKBPs are best known as the mediators of the effects of the immunosuppressive drugs, FK506 and rapamycin (Blackburn & Walkinshaw, 2011; Tong & Jiang, 2015; Van Duyne, Standaert, Karplus, Schreiber, & Clardy, 1993). A classical FKBP fold, exemplified by the structure of FKBP12, is comprised of a half-conical six-stranded β-sheet surrounding a short central α-helix (Van Duyne et al., 1993). FK506 and rapamycin bind within a conserved hydrophobic cavity between the β-sheet and the α-helix, which also overlaps with the PPIase catalytic site (Alag et al., 2013). A short hairpin-like loop linking the β-strands β5 and β6 at the periphery above the cavity also contributes to the drug-binding (Van Duyne et al., 1993). In contrast to conventional FKBPs, AIPL1 and AIP do not possess PPIase activity and do not bind FK506 or rapamycin. The structural basis for these unusual properties was elucidated with the NMR structure of AIP (Linnert et al., 2013). In AIP (and AIPL1) the hairpin loop is replaced by a longer "insert" region. The AIP structure showed that the insert region overhangs the hydrophobic cavity and completely blocks a surface access to the cavity (Linnert et al., 2013), leaving no plausible binding site(s) for hydrophobic ligand(s). The mystery of the prenyl-lipid binding of AILP1 was solved when the crystal structures of the FKBP-domain of AILP1 were determined in the apo form and in complex with isoprenyl moieties (Yadav, Gakhar, Yu, & Artemyev, 2017). These structures exposed two principal differences between AIP and AIPL1 enabling the latter to bind prenyl lipids (Yadav et al., 2017). In AIP, the β4-α1 loop preceding the central α1-helix interacts with the insert region and helps to seal the cavity. In AIPL1, the β4-α1 loop moves away from the insert region and adopts a

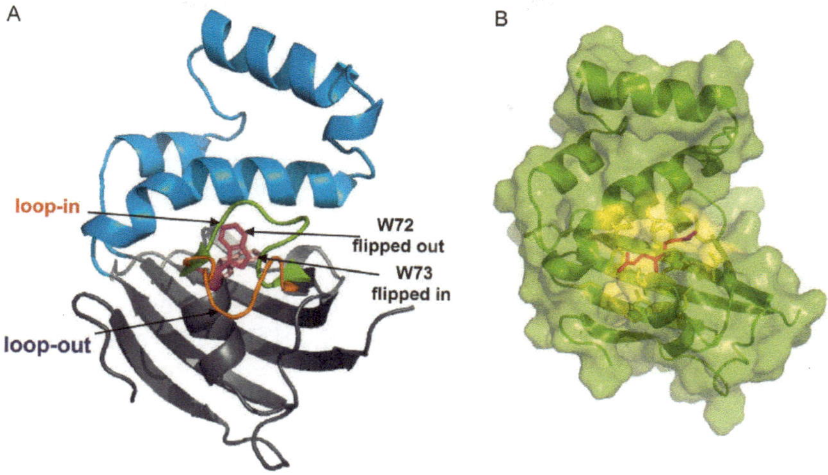

Fig. 4 Structural features of AIPL1 underlying prenyl-lipid binding. (A) The crystal structure of apo AIPL1-FKBP (PDB ID: 5U9A). The core FKBP fold is shown in gray and the "insert" region is in blue. The β4-α1 loop in the "loop-out" conformation (orange) and the W72 residue in the "flip-out" conformation (magenta). The AIP β4-α1 loop in the "loop-in" conformation (green) and the W73 in the "flip-in" conformation (light pink) are superimposed onto the structure of AIPL1-FKBP from the NMR structure of AIP (PDB ID: 2LKN). (B) The crystal structure of AIPL1-FKBP in complex with geranylgeranyl moiety (PDB ID: 5U9J). The geranylgeranyl moiety is bound to a deep hydrophobic pocket between the core FKBP and the "insert" region. Hydrophobic residues lining up the binding pocket are shown as yellow sticks.

distinct "loop-out" conformation thereby opening an access into the hydrophobic cavity (Yadav et al., 2017) (Fig. 4A). The second important and unique feature of AIPL1 is the orientation of the side chain of Trp72. The side chains of the counterpart Trp residues in AIP and FKPB12 are perpendicular to the α1 axis and constitute the base of the cavity (Linnert et al., 2013; Van Duyne et al., 1993). In AIPL1, the side chain of Trp72 flips out of the cavity to a position parallel to the α1 axis (Fig. 4A). A "loop-out" conformation of β4-α1 combined with the "flip-out" position of Trp72 creates an accessible and deep hydrophobic pocket, which is sandwiched between the core FKBP fold and the α2 helix of the insert region (Yadav et al., 2017). This hydrophobic pocket accommodates the prenyl moieties by slight rearrangement of the side chains lining up the pocket (Fig. 4B).

5.3 Conformational dynamics of AIPL1 explain pathogenicity of LCA mutations

The crystal structures of AIPL1 FKBP provided a clear explanation for pathogenicity of certain LCA4 mutations such as C89R variant. Cys89 is a part of the hydrophobic pocket, and the mutation disrupts the pocket and the prenyl-binding to AIPL1-FKBP (Majumder et al., 2013; Yadav et al., 2017). However, the AIPL1-FKBP structures did not offer obvious mechanisms for pathogenicity of at least two other LCA4 variants, V71F and W72S, which retained the ability to bind prenyl moieties (Gopalakrishna et al., 2016; Majumder et al., 2013; Yadav et al., 2017). Furthermore, the crystal structure of liganded V17F has been solved, and it was nearly identical to the structure of AIPL1-FKBP (Yadav et al., 2017). The clues were uncovered by subsequent analysis of the protein dynamics in solution by NMR (Yadav et al., 2017; Yu, Yadav & Artemyev, 2017). This analysis revealed that apo APL1-FKBP in solution is present in two major conformations referred to as "open" and "closed" states. Upon binding of FC, the two conformations converged into a single liganded state (Yadav et al., 2017). The two conformations of apo AIPL1-FKBP apparently resulted from movement of the side chain of Trp72 in and out of the ligand binding pocket (Yadav et al., 2017). The "flip-out" conformation of Trp72 in the open state is thought to be similar to that observed in the crystal structure of AIPL1-FKBP, whereas the "flip-in" conformation may resemble positions of the counterpart Trp residues in AIP and FKBP12 (Fig. 4A). Interestingly, because Trp72 in the closed state occupies the binding pocket and mimics the ligand, the positions of NMR peaks of the closed state of apo AIPL1-FKBP and those in the liganded state were similar (Yadav et al., 2017). The existence of the closed state of AIPL1-FKBP in solution aids in understanding the pathogenic nature of V71F and W72S LCA mutations. The closed state is predicted to stabilize apo AIPL1-FKBP. This state would not be stable if the bulky Trp is replaced by small hydrophilic Ser. The protein instability of W72S has been confirmed experimentally (Yadav et al., 2017). The V71F mutation is problematic for the AIPL1 function, because this mutant assumes the closed conformation in solution, which interferes with ligand binding (Yadav et al., 2017).

6. UNC119 and PDE6D as trafficking chaperones of transducin and PDE6

6.1 UNC119 and PDE6D aid trafficking of photoreceptor proteins by sequestering lipid modifications

Following folding and assembly, heterotrimeric transducin is transported from the IS, a site of the protein synthesis, to the OS, a site of phototransduction. Furthermore, transducin massively translocates from OS to the IS in response to bright light, and it returns to the OS during dark-adaptation (Artemyev, 2008; Sokolov et al., 2002). This light-dependent translocation of transducin is thought to contribute to light-adaptation of photoreceptors. The dual lipidation of $G\alpha_{t1}$ and $G\gamma_1$ provides for a relatively stable association of newly synthesized or translocated transducin with the IS endomembranes. Therefore, the IS→OS transport mechanisms of transducin may rely on vesicular trafficking, and/or diffusion coupled with sequestration of the lipid modifications.

UNC119, a mammalian ortholog of *C. elegans unc-119* (Maduro & Pilgrim, 1995), also known as Retina Gene 4 protein (RG4) (Higashide, Murakami, McLaren, & Inana, 1996) has been identified as a major protein partner of $G\alpha_{t1}$ and an essential trafficking chaperone of transducin (Gopalakrishna et al., 2011; Sinha, Majumder, Belcastro, Sokolov, & Artemyev, 2013; Zhang et al., 2011). UNC119 is now generally viewed as a carrier protein for myristoylated cargo with a preference for cargo proteins targeted to primary cilia (Constantine, Zhang, Gerstner, Frederick, & Baehr, 2012; Fansa & Wittinghofer, 2016). UNC119 (isoform a) is most abundant in photoreceptor cells, specifically the rod IS (Higashide, McLaren, & Inana, 1998), and to a much lesser extent in lung fibroblasts, adrenal glands, cerebellum, and kidney (Karim, Vepachedu, Gorska, & Alam, 2010; Swanson, Chang, Campochiaro, Zack, & Valle, 1998). Therefore, the role of UNC119a in rods is arguably best understood. A second isoform of UNC119, UNC119b, exists in vertebrates, but its expression in photoreceptor cells has not been investigated (Zhang et al., 2011). Truncation mutation in UNC119a has been linked to cone-rod dystrophy in human patients (Kobayashi et al., 2000). Knockout of UNC119 in mice caused slow retinal degeneration (Ishiba et al., 2007). Importantly, in these mice return of light-translocated $G\alpha_{t1}$ to the OS during dark adaptation was impaired (Zhang et al., 2011). However, the trafficking impairment of $G\alpha_{t1}$ in mice lacking UNC119 is not complete, suggesting that transducin utilizes additional UNC119-independent transport mechanisms. The lipid-binding module of UNC119 is an immunoglobulin-like β-sandwich made up of two opposed β-sheets (Zhang et al., 2011) (Fig. 5A). One β-sheet is comprised of four antiparallel β strands, whereas the opposite β-sheet contains five β strands, four of which are antiparallel. The two β-sheets at the center harbor a narrow and deep cavity lined with hydrophobic residues. The crystal structure of the complex of UNC119 with bound lauroylated $G\alpha_{t1}$ N-terminal peptide revealed that this cavity is utilized for binding of the acyl moiety and the first six residues of $G\alpha_{t1}$ (Zhang et al., 2011). Besides interacting with the

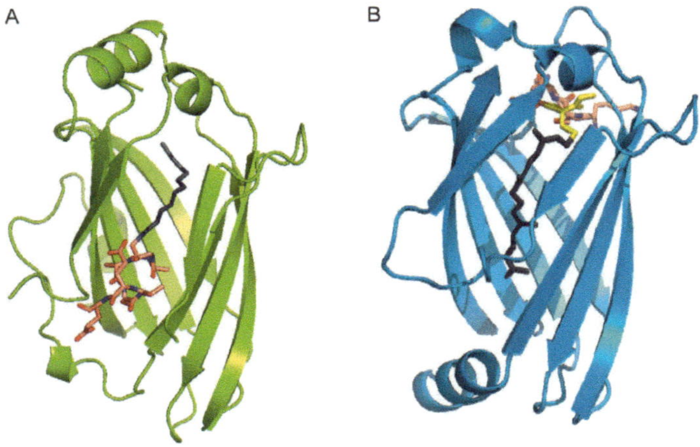

Fig. 5 Crystal structures of cargo-bound trafficking chaperons UNC119 and PDE6D. (A) The crystal structure of the complex of UNC119 with bound lauroylated $G\alpha_{t1}$ N-terminal peptide (PDB ID: 3RBQ). (B) The structure of PDE6D complexed with the geranylgeranylated PDE6C peptide (PDB ID: 5E8F). The peptide portions of cargo are shown as sticks colored by element; only the prenylated Cys is shown in yellow. The lipid modifications are shown as gray sticks. UNC119 and PDE6D share the similar immunoglobulin-like β-sandwich fold, but the points of insertion of lipidated cargo into the hydrophobic pocket are located on the opposite sides.

acylated (myristoylated) N-terminus of $G\alpha_{t1}$ (Zhang et al., 2011), UNC119a can interact with heterotrimeric $G\alpha_{t1}\beta_1\gamma_1$, promote dissociation of $G\alpha_{t1}$ from $G\beta_1\gamma_1$, and release them from the membrane (Gopalakrishna et al., 2011). Analysis of the complex between UNC119a and the full-length $G\alpha_{t1}$ by Small Angle X-ray Scattering (SAXS) and chemical crosslinking suggested an additional interface between the proteins involving the switch II/α3-β5 surface of $G\alpha_{t1}$, which overlaps with the $G\beta_1\gamma_1$ binding site (Cheguru et al., 2015). Thus, UNC119 apparently dissociates transducin subunits by disrupting and sterically occluding the $G\beta_1\gamma_1$-binding sites on $G\alpha_{t1}$ (Cheguru et al., 2015). The renal ciliopathy nephronophthisis (NPHP) protein nephrocystin-3 (NPHP3) is also among the critical myristoylation-dependent partners of UNC119 in photoreceptor cells (Wright et al., 2011). Mutations in the *NPHP3* gene cause NPHP and associated renal-retinal SLS (Ronquillo et al., 2012). NPHP3 is localized to the primary cilium where it interacts with other NPHP proteins (Wright et al., 2011). The precise function of NPHP3 in the photoreceptor connecting cilium remains unknown.

UNC119 shares significant sequence and structural homology with the prenyl-binding protein PDE6D (Gillespie, Prusti, Apel, & Beavo, 1989;

Norton et al., 2005; Zhang et al., 2011). PDE6D was originally identified based on co-purification with PDE6 from retina extracts, and hence it was first regarded as an auxiliary subunit of PDE6 (Gillespie et al., 1989). Subsequently, PDE6D has been recognized as an ubiquitous prenyl-binding protein and a trafficking chaperone of small GTPases of the Ras-superfamily and prenylated phototransduction proteins including farnesylated $G\beta_1\gamma_1$ (Baehr, 2014; Fansa & Wittinghofer, 2016). Featuring the equivalent immunoglobulin-like β-sandwich fold, PDE6D differs from UNC119 in loops connecting the β sheets, and notably, in the point of insertion of lipidated cargo into the hydrophobic pocket, which is located on the opposite side of the β-sandwich compared to that in UNC119 (Baehr, 2014; Hanzal-Bayer, Renault, Roversi, Wittinghofer, & Hillig, 2002) (Fig. 5A and B). PDE6D knockout mice reveal slow retinal degeneration and varying degrees of mislocalization of prenylated proteins including PDE6, GRK1, and $G\gamma_1$ (Zhang et al., 2007). A homozygous splice mutation was identified in *PDE6D* gene in patients with Joubert syndrome (JS). This mutation caused in-frame deletion of exon 3, thus essentially disrupting the entire function of PDE6D and leading to severe multiorgan abnormalities including retinal dysfunction (Thomas et al., 2014). In addition to prenylated phototransduction proteins, PDE6D is involved in trafficking proteins that are important in the development and stability of the primary cilium. Critical cargos of PDE6D include inositol polyphosphate 5-phosphatase E (INPP5E) as well as retinitis pigmentosa GTPase regulator (RPGR) (Fansa, Kosling, Zent, Wittinghofer, & Ismail, 2016; Humbert et al., 2012; Lee & Seo, 2015; Rao, Zhang, Li, Anand, & Khanna, 2016; Thomas et al., 2014). INPP5E itself has been linked to JS (Conduit, Dyson, & Mitchell, 2012). However, it is unclear if INPP5E truly traffics to the OS since a recent study localized INPP5E to Golgi in mouse photoreceptors (Hanke-Gogokhia et al., 2016). Mutations in RPGR affect 10–20% of people with RP and account for 70–90% of the X-linked form of RP (XLRP) (Lyraki, Megaw, & Hurd, 2016; Megaw, Soares, & Wright, 2015).

6.2 Ciliary release of photoreceptor cargo is tightly regulated and is impaired in retinal diseases

UNC119 and PDE6D are the effectors of small ARF-like GTPases ARL2 and ARL3, which regulate cargo release from carrier proteins. Low-affinity cargo molecules, including myristoylated Src family kinases or prenylated Ras and Rheb, are released in the cytosol by ARL2-GTP (Baehr, 2014; Fansa & Wittinghofer, 2016). However, only ARL3-GTP can release high-affinity cargo from UNC119 and PDE6D (Fansa & Wittinghofer, 2016). In the UNC119 complex with ARL2/ARL3-GTP, the lipid-binding pocket of UNC119 is expanded thereby weakening the interaction with cargo (Pison et al., 1990). In contrast, the hydrophobic pocket of PDE6D shrinks upon binding of ARL2/ARL3 thus squeezing the lipid out of the binding site (Ismail et al., 2011). Targeting of the cargo to the cilium is achieved because only ARL3-GTP is concentrated in the cilia. Accumulation of ARL3-GTP in the cilium occurs since the ARL3 GEF protein, ARL13B, is localized inside the cilium, while the ARL3-specific GAP protein retinitis pigmentosa protein 2 (RP2) is found primarily at the base of the cilium and near the basal body (Fansa & Wittinghofer, 2016). The structures of the complexes of ARL3 with GAP and GEF have been solved paving the way to elucidating the mechanisms of ARL3-dependent ciliary transport, which is critical to understanding many retinal diseases (Gotthardt et al., 2015; Veltel, Gasper, Eisenacher, & Wittinghofer, 2008). Given the role of ARL3 in the ciliated cells, it is not surprising that a germline knockout of the protein leads to syndromic ciliopathy akin to JS and BBS (Schrick, Vogel, Abuin, Hampton, & Rice, 2006). Rod specific knockout of ARL3 supported its role as a regulator of ciliary trafficking of lipidated phototransduction proteins, whereas retina-wide deletion of ARL3 during embryonic development highlighted the protein role in ciliogenesis (Hanke-Gogokhia et al., 2016). A rare pathogenic variant of ARL3, Y90C, has been recently identified in patients with adRP (Strom et al., 2016). The specific mechanism of this dominant mutation is unknown, but it may parallel that of the dominant active form ARL3-QL (Wright et al., 2016). The ARL3-QL form caused rod cell death in a transgenic mouse model apparently by sequestering PDE6D and leading to mistrafficking of prenylated phototransduction proteins (Wright et al., 2016). Compared to ARL3, mutations in RP2 are much more commonly linked to retinal diseases and account for up to 20% of XLRP (Lyraki et al., 2016). Many pathogenic mutations of RP2 lie in the N-terminal β-helix domain that donates an Arg118 residue critical to the ARL3 GAP activity of RP3. RP2Arg118 positions the key catalytic Gln residue of ARL3 and neutralizes the negative charge in the transition state of the GTP hydrolysis, similarly to the arginine finger mechanism of the Ras GAPs (Veltel et al., 2008). Thus, the impaired GAP activity underlies RP2-linked XLRP. An RP-2 knockout mouse model revealed mistrafficking of PDE6 and GRK1 accompanied by rod-cone dystrophy (Zhang et al., 2015). While malfunction of RP2 elevates active ARL3-GTP, deficiency of the GEF

ARL13B leads to abnormally low levels of the activated ARL3. Retina-specific deletion of ARL13B in mouse revealed its requirement for elaboration of photoreceptor OS (Hanke-Gogokhia et al., 2017). Mutations in ARL13B have been linked to JS (Thomas et al., 2015), underscoring the need for precise regulation of active ARL3 inside the cilium.

6.3 Affinities of transducin and PDE6 for the carrier proteins PDE6D and UNC119 are essential to their ciliary trafficking in photoreceptors

Exclusive gradient of activated ARL3 inside the cilium, and the ability of ARL3GTP but not active ARL2 to release high affinity cargo from both UNC119a and PDE6D underlie the sorting of lipidated cargo to the cilium, including protein sorting into the photoreceptor OS (Fansa & Wittinghofer, 2016). Thus, affinities of PDE6 and $G\beta_1\gamma_1$ for PDE6D, as well as the affinity of $G\alpha_{t1}$ for UNC119a may define the significance of PDE6D and UNC119 as trafficking chaperones for these phototransduction proteins. Structures of PDE6D with its cargo Rheb or the prenylated C-terminal peptides of INPP5E and PDE6C suggested that residues at positions −1 and −3 relative to the farnesylated cysteine are the key determinants of the PDE6D cargo affinity (Fansa et al., 2016; Fansa, O'Reilly, Ismail, & Wittinghofer, 2015; Ismail et al., 2011). In particular, all experimentally verified high-affinity prenylated cargo peptides including those from PDE6C, GRK1, and INPP5E feature Ser at −3 position, whereas low-affinity cargo molecules such as Rheb, KRas, and $G\gamma_1$ contain Lys at equivalent position (Fansa et al., 2016, 2015). This may explain why PDE6D knockout mice revealed more severe mislocalization of PDE6 and GRK1 compared to transducin $G\gamma_1$ (Zhang et al., 2007). Structures of UNC119 with acylated cargo fragments of $G\alpha_{t1}$ and NPHP3 suggested the role of residues at positions +2 and +3 relative to the modified N-terminal Gly (Jaiswal et al., 2016; Zhang et al., 2011). Only residues with small side chains at these positions allow for high affinity interactions with the myristoylated N-termini (Jaiswal et al., 2016). $G\alpha_{t1}$ satisfies this criterion and indeed binds UNC119 with high affinity (Jaiswal et al., 2016). UNC119-dependent transport of transducin implies trafficking of its individual subunits since UNC119 causes dissociation of heterotrimeric G proteins (Gopalakrishna et al., 2011). However, transducin can clearly traffic to the outer segment even in the absence of UNC119 (Zhang et al., 2011). Therefore, an alternative trafficking mechanism for transducin must exist that is independent of UNC119 and PDE6D. This mechanism may be similar to the commonly accepted route of G proteins to the plasma membrane, which involves assembly of the lipid-modified heterotrimeric G protein on the cytoplasmic surface of the ER followed by vesicular trafficking (Wedegaertner, 2012). The UNC119-dependent mechanism appears to be utilized when the IS → OS transport of massive amounts of transducin is required as it is the case following its light-dependent translocation to the rod IS (Zhang et al., 2011).

7. Outlook

Significant progress has been made in understanding molecular mechanisms of retinal diseases related to misfolding and/or mistrafficking of photoreceptor proteins. New atomic structures solved by classical X-crystallography approaches shed light on the PDE6-specific chaperone AIPL1 and the trafficking complexes of transducin and PDE6 cargo fragments with UNC119 and PDE6D. Remarkable insights into the mechanisms of ciliary trafficking have been gained from structures of complexes of ARL3 and its regulators RP2 and ARL13B. Cryo-EM is emerging as a powerful tool to study multimeric assembly complexes in photoreceptors, such as the complex of the CCT-chaperonin with PhLP1 and transducin β or the core BBSome (Klink et al., 2017; Plimpton et al., 2015). An anticipated near atomic reconstruction of large and complex photoreceptor assemblies holds a promise of developing novel therapeutic strategies for a variety of blinding diseases.

Acknowledgment

N.O.A.'s laboratory is supported by the National Institutes of Health grants EY-10843 and EY-12682.

References

Alag, R., Balakrishna, A. M., Rajan, S., Qureshi, I. A., Shin, J., Lescar, J., et al. (2013). Structural insights into substrate binding by PvFKBP35, a peptidylprolyl cis-trans isomerase from the human malarial parasite Plasmodium vivax. *Eukaryotic Cell*, *12*(4), 627–634. https://doi.org/10.1128/EC.00016-13.

Anant, J. S., Ong, O. C., Xie, H. Y., Clarke, S., O'Brien, P. J., & Fung, B. K. (1992). In vivo differential prenylation of retinal cyclic GMP phosphodiesterase catalytic subunits. *Journal of Biological Chemistry*, *267*(2), 687–690.

Arshavsky, V. Y., Lamb, T. D., & Pugh, E. N., Jr. (2002). G proteins and phototransduction. *Annual Review of Physiology*, *64*, 153–187. https://doi.org/10.1146/annurev.physiol.64.082701.102229.

Artemyev, N. O. (2008). Light-dependent compartmentalization of transducin in rod photoreceptors. *Molecular Neurobiology*, *37*(1), 44–51. https://doi.org/10.1007/s12035-008-8015-2.

Athanasiou, D., Aguila, M., Bellingham, J., Li, W., McCulley, C., Reeves, P. J., et al. (2018). The molecular and cellular basis of rhodopsin retinitis pigmentosa reveals potential strategies for therapy. *Progress in Retinal and Eye Research*, *62*, 1–23. https://doi.org/10.1016/j.preteyeres.2017.10.002.

Athanasiou, D., Aguila, M., Bevilacqua, D., Novoselov, S. S., Parfitt, D. A., & Cheetham, M. E. (2013). The cell stress machinery and retinal degeneration. *FEBS Letters*, *587*(13), 2008–2017. https://doi.org/10.1016/j.febslet.2013.05.020.

Baehr, W. (2014). Membrane protein transport in photoreceptors: The function of PDEdelta: The Proctor lecture. *Investigative Ophthalmology & Visual Science*, *55*(12), 8653–8666. https://doi.org/10.1167/iovs.14-16066.

Belcastro, M., Song, H., Sinha, S., Song, C., Mathers, P. H., & Sokolov, M. (2012). Phosphorylation of phosducin accelerates rod recovery from transducin translocation. *Investigative Ophthalmology & Visual Science*, *53*(6), 3084–3091. https://doi.org/10.1167/iovs.11-8798.

Blackburn, E. A., & Walkinshaw, M. D. (2011). Targeting FKBP isoforms with small-molecule ligands. *Current Opinion in Pharmacology*, *11*(4), 365–371. https://doi.org/10.1016/j.coph.2011.04.007.

Bowes, C., Li, T., Danciger, M., Baxter, L. C., Applebury, M. L., & Farber, D. B. (1990). Retinal degeneration in the rd mouse is caused by a defect in the beta subunit of rod cGMP-phosphodiesterase. *Nature*, *347*(6294), 677–680. https://doi.org/10.1038/347677a0.

Brooks, C., Murphy, J., Belcastro, M., Heller, D., Kolandaivelu, S., Kisselev, O., et al. (2018). Farnesylation of the transducin G protein gamma subunit is a prerequisite for its ciliary targeting in rod photoreceptors. *Frontiers in Molecular Neuroscience*, *11*, 16. https://doi.org/10.3389/fnmol.2018.00016.

Bujakowska, K. M., Liu, Q., & Pierce, E. A. (2017). Photoreceptor cilia and retinal ciliopathies. *Cold Spring Harbor Perspectives in Biology*, *9*(10), 1–28. https://doi.org/10.1101/cshperspect.a028274.

Calvert, P. D., Krasnoperova, N. V., Lyubarsky, A. L., Isayama, T., Nicolo, M., Kosaras, B., et al. (2000). Phototransduction in transgenic mice after targeted deletion of the rod transducin alpha-subunit. *Proceedings of the National Academy of Sciences of the United States of America*, *97*(25), 13913–13918. https://doi.org/10.1073/pnas.250478897.

Chan, P., Gabay, M., Wright, F. A., Kan, W., Oner, S. S., Lanier, S. M., et al. (2011). Purification of heterotrimeric G protein alpha subunits by GST-Ric-8 association: Primary characterization of purified G alpha(olf). *The Journal of Biological Chemistry*, *286*(4), 2625–2635. https://doi.org/10.1074/jbc.M110.178897.

Chan, P., Thomas, C. J., Sprang, S. R., & Tall, G. G. (2013). Molecular chaperoning function of Ric-8 is to fold nascent heterotrimeric G protein alpha subunits. *Proceedings of the National Academy of Sciences of the United States of America*, *110*(10), 3794–3799. https://doi.org/10.1073/pnas.1220943110.

Chang, B., Grau, T., Dangel, S., Hurd, R., Jurklies, B., Sener, E. C., et al. (2009). A homologous genetic basis of the murine cpfl1 mutant and human achromatopsia linked to mutations in the PDE6C gene. *Proceedings of the National Academy of Sciences of the United States of America*, *106*(46), 19581–19586. https://doi.org/10.1073/pnas.0907720106.

Cheguru, P., Majumder, A., Yadav, R., Gopalakrishna, K. N., Gakhar, L., & Artemyev, N. O. (2015). The solution structure of the transducin-alpha-uncoordinated 119 protein complex suggests occlusion of the Gbeta(1)gamma(1)-binding sites. *The FEBS Journal*, *282*(3), 550–561. https://doi.org/10.1111/febs.13161.

Christiansen, J. R., & Ramamurthy, V. (2012). Greasing the protein biosynthesis machinery of photoreceptor neurons: Role for postprenylation processing of proteins. *Cellular Logistics*, *2*(1), 15–19. https://doi.org/10.4161/cl.19804.

Conduit, S. E., Dyson, J. M., & Mitchell, C. A. (2012). Inositol polyphosphate 5-phosphatases; new players in the regulation of cilia and ciliopathies. *FEBS Letters*, *586*(18), 2846–2857. https://doi.org/10.1016/j.febslet.2012.07.037.

Constantine, R., Zhang, H., Gerstner, C. D., Frederick, J. M., & Baehr, W. (2012). Uncoordinated (UNC)119: Coordinating the trafficking of myristoylated proteins. *Vision Research*, *75*, 26–32. https://doi.org/10.1016/j.visres.2012.08.012.

Cote, R. H. (2004). Characteristics of photoreceptor PDE (PDE6): Similarities and differences to PDE5. *International Journal of Impotence Research*, *16*(Suppl. 1), S28–S33. https://doi.org/10.1038/sj.ijir.3901212.

Dekker, C., Roe, S. M., McCormack, E. A., Beuron, F., Pearl, L. H., & Willison, K. R. (2011). The crystal structure of yeast CCT reveals intrinsic asymmetry of eukaryotic cytosolic chaperonins. *The EMBO Journal*, *30*(15), 3078–3090. https://doi.org/10.1038/emboj.2011.208.

den Hollander, A. I., Roepman, R., Koenekoop, R. K., & Cremers, F. P. (2008). Leber congenital amaurosis: Genes, proteins and disease mechanisms. *Progress in Retinal and Eye Research*, *27*(4), 391–419. https://doi.org/10.1016/j.preteyeres.2008.05.003.

Dryja, T. P., Rucinski, D. E., Chen, S. H., & Berson, E. L. (1999). Frequency of mutations in the gene encoding the alpha subunit of rod cGMP-phosphodiesterase in autosomal recessive retinitis pigmentosa. *Investigative Ophthalmology and Visual Science*, *40*(8), 1859–1865.

Fanghanel, J., & Fischer, G. (2004). Insights into the catalytic mechanism of peptidyl prolyl cis/trans isomerases. *Frontiers in Bioscience*, *9*, 3453–3478.

Fansa, E. K., Kosling, S. K., Zent, E., Wittinghofer, A., & Ismail, S. (2016). PDE6delta-mediated sorting of INPP5E into the cilium is determined by cargo-carrier affinity. *Nature Communications*, 7, 11366. https://doi.org/10.1038/ncomms11366.

Fansa, E. K., O'Reilly, N. J., Ismail, S., & Wittinghofer, A. (2015). The N- and C-terminal ends of RPGR can bind to PDE6delta. *EMBO Reports*, *16*(12), 1583–1585. https://doi.org/10.15252/embr.201541404.

Fansa, E. K., & Wittinghofer, A. (2016). Sorting of lipidated cargo by the Arl2/Arl3 system. *Small GTPases*, 7(4), 222–230. https://doi.org/10.1080/21541248.2016.1224454.

Farber, D. B., & Lolley, R. N. (1974). Cyclic guanosine monophosphate: Elevation in degenerating photoreceptor cells of the C3H mouse retina. *Science*, *186*(4162), 449–451.

Fenech, C., Patrikainen, L., Kerr, D. S., Grall, S., Liu, Z., Laugerette, F., et al. (2009). Ric-8A, a Galpha protein guanine nucleotide exchange factor potentiates taste receptor signaling. *Frontiers in Cellular Neuroscience*, *3*, 11. https://doi.org/10.3389/neuro.03.011.2009.

Figueroa, M., Hinrichs, M. V., Bunster, M., Babbitt, P., Martinez-Oyanedel, J., & Olate, J. (2009). Biophysical studies support a predicted superhelical structure with armadillo repeats for Ric-8. *Protein Science*, *18*(6), 1139–1145. https://doi.org/10.1002/pro.124.

Forsythe, E., & Beales, P. L. (2013). Bardet-Biedl syndrome. *European Journal of Human Genetics*, *21*(1), 8–13. https://doi.org/10.1038/ejhg.2012.115.

Frydman, J., Nimmesgern, E., Erdjument-Bromage, H., Wall, J. S., Tempst, P., & Hartl, F. U. (1992). Function in protein folding of TRiC, a cytosolic ring complex containing TCP-1 and structurally related subunits. *The EMBO Journal*, *11*(13), 4767–4778.

Gabay, M., Pinter, M. E., Wright, F. A., Chan, P., Murphy, A. J., Valenzuela, D. M., et al. (2011). Ric-8 proteins are molecular chaperones that direct nascent G protein alpha subunit membrane association. *Science Signaling*, *4*(200), ra79. https://doi.org/10.1126/scisignal.2002223.

Gao, X., Sinha, S., Belcastro, M., Woodard, C., Ramamurthy, V., Stoilov, P., et al. (2013). Splice isoforms of phosducin-like protein control the expression of heterotrimeric G proteins. *The Journal of Biological Chemistry*, *288*(36), 25760–25768. https://doi.org/10.1074/jbc.M113.486258.

Gao, Y., Thomas, J. O., Chow, R. L., Lee, G. H., & Cowan, N. J. (1992). A cytoplasmic chaperonin that catalyzes beta-actin folding. *Cell*, *69*(6), 1043–1050.

Garcia-Higuera, I., Gaitatzes, C., Smith, T. F., & Neer, E. J. (1998). Folding a WD repeat propeller. Role of highly conserved aspartic acid residues in the G protein beta subunit and Sec13. *The Journal of Biological Chemistry*, *273*(15), 9041–9049.

Gaudet, R., Bohm, A., & Sigler, P. B. (1996). Crystal structure at 2.4 angstroms resolution of the complex of transducin betagamma and its regulator, phosducin. *Cell*, *87*(3), 577–588.

Gillespie, P. G., Prusti, R. K., Apel, E. D., & Beavo, J. A. (1989). A soluble form of bovine rod photoreceptor phosphodiesterase has a novel 15-kDa subunit. *The Journal of Biological Chemistry*, *264*(21), 12187–12193.

Gopalakrishna, K. N., Boyd, K., & Artemyev, N. O. (2017). Mechanisms of mutant PDE6 proteins underlying retinal diseases. *Cellular Signalling*, *37*, 74–80. https://doi.org/10.1016/j.cellsig.2017.06.002.

Gopalakrishna, K. N., Boyd, K., Yadav, R. P., & Artemyev, N. O. (2016). Aryl hydrocarbon receptor-interacting protein-like 1 is an obligate chaperone of phosphodiesterase 6 and is assisted by the gamma-subunit of its client. *The Journal of Biological Chemistry*, *291*(31), 16282–16291. https://doi.org/10.1074/jbc.M116.737593.

Gopalakrishna, K. N., Doddapuneni, K., Boyd, K. K., Masuho, I., Martemyanov, K. A., & Artemyev, N. O. (2011). Interaction of transducin with uncoordinated 119 protein (UNC119): Implications for the model of transducin trafficking in rod photoreceptors. *The Journal of Biological Chemistry*, *286*(33), 28954–28962. https://doi.org/10.1074/jbc.M111.268821.

Gotthardt, K., Lokaj, M., Koerner, C., Falk, N., Giessl, A., & Wittinghofer, A. (2015). A G-protein activation cascade from Arl13B to Arl3 and implications for ciliary targeting of lipidated proteins. *eLife*, *4*. https://doi.org/10.7554/eLife.11859.

Grau, T., Artemyev, N. O., Rosenberg, T., Dollfus, H., Haugen, O. H., Cumhur Sener, E., et al. (2011). Decreased catalytic activity and altered activation properties of PDE6C mutants associated with autosomal recessive achromatopsia. *Human Molecular Genetics*, *20*(4), 719–730. https://doi.org/10.1093/hmg/ddq517.

Gutsche, I., Essen, L. O., & Baumeister, W. (1999). Group II chaperonins: New TRiC(k)s and turns of a protein folding machine. *Journal of Molecular Biology*, *293*(2), 295–312. https://doi.org/10.1006/jmbi.1999.3008.

Hamel, C. (2006). Retinitis pigmentosa. *Orphanet Journal of Rare Diseases*, *1*, 40. https://doi.org/10.1186/1750-1172-1-40.

Hamel, C. P. (2007). Cone rod dystrophies. *Orphanet Journal of Rare Diseases*, *2*, 7. https://doi.org/10.1186/1750-1172-2-7.

Hanke-Gogokhia, C., Wu, Z., Gerstner, C. D., Frederick, J. M., Zhang, H., & Baehr, W. (2016). Arf-like protein 3 (ARL3) regulates protein trafficking and ciliogenesis in mouse photoreceptors. *The Journal of Biological Chemistry*, *291*(13), 7142–7155. https://doi.org/10.1074/jbc.M115.710954.

Hanke-Gogokhia, C., Wu, Z., Sharif, A., Yazigi, H., Frederick, J. M., & Baehr, W. (2017). The guanine nucleotide exchange factor Arf-like protein 13b is essential for assembly of the mouse photoreceptor transition zone and outer segment. *The Journal of Biological Chemistry*, *292*(52), 21442–21456. https://doi.org/10.1074/jbc.RA117.000141.

Hanzal-Bayer, M., Renault, L., Roversi, P., Wittinghofer, A., & Hillig, R. C. (2002). The complex of Arl2-GTP and PDE delta: From structure to function. *The EMBO Journal*, *21*(9), 2095–2106. https://doi.org/10.1093/emboj/21.9.2095.

Hartong, D. T., Berson, E. L., & Dryja, T. P. (2006). Retinitis pigmentosa. *Lancet*, *368*(9549), 1795–1809. https://doi.org/10.1016/S0140-6736(06)69740-7.

Hausch, F. (2015). FKBPs and their role in neuronal signaling. *Biochimica et Biophysica Acta*, *1850*(10), 2035–2040. https://doi.org/10.1016/j.bbagen.2015.01.012.

Hemmingsen, S. M., Woolford, C., van der Vies, S. M., Tilly, K., Dennis, D. T., Georgopoulos, C. P., et al. (1988). Homologous plant and bacterial proteins chaperone oligomeric protein assembly. *Nature, 333*(6171), 330–334. https://doi.org/10.1038/333330a0.

Higashide, T., McLaren, M. J., & Inana, G. (1998). Localization of HRG4, a photoreceptor protein homologous to Unc-119, in ribbon synapse. *Investigative Ophthalmology & Visual Science, 39*(5), 690–698.

Higashide, T., Murakami, A., McLaren, M. J., & Inana, G. (1996). Cloning of the cDNA for a novel photoreceptor protein. *The Journal of Biological Chemistry, 271*(3), 1797–1804.

Hirji, N., Aboshiha, J., Georgiou, M., Bainbridge, J., & Michaelides, M. (2018). Achromatopsia: Clinical features, molecular genetics, animal models and therapeutic options. *Ophthalmic Genetics, 39*(2), 149–157. https://doi.org/10.1080/13816810.2017.1418389.

Humbert, M. C., Weihbrecht, K., Searby, C. C., Li, Y., Pope, R. M., Sheffield, V. C., et al. (2012). ARL13B, PDE6D, and CEP164 form a functional network for INPP5E ciliary targeting. *Proceedings of the National Academy of Sciences of the United States of America, 109*(48), 19691–19696. https://doi.org/10.1073/pnas.1210916109.

Humrich, J., Bermel, C., Bunemann, M., Harmark, L., Frost, R., Quitterer, U., et al. (2005). Phosducin-like protein regulates G-protein betagamma folding by interaction with tailless complex polypeptide-1alpha: Dephosphorylation or splicing of PhLP turns the switch toward regulation of Gbetagamma folding. *The Journal of Biological Chemistry, 280*(20), 20042–20050. https://doi.org/10.1074/jbc.M409233200.

Ishiba, Y., Higashide, T., Mori, N., Kobayashi, A., Kubota, S., McLaren, M. J., et al. (2007). Targeted inactivation of synaptic HRG4 (UNC119) causes dysfunction in the distal photoreceptor and slow retinal degeneration, revealing a new function. *Experimental Eye Research, 84*(3), 473–485. https://doi.org/10.1016/j.exer.2006.10.016.

Ismail, S. A., Chen, Y. X., Rusinova, A., Chandra, A., Bierbaum, M., Gremer, L., et al. (2011). Arl2-GTP and Arl3-GTP regulate a GDI-like transport system for farnesylated cargo. *Nature Chemical Biology*, 7(12), 942–949. https://doi.org/10.1038/nchembio.686.

Jaiswal, M., Fansa, E. K., Kosling, S. K., Mejuch, T., Waldmann, H., & Wittinghofer, A. (2016). Novel biochemical and structural insights into the interaction of myristoylated cargo with Unc119 protein and their release by Arl2/3. *The Journal of Biological Chemistry, 291*(39), 20766–20778. https://doi.org/10.1074/jbc.M116.741827.

Jin, H., White, S. R., Shida, T., Schulz, S., Aguiar, M., Gygi, S. P., et al. (2010). The conserved Bardet-Biedl syndrome proteins assemble a coat that traffics membrane proteins to cilia. *Cell, 141*(7), 1208–1219. https://doi.org/10.1016/j.cell.2010.05.015.

Kant, R., Zeng, B., Thomas, C. J., Bothner, B., & Sprang, S. R. (2016). Ric-8A, a G protein chaperone with nucleotide exchange activity induces long-range secondary structure changes in Galpha. *eLife, 5*. https://doi.org/10.7554/eLife.19238.

Karan, S., Zhang, H., Li, S., Frederick, J. M., & Baehr, W. (2008). A model for transport of membrane-associated phototransduction polypeptides in rod and cone photoreceptor inner segments. *Vision Research, 48*(3), 442–452. https://doi.org/10.1016/j.visres.2007.08.020.

Karim, Z., Vepachedu, R., Gorska, M., & Alam, R. (2010). UNC119 inhibits dynamin and dynamin-dependent endocytic processes. *Cellular Signalling, 22*(1), 128–137. https://doi.org/10.1016/j.cellsig.2009.09.022.

Klink, B. U., Zent, E., Juneja, P., Kuhlee, A., Raunser, S., & Wittinghofer, A. (2017). A recombinant BBSome core complex and how it interacts with ciliary cargo. *eLife, 6*. https://doi.org/10.7554/eLife.27434.

Klumpp, M., Baumeister, W., & Essen, L. O. (1997). Structure of the substrate binding domain of the thermosome, an archaeal group II chaperonin. *Cell, 91*(2), 263–270.

Kobayashi, A., Higashide, T., Hamasaki, D., Kubota, S., Sakuma, H., An, W., et al. (2000). HRG4 (UNC119) mutation found in cone-rod dystrophy causes retinal degeneration in a transgenic model. *Investigative Ophthalmology & Visual Science, 41*(11), 3268–3277.

Koenekoop, R. K. (2004). An overview of Leber congenital amaurosis: A model to understand human retinal development. *Survey of Ophthalmology, 49*(4), 379–398. https://doi.org/10.1016/j.survophthal.2004.04.003.

Kosmaoglou, M., Schwarz, N., Bett, J. S., & Cheetham, M. E. (2008). Molecular chaperones and photoreceptor function. *Progress in Retinal and Eye Research, 27*(4), 434–449. https://doi.org/10.1016/j.preteyeres.2008.03.001.

Krispel, C. M., Sokolov, M., Chen, Y. M., Song, H., Herrmann, R., Arshavsky, V. Y., et al. (2007). Phosducin regulates the expression of transducin betagamma subunits in rod photoreceptors and does not contribute to phototransduction adaptation. *The Journal of General Physiology, 130*(3), 303–312. https://doi.org/10.1085/jgp.200709812.

Kubota, H., Hynes, G., Carne, A., Ashworth, A., & Willison, K. (1994). Identification of six Tcp-1-related genes encoding divergent subunits of the TCP-1-containing chaperonin. *Current Biology, 4*(2), 89–99.

Kumaran, N., Moore, A. T., Weleber, R. G., & Michaelides, M. (2017). Leber congenital amaurosis/early-onset severe retinal dystrophy: Clinical features, molecular genetics and therapeutic interventions. *The British Journal of Ophthalmology, 101*(9), 1147–1154. https://doi.org/10.1136/bjophthalmol-2016-309975.

Lai, C. W., Kolesnikov, A. V., Frederick, J. M., Blake, D. R., Jiang, L., Stewart, J. S., et al. (2013). Phosducin-like protein 1 is essential for G-protein assembly and signaling in retinal rod photoreceptors. *The Journal of Neuroscience, 33*(18), 7941–7951. https://doi.org/10.1523/JNEUROSCI.5001-12.2013.

Lai, R. K., Perez-Sala, D., Canada, F. J., & Rando, R. R. (1990). The gamma subunit of transducin is farnesylated. *Proceedings of the National Academy of Sciences of the United States of America, 87*(19), 7673–7677.

Lamb, T. D., & Pugh, E. N., Jr. (2006). Phototransduction, dark adaptation, and rhodopsin regeneration the proctor lecture. *Investigative Ophthalmology & Visual Science, 47*(12), 5137–5152. https://doi.org/10.1167/iovs.06-0849.

Lee, R. H., Brown, B. M., & Lolley, R. N. (1984). Light-induced dephosphorylation of a 33K protein in rod outer segments of rat retina. *Biochemistry, 23*(9), 1972–1977.

Lee, R. H., Lieberman, B. S., & Lolley, R. N. (1987). A novel complex from bovine visual cells of a 33,000-dalton phosphoprotein with beta- and gamma-transducin: Purification and subunit structure. *Biochemistry, 26*(13), 3983–3990.

Lee, J. J., & Seo, S. (2015). PDE6D binds to the C-terminus of RPGR in a prenylation-dependent manner. *EMBO Reports, 16*(12), 1581–1582. https://doi.org/10.15252/embr.201541220.

Lee, R. H., Ting, T. D., Lieberman, B. S., Tobias, D. E., Lolley, R. N., & Ho, Y. K. (1992). Regulation of retinal cGMP cascade by phosducin in bovine rod photoreceptor cells. Interaction of phosducin and transducin. *The Journal of Biological Chemistry, 267*(35), 25104–25112.

Linnert, M., Lin, Y. J., Manns, A., Haupt, K., Paschke, A. K., Fischer, G., et al. (2013). The FKBP-type domain of the human aryl hydrocarbon receptor-interacting protein reveals an unusual Hsp90 interaction. *Biochemistry, 52*(12), 2097–2107. https://doi.org/10.1021/bi301649m.

Liu, X., Bulgakov, O. V., Wen, X. H., Woodruff, M. L., Pawlyk, B., Yang, J., et al. (2004). AIPL1, the protein that is defective in Leber congenital amaurosis, is essential for the biosynthesis of retinal rod cGMP phosphodiesterase. *Proceedings of the National Academy of Sciences of the United States of America, 101*(38), 13903–13908. https://doi.org/10.1073/pnas.0405160101.

Llorca, O., Martin-Benito, J., Ritco-Vonsovici, M., Grantham, J., Hynes, G. M., Willison, K. R., et al. (2000). Eukaryotic chaperonin CCT stabilizes actin and tubulin folding intermediates in open quasi-native conformations. *The EMBO Journal, 19*(22), 5971–5979. https://doi.org/10.1093/emboj/19.22.5971.

Lobanova, E. S., Finkelstein, S., Herrmann, R., Chen, Y. M., Kessler, C., Michaud, N. A., et al. (2008). Transducin gamma-subunit sets expression levels of alpha- and beta-subunits and is crucial for rod viability. *The Journal of Neuroscience, 28*(13), 3510–3520. https://doi.org/10.1523/JNEUROSCI.0338-08.2008.

Lobanova, E. S., Finkelstein, S., Skiba, N. P., & Arshavsky, V. Y. (2013). Proteasome overload is a common stress factor in multiple forms of inherited retinal degeneration. *Proceedings of the National Academy of Sciences of the United States of America, 110*(24), 9986–9991. https://doi.org/10.1073/pnas.1305521110.

Loew, A., Ho, Y. K., Blundell, T., & Bax, B. (1998). Phosducin induces a structural change in transducin beta gamma. *Structure, 6*(8), 1007–1019.

Loktev, A. V., Zhang, Q., Beck, J. S., Searby, C. C., Scheetz, T. E., Bazan, J. F., et al. (2008). A BBSome subunit links ciliogenesis, microtubule stability, and acetylation. *Developmental Cell, 15*(6), 854–865. https://doi.org/10.1016/j.devcel.2008.11.001.

Lukov, G. L., Hu, T., McLaughlin, J. N., Hamm, H. E., & Willardson, B. M. (2005). Phosducin-like protein acts as a molecular chaperone for G protein betagamma dimer assembly. *The EMBO Journal, 24*(11), 1965–1975. https://doi.org/10.1038/sj.emboj.7600673.

Lyraki, R., Megaw, R., & Hurd, T. (2016). Disease mechanisms of X-linked retinitis pigmentosa due to RP2 and RPGR mutations. *Biochemical Society Transactions, 44*(5), 1235–1244. https://doi.org/10.1042/BST20160148.

MacMillan, D. (2013). FK506 binding proteins: Cellular regulators of intracellular Ca2 + signalling. *European Journal of Pharmacology, 700*(1–3), 181–193. https://doi.org/10.1016/j.ejphar.2012.12.029.

Maduro, M., & Pilgrim, D. (1995). Identification and cloning of unc-119, a gene expressed in the Caenorhabditis elegans nervous system. *Genetics, 141*(3), 977–988.

Mahoney, J. P., & Sunahara, R. K. (2016). Mechanistic insights into GPCR-G protein interactions. *Current Opinion in Structural Biology, 41*, 247–254. https://doi.org/10.1016/j.sbi.2016.11.005.

Majumder, A., Gopalakrishna, K. N., Cheguru, P., Gakhar, L., & Artemyev, N. O. (2013). Interaction of aryl hydrocarbon receptor-interacting protein-like 1 with the farnesyl moiety. *The Journal of Biological Chemistry, 288*(29), 21320–21328. https://doi.org/10.1074/jbc.M113.476242.

Malanson, K. M., & Lem, J. (2009). Rhodopsin-mediated retinitis pigmentosa. *Progress in Molecular Biology and Translational Science, 88*, 1–31. https://doi.org/10.1016/S1877-1173(09)88001-0.

Matsuzaki, F. (2005). Drosophila G-protein signalling: Intricate roles for Ric-8? *Nature Cell Biology*, 7(11), 1047–1049.

McLaughlin, M. E., Ehrhart, T. L., Berson, E. L., & Dryja, T. P. (1995). Mutation spectrum of the gene encoding the beta subunit of rod phosphodiesterase among patients with autosomal recessive retinitis pigmentosa. *Proceedings of the National Academy of Sciences of the United States of America, 92*(8), 3249–3253.

Megaw, R. D., Soares, D. C., & Wright, A. F. (2015). RPGR: Its role in photoreceptor physiology, human disease, and future therapies. *Experimental Eye Research, 138*, 32–41. https://doi.org/10.1016/j.exer.2015.06.007.

Mourao, A., Nager, A. R., Nachury, M. V., & Lorentzen, E. (2014). Structural basis for membrane targeting of the BBSome by ARL6. *Nature Structural & Molecular Biology, 21*(12), 1035–1041. https://doi.org/10.1038/nsmb.2920.

Nachury, M. V., Loktev, A. V., Zhang, Q., Westlake, C. J., Peranen, J., Merdes, A., et al. (2007). A core complex of BBS proteins cooperates with the GTPase Rab8 to promote ciliary membrane biogenesis. *Cell, 129*(6), 1201–1213. https://doi.org/10.1016/j.cell.2007.03.053.

Neubert, T. A., Johnson, R. S., Hurley, J. B., & Walsh, K. A. (1992). The rod transducin alpha subunit amino terminus is heterogeneously fatty acylated. *The Journal of Biological Chemistry*, *267*(26), 18274–18277.

Norton, A. W., Hosier, S., Terew, J. M., Li, N., Dhingra, A., Vardi, N., et al. (2005). Evaluation of the 17-kDa prenyl-binding protein as a regulatory protein for phototransduction in retinal photoreceptors. *The Journal of Biological Chemistry*, *280*(2), 1248–1256. https://doi.org/10.1074/jbc.M410475200.

Obin, M., Lee, B. Y., Meinke, G., Bohm, A., Lee, R. H., Gaudet, R., et al. (2002). Ubiquitylation of the transducin betagamma subunit complex. Regulation by phosducin. *The Journal of Biological Chemistry*, *277*(46), 44566–44575. https://doi.org/10.1074/jbc.M205308200.

Palczewski, K., Kumasaka, T., Hori, T., Behnke, C. A., Motoshima, H., Fox, B. A., et al. (2000). Crystal structure of rhodopsin: A G protein-coupled receptor. *Science*, *289*(5480), 739–745.

Papasergi, M. M., Patel, B. R., & Tall, G. G. (2015). The G protein alpha chaperone Ric-8 as a potential therapeutic target. *Molecular Pharmacology*, *87*(1), 52–63. https://doi.org/10.1124/mol.114.094664.

Pearring, J. N., Salinas, R. Y., Baker, S. A., & Arshavsky, V. Y. (2013). Protein sorting, targeting and trafficking in photoreceptor cells. *Progress in Retinal and Eye Research*, *36*, 24–51. https://doi.org/10.1016/j.preteyeres.2013.03.002.

Pison, U., Obertacke, U., Brand, M., Seeger, W., Joka, T., Bruch, J., et al. (1990). Altered pulmonary surfactant in uncomplicated and septicemia-complicated courses of acute respiratory failure. *The Journal of Trauma*, *30*(1), 19–26.

Pittler, S. J., & Baehr, W. (1991). Identification of a nonsense mutation in the rod photoreceptor cGMP phosphodiesterase beta-subunit gene of the rd mouse. *Proceedings of the National Academy of Sciences of the United States of America*, *88*(19), 8322–8326.

Plimpton, R. L., Cuellar, J., Lai, C. W., Aoba, T., Makaju, A., Franklin, S., et al. (2015). Structures of the Gbeta-CCT and PhLP1-Gbeta-CCT complexes reveal a mechanism for G-protein beta-subunit folding and Gbetagamma dimer assembly. *Proceedings of the National Academy of Sciences of the United States of America*, *112*(8), 2413–2418. https://doi.org/10.1073/pnas.1419595112.

Posokhova, E., Song, H., Belcastro, M., Higgins, L., Bigley, L. R., Michaud, N. A., et al. (2011). Disruption of the chaperonin containing TCP-1 function affects protein networks essential for rod outer segment morphogenesis and survival. *Molecular and Cellular Proteomics*, *10*(1). https://doi.org/10.1074/mcp.M110.000570. M110 000570.

Preininger, A. M., Meiler, J., & Hamm, H. E. (2013). Conformational flexibility and structural dynamics in GPCR-mediated G protein activation: A perspective. *Journal of Molecular Biology*, *425*(13), 2288–2298. https://doi.org/10.1016/j.jmb.2013.04.011.

Rachel, R. A., Li, T., & Swaroop, A. (2012). Photoreceptor sensory cilia and ciliopathies: Focus on CEP290, RPGR and their interacting proteins. *Cilia*, *1*(1), 22. https://doi.org/10.1186/2046-2530-1-22.

Ramamurthy, V., Niemi, G. A., Reh, T. A., & Hurley, J. B. (2004). Leber congenital amaurosis linked to AIPL1: A mouse model reveals destabilization of cGMP phosphodiesterase. *Proceedings of the National Academy of Sciences of the United States of America*, *101*(38), 13897–13902. https://doi.org/10.1073/pnas.0404197101.

Ramamurthy, V., Roberts, M., van den Akker, F., Niemi, G., Reh, T. A., & Hurley, J. B. (2003). AIPL1, a protein implicated in Leber's congenital amaurosis, interacts with and aids in processing of farnesylated proteins. *Proceedings of the National Academy of Sciences of the United States of America*, *100*(22), 12630–12635. https://doi.org/10.1073/pnas.2134194100.

Rao, K. N., Zhang, W., Li, L., Anand, M., & Khanna, H. (2016). Prenylated retinal ciliopathy protein RPGR interacts with PDE6delta and regulates ciliary localization of Joubert syndrome-associated protein INPP5E. *Human Molecular Genetics*, *25*(20), 4533–4545. https://doi.org/10.1093/hmg/ddw281.

Ronquillo, C. C., Bernstein, P. S., & Baehr, W. (2012). Senior-Loken syndrome: A syndromic form of retinal dystrophy associated with nephronophthisis. *Vision Research*, *75*, 88–97. https://doi.org/10.1016/j.visres.2012.07.003.

Russmann, F., Stemp, M. J., Monkemeyer, L., Etchells, S. A., Bracher, A., & Hartl, F. U. (2012). Folding of large multidomain proteins by partial encapsulation in the chaperonin TRiC/CCT. *Proceedings of the National Academy of Sciences of the United States of America*, *109*(52), 21208–21215. https://doi.org/10.1073/pnas.1218836109.

Schrick, J. J., Vogel, P., Abuin, A., Hampton, B., & Rice, D. S. (2006). ADP-ribosylation factor-like 3 is involved in kidney and photoreceptor development. *The American Journal of Pathology*, *168*(4), 1288–1298. https://doi.org/10.2353/ajpath.2006.050941.

Seo, S., Baye, L. M., Schulz, N. P., Beck, J. S., Zhang, Q., Slusarski, D. C., et al. (2010). BBS6, BBS10, and BBS12 form a complex with CCT/TRiC family chaperonins and mediate BBSome assembly. *Proceedings of the National Academy of Sciences of the United States of America*, *107*(4), 1488–1493. https://doi.org/10.1073/pnas.0910268107.

Sinha, S., Belcastro, M., Datta, P., Seo, S., & Sokolov, M. (2014). Essential role of the chaperonin CCT in rod outer segment biogenesis. *Investigative Ophthalmology & Visual Science*, *55*(6), 3775–3785. https://doi.org/10.1167/iovs.14-13889.

Sinha, S., Majumder, A., Belcastro, M., Sokolov, M., & Artemyev, N. O. (2013). Expression and subcellular distribution of UNC119a, a protein partner of transducin alpha subunit in rod photoreceptors. *Cellular Signalling*, *25*(1), 341–348. https://doi.org/10.1016/j.cellsig.2012.10.005.

Skiba, N. P., Bae, H., & Hamm, H. E. (1996). Mapping of effector binding sites of transducin alpha-subunit using G alpha t/G alpha i1 chimeras. *The Journal of Biological Chemistry*, *271*(1), 413–424.

Sohocki, M. M., Bowne, S. J., Sullivan, L. S., Blackshaw, S., Cepko, C. L., Payne, A. M., et al. (2000). Mutations in a new photoreceptor-pineal gene on 17p cause Leber congenital amaurosis. *Nature Genetics*, *24*(1), 79–83. https://doi.org/10.1038/71732.

Sokolov, M., Lyubarsky, A. L., Strissel, K. J., Savchenko, A. B., Govardovskii, V. I., Pugh, E. N., Jr., et al. (2002). Massive light-driven translocation of transducin between the two major compartments of rod cells: A novel mechanism of light adaptation. *Neuron*, *34*(1), 95–106.

Sokolov, M., Strissel, K. J., Leskov, I. B., Michaud, N. A., Govardovskii, V. I., & Arshavsky, V. Y. (2004). Phosducin facilitates light-driven transducin translocation in rod photoreceptors. Evidence from the phosducin knockout mouse. *The Journal of Biological Chemistry*, *279*(18), 19149–19156. https://doi.org/10.1074/jbc.M311058200.

Song, H., Belcastro, M., Young, E. J., & Sokolov, M. (2007). Compartment-specific phosphorylation of phosducin in rods underlies adaptation to various levels of illumination. *The Journal of Biological Chemistry*, *282*(32), 23613–23621. https://doi.org/10.1074/jbc.M701974200.

Sprang, S. R., Chen, Z., & Du, X. (2007). Structural basis of effector regulation and signal termination in heterotrimeric Galpha proteins. *Advances in Protein Chemistry*, *74*, 1–65. https://doi.org/10.1016/S0065-3233(07)74001-9.

Stoetzel, C., Laurier, V., Davis, E. E., Muller, J., Rix, S., Badano, J. L., et al. (2006). BBS10 encodes a vertebrate-specific chaperonin-like protein and is a major BBS locus. *Nature Genetics*, *38*(5), 521–524. https://doi.org/10.1038/ng1771.

Stone, E. M. (2007). Leber congenital amaurosis—A model for efficient genetic testing of heterogeneous disorders: LXIV Edward Jackson Memorial Lecture. *American Journal of Ophthalmology*, *144*(6), 791–811. https://doi.org/10.1016/j.ajo.2007.08.022.

Storer, C. L., Dickey, C. A., Galigniana, M. D., Rein, T., & Cox, M. B. (2011). FKBP51 and FKBP52 in signaling and disease. *Trends in Endocrinology and Metabolism: TEM*, *22*(12), 481–490. https://doi.org/10.1016/j.tem.2011.08.001.

Strom, S. P., Clark, M. J., Martinez, A., Garcia, S., Abelazeem, A. A., Matynia, A., et al. (2016). De novo occurrence of a variant in ARL3 and apparent autosomal dominant transmission of retinitis pigmentosa. *PLoS One*, *11*(3), e0150944. https://doi.org/10.1371/journal.pone.0150944.

Swanson, D. A., Chang, J. T., Campochiaro, P. A., Zack, D. J., & Valle, D. (1998). Mammalian orthologs of C. elegans unc-119 highly expressed in photoreceptors. *Investigative Ophthalmology & Visual Science*, *39*(11), 2085–2094.

Tall, G. G. (2013). Ric-8 regulation of heterotrimeric G proteins. *Journal of Receptor and Signal Transduction Research*, *33*(3), 139–143. https://doi.org/10.3109/10799893.2013.763828.

Tall, G. G., Krumins, A. M., & Gilman, A. G. (2003). Mammalian Ric-8A (synembryn) is a heterotrimeric Galpha protein guanine nucleotide exchange factor. *The Journal of Biological Chemistry*, *278*(10), 8356–8362. https://doi.org/10.1074/jbc.M211862200.

Thiadens, A. A., den Hollander, A. I., Roosing, S., Nabuurs, S. B., Zekveld-Vroon, R. C., Collin, R. W., et al. (2009). Homozygosity mapping reveals PDE6C mutations in patients with early-onset cone photoreceptor disorders. *American Journal of Human Genetics*, *85*(2), 240–247. https://doi.org/10.1016/j.ajhg.2009.06.016.

Thomas, C. J., Briknarova, K., Hilmer, J. K., Movahed, N., Bothner, B., Sumida, J. P., et al. (2011). The nucleotide exchange factor Ric-8A is a chaperone for the conformationally dynamic nucleotide-free state of Galphai1. *PLoS One*, *6*(8), e23197. https://doi.org/10.1371/journal.pone.0023197.

Thomas, S., Cantagrel, V., Mariani, L., Serre, V., Lee, J. E., Elkhartoufi, N., et al. (2015). Identification of a novel ARL13B variant in a Joubert syndrome-affected patient with retinal impairment and obesity. *European Journal of Human Genetics*, *23*(5), 621–627. https://doi.org/10.1038/ejhg.2014.156.

Thomas, S., Wright, K. J., Le Corre, S., Micalizzi, A., Romani, M., Abhyankar, A., et al. (2014). A homozygous PDE6D mutation in Joubert syndrome impairs targeting of farnesylated INPP5E protein to the primary cilium. *Human Mutation*, *35*(1), 137–146. https://doi.org/10.1002/humu.22470.

Tong, M., & Jiang, Y. (2015). FK506-binding proteins and their diverse functions. *Current Molecular Pharmacology*, *9*(1), 48–65.

Tzekov, R., Stein, L., & Kaushal, S. (2011). Protein misfolding and retinal degeneration. *Cold Spring Harbor Perspectives in Biology*, *3*(11), a007492. https://doi.org/10.1101/cshperspect.a007492.

Valpuesta, J. M., Martin-Benito, J., Gomez-Puertas, P., Carrascosa, J. L., & Willison, K. R. (2002). Structure and function of a protein folding machine: The eukaryotic cytosolic chaperonin CCT. *FEBS Letters*, *529*(1), 11–16.

van der Spuy, J., Chapple, J. P., Clark, B. J., Luthert, P. J., Sethi, C. S., & Cheetham, M. E. (2002). The Leber congenital amaurosis gene product AIPL1 is localized exclusively in rod photoreceptors of the adult human retina. *Human Molecular Genetics*, *11*(7), 823–831.

Van Duyne, G. D., Standaert, R. F., Karplus, P. A., Schreiber, S. L., & Clardy, J. (1993). Atomic structures of the human immunophilin FKBP-12 complexes with FK506 and rapamycin. *Journal of Molecular Biology*, *229*(1), 105–124. https://doi.org/10.1006/jmbi.1993.1012.

Van Eps, N., Thomas, C. J., Hubbell, W. L., & Sprang, S. R. (2015). The guanine nucleotide exchange factor Ric-8A induces domain separation and Ras domain plasticity in Galphai1. *Proceedings of the National Academy of Sciences of the United States of America*, *112*(5), 1404–1409. https://doi.org/10.1073/pnas.1423878112.

Veltel, S., Gasper, R., Eisenacher, E., & Wittinghofer, A. (2008). The retinitis pigmentosa 2 gene product is a GTPase-activating protein for Arf-like 3. *Nature Structural & Molecular Biology*, *15*(4), 373–380. https://doi.org/10.1038/nsmb.1396.

Von Dannecker, L. E., Mercadante, A. F., & Malnic, B. (2005). Ric-8B, an olfactory putative GTP exchange factor, amplifies signal transduction through the olfactory-specific G-protein Galphaolf. *The Journal of Neuroscience*, *25*(15), 3793–3800. https://doi.org/10.1523/JNEUROSCI.4595-04.2005.

Wang, T., Tsang, S. H., & Chen, J. (2017). Two pathways of rod photoreceptor cell death induced by elevated cGMP. *Human Molecular Genetics*, *26*(12), 2299–2306. https://doi.org/10.1093/hmg/ddx121.

Wedegaertner, P. B. (2012). G protein trafficking. *Subcellular Biochemistry*, *63*, 193–223. https://doi.org/10.1007/978-94-007-4765-4_11.

Wells, C. A., Dingus, J., & Hildebrandt, J. D. (2006). Role of the chaperonin CCT/TRiC complex in G protein betagamma-dimer assembly. *The Journal of Biological Chemistry*, *281*(29), 20221–20232. https://doi.org/10.1074/jbc.M602409200.

Wilkie, T. M., & Kinch, L. (2005). New roles for Galpha and RGS proteins: Communication continues despite pulling sisters apart. *Current Biology*, *15*(20), R843–R854. https://doi.org/10.1016/j.cub.2005.10.008.

Wilkins, J. F., Bitensky, M. W., & Willardson, B. M. (1996). Regulation of the kinetics of phosducin phosphorylation in retinal rods. *The Journal of Biological Chemistry*, *271*(32), 19232–19237.

Willardson, B. M., & Howlett, A. C. (2007). Function of phosducin-like proteins in G protein signaling and chaperone-assisted protein folding. *Cellular Signalling*, *19*(12), 2417–2427. https://doi.org/10.1016/j.cellsig.2007.06.013.

Willison, K. R. (2018). The substrate specificity of eukaryotic cytosolic chaperonin CCT. *Philosophical Transactions of the Royal Society of London. Series B, Biological Sciences*, *373*(1749). https://doi.org/10.1098/rstb.2017.0192.

Wright, K. J., Baye, L. M., Olivier-Mason, A., Mukhopadhyay, S., Sang, L., Kwong, M., et al. (2011). An ARL3-UNC119-RP2 GTPase cycle targets myristoylated NPHP3 to the primary cilium. *Genes & Development*, *25*(22), 2347–2360. https://doi.org/10.1101/gad.173443.111.

Wright, Z. C., Singh, R. K., Alpino, R., Goldberg, A. F., Sokolov, M., & Ramamurthy, V. (2016). ARL3 regulates trafficking of prenylated phototransduction proteins to the rod outer segment. *Human Molecular Genetics*, *25*(10), 2031–2044. https://doi.org/10.1093/hmg/ddw077.

Yadav, R. P., & Artemyev, N. O. (2017). AIPL1: A specialized chaperone for the phototransduction effector. *Cellular Signalling*, *40*, 183–189. https://doi.org/10.1016/j.cellsig.2017.09.014.

Yadav, R. P., Gakhar, L., Yu, L., & Artemyev, N. O. (2017). Unique structural features of the AIPL1-FKBP domain that support prenyl lipid binding and underlie protein malfunction in blindness. *Proceedings of the National Academy of Sciences of the United States of America*, *114*(32), E6536–E6545. https://doi.org/10.1073/pnas.1704782114.

Yadav, R. P., Majumder, A., Gakhar, L., & Artemyev, N. O. (2015). Extended conformation of the proline-rich domain of human aryl hydrocarbon receptor-interacting protein-like 1: Implications for retina disease. *Journal of Neurochemistry*, *135*(1), 165–175. https://doi.org/10.1111/jnc.13223.

Yoshida, T., Willardson, B. M., Wilkins, J. F., Jensen, G. J., Thornton, B. D., & Bitensky, M. W. (1994). The phosphorylation state of phosducin determines its ability to block transducin subunit interactions and inhibit transducin binding to activated rhodopsin. *The Journal of Biological Chemistry*, *269*(39), 24050–24057.

Yu, L., Yadav, R. P., & Artemyev, N. O. (2017). NMR resonance assignments of the FKBP domain of human aryl hydrocarbon receptor-interacting protein-like 1 (AIPL1) in complex with a farnesyl ligand. *Biomolecular NMR Assignments*, *11*(1), 111–115. https://doi.org/10.1007/s12104-017-9730-2.

Zaghloul, N. A., & Katsanis, N. (2009). Mechanistic insights into Bardet-Biedl syndrome, a model ciliopathy. *The Journal of Clinical Investigation, 119*(3), 428–437. https://doi.org/10.1172/JCI37041.

Zhang, J., Baker, M. L., Schroder, G. F., Douglas, N. R., Reissmann, S., Jakana, J., et al. (2010). Mechanism of folding chamber closure in a group II chaperonin. *Nature, 463*(7279), 379–383. https://doi.org/10.1038/nature08701.

Zhang, H., Constantine, R., Vorobiev, S., Chen, Y., Seetharaman, J., Huang, Y. J., et al. (2011). UNC119 is required for G protein trafficking in sensory neurons. *Nature Neuroscience, 14*(7), 874–880. https://doi.org/10.1038/nn.2835.

Zhang, H., Hanke-Gogokhia, C., Jiang, L., Li, X., Wang, P., Gerstner, C. D., et al. (2015). Mistrafficking of prenylated proteins causes retinitis pigmentosa 2. *The FASEB Journal, 29*(3), 932–942. https://doi.org/10.1096/fj.14-257915.

Zhang, H., Li, S., Doan, T., Rieke, F., Detwiler, P. B., Frederick, J. M., et al. (2007). Deletion of PrBP/delta impedes transport of GRK1 and PDE6 catalytic subunits to photoreceptor outer segments. *Proceedings of the National Academy of Sciences of the United States of America, 104*(21), 8857–8862. https://doi.org/10.1073/pnas.0701681104.

Zhang, Q., Yu, D., Seo, S., Stone, E. M., & Sheffield, V. C. (2012). Intrinsic protein-protein interaction-mediated and chaperonin-assisted sequential assembly of stable Bardet-Biedl syndrome protein complex, the BBSome. *The Journal of Biological Chemistry, 287*(24), 20625–20635. https://doi.org/10.1074/jbc.M112.341487.

CHAPTER FOUR

Structural and functional insights on the roles of molecular chaperones in the mistargeting and aggregation phenotypes associated with primary hyperoxaluria type I

José Ángel Fernández-Higuero[a,†], Isabel Betancor-Fernández[b,†], Noel Mesa-Torres[c,†], Arturo Muga[a], Eduardo Salido[b], Angel L. Pey[c,*]

[a]Biofisika Institute (CSIC, UPV/EHU) and Department of Biochemistry and Molecular Biology, Faculty of Science and Technology, University of the Basque Country (UPV/EHU), Bilbao, Spain
[b]Centre for Biomedical Research on Rare Diseases (CIBERER), Hospital Universitario de Canarias, ITB, University of La Laguna, Tenerife, Spain
[c]Department of Physical Chemistry, University of Granada, Granada, Spain
*Corresponding author: e-mail address: angelpey@ugr.es

Contents

† These authors equally contributed to this work.

Advances in Protein Chemistry and Structural Biology, Volume 114
ISSN 1876-1623
https://doi.org/10.1016/bs.apcsb.2018.09.003

Abstract

To carry out their biological function in cells, proteins must be folded and targeted to the appropriate subcellular location. These processes are controlled by a vast collection of interacting proteins collectively known as the protein homeostasis network, in which molecular chaperones play a prominent role. Protein homeostasis can be impaired by inherited mutations leading to genetic diseases. In this chapter, we focus on a particular disease, primary hyperoxaluria type 1 (PH1), in which disease-associated mutations exacerbate protein aggregation in the cell and mistarget the peroxisomal alanine: glyoxylate aminotransferase (AGT) protein to mitochondria, in part due to native state destabilization and enhanced interaction with Hsp60, 70 and 90 chaperone systems. After a general introduction of molecular chaperones and PH1, we review our current knowledge on the structural and energetic features of PH1-causing mutants that lead to these particular pathogenic mechanisms. From this perspective, and in the context of the key role of molecular chaperones in PH1 pathogenesis, we present and discuss current and future perspectives for pharmacological treatments for this disease.

1. Molecular chaperones in protein homeostasis

1.1 A general perspective of protein homeostasis

Proteins are essential players in all cellular processes, being the most versatile and complex macromolecules from the structural viewpoint. They are synthesized on ribosomes as linear chains that must fold into their native states, a set of closely related three-dimensional structures (Bartlett & Radford, 2009; Dobson, Šali, & Karplus, 1998). Translation of this linear sequence into the functional protein is a slow process in the eukaryotic cell, where the polypeptide chain, based on an mRNA template, is assembled at the ribosome (Vabulas, Raychaudhuri, Hayer-Hartl, & Hartl, 2010). Although some newly translated proteins are able to fold spontaneously, a significant fraction of proteins, particularly large ones with complex structures, require the assistance of chaperones, which also regulate conformational integrity and control aggregation under stress conditions (Chiti & Dobson, 2017). The partially folded or misfolded states that are populated during the folding process often expose stretches of hydrophobic amino acid residues that in the native protein are buried from the solvent (Hartl & Hayer-Hartl, 2009; Vabulas et al., 2010). This favors unspecific intermolecular interactions, which are promoted by the crowded intracellular milieu, that interfere with protein folding and leads to protein aggregation. Furthermore, achieving a folded state does not ensure that a given protein is safe as, due to the flexibility requirement to function, proteins are only marginally stable in their physiological environment and can be easily denatured when they are subjected to stress conditions, becoming susceptible to aggregation. All these protein products are not functional and potentially toxic, and thus their production has to be inhibited or, if it cannot be avoided, they must be reactivated or recycled by the ubiquitin proteasome system or autophagic pathways (Fig. 1A).

To prevent the generation of protein toxic species and to ensure proteome integrity, cells invest in protein quality control pathways that refold or degrade these protein products (Brehme et al., 2014; Hipp, Park, & Hartl, 2014). However, when cells are exposed to harsh conditions such as thermal, oxidative or metabolic stresses, the cellular refolding and degradation capacity may become overloaded and global protein misfolding cannot be avoided. The cellular capacity to maintain a healthy proteome also becomes a challenging task when disease-associated mutations occur and when the overall activity of the protein quality control pathways declines during aging (Labbadia & Morimoto, 2015; Walther et al., 2015). Under these conditions, the higher concentration of misfolded protein species leads to protein aggregation and to chaperone-induced formation of protein quality control inclusions. These inclusions usually contain small Heat shock proteins (sHsps), which can function as aggregases that interacting with exposed hydrophobic regions of misfolded proteins prevent their undesired binding to other cellular components (Mogk & Bukau, 2017).

Avoiding protein misfolding and aggregation is so important for cell viability that nature employs several classes of chaperones in the regulation of protein folding, the degradation of misfolded proteins and maintenance of the protein native states (Fig. 1A). Proper balance between protein folding and degradation is maintained by the protein quality control system and is one of the processes influencing protein homeostasis (proteostasis) (Balch, Morimoto, Dillin, & Kelly, 2008). The proteostasis network preserves protein homeostasis and avoids formation of toxic products by controlling concentration, conformation, location, and degradation of individual proteins. The loss of capacity to restore proteostasis after a stress condition often leads to disorders that are classified into two groups: loss-of-function (LOF) and gain-of-function (GOF) disorders (Balch et al., 2008). In this chapter we will focus on LOF mechanisms associated with disease. Excellent reviews on GOF disorders can be found elsewhere (Cox, Raeburn, Sui, & Hatters, 2018; Scannevin, 2018). LOF due to protein misfolding and degradation by the proteasome system or the lysosome is one of the most important mechanisms associated with an increasing number of inborn metabolic

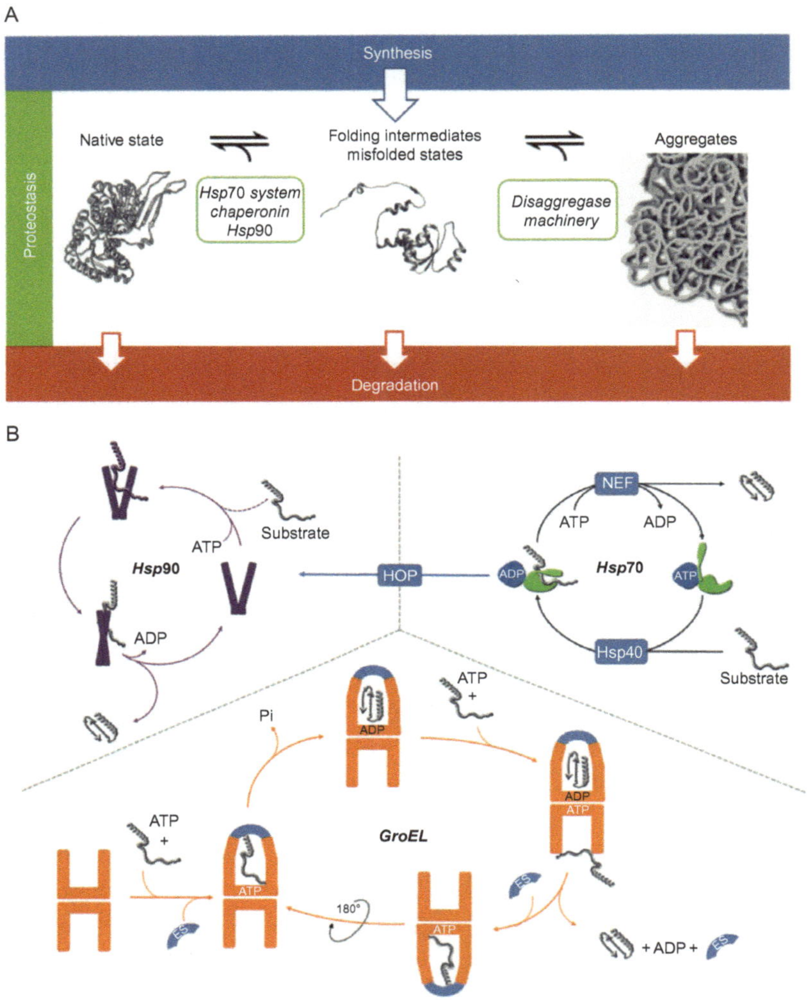

Fig. 1 Role of molecular chaperones in protein homeostasis. (A) The joint effort of different chaperone systems maintains proteome integrity through their involvement in the folding of newly synthesized proteins, remodeling of misfolded states, reactivation of protein aggregates, and targeting of misfolded polypeptides to protein degradation. (B) Schematic representation of the functional cycles of Hsp70, Hsp90 and chaperonins. They all are governed by their ATPase cycle that regulates the interaction with accessory and client proteins. The ATPase cycle of *Hsp70* starts with the chaperone in the ATP state (open conformation), which forms a ternary complex with a misfolded substrate protein and the cochaperone Hsp40. The interaction of the Hsp40 J-domain with Hsp70 accelerates ATP hydrolysis. The affinity of Hsp70 for protein substrates increases in the ADP state (close conformation) facilitating substrate transfer from Hsp40. In the next step, a nucleotide exchange factor (NEF) promotes ADP/ATP exchange lowering the affinity of

diseases (Gregersen, Bross, Vang, & Christensen, 2006; Jorge-Finnigan et al., 2010; Parenti, Pignata, Vajro, & Salerno, 2013). Native-state destabilization due to mutations seems to be a general feature involved in many LOF genetic diseases (Oppici, Dindo, Conter, Borri Voltattorni, & Cellini, 2018; Pey, Stricher, Serrano, & Martinez, 2007). Treatment for these conformational diseases is currently based on the use of proteostasis regulators that enhance the proteostasis capacity and small chemical molecules or pharmacological chaperones that specifically bind and stabilize the native conformation of the target protein (Tao & Conn, 2018).

1.2 The pivotal role of molecular chaperones in protein homeostasis

Molecular chaperones are proteins that interact cotranslationally with the nascent polypeptide chain or folding intermediates to stabilize or help a non-native protein to acquire its native conformation, inhibiting premature misfolding and aggregation, but are not present in the final functional structure (Fig. 1A) (Hartl & Hayer-Hartl, 2009; Vabulas et al., 2010). Different classes of structurally unrelated molecular chaperones have been described, often known as heat shock proteins (Hsp). They are classified according to their molecular weight indicated in the name of the chaperone, Hsp40, Hsp60, Hsp70, Hsp90, Hsp100, whereas small heat shock proteins exhibit a molecular weight in the range between 15 and 40 kDa. Some chaperones promote folding (foldases), others maintain solubility (holdases), and a last type of chaperones is responsible for the disaggregation of protein aggregates (disaggregases) (Gregersen et al., 2006). Although the core chaperone

Hsp70 for its substrate. The released substrate either folds into its native structure or is transferred to other chaperone (Hsp90 or chaperonin) for further processing. The functional cycle of the Hsp90 is regulated by ATP binding and hydrolysis, and interaction with accessory proteins. Client proteins are recruited from the upstream folding machinery via adaptor proteins such as HOP, which links Hsp70 and Hsp90 and inhibit their ATPase activity. In the absence of ATP, Hsp90 adopts mainly a V shaped open conformation. ATP binding promotes interaction with substrates and accessory proteins and induces a structural rearrangement that leads to the closed state via intermediate steps. Finally, ATP hydrolysis promotes the release of the bound substrate. Functional cycle of the chaperonin *GroEL*. The cycle starts by binding of ATP and unfolded client proteins to one GroEL ring. Subsequent interaction of the cochaperone GroES encapsulates the substrate into the enlarged central cavity of the *cis*-ring where it folds. Nucleotide hydrolysis and binding of ATP and client protein to the opposite *trans*-ring trigger substrate release form the *cis*-ring. This traditional "bullet cycle" that assumes that only one folding chamber is active at a given time has been recently challenged with an alternative "football cycle" in which both GroEL rings can accept substrates at the same time.

machineries (Hsp70s, Hsp90s, Hsp100s, chaperonins, and small Hsps [sHsps]) are present in prokaryotes, evolution has significantly increased the number of their regulatory cofactors (Hsp40s, tetratricopeptide repeat proteins) (Brehme et al., 2014; Genevaux, Georgopoulos, & Kelley, 2007; Rizzolo et al., 2017; Vos, Hageman, Carra, & Kampinga, 2008). The four large and ubiquitous families of ATP-dependent molecular chaperones, namely, Hsp100, Hsp90, Hsp70 and Hsp60 proteins, may be represented in an organism by different members. The diversity of molecular chaperones is evidenced by the existence, besides ATP-dependent ones, of nucleotide-independent chaperones, such as small Hsps (sHsps), and others that are responsible for the remodeling of specific substrate proteins. In the next sections we will review the current knowledge of three of the main chaperone families, Hsp60, Hsp70 and Hsp90.

1.2.1 The Hsp70 system

This system is formed by a central chaperone, named Hsp70, and two accessory proteins, one belonging to the Hsp40 family and the other functioning as a nucleotide exchange factor (NEF). The functional versatility of this system comes from the ability of Hsp70 to interact with many different conformations of substrate proteins, including largely unfolded (Hartl, Bracher, & Hayer-Hartl, 2011; Neupert & Herrmann, 2007) and native conformations (Sousa & Lafer, 2015).

Hsp70 chaperones are highly conserved in all kingdoms, with the bacterial representative, DnaK, being the best characterized both functionally and structurally. DnaK builds a system with an Hsp40 protein, DnaJ, and a NEF, GrpE. In eukaryotes this system is more complex, with 13 Hsp70s, 50 J-proteins and 7 NEFs (Kampinga et al., 2009). This diversity in eukaryotes reflects the requirements of different cellular compartments, their function in signaling and trafficking, and their cross-talk with proteosomal and autophagosomal degradation systems and the Hsp90 system through accessory proteins.

Hsp70s consist of an N-terminal 43-kDa nucleotide binding domain (NBD) and a C-terminal 27-kDa substrate binding domain (SBD), which are allosterically connected by a highly conserved linker. Human Hsp70 proteins closely resemble this domain distribution. The affinity of Hsp70 for substrates varies during the ATPase cycle of the protein due to nucleotide-driven conformational changes (Fig. 1B). In the ATP state, the chaperone binds exposed hydrophobic regions of misfolded/unfolded proteins with a weak affinity (Kityk, Kopp, Sinning, & Mayer, 2012). In the ADP state, its affinity for misfolded/unfolded proteins increases 10–50 times (Schmid, Baici, Gehring, & Christen, 1994). Hsp70s bind a motif formed by a core of five preferentially hydrophobic amino acids flanked by positively charged regions, which occur frequently in most proteins (Rüdiger, Buchberger, & Bukau, 1997). Unfolded or partially folded proteins expose these motifs which can bind to Hsp70, in contrast to native proteins that bury them in their hydrophobic core. Recognition of such short sequences in protein substrates avoids size limitations, in contrast to chaperonins (see below), and allows Hsp70 to interact with large proteins, protein complexes and protein aggregates.

Although classification of J-proteins is not straightforward because of their variable structure outside of the characteristic J-domain (Craig & Marszalek, 2017), these multidomain proteins are divided into three main classes (A, B and C) considering their domain composition. All contain a J-domain, a helical hairpin structure of 70–75 residues, which is at the N terminus of class A and class B, and anywhere in class C. Members of class A share the domain architecture of *Escherichia coli* DnaJ: the N-terminal J-domain is followed by a glycine/phenylalanine (G/F)-rich, flexible region of 30 residues, two homologous β sandwich domains, with a Zn^{2+}-finger domain inserted into the first one, and a C-terminal dimerization domain. Class B representatives contain all these domains, except the Zn^{2+} finger, and class C JDPs only have the J-domain in common and can have a wide variety of different domains or functional motifs. Regarding function and specificity, class A and class B behave as general cochaperones that interact with unfolded, misfolded, or aggregated proteins and transfer them to Hsp70, whereas members of class C are believed to be more specific, as they seem to interact with a small number of substrates. The different representatives of the J proteins can bind themselves to protein substrates or to protein complexes with a wide range of specificity (Hageman et al., 2010; Kakkar, Kuiper, Pandey, Braakman, & Kampinga, 2016), and recruit Hsp70 for specific protein folding tasks.

NEFs activate the Hsp70 folding cycle by replacing Hsp70-bound ADP by ATP. Bacteria and organelles of bacterial origin have only one known NEF for Hsp70, GrpE. While in mitochondria and chloroplasts GrpE homologs are preserved, the cytosol and endoplasmic reticulum (ER) of eukaryotes contain three divergent families of NEFs: Hsp110/Grp170, HspBP1/Sil1 homologs and BAG-domain proteins (Dragovic, Broadley, Shomura, Bracher, & Hartl, 2006; Höhfeld & Jentsch, 1997; Kabani, McLellan, Raynes, Guerriero, & Brodsky, 2002; Raviol, Sadlish, Rodriguez, Mayer, & Bukau, 2006; Steel, Fullerton, Tyson, & Stirling, 2004). The human genome encodes four Hsp110/Grp170, two HspBP1/Sil1 homologues, five BAG-domain proteins and two mitochondrial GrpE species. Despite the remarkable structural diversity found among NEFs, it is interesting to highlight that all of them target the same region in the Hsp70 NBD, suggesting convergent evolution to activate the molecular switch built into the Hsp70 NBD. This NEF diversity might enable adaptation of the Hsp70 ATPase cycle to the folding needs of specific substrate proteins (Rauch & Gestwicki, 2014; Tzankov, Wong, Shi, Nassif, & Young, 2008), and could direct Hsp70s to different functions (Bracher & Verghese, 2015).

The most ancient and universal type of eukaryotic NEF, Hsp110, was first identified as a divergent member of the Hsp70 family, with a sequence identity of <30%. From the structural viewpoint, both contain two domains, an N-terminal NBD and a C-terminal SBD, and differ by the presence of an inserted acidic sequence in the SBD, and an intrinsically disordered C-terminal extension in Hsp110 (Aguado, Fernández-Higuero, Moro, & Muga, 2015). It remains unclear whether Hsp110/Grp170 and Hsp70 similarly recognize the substrate or follow a similar ATP-dependent functional cycle (Rampelt et al., 2012; Shorter, 2011; Xu et al., 2012). Hsp110s function as NEFs for Hsp70 and display a potent holdase activity, binding denatured proteins and preventing their aggregation (Xu et al., 2012). It is also part of the disaggregase machinery of higher eukaryotes that extracts unfolded proteins from aggregates and reactivates them through an energy-dependent mechanism (Aguado et al., 2015; Rampelt et al., 2012; Shorter, 2011), in collaboration with Hsc70 and J-proteins (Nillegoda et al., 2015; Rampelt et al., 2012).

1.2.2 Hsp60 (chaperonins)

The chaperonins are double-ring complexes of 800–1000 kDa built of 7–9 subunits per ring. They facilitate protein folding by providing a protective chamber where non-native substrate proteins can (re)fold in isolation from the cell environment to avoid undesired intermolecular interactions (Bukau & Horwich, 1998; Hayer-Hartl, Bracher, & Hartl, 2016). Chaperonins are classified into two distantly related structural groups: Group I is found in bacteria (GroEL) and eukaryotic organelles (Hsp60), and Group II is expressed in Archaea (thermosome) and in the eukaryotic cytosol (chaperonin containing TCP1 (CCT), or TCP1 ring complex (TRiC)) (Kim, Hipp, Bracher, Hayer-Hartl, & Hartl, 2013). Their gene family is extensive and complex (Dekker, Willison, & Taylor, 2011), with high structural similarity despite their low sequence conservation (Skjærven, Cuellar, Martinez, & Valpuesta, 2015).

GroEL is a large double-ring cylinder with ATPase activity that binds non-native substrate protein via hydrophobic residues exposed toward the central cavity. Binding of the cochaperone, dome-shaped GroES to GroEL displaces the bound protein into an enlarged chamber, allowing folding to occur unimpaired by aggregation (Hayer-Hartl et al., 2016). GroES and substrate proteins undergo cycles of binding and release, regulated allosterically by the GroEL ATPase. Each GroEL monomer folds into three distinct structural domains: the equatorial, intermediate, and apical domain (Braig et al., 1994; Xu, Horwich, & Sigler, 1997) (Fig. 1B). The intermediate domain acts as a hinge between the apical and equatorial domains, providing flexibility to the GroEL assembly and mediating large-scale conformational changes (Xu et al., 1997), which drive the functional cycle of productive protein folding (Ranson et al., 2001).

The structure of Group II chaperonins is more complex. This group includes the archaeal chaperonins or thermosomes, which are composed of octameric or nonameric rings made up of 1–3 different subunits (Bigotti & Clarke, 2008), and CCT, which is found in the cytosol of eukaryotic organisms, and is composed of eight distinct subunits (Leitner et al., 2012; Muñoz et al., 2011). The monomer of Group II chaperonins also contains equatorial, intermediate and apical domains but differs in the inter-ring structure and in the lid arrangement that covers the central chamber, which frees this group of the cochaperone need (Skjaerven et al., 2015, 2011). Hsp70 interacts directly with CCT, transferring substrates to the chaperonin (Cuéllar et al., 2008) and cooperating in cotranslational folding of nascent-polypeptide chains.

Substrate recognition ranges from the simple and promiscuous GroEL to the much more sophisticated CCT: in the latter, each of the eight distinct subunits interacts differently with client proteins. In contrast to GroEL, differences in ATP binding affinity of each of the eight CCT monomers may cause timely delayed conformational changes and facilitate an ordered release of client segments into the chamber, potentially avoiding misfolding of topologically complex client proteins (Reissmann et al., 2012). Additionally, simultaneous binding of several hydrophobic segments of the substrate to different apical domains of the chaperonin can partially unfold the client protein (Farr et al., 2000; Priya et al., 2013), which can be further fostered through ATP-driven conformational changes (Lin, Madan, & Rye, 2008). Many substrates have complex folding landscapes in which kinetically trapped folding intermediates and misfolded states are frequently populated (Kerner et al., 2005; Yam et al., 2008). It has been proposed that unfolding could pull the substrate protein out of kinetically trapped misfolded state(s) (Priya et al., 2013).

1.2.3 The Hsp90 system

Hsp90 is an essential protein under normal growth conditions in all eukaryotes with observable increase in expression under thermal stress. Many eukaryotes contain two copies of cytosolic Hsp90 and additional homologs, which encode mitochondrial or ER-localized variants. The function of this chaperone has evolved from a dispensable bacterial heat shock protein and to a mandatory chaperone for many essential clients. At the same time, this chaperone system has expanded to include about 20 cofactors that interact with the Hsp90 dimer and prime it for client-specific activities (Prodromou, 2017; Schopf, Biebl, & Buchner, 2017). While the cytosolic Hsp90 protein itself is promiscuous, its cofactors are much more specific. Some of these proteins simultaneously interact with Hsp90 and specific client proteins, forming multimeric protein complexes. Hsp90 is associated with many human diseases, which include cancer, neurodegenerative diseases and infectious diseases (Schopf et al., 2017).

Hsp90 functions as a homodimer, and dimerization is essential for its function in vivo (Mayer & Le Breton, 2015; Wayne & Bolon, 2007). An Hsp90 monomer consists of three highly conserved domains: the amino-terminal domain (NTD), which binds ATP (Prodromou et al., 1997); the middle domain (MD), which is important for ATP hydrolysis and binding to clients; and the carboxy-terminal domain (CTD), which mediates Hsp90 dimerization. It also contains a C-terminal MEEVD motif, which is involved in the interaction with co-chaperones that contain tetratricopeptide repeat (TPR) domains (Buchner, 1999). The NTD and the MD are connected by a long, flexible, charged linker that modulates NTD–MD contacts and affect Hsp90 function (Jahn et al., 2014; Tsutsumi et al., 2012). Analysis of the binding sites for co-chaperones, which are detected in all three domains of Hsp90, suggests that some co-chaperones compete with each other for binding to Hsp90, whereas others can bind simultaneously or even synergistically. Some co-chaperones are modulators of the Hsp90 chaperone cycle and others act as adaptor proteins for client recruitment, although there are co-chaperones that display both functions. Hsp90 also undergoes large conformational changes during the ATPase cycle of the protein, which are used to properly chaperone its client proteins (Shiau, Harris, Southworth, & Agard, 2006) (Fig. 1B). In the absence of ATP, Hsp90 adopts mainly a V-shaped open conformation. ATP binding induces a structural rearrangement that leads to an N-terminally closed state via intermediate steps (Hellenkamp, Wortmann, Kandzia, Zacharias, & Hugel, 2017).

Hsp90 fosters the folding, maturation and degradation of a large number of often structurally and sequentially unrelated client proteins (Chiosis, Dickey, & Johnson, 2013; Johnson, 2012; Taipale et al., 2012), typically acting downstream of Hsp70 (Alvira et al., 2014; Genest, Hoskins, Camberg, Doyle, & Wickner, 2011; Morán Luengo, Kityk, Mayer, & Rüdiger, 2018). In contrast to Hsp70 that recognizes short, hydrophobic patches, Hsp90 usually binds partially folded, less hydrophobic intermediates (Karagöz et al., 2014). Hsp90 clients include many aggregation-prone and unstable regulatory proteins, such as kinases; transcription factors, including steroid hormone receptors; and surprisingly, intrinsically disordered proteins such as Tau or α-synuclein. The comparison of the Hsp90 binding sites of representatives of these three types of substrates reveals some features that may be general trademarks for Hsp90 clients (Karagöz et al., 2014; Lorenz et al., 2014; Radli & Rüdiger, 2018; Verba et al., 2016). Hsp90 recognizes extended (50–170 amino acids) regions of the client, in contrast to the short peptides bound by Hsp70. This suggests that Hsp90 binds substrate proteins through a large number of weak contacts allowing for dynamic, transient interactions, and avoids binding of hydrophobic, aggregation-prone short stretches that typically form the hydrophobic core of proteins. This observation indicates that it preferentially binds to late folding intermediates in which these sites are already buried. The nature of the binding sites of co-chaperones is more variable (Ali et al., 2006; Meyer et al., 2004; Roe et al., 2004), suggesting that these interfaces are designed for a specific interaction, whereas the client interaction surface requires versatility to face structural diversity. Post-translational modifications (PTM), including phosphorylation, S-nitrosylation, acetylation and methylation, of Hsp90 provide further regulation to its function (Mayer & Le Breton, 2015; Röhl, Rohrberg, & Buchner, 2013). The number of phosphorylation sites and other PTMs on Hsp90 and on the regulatory cofactors are surprisingly high and they might be used to adjust the general Hsp90 functionality toward the individual needs of specific clients.

1.3 Molecular chaperones and protein intracellular traffic

The Hsp70 and Hsp90 systems collaborate in all stages of intracellular traffic of proteins from cytoplasm to organelles. The interplay between these chaperone machineries links protein trafficking to protein quality control and determines whether client proteins achieve their final location or are deflected to degradation pathways. Both chaperone systems can regulate protein turnover by opposite effects on CHIP-dependent ubiquitination, which is promoted or inhibited by interaction with Hsp70 or Hsp90, respectively (Kevei, Pokrzywa, & Hoppe, 2017).

Newly synthetized proteins can be matured in the cytosol or delivered to an organelle. In the latter case, organelle specific import receptors recruit protein-chaperone heterocomplexes through the binding of multiple tetratricopeptide repeats (TPR) to C-terminal acceptor sites of Hsp70 and Hsp90. When the targeting compartment is the mitochondria, carrier complexes are recognized by the Translocase of the Outer Mitochondrial Membrane (TOM) through different receptors (Tom70, Tom20 and Tom22) (Wiedemann & Pfanner, 2017). Once in the intermembrane space (IMS), mitochondrial membrane proteins are stabilized by chaperones prior to their transference to the outer or inner membrane by the Sorting and Assembly Machinery (SAM) or the Translocase of the Inner Membrane 22 (TIM22 complex), respectively (Dudek, Rehling, & van der Laan, 2013). Mitochondrial matrix proteins containing the N-terminal cleavable presequence are bound to the Tim50 receptor of the Translocase of the Inner membrane 23 (TIM23) complex. Complete protein translocation into the matrix requires the participation of mitochondrial Hsp70 (mtHsp70) that is coupled to the translocase complex through the peripheral membrane protein Tim44, and acts as the ATP driven import motor. In this process two membrane-bound J-proteins (Tim14 and Tim16) and the nucleotide exchange factor Mge1 regulate the Hsp70 functional cycle. After the removal of the presequences by mitochondrial peptidases, the Hsp60 machinery promotes the folding of imported proteins (Wiedemann & Pfanner, 2017). A similar mechanism has been found for the import of proteins into chloroplasts (Gray, Rustgi, von Wettstein, Reinbothe, & Reinbothe, 2016; Jarvis & Soll, 2001).

Protein import into the peroxisomal matrix may also require Hsp70/Hsp40 (Hettema et al., 1998; Walton, Wendland, Subramani, Rachubinski, & Welch, 1994) and Hsp90 (Crookes & Olsen, 1998) systems. The lack of evidence for the presence of Hsp70 or Hsp90 proteins in the peroximal matrix may indicate that proteins are imported in a near native conformation assisted by cytosolic chaperones and once into the organelle they are able to spontaneously acquire their functional state. However, the presence of proteins with chaperone activity in peroxisomes of other organism as plants and fungus (Bartoszewska et al., 2012; Du et al., 2015) suggests that other matrix proteins could perform chaperone-like actions.

2. Primary hyperoxaluria type I (PH1): A LOF disease associated with peroxisomal aggregation and mitochondrial mistargeting

2.1 PH1: A metabolic disorder with complex genotype-phenotype relationships and disease mechanisms

Primary hyperoxalurias (PH) are a group of recessive inborn errors of glyoxylate metabolism characterized by overproduction of oxalate mainly in the liver and massive excretion of oxalate in the urine (reviewed in Salido, Pey, Rodriguez, & Lorenzo, 2012). Oxalate, a dicarboxylic acid with low solubility in the presence of calcium (K_{ps} of 2.3×10^{-9}), is an end-product of metabolism without a biological role in mammals. Although oxalate can be also absorbed in the gut from the diet, most of the oxalate produced by PH patients is generated by hepatocytes from glyoxylate, hydroxyproline, glycolate and other precursors. Under normal circumstances oxalate is efficiently removed by the kidneys and excreted in the urine without causing adverse effects. But the high oxalate excretion in PH patients causes frequent calcium oxalate stone formation (urolithiasis) and deposits in the renal parenchyma (nephrocalcinosis), which result in kidney injury and eventual loss of renal function (end stage renal disease, ESRD). At this point, oxalate levels increase dramatically in plasma and widespread oxalate deposition (oxalosis) can be lethal unless aggressive measures are taken, such as double liver-kidney transplantation.

Primary hyperoxaluria type 1 (PH1) accounts for approximately 80% of diagnosed PH patients (Hopp et al., 2015; Mandrile et al., 2014) and is caused by mutations of the AGXT gene encoding liver peroxisomal alanine-glyoxylate aminotransferase (AGT) (Purdue, Takada, & Danpure, 1990; OMIM: #259900). AGT forms functional homodimers, each monomer containing: (i) an unstructured N-terminal tail (NTT; residues 1–21) that contacts the other subunit of the dimer; (ii) a large N-terminal domain (NTD; residues 22–282), with a pyridoxal 5′-phosphate molecule (PLP) covalently bound to Lys209 forming a Schiff base; and (iii) a smaller C-terminal domain (CTD; residues 283–392; containing a PTS1 peroxisomal targeting sequence at the end) (Fig. 2A). The primary function of AGT is the amino transfer from alanine to glyoxylate to form pyruvate and glycine. Since glyoxylate cannot be converted to glycine, it is oxidized to oxalate by lactate dehydrogenase. Over 200 pathogenic mutations have been described, involving all exons, and the majority are point are missense

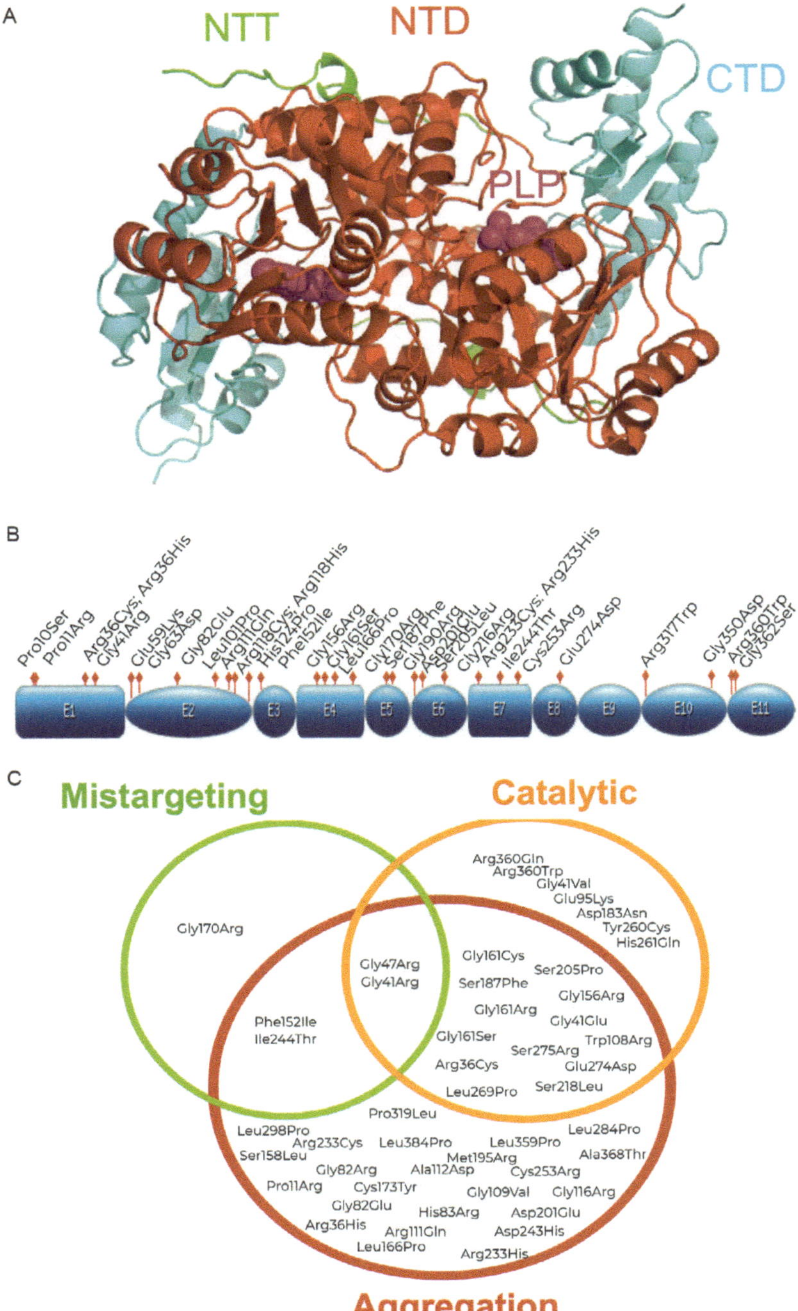

Fig. 2 AGT and Primary hyperoxaluria type I. (A) X-ray crystallographic structure (PDB entry 2F1O) for the AGT dimer. Different structural regions are displayed in different colors: N-terminal tail (NTT, in green), N-terminal domain (NTD, in red) and C-terminal domain (CTD, in cyan). The PLP bound is shown in magenta sphere representation. (B) Representation of AGXT exons and the main PH1-associated mutations; (C) PH1-associated mutations classified according to their pathogenic mechanism.

mutations (https://www.uclh.nhs.uk/OurServices/ServiceA-Z/PATH/PATHBIOMED/CBIO/Pages/Phmdatabase.aspx, Fig. 2B). Most AGXT missense mutations are now considered examples of conformational or protein homeostasis disease, where a decrease in folding efficiency is thought to be the main mechanism involved, resulting in very low levels of active protein and/or its aberrant targeting to the mitochondria (Pey, Albert, & Salido, 2013) (Fig. 2C).

No universal genotype-phenotype correlation has been found (Frishberg et al., 2005; Lorenzo et al., 2006) except for patients homozygous for the G170R mutation on the minor allele (LM). This genotype appears to predict pyridoxine responsiveness (Monico, Rossetti, Olson, & Milliner, 2005; van Woerden et al., 2004), to correlate with slower progression to ESRD (OxalEurope registry) and to increase renal survival (data from the Rare Kidney Stone Consortium, RKSC, and compared to other PH1 mutations; Hopp et al., 2015). The benefit of G170R homozygosity may relate to lower oxalate excretion rates in this population (Hopp et al., 2015), in addition to marked reduction in urinary oxalate excretion when pyridoxine treatment is successful (Monico et al., 2005)

The LM haplotype (allelic frequency 0.05–0.1) contains a number of polymorphic changes, including P11L and I340M. Given the short distance between these points in the genome, these two polymorphisms are inherited in very strong linkage disequilibrium (P11-I340—major haplotype—vs. L11-M340—minor haplotype) and it is very rare to find one without the other. The minor AGT allele has reduced enzymatic activity (Lumb & Danpure, 2000). The I340M change by itself has no deleterious consequences and in vitro studies demonstrate a beneficial effect on enzymatic activity (Mesa-Torres, Yunta, et al., 2014). On the other hand, the P11L substitution has a synergistic effect with various missense mutations, playing important roles in PH1 protein homeostasis defects. Indeed, P11L introduces a weak mitochondrial targeting signal that, in conjunction with the very common pathogenic variant G170R, causes peroxisome-to-mitochondrion mistargeting of the AGT enzyme (Purdue et al., 1990; Purdue, Allsop, Isaya, Rosenberg, & Danpure, 1991). Similarly, P11L is required in *cis* for the common I244T mutation to result in AGT misfolding and protein aggregation (Santana, Salido, Torres, & Shapiro, 2003). The strong association to the minor haplotype of a large proportion of missense mutations suggests that P11L might have a significant role in many of these mutations, although experimental evidence is lacking for most of them (Mesa-Torres, Salido, & Pey, 2014).

2.2 Molecular pathogenesis

Over the 92 PH1-associated missense mutations, 47 have been characterized, to some extent, in terms of their pathogenic mechanism (Fig. 2C). Most of them clearly affect protein homeostasis in different manners (Fig. 2C). In this section, we will discuss the links between mutation-induced perturbation of the native state and folding/unfolding pathways and the different pathogenic mechanisms in PH1. This will bring us to discuss current and future directions to further understand this disease, with the hope of developing new therapeutic strategies specifically targeting different disease mechanisms.

2.2.1 Mitochondrial mistargeting

The subcellular localization of AGT varies among mammals. While human AGT is exclusively peroxisomal, the orthologous enzyme is mitochondrial in many carnivorous animals and others, such as mice, have both peroxisomal and mitochondrial variants (Birdsey et al., 2005; Mesa-Torres et al., 2016). In evolutionary terms, it has been suggested that proteins obtain novel compartmentation variants by "minor mistargeting" mechanisms (Martin, 2010). This argument could help rationalize the propensity of human AGT to end up in the wrong organelle (mitochondria) as a consequence of missense mutations in the minor haplotype. One of the most common PH1 mutations, G170R, is the best example of mistargeting of a peroxisomal protein to the mitochondria, where AGT cannot detoxify glyoxylate in humans (Purdue et al., 1991, 1990). As for the mechanism involved in such mistargeting, it is known that the P11L amino acid replacement encoded by the LM may create a potential N-terminal mitochondrial targeting sequence that results in mistargeting if the G170R mutation is also present in the same allele. Mistargeting has also been observed with other mutations that segregate with the LM (I244T, F152I, and G41R) but only under some experimental conditions (Fargue, Lewin, Rumsby, & Danpure, 2013; Santana et al., 2003) suggesting a dynamic intracellular interplay between normal peroxisomal targeting and mitochondrial mistargeting of PH1 variants.

2.2.2 Protein aggregation

Intracellularly, the balance of AGT protein between soluble and aggregated states for PH1-associated mutants is delicate (Mesa-Torres et al., 2013). In some cases, LM can severely reduce steady-state protein levels and exacerbate protein aggregation (Mesa-Torres et al., 2013) and these effects are magnified in the presence of some PH1 mutations (e.g., G41R, G47R, G170R, F152I, I244T or P319L) (Fargue, Lewin, et al., 2013; Mesa-Torres et al., 2013; Montioli et al., 2015). Interestingly, mutations associated with either aggregation or mitochondrial mistargeting are scattered though the dimeric structure of AGT (Fig. 3A), suggesting that significantly unfolded intermediate(s) are associated with either of these two pathogenic mechanisms.

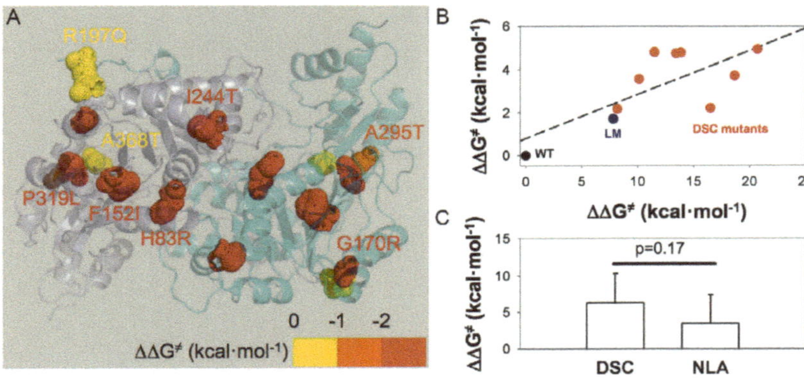

Fig. 3 PH1-causing mutations often destabilize the native state. (A) Structural location of several PH1-causing mutations known to exacerbated intracellular aggregation of AGT due to kinetic destabilization of the apo-protein. The color code indicates the extent of kinetic destabilization of the apo-AGT determined by differential scanning calorimetry (DSC). (B) Correlation between thermodynamic (*x*-axis, change in unfolding free energy, $\Delta\Delta G$ as determined by FoldX; Schymkowitz et al., 2005) and kinetic (*y*-axis, change in activation free energy, $\Delta\Delta G^{\ddagger}$ as determined by DSC measurements). Values are those for the free energy difference between WT and the disease-associated variant (LM or the mutations on the LM background; DSC mutants). The slope of the plot is 0.20 ± 0.05, supporting that about 20% of the thermodynamic destabilization of the native state is translated to a decrease in the denaturation free energy barrier. FoldX calculations were determined as a plug-in of Yasara software (www.yasara.org/) at 298.15K, 100mM ionic strength and pH 7. (C) Thermodynamic destabilization of the native state for some PH1-causing mutations for which the extent of aggregation/mistargeting in cells has been compared with their kinetic destabilization (DSC mutants; H83R, F152I, G170R, R197Q, I244T, A295T, P319L and A368T) and those proposed to cause native-like aggregation (NLA mutants; E95K, D129H, E141D, R197Q, R233H, C253R, Q282R, R289H and R289C). A statistical analysis using a Mann-Whitney test provided no significantly difference between the thermodynamic destabilization of these two groups of mutants. *DSC data in panel (B) is from Mesa-Torres, N., Salido, E., & Pey, A. L. (2014). The lower limits for protein stability and foldability in primary hyperoxaluria type I.* Biochimica et Biophysica Acta, 1844*(12), 2355–2365. https://doi.org/10.1016/j.bbapap.2014.10.010.*

2.2.3 The role of native state stability and unfolding intermediates in mistargeting and aggregation phenotypes

The intracellular process by which AGT is synthesized, folded and assembled into functional dimers and transported into peroxisomes is complex by nature (Oppici et al., 2018; Pey et al., 2013). Intrinsic physicochemical features of the AGT polypeptide as well the interaction of partially and fully folded AGT monomers and dimers between them and with molecular chaperones (Pey et al., 2013), as well as direct channeling of PLP cofactor from enzymes (di Salvo, Contestabile, & Safo, 2011; di Salvo, Safo, & Contestabile, 2012), may play key roles in correct folding and peroxisomal transport.

The stability of the folded dimer of AGT is key in PH1 pathogenesis. The native holo-AGT dimer is a highly stable protein in vitro, while removal of the PLP (apo-AGT) leads to much lower stability (Cellini et al., 2010; Mesa-Torres et al., 2013; Mesa-Torres, Salido, et al., 2014; Pey, Salido, & Sanchez-Ruiz, 2011). Biophysical studies supported a simple in vitro denaturation model for AGT (Mesa-Torres et al., 2013; Pey, 2013; Pey et al., 2011) (Eq. 1):

$$N \overset{K_U}{\longleftrightarrow} U \overset{k_I}{\longrightarrow} F \tag{1}$$

where N stands for the native AGT dimer, U for an unfolded state and F for the irreversible denatured protein. The N to U step is labelled as reversible and thus described by an unfolding equilibrium constant $K_U = \frac{[U]}{[N]} = \frac{k_U}{k_F}$, in which k_U and k_F are the relevant kinetic rate constants for unfolding and folding, respectively. The step U to F is a kinetic process described by a first-order rate constant k_I (Mesa-Torres, Salido, et al., 2014). In this context, PH1-associated mutants primarily affect the kinetic stability of apo-AGT (Fig. 3A), synergizing with the destabilizing effect of the LM, the latter representing a non-pathogenic lower limit for stability and foldability in AGT (Mesa-Torres, Salido, et al., 2014).

Structure-energetic analyses have shown that the N to U step may involve significant loss of native structure, and the U to F step mostly reflects "docking" or "assembly" of partially unfolded states into the final (aggregated) states. For AGT mutants studied so far, the irreversible step is not-rate limiting, and therefore, we can gain insight into structural properties of partially unfolded states involved in both aggregation and mistargeting phenotypes. Indeed, physiologically relevant partially unfolded states are likely

AGT dimers with low residual native structure resembling a *molten globule* (Mesa-Torres et al., 2013; Mesa-Torres, Salido, et al., 2014; Pey et al., 2011). PH1-causing mutations reduce the thermodynamic stability of the native state thus increasing the population of partially unfolded states according to their effects on K_U (Mesa-Torres, Salido, et al., 2014). This interpretation is consistent with structure-based energy calculations for the impact of disease-mutations on the stability of the native state against global unfolding, showing a link between mutation-induced destabilization of the native state in thermodynamic and kinetic terms (Fig. 3B). Importantly, this propensity of disease-associated variants to populate partially unfolded states correlate to some extent with their ability to interact with Hsc70 and Hsp90 chaperones, and with the overall yield of protein properly folded and imported into peroxisomes (Mesa-Torres et al., 2013; Mesa-Torres, Salido, et al., 2014; Pey et al., 2011). Overall, all these evidences support that the stability of AGT toward partially unfolding as well as the accumulation of partially folded states during folding intracellularly are likely associated, and may strongly determine the partition of AGT between properly folded and peroxisomally targeted protein, aggregation and mitochondrially mistargeted protein.

A recent study has proposed a native-like and electrostatic aggregation (NLA) mechanism that may contribute to the enhanced aggregation of certain PH1-associated mutants (Dindo, Conter, & Cellini, 2017). This mechanism implies the following two steps:

$$A \xrightarrow{k_1} B \tag{2}$$

$$A + B \xrightarrow{k_2} B \tag{3}$$

in which A represents the AGT dimer and B is a multimeric aggregated species. k_1 describes a first-order conversion (although the oligomeric status of A and B are by definition different) and k_2 must characterize a second-order aggregation process to yield B species (note that this step appears to violate law of mass conservation). Although the first step (Eq. 2) is defined as first-order, its value drastically changes (by up to 2000-fold) with a 10-fold increase in protein concentration (Dindo et al., 2017). In addition, a substantial fraction of secondary structure is lost due to aggregation deviating from the proposed native-like aggregation species, and mutational effects on aggregation kinetics follow closely their thermal stabilities at low protein concentrations while no remarkable differences are shown at higher concentrations (Dindo et al., 2017; Mesa-Torres et al., 2013). This mechanism is also invoked to predict deleterious effects of mutations seemingly as destabilizing as those for which thermodynamic and kinetic destabilization seems to be the main molecular consequence associated with aggregation (or mistargeting) (Fig. 3C). Thus, further research is required to assess the impact of this intriguing mechanism on the loss-of-function aggregation phenotypes in PH1.

2.2.4 Interaction of AGT with molecular chaperones

Human AGT is synthesized in the cytoplasmic free ribosomes, and likely assisted by different chaperone pathways working in a stepwise fashion (Hartl & Hayer-Hartl, 2002): in a first step, ribosome-associated chaperones may interact with the nascent polypeptide preventing its aggregation, stabilizing it and helping its correct folding, and second, downstream chaperones can take on the task of completing folding and assembly of AGT dimers. AGT dimer can then interact in the cytosol with the peroxin 5 (Pex5p) receptor via its C-terminal PTS1 sequence to be imported into the peroxisome.

While the interaction of nascent AGT with the ribosome-associated complex (RAC) has not been particularly studied, a series of experiments has underscored the importance of the interaction of mutant AGT with the second tier of cytosolic chaperones. WT AGT shows transient interactions with cytosolic chaperones, many PH1 mutants of the LM strongly interact with Hsc70 chaperones, adding to our previous work on I244T and G170R that also showed enhanced interactions with Hsp90 (Mesa-Torres et al., 2013; Pey et al., 2011; Santana et al., 2003). AGT WT interacts with Hsc70 and Hsp90 chaperones through a molten globule folding intermediate, and this interaction is notably enhanced by LM-G170R upon in vitro synthesis and folding in eukaryotic cell-free systems (Pey et al., 2011). Correct folding of LM-I244T is enhanced upon coexpression of with bacterial GroEL/ES and forms stable complexes with Hsp60 chaperones in partially folded monomeric state with the folded N-terminal and C-terminal domains in an open conformation (Albert et al., 2010). In vitro refolding experiments of AGT WT in the presence of bacterial Hsp40 protein DnaJ also revealed that this chaperone successfully prevents its aggregation (Pey et al., 2013). We thus propose that the last steps of AGT folding are crucial for proper folding and involve docking of tertiary structure elements and acquisition of the dimeric quaternary structure. The enhanced interaction of PH1 mutants with cytosolic molecular chaperones suggests a rougher folding landscape for these mutants (with a higher or longer lasting population of kinetic/equilibrium intermediates). Since Hsp70/40, Hsp60/10 and Hsp90 chaperone systems are known to cooperate in assisting protein folding, the regulation of chaperone activity by co-chaperones and regulatory proteins may lead to different fates (i.e., folding vs. degradation) for the client proteins (Hartl et al., 2011).

Thus, molecular chaperones may represent important checkpoints in the folding and misfolding of PH1 mutants, likely by partitioning AGT folding intermediates into productive formation of native dimers and peroxisomal import, presentation of partially folded states to the mitochondrial import systems, aggregation and cytosolic/peroxisomal degradation (Mesa-Torres et al., 2013; Salido et al., 2012).

2.2.5 Interaction with peroxisomal import machineries

AGT contains one of weakest PTS1 sequences reported so far, with K_d values of about 2 and 20 μM, for the full-length protein and the C-terminal PTS1 sequence, respectively (Mesa-Torres, Tomic, Albert, Salido, & Pey, 2015), although this low affinity does not prevent its targeting to peroxisomes due to the high expression levels of AGT (i.e., high concentration of folded dimers; Ghosh & Berg, 2010; Mesa-Torres et al., 2013, 2015). We reported that enhancing the interaction of AGT with the Pex5p receptor via a 100-fold stronger PTS1 sequences partially reverted mitochondrial mistargeting of the LM-G170R mutant (Mesa-Torres et al., 2015). These analyses also revealed a complex interplay between the structure of the complex and the underlying binding thermodynamics, including unexpected sequence-dependent structural rearrangements and a 10-fold enhancement of the interaction by ancillary regions beyond the PTS1 sequence (Fodor et al., 2015; Mesa-Torres et al., 2015). Importantly, peroxisomal rescue of LM-G170R was in part associated to more stable AGT: Pex5p complexes (Mesa-Torres et al., 2015). In the case of LM-G170R, the formation of the complex with Pex5p is likely prevented due to its trapping along the folding process with molecular chaperones that thus promote its interaction with mitochondrial import machineries (Mesa-Torres et al., 2013; Pey et al., 2011). Consequently, combining native state stabilization with enhanced folding and peroxisomal import of PH1 mutants would in principle lead to full correction of mitochondrial mistargeting in PH1 patients.

3. Small molecule correction of protein homeostasis defects in PH1: Pyridoxine responsiveness and modulators of protein homeostasis and mitochondrial import

So far, we have shown that thermodynamic and kinetic factors governing folding and stability of the native dimer of AGT seem to be critical to determine two of the main pathogenic mechanisms in PH1, aggregation and mistargeting. Population of partially unfolded states may promote mitochondrial mistargeting enhancing interaction with molecular chaperones of the Hsp70, Hsp60 and Hsp90 families, while the fraction of stably folded dimeric AGT is key for its import to peroxisomes mediated by Pex5p-dependent pathway. Consequently, a critical component in these scenarios is binding of the PLP cofactor, which boosts AGT stability in thermodynamic and kinetic terms. Thus, we are ready to describe in some detail a vitamin based therapy and current research in other strategies to modulate protein homeostasis in PH1.

Pyridoxine (vitamin B6 or PN) is the only medication that is effective in lowering urinary oxalate in primary hyperoxaluria, an effect that has been known for decades (McLaurin, Beisel, McCormick, Scalettar, & Herman, 1961). However, it is only efficacious in about a third of PH1 patients, after a dose escalation trial ranging between 5 and 20 $mg\,kg^{-1}\,day^{-1}$. The molecular basis for its effect is only partially understood (Fargue, Rumsby, & Danpure, 2013). PN is a precursor to the pyridoxal phosphate (PLP) cofactor for AGT. PN sensitivity is associated with specific mutations, particularly in patients homozygous for the LM-G170R mutation of AGXT who demonstrate reduction in urine oxalate to normal or near normal when receiving pharmacologic doses of PN.

In the last years, significant advances have been reported explaining the molecular basis of PN responsiveness and expanding, on theoretical grounds, the number of mutations that could potentially respond to this therapy. PN administration to cells in culture increases intracellular PLP concentration, and this cofactor plays a dual role as prosthetic molecule and chaperone (Cellini, Montioli, Oppici, Astegno, & Voltattorni, 2014), partially rescuing the misfolding of the apoenzyme. Thus, PN treatment favors the formation of holoproteins with higher stability and activity, and partially prevents mitochondrial mistargeting (Fargue, Rumsby, et al., 2013). While only PLP and pyridoxamine 5′-phosphate (PMP) are the active forms of the vitamin, there are several vitamer forms, differing in phosphorylation state and modification of the 4′ carbon. Vertebrates (including humans) cannot synthesize PLP and they depend on the salvage pathway to convert dietary B6 vitamers into the enzyme cofactor (di Salvo et al., 2011). Since the salvage pathway enzymes are cytosolic, specific transport systems exist for cellular uptake of B6 vitamers (Whittaker, 2016). Vitamin B6 is administered orally to patients as PN, which is absorbed and metabolized to PLP in the hepatocytes. In addition, two other non-phosphorylated vitamin B6 forms, pyridoxamine (PM) and pyridoxal (PL), are absorbed by the intestine and converted to PLP. PM and PL were shown to be more effective than PN in rescuing the AGT mutants G170R and F152I, inducing an intracellular accumulation of PLP and PMP, which exert their chaperone and a prosthetic role without any accumulation of pyridoxine phosphate (PNP) (Oppici et al., 2015). PNP is known to compete with PLP for apoenzyme binding, resulting in an inactive AGT-PNP complex. An interesting but yet unexplored aspect of the physiology of the AGT enzyme as well as of the responsiveness to vitamin B6 supplementation in PH1 patients is how the PLP cofactor is physically channelled to apo-AGT enzymes in the cytosol. Due to its high reactivity, free PLP concentrations are kept low (di Salvo et al., 2011, 2012), so to avoid indesired side reactions, PLP could be channeled from biosynthetic enzymes such as pyridoxal kinase (PLK) and pyridoxine (pyridoxamine) 5′-phosphate oxidase (PNPox) to apo-enzymes (di Salvo et al., 2011, 2012).

Using CHO cells expressing the mitochondrial mistargeting mutation (G170R), a compound previously identified in a screening for molecules capable of inhibiting mitochondrial import has been shown to blunt the mitochondrial mistargeting of mutant AGT, while enhancing the peroxisomal distribution, and contribute to reduced oxalate production (Miyata et al., 2014), underscoring the potential therapeutic benefit of modifying the balance between mitochondrial and peroxisomal targeting in favor of the latter.

4. Outlook and future directions

Among the thousands of human diseases involving defects in protein homeostasis, PH1 emerges as a particularly interesting case. AGT folding and assembly into stable dimers, the shuttling of PLP to form the holoenzyme and its transport across peroxisomal membranes likely involve the interaction of the protein with several protein homeostasis machineries. In addition, the presence of a polymorphic variant (the LM) predisposes toward different types of conformational defects, the most prominent being aggregation and mitochondrial, which are likely modulated by chaperones, mitochondrial and peroxisomal import machineries. Accordingly, several checkpoints in the folding, misfolding and mistargeting of AGT emerge as potential targets for pharmacological correction of PH1.

Most evidence collected from *in vitro* and *in cellulo* studies have supported that loading with PLP precursor may alleviate AGT misfolding and mistargeting in PH1 (Dindo, Oppici, Dell'Orco, Montone, & Cellini, 2018; Fargue, Rumsby, et al., 2013; Montioli et al., 2015; Oppici et al., 2015). From these studies, a model centered on the stabilizing effect of PLP binding on the native structure of AGT has emerged (Mesa-Torres et al., 2013; Pey et al., 2013). However, to explain the correction of disease phenotypes in PH1 at the molecular level, we will require knowledge on how PLP is delivered to the apo-enzymes, and potentially on how PH1 mutations may interfere with this process.

From the viewpoint of the proteostasis machineries, we must learn on the specific proteostasis components or circuits modulating phenotypic traits (e.g., aggregation vs. mistargeting) in PH1. First, those proteostasis elements responsible for proper folding of AGT into folded dimers as well as their reconstitution as holo-enzymes. Second, whether mutations interfere with proper folding and interaction with the protein homeostasis network, increasing the population of partially folded states associated with chaperones and leading to aggregation or presentation to mitochondrial import machineries. Several evidences support that modulation of such proteostasis elements could overcome for instance mitochondrial mistargeting (Mesa-Torres et al., 2015; Miyata et al., 2014). Third, whether components of human protein homeostasis networks can reactivate aggregated AGT species found for PH1 variants in cells (Dindo et al., 2018; Mesa-Torres et al., 2013, 2015; Mesa-Torres, Salido, et al., 2014; Nishiyama et al., 1991). Fourth, little is known on the proteostasis elements that lead to degradation of aberrantly folded AGT, and what is the contribution from cytosolic degradation (i.e., the ubiquitin-proteasome system; Coulter-Mackie & Lian, 2006) or peroxisomal (i.e., LonP2 protease; Walker, Pomatto, Tripathi, & Davies, 2018) degradation machineries, ultimately decreasing steady-state levels of many PH1 mutants (Dindo et al., 2018; Fargue, Lewin, et al., 2013; Mesa-Torres et al., 2013). Fifth, are there specific features, in terms of the magnitude of the energetic impact or particular structural perturbations due to PH1 mutations, that balance the partition between peroxisomal and mitochondrial targeting or cytosolic and peroxisomal degradation (Mesa-Torres, Salido, et al., 2014).

Acknowledgments

This work was supported by Grants from Junta de Andalucía (P11-CTS-07187 to A.L.P.), from MINECO/FEDER, UE (BFU2016-75983 to A.M. and SAF2015-69796 to E.S.), and the Basque Government (IT709-13 to A.M.).

References

Aguado, A., Fernández-Higuero, J. A., Moro, F., & Muga, A. (2015). Chaperone-assisted protein aggregate reactivation: Different solutions for the same problem. *Archives of Biochemistry and Biophysics*, *580*, 121–134. https://doi.org/10.1016/j.abb.2015.07.006.

Albert, A., Yunta, C., Arranz, R., Peña, A., Salido, E., Valpuesta, J. M., et al. (2010). Structure of GroEL in complex with an early folding intermediate of alanine glyoxylate aminotransferase. *The Journal of Biological Chemistry*, *285*(9), 6371–6376. https://doi.org/10.1074/jbc.M109.062471.

Ali, M. M. U., Roe, S. M., Vaughan, C. K., Meyer, P., Panaretou, B., Piper, P. W., et al. (2006). Crystal structure of an Hsp90-nucleotide-p23/Sba1 closed chaperone complex. *Nature*, *440*(7087), 1013–1017. https://doi.org/10.1038/nature04716.

Alvira, S., Cuéllar, J., Röhl, A., Yamamoto, S., Itoh, H., Alfonso, C., et al. (2014). Structural characterization of the substrate transfer mechanism in Hsp70/Hsp90 folding machinery mediated by hop. *Nature Communications*, *5*, 5484. https://doi.org/10.1038/ncomms6484.

Balch, W. E., Morimoto, R. I., Dillin, A., & Kelly, J. W. (2008). Adapting proteostasis for disease intervention. *Science (New York, N.Y.)*, *319*(5865), 916–919. https://doi.org/10.1126/science.1141448.

Bartlett, A. I., & Radford, S. E. (2009). An expanding arsenal of experimental methods yields an explosion of insights into protein folding mechanisms. *Nature Structural & Molecular Biology*, *16*(6), 582–588. https://doi.org/10.1038/nsmb.1592.

Bartoszewska, M., Williams, C., Kikhney, A., Opaliński, Ł., van Roermund, C. W. T., de Boer, R., et al. (2012). Peroxisomal proteostasis involves a lon family protein that functions as protease and chaperone. *The Journal of Biological Chemistry*, *287*(33), 27380–27395. https://doi.org/10.1074/jbc.M112.381566.

Bigotti, M. G., & Clarke, A. R. (2008). Chaperonins: The hunt for the group II mechanism. *Archives of Biochemistry and Biophysics*, *474*(2), 331–339. https://doi.org/10.1016/j.abb.2008.03.015.

Birdsey, G. M., Lewin, J., Holbrook, J. D., Simpson, V. R., Cunningham, A. A., & Danpure, C. J. (2005). A comparative analysis of the evolutionary relationship between diet and enzyme targeting in bats, marsupials and other mammals. *Proceedings. Biological Sciences*, *272*(1565), 833–840. https://doi.org/10.1098/rspb.2004.3011.

Bracher, A., & Verghese, J. (2015). The nucleotide exchange factors of Hsp70 molecular chaperones. *Frontiers in Molecular Biosciences*, *2*, 10. https://doi.org/10.3389/fmolb.2015.00010.

Braig, K., Otwinowski, Z., Hegde, R., Boisvert, D. C., Joachimiak, A., Horwich, A. L., et al. (1994). The crystal structure of the bacterial chaperonin GroEL at 2.8 A. *Nature*, *371*(6498), 578–586. https://doi.org/10.1038/371578a0.

Brehme, M., Voisine, C., Rolland, T., Wachi, S., Soper, J. H., Zhu, Y., et al. (2014). A chaperome subnetwork safeguards proteostasis in aging and neurodegenerative disease. *Cell Reports*, *9*(3), 1135–1150. https://doi.org/10.1016/j.celrep.2014.09.042.

Buchner, J. (1999). Hsp90 & Co.—A holding for folding. *Trends in Biochemical Sciences*, *24*(4), 136–141. Retrieved from http://www.ncbi.nlm.nih.gov/pubmed/10322418.

Bukau, B., & Horwich, A. L. (1998). The Hsp70 and Hsp60 chaperone machines. *Cell*, *92*(3), 351–366. Retrieved from http://www.ncbi.nlm.nih.gov/pubmed/9476895.

Cellini, B., Montioli, R., Oppici, E., Astegno, A., & Voltattorni, C. B. (2014). The chaperone role of the pyridoxal 5′-phosphate and its implications for rare diseases involving B6-dependent enzymes. *Clinical Biochemistry*, *47*(3), 158–165. https://doi.org/10.1016/j.clinbiochem.2013.11.021.

Cellini, B., Montioli, R., Paiardini, A., Lorenzetto, A., Maset, F., Bellini, T., et al. (2010). Molecular defects of the glycine 41 variants of alanine glyoxylate aminotransferase associated with primary hyperoxaluria type I. *Proceedings of the National Academy of Sciences of the United States of America*, *107*(7), 2896–2901. https://doi.org/10.1073/pnas.0908565107.

Chiosis, G., Dickey, C. A., & Johnson, J. L. (2013). A global view of Hsp90 functions. *Nature Structural & Molecular Biology*, *20*(1), 1–4. https://doi.org/10.1038/nsmb.2481.

Chiti, F., & Dobson, C. M. (2017). Protein misfolding, amyloid formation, and human disease: A summary of progress over the last decade. *Annual Review of Biochemistry*, *86*, 27–68. https://doi.org/10.1146/annurev-biochem-061516-045115.

Coulter-Mackie, M. B., & Lian, Q. (2006). Consequences of missense mutations for dimerization and turnover of alanine: Glyoxylate aminotransferase: Study of a spectrum of mutations. *Molecular Genetics and Metabolism*, *89*(4), 349–359. https://doi.org/10.1016/j.ymgme.2006.07.013.

Cox, D., Raeburn, C., Sui, X., & Hatters, D. M. (2018). Protein aggregation in cell biology: An aggregomics perspective of health and disease. *Seminars in Cell & Developmental Biology*, (in press). https://doi.org/10.1016/j.semcdb.2018.05.003.

Craig, E. A., & Marszalek, J. (2017). How do J-proteins get Hsp70 to do so many different things? *Trends in Biochemical Sciences*, *42*(5), 355–368. https://doi.org/10.1016/j.tibs.2017.02.007.

Crookes, W. J., & Olsen, L. J. (1998). The effects of chaperones and the influence of protein assembly on peroxisomal protein import. *The Journal of Biological Chemistry*, *273*(27), 17236–17242. Retrieved from http://www.ncbi.nlm.nih.gov/pubmed/9642294.

Cuéllar, J., Martín-Benito, J., Scheres, S. H. W., Sousa, R., Moro, F., López-Viñas, E., et al. (2008). The structure of CCT-Hsc70 NBD suggests a mechanism for Hsp70 delivery of substrates to the chaperonin. *Nature Structural & Molecular Biology*, *15*(8), 858–864. https://doi.org/10.1038/nsmb.1464.

Dekker, C., Willison, K. R., & Taylor, W. R. (2011). On the evolutionary origin of the chaperonins. *Proteins*, *79*(4), 1172–1192. https://doi.org/10.1002/prot.22952.

di Salvo, M. L., Contestabile, R., & Safo, M. K. (2011). Vitamin B(6) salvage enzymes: Mechanism, structure and regulation. *Biochimica et Biophysica Acta*, *1814*(11), 1597–1608. https://doi.org/10.1016/j.bbapap.2010.12.006.

di Salvo, M. L., Safo, M. K., & Contestabile, R. (2012). Biomedical aspects of pyridoxal 5′-phosphate availability. *Frontiers in Bioscience (Elite Edition)*, *4*, 897–913. Retrieved from http://www.ncbi.nlm.nih.gov/pubmed/22201923.

Dindo, M., Conter, C., & Cellini, B. (2017). Electrostatic interactions drive native-like aggregation of human alanine: Glyoxylate aminostransferase. *The FEBS Journal*, *284*(21), 3739–3764. https://doi.org/10.1111/febs.14269.

Dindo, M., Oppici, E., Dell'Orco, D., Montone, R., & Cellini, B. (2018). Correlation between the molecular effects of mutations at the dimer interface of alanine-glyoxylate aminotransferase leading to primary hyperoxaluria type I and the cellular response to vitamin B6. *Journal of Inherited Metabolic Disease*, *41*(2), 263–275. https://doi.org/10.1007/s10545-017-0105-8.

Dobson, C. M., Šali, A., & Karplus, M. (1998). Protein folding: A perspective from theory and experiment. *Angewandte Chemie (International Ed. in English)*, *37*(7), 868–893. https://doi.org/10.1002/(SICI)1521-3773(19980420)37:7< 868::AID-ANIE868 > 3.0.CO;2-H.

Dragovic, Z., Broadley, S. A., Shomura, Y., Bracher, A., & Hartl, F. U. (2006). Molecular chaperones of the Hsp110 family act as nucleotide exchange factors of Hsp70s. *The EMBO Journal*, *25*(11), 2519–2528. https://doi.org/10.1038/sj.emboj.7601138.

Du, H., Kim, S., Hur, Y.-S., Lee, M.-S., Lee, S.-H., & Cheon, C.-I. (2015). A cytosolic thioredoxin acts as a molecular chaperone for peroxisome matrix proteins as well as antioxidant in peroxisome. *Molecules and Cells*, *38*(2), 187–194. https://doi.org/10.14348/molcells.2015.2255.

Dudek, J., Rehling, P., & van der Laan, M. (2013). Mitochondrial protein import: Common principles and physiological networks. *Biochimica et Biophysica Acta*, *1833*(2), 274–285. https://doi.org/10.1016/j.bbamcr.2012.05.028.

Fargue, S., Lewin, J., Rumsby, G., & Danpure, C. J. (2013). Four of the most common mutations in primary hyperoxaluria type 1 unmask the cryptic mitochondrial targeting sequence of alanine: Glyoxylate aminotransferase encoded by the polymorphic minor allele. *The Journal of Biological Chemistry*, *288*(4), 2475–2484. https://doi.org/10.1074/jbc.M112.432617.

Fargue, S., Rumsby, G., & Danpure, C. J. (2013). Multiple mechanisms of action of pyridoxine in primary hyperoxaluria type 1. *Biochimica et Biophysica Acta*, *1832*(10), 1776–1783. https://doi.org/10.1016/j.bbadis.2013.04.010.

Farr, G. W., Furtak, K., Rowland, M. B., Ranson, N. A., Saibil, H. R., Kirchhausen, T., et al. (2000). Multivalent binding of nonnative substrate proteins by the chaperonin GroEL. *Cell*, *100*(5), 561–573. Retrieved from http://www.ncbi.nlm.nih.gov/pubmed/10721993.

Fodor, K., Wolf, J., Reglinski, K., Passon, D. M., Lou, Y., Schliebs, W., et al. (2015). Ligand-induced compaction of the PEX5 receptor-binding cavity impacts protein import efficiency into peroxisomes. *Traffic (Copenhagen, Denmark)*, *16*(1), 85–98. https://doi.org/10.1111/tra.12238.

Frishberg, Y., Rinat, C., Shalata, A., Khatib, I., Feinstein, S., Becker-Cohen, R., et al. (2005). Intra-familial clinical heterogeneity: Absence of genotype-phenotype correlation in primary hyperoxaluria type 1 in Israel. *American Journal of Nephrology*, *25*(3), 269–275. https://doi.org/10.1159/000086357.

Genest, O., Hoskins, J. R., Camberg, J. L., Doyle, S. M., & Wickner, S. (2011). Heat shock protein 90 from *Escherichia coli* collaborates with the DnaK chaperone system in client protein remodeling. *Proceedings of the National Academy of Sciences of the United States of America*, *108*(20), 8206–8211. https://doi.org/10.1073/pnas.1104703108.

Genevaux, P., Georgopoulos, C., & Kelley, W. L. (2007). The Hsp70 chaperone machines of *Escherichia coli*: A paradigm for the repartition of chaperone functions. *Molecular Microbiology*, *66*(4), 840–857. https://doi.org/10.1111/j.1365-2958.2007.05961.x.

Ghosh, D., & Berg, J. M. (2010). A proteome-wide perspective on peroxisome targeting signal 1(PTS1)-Pex5p affinities. *Journal of the American Chemical Society*, *132*(11), 3973–3979. https://doi.org/10.1021/ja9109049.

Gray, J., Rustgi, S., von Wettstein, D., Reinbothe, C., & Reinbothe, S. (2016). Common functions of the chloroplast and mitochondrial co-chaperones cpDnaJL (CDF1) and mtDnaJ (PAM16) in protein import and ROS scavenging in *Arabidopsis thaliana*. *Communicative & Integrative Biology*, *9*(5), e1119343. https://doi.org/10.1080/19420889.2015.1119343.

Gregersen, N., Bross, P., Vang, S., & Christensen, J. H. (2006). Protein misfolding and human disease. *Annual Review of Genomics and Human Genetics*, 7, 103–124. https://doi.org/10.1146/annurev.genom.7.080505.115737.

Hageman, J., Rujano, M. A., van Waarde, M. A. W. H., Kakkar, V., Dirks, R. P., Govorukhina, N., et al. (2010). A DNAJB chaperone subfamily with HDAC-dependent activities suppresses toxic protein aggregation. *Molecular Cell*, *37*(3), 355–369. https://doi.org/10.1016/j.molcel.2010.01.001.

Hartl, F. U., Bracher, A., & Hayer-Hartl, M. (2011). Molecular chaperones in protein folding and proteostasis. *Nature*, *475*(7356), 324–332. https://doi.org/10.1038/nature10317.

Hartl, F. U., & Hayer-Hartl, M. (2002). Molecular chaperones in the cytosol: From nascent chain to folded protein. *Science (New York, N.Y.)*, *295*(5561), 1852–1858. https://doi.org/10.1126/science.1068408.

Hartl, F. U., & Hayer-Hartl, M. (2009). Converging concepts of protein folding in vitro and in vivo. *Nature Structural & Molecular Biology*, *16*(6), 574–581. https://doi.org/10.1038/nsmb.1591.

Hayer-Hartl, M., Bracher, A., & Hartl, F. U. (2016). The GroEL-GroES chaperonin machine: A Nano-cage for protein folding. *Trends in Biochemical Sciences*, *41*(1), 62–76. https://doi.org/10.1016/j.tibs.2015.07.009.

Hellenkamp, B., Wortmann, P., Kandzia, F., Zacharias, M., & Hugel, T. (2017). Multidomain structure and correlated dynamics determined by self-consistent FRET networks. *Nature Methods*, *14*(2), 174–180. https://doi.org/10.1038/nmeth.4081.

Hettema, E. H., Ruigrok, C. C., Koerkamp, M. G., van den Berg, M., Tabak, H. F., Distel, B., et al. (1998). The cytosolic DnaJ-like protein djp1p is involved specifically in peroxisomal protein import. *The Journal of Cell Biology*, *142*(2), 421–434. Retrieved from http://www.ncbi.nlm.nih.gov/pubmed/9679141.

Hipp, M. S., Park, S.-H., & Hartl, F. U. (2014). Proteostasis impairment in protein-misfolding and -aggregation diseases. *Trends in Cell Biology*, *24*(9), 506–514. https://doi.org/10.1016/j.tcb.2014.05.003.

Höhfeld, J., & Jentsch, S. (1997). GrpE-like regulation of the hsc70 chaperone by the anti-apoptotic protein BAG-1. *The EMBO Journal*, *16*(20), 6209–6216. https://doi.org/10.1093/emboj/16.20.6209.

Hopp, K., Cogal, A. G., Bergstralh, E. J., Seide, B. M., Olson, J. B., Meek, A. M., et al. (2015). Phenotype-genotype correlations and estimated carrier frequencies of primary hyperoxaluria. *Journal of the American Society of Nephrology: JASN*, *26*(10), 2559–2570. https://doi.org/10.1681/ASN.2014070698.

Jahn, M., Rehn, A., Pelz, B., Hellenkamp, B., Richter, K., Rief, M., et al. (2014). The charged linker of the molecular chaperone Hsp90 modulates domain contacts and biological function. *Proceedings of the National Academy of Sciences of the United States of America*, *111*(50), 17881–17886. https://doi.org/10.1073/pnas.1414073111.

Jarvis, P., & Soll, J. (2001). Toc, tic, and chloroplast protein import. *Biochimica et Biophysica Acta*, *1541*(1–2), 64–79. Retrieved from http://www.ncbi.nlm.nih.gov/pubmed/11750663.

Johnson, J. L. (2012). Evolution and function of diverse Hsp90 homologs and cochaperone proteins. *Biochimica et Biophysica Acta*, *1823*(3), 607–613. https://doi.org/10.1016/j.bbamcr.2011.09.020.

Jorge-Finnigan, A., Aguado, C., Sánchez-Alcudia, R., Abia, D., Richard, E., Merinero, B., et al. (2010). Functional and structural analysis of five mutations identified in methylmalonic aciduria cblB type. *Human Mutation*, *31*(9), 1033–1042. https://doi.org/10.1002/humu.21307.

Kabani, M., McLellan, C., Raynes, D. A., Guerriero, V., & Brodsky, J. L. (2002). HspBP1, a homologue of the yeast Fes1 and Sls1 proteins, is an Hsc70 nucleotide exchange factor. *FEBS Letters*, *531*(2), 339–342. Retrieved from http://www.ncbi.nlm.nih.gov/pubmed/12417338.

Kakkar, V., Kuiper, E. F. E., Pandey, A., Braakman, I., & Kampinga, H. H. (2016). Versatile members of the DNAJ family show Hsp70 dependent anti-aggregation activity on RING1 mutant parkin C289G. *Scientific Reports*, *6*, 34830. https://doi.org/10.1038/srep34830.

Kampinga, H. H., Hageman, J., Vos, M. J., Kubota, H., Tanguay, R. M., Bruford, E. A., et al. (2009). Guidelines for the nomenclature of the human heat shock proteins. *Cell Stress & Chaperones*, *14*(1), 105–111. https://www.ncbi.nlm.nih.gov/pubmed/18663603.

Karagöz, G. E., Duarte, A. M. S., Akoury, E., Ippel, H., Biernat, J., Morán Luengo, T., et al. (2014). Hsp90-Tau complex reveals molecular basis for specificity in chaperone action. *Cell, 156*(5), 963–974. https://doi.org/10.1016/j.cell.2014.01.037.

Kerner, M. J., Naylor, D. J., Ishihama, Y., Maier, T., Chang, H.-C., Stines, A. P., et al. (2005). Proteome-wide analysis of chaperonin-dependent protein folding in *Escherichia coli*. *Cell, 122*(2), 209–220. https://doi.org/10.1016/j.cell.2005.05.028.

Kevei, É., Pokrzywa, W., & Hoppe, T. (2017). Repair or destruction-an intimate liaison between ubiquitin ligases and molecular chaperones in proteostasis. *FEBS Letters, 591*(17), 2616–2635. https://doi.org/10.1002/1873-3468.12750.

Kim, Y. E., Hipp, M. S., Bracher, A., Hayer-Hartl, M., & Hartl, F. U. (2013). Molecular chaperone functions in protein folding and proteostasis. *Annual Review of Biochemistry, 82*, 323–355. https://doi.org/10.1146/annurev-biochem-060208-092442.

Kityk, R., Kopp, J., Sinning, I., & Mayer, M. P. (2012). Structure and dynamics of the ATP-bound open conformation of Hsp70 chaperones. *Molecular Cell, 48*(6), 863–874. https://doi.org/10.1016/j.molcel.2012.09.023.

Labbadia, J., & Morimoto, R. I. (2015). The biology of proteostasis in aging and disease. *Annual Review of Biochemistry, 84*, 435–464. https://doi.org/10.1146/annurev-biochem-060614-033955.

Leitner, A., Joachimiak, L. A., Bracher, A., Mönkemeyer, L., Walzthoeni, T., Chen, B., et al. (2012). The molecular architecture of the eukaryotic chaperonin TRiC/CCT. *Structure (London, England: 1993), 20*(5), 814–825. https://doi.org/10.1016/j.str.2012.03.007.

Lin, Z., Madan, D., & Rye, H. S. (2008). GroEL stimulates protein folding through forced unfolding. *Nature Structural & Molecular Biology, 15*(3), 303–311. https://doi.org/10.1038/nsmb.1394.

Lorenz, O. R., Freiburger, L., Rutz, D. A., Krause, M., Zierer, B. K., Alvira, S., et al. (2014). Modulation of the Hsp90 chaperone cycle by a stringent client protein. *Molecular Cell, 53*(6), 941–953. https://doi.org/10.1016/j.molcel.2014.02.003.

Lorenzo, V., Alvarez, A., Torres, A., Torregrosa, V., Hernández, D., & Salido, E. (2006). Presentation and role of transplantation in adult patients with type 1 primary hyperoxaluria and the I244T AGXT mutation: Single-center experience. *Kidney International, 70*(6), 1115–1119. https://doi.org/10.1038/sj.ki.5001758.

Lumb, M. J., & Danpure, C. J. (2000). Functional synergism between the most common polymorphism in human alanine: Glyoxylate aminotransferase and four of the most common disease-causing mutations. *The Journal of Biological Chemistry, 275*(46), 36415–36422. https://doi.org/10.1074/jbc.M006693200.

Mandrile, G., van Woerden, C. S., Berchialla, P., Beck, B. B., Acquaviva Bourdain, C., Hulton, S.-A., et al. (2014). Data from a large European study indicate that the outcome of primary hyperoxaluria type 1 correlates with the AGXT mutation type. *Kidney International, 86*(6), 1197–1204. https://doi.org/10.1038/ki.2014.222.

Martin, W. (2010). Evolutionary origins of metabolic compartmentalization in eukaryotes. *Philosophical Transactions of the Royal Society of London. Series B, Biological Sciences, 365*(1541), 847–855. https://doi.org/10.1098/rstb.2009.0252.

Mayer, M. P., & Le Breton, L. (2015). Hsp90: Breaking the symmetry. *Molecular Cell, 58*(1), 8–20. https://doi.org/10.1016/j.molcel.2015.02.022.

McLaurin, A. W., Beisel, W. R., McCormick, G. J., Scalettar, R., & Herman, R. H. (1961). Primary hyperoxaluria. *Annals of Internal Medicine, 55*, 70–80. Retrieved from http://www.ncbi.nlm.nih.gov/pubmed/13774084.

Mesa-Torres, N., Calvo, A. C., Oppici, E., Titelbaum, N., Montioli, R., Miranda-Vizuete, A., et al. (2016). Caenorhabditis elegans AGXT-1 is a mitochondrial and temperature-adapted ortholog of peroxisomal human AGT1: New insights into between-species divergence in glyoxylate metabolism. *Biochimica et Biophysica Acta, 1864*(9), 1195–1205. https://doi.org/10.1016/j.bbapap.2016.05.004.

Mesa-Torres, N., Fabelo-Rosa, I., Riverol, D., Yunta, C., Albert, A., Salido, E., et al. (2013). The role of protein denaturation energetics and molecular chaperones in the aggregation and mistargeting of mutants causing primary hyperoxaluria type I. *PLoS One*, *8*(8), e71963. https://doi.org/10.1371/journal.pone.0071963.

Mesa-Torres, N., Salido, E., & Pey, A. L. (2014). The lower limits for protein stability and foldability in primary hyperoxaluria type I. *Biochimica et Biophysica Acta*, *1844*(12), 2355–2365. https://doi.org/10.1016/j.bbapap.2014.10.010.

Mesa-Torres, N., Tomic, N., Albert, A., Salido, E., & Pey, A. L. (2015). Molecular recognition of PTS-1 cargo proteins by Pex5p: Implications for protein mistargeting in primary hyperoxaluria. *Biomolecules*, *5*(1), 121–141. https://doi.org/10.3390/biom5010121.

Mesa-Torres, N., Yunta, C., Fabelo-Rosa, I., Gonzalez-Rubio, J. M., Sánchez-Ruiz, J. M., Salido, E., et al. (2014). The consensus-based approach for gene/enzyme replacement therapies and crystallization strategies: The case of human alanine-glyoxylate aminotransferase. *The Biochemical Journal*, *462*(3), 453–463. https://doi.org/10.1042/BJ20140250.

Meyer, P., Prodromou, C., Liao, C., Hu, B., Mark Roe, S., Vaughan, C. K., et al. (2004). Structural basis for recruitment of the ATPase activator Aha1 to the Hsp90 chaperone machinery. *The EMBO Journal*, *23*(3), 511–519. https://doi.org/10.1038/sj.emboj.7600060.

Miyata, N., Steffen, J., Johnson, M. E., Fargue, S., Danpure, C. J., & Koehler, C. M. (2014). Pharmacologic rescue of an enzyme-trafficking defect in primary hyperoxaluria 1. *Proceedings of the National Academy of Sciences of the United States of America*, *111*(40), 14406–14411. https://doi.org/10.1073/pnas.1408401111.

Mogk, A., & Bukau, B. (2017). Role of sHsps in organizing cytosolic protein aggregation and disaggregation. *Cell Stress & Chaperones*, *22*(4), 493–502. https://doi.org/10.1007/s12192-017-0762-4.

Monico, C. G., Rossetti, S., Olson, J. B., & Milliner, D. S. (2005). Pyridoxine effect in type I primary hyperoxaluria is associated with the most common mutant allele. *Kidney International*, *67*(5), 1704–1709. https://doi.org/10.1111/j.1523-1755.2005.00267.x.

Montioli, R., Oppici, E., Dindo, M., Roncador, A., Gotte, G., Cellini, B., et al. (2015). Misfolding caused by the pathogenic mutation G47R on the minor allele of alanine: Glyoxylate aminotransferase and chaperoning activity of pyridoxine. *Biochimica et Biophysica Acta*, *1854*(10 Pt. A), 1280–1289. https://doi.org/10.1016/j.bbapap.2015.07.002.

Morán Luengo, T., Kityk, R., Mayer, M. P., & Rüdiger, S. G. D. (2018). Hsp90 breaks the deadlock of the Hsp70 chaperone system. *Molecular Cell*, *70*(3), 545–552.e9. https://doi.org/10.1016/j.molcel.2018.03.028.

Muñoz, I. G., Yébenes, H., Zhou, M., Mesa, P., Serna, M., Park, A. Y., et al. (2011). Crystal structure of the open conformation of the mammalian chaperonin CCT in complex with tubulin. *Nature Structural & Molecular Biology*, *18*(1), 14–19. https://doi.org/10.1038/nsmb.1971.

Neupert, W., & Herrmann, J. M. (2007). Translocation of proteins into mitochondria. *Annual Review of Biochemistry*, *76*, 723–749. https://doi.org/10.1146/annurev.biochem.76.052705.163409.

Nillegoda, N. B., Kirstein, J., Szlachcic, A., Berynskyy, M., Stank, A., Stengel, F., et al. (2015). Crucial HSP70 co-chaperone complex unlocks metazoan protein disaggregation. *Nature*, *524*(7564), 247–251. https://doi.org/10.1038/nature14884.

Nishiyama, K., Funai, T., Katafuchi, R., Hattori, F., Onoyama, K., & Ichiyama, A. (1991). Primary hyperoxaluria type I due to a point mutation of T to C in the coding region of the serine: Pyruvate aminotransferase gene. *Biochemical and Biophysical Research Communications*, *176*(3), 1093–1099. Retrieved from http://www.ncbi.nlm.nih.gov/pubmed/2039493.

Oppici, E., Dindo, M., Conter, C., Borri Voltattorni, C., & Cellini, B. (2018). Folding defects leading to primary hyperoxaluria. *Handbook of Experimental Pharmacology, 245*, 313–343. https://doi.org/10.1007/164_2017_59.

Oppici, E., Fargue, S., Reid, E. S., Mills, P. B., Clayton, P. T., Danpure, C. J., et al. (2015). Pyridoxamine and pyridoxal are more effective than pyridoxine in rescuing folding-defective variants of human alanine: Glyoxylate aminotransferase causing primary hyperoxaluria type I. *Human Molecular Genetics, 24*(19), 5500–5511. https://doi.org/10.1093/hmg/ddv276.

Parenti, G., Pignata, C., Vajro, P., & Salerno, M. (2013). New strategies for the treatment of lysosomal storage diseases (Review). *International Journal of Molecular Medicine, 31*(1), 11–20. https://doi.org/10.3892/ijmm.2012.1187.

Pey, A. L. (2013). Protein homeostasis disorders of key enzymes of amino acids metabolism: Mutation-induced protein kinetic destabilization and new therapeutic strategies. *Amino Acids, 45*(6), 1331–1341. https://doi.org/10.1007/s00726-013-1609-7.

Pey, A. L., Albert, A., & Salido, E. (2013). Protein homeostasis defects of alanine-glyoxylate aminotransferase: New therapeutic strategies in primary hyperoxaluria type I. *BioMed Research International, 2013*, 687658. https://doi.org/10.1155/2013/687658.

Pey, A. L., Salido, E., & Sanchez-Ruiz, J. M. (2011). Role of low native state kinetic stability and interaction of partially unfolded states with molecular chaperones in the mitochondrial protein mistargeting associated with primary hyperoxaluria. *Amino Acids, 41*(5), 1233–1245. https://doi.org/10.1007/s00726-010-0801-2.

Pey, A. L., Stricher, F., Serrano, L., & Martinez, A. (2007). Predicted effects of missense mutations on native-state stability account for phenotypic outcome in phenylketonuria, a paradigm of misfolding diseases. *American Journal of Human Genetics, 81*(5), 1006–1024. https://doi.org/10.1086/521879.

Priya, S., Sharma, S. K., Sood, V., Mattoo, R. U. H., Finka, A., Azem, A., et al. (2013). GroEL and CCT are catalytic unfoldases mediating out-of-cage polypeptide refolding without ATP. *Proceedings of the National Academy of Sciences of the United States of America, 110*(18), 7199–7204. https://doi.org/10.1073/pnas.1219867110.

Prodromou, C. (2017). Regulatory mechanisms of Hsp90. *Biochemistry & Molecular Biology Journal, 3*(1), 2. https://doi.org/10.21767/2471-8084.100030.

Prodromou, C., Roe, S. M., O'Brien, R., Ladbury, J. E., Piper, P. W., & Pearl, L. H. (1997). Identification and structural characterization of the ATP/ADP-binding site in the Hsp90 molecular chaperone. *Cell, 90*(1), 65–75. Retrieved from http://www.ncbi.nlm.nih.gov/pubmed/9230303.

Purdue, P. E., Allsop, J., Isaya, G., Rosenberg, L. E., & Danpure, C. J. (1991). Mistargeting of peroxisomal L-alanine: Glyoxylate aminotransferase to mitochondria in primary hyperoxaluria patients depends upon activation of a cryptic mitochondrial targeting sequence by a point mutation. *Proceedings of the National Academy of Sciences of the United States of America, 88*(23), 10900–10904. Retrieved from http://www.ncbi.nlm.nih.gov/pubmed/1961759.

Purdue, P. E., Takada, Y., & Danpure, C. J. (1990). Identification of mutations associated with peroxisome-to-mitochondrion mistargeting of alanine/glyoxylate aminotransferase in primary hyperoxaluria type 1. *The Journal of Cell Biology, 111*(6 Pt. 1), 2341–2351. Retrieved from http://www.ncbi.nlm.nih.gov/pubmed/1703535.

Radli, M., & Rüdiger, S. G. D. (2018). Dancing with the diva: Hsp90-client interactions. *Journal of Molecular Biology, 430*(18 Pt. B), 3029–3040. https://doi.org/10.1016/j.jmb.2018.05.026.

Rampelt, H., Kirstein-Miles, J., Nillegoda, N. B., Chi, K., Scholz, S. R., Morimoto, R. I., et al. (2012). Metazoan Hsp70 machines use Hsp110 to power protein disaggregation. *The EMBO Journal, 31*(21), 4221–4235. https://doi.org/10.1038/emboj.2012.264.

Ranson, N. A., Farr, G. W., Roseman, A. M., Gowen, B., Fenton, W. A., Horwich, A. L., et al. (2001). ATP-bound states of GroEL captured by cryo-electron microscopy. *Cell, 107*(7), 869–879. https://doi.org/11779463.

Rauch, J. N., & Gestwicki, J. E. (2014). Binding of human nucleotide exchange factors to heat shock protein 70 (Hsp70) generates functionally distinct complexes in vitro. *The Journal of Biological Chemistry, 289*(3), 1402–1414. https://doi.org/10.1074/jbc.M113.521997.

Raviol, H., Sadlish, H., Rodriguez, F., Mayer, M. P., & Bukau, B. (2006). Chaperone network in the yeast cytosol: Hsp110 is revealed as an Hsp70 nucleotide exchange factor. *The EMBO Journal, 25*(11), 2510–2518. https://doi.org/10.1038/sj.emboj.7601139.

Reissmann, S., Joachimiak, L. A., Chen, B., Meyer, A. S., Nguyen, A., & Frydman, J. (2012). A gradient of ATP affinities generates an asymmetric power stroke driving the chaperonin TRIC/CCT folding cycle. *Cell Reports, 2*(4), 866–877. https://doi.org/10.1016/j.celrep.2012.08.036.

Rizzolo, K., Huen, J., Kumar, A., Phanse, S., Vlasblom, J., Kakihara, Y., et al. (2017). Features of the chaperone cellular network revealed through systematic interaction mapping. *Cell Reports, 20*(11), 2735–2748. https://doi.org/10.1016/j.celrep.2017.08.074.

Roe, S. M., Ali, M. M. U., Meyer, P., Vaughan, C. K., Panaretou, B., Piper, P. W., et al. (2004). The mechanism of Hsp90 regulation by the protein kinase-specific cochaperone p50(cdc37). *Cell, 116*(1), 87–98. Retrieved from http://www.ncbi.nlm.nih.gov/pubmed/14718169.

Röhl, A., Rohrberg, J., & Buchner, J. (2013). The chaperone Hsp90: Changing partners for demanding clients. *Trends in Biochemical Sciences, 38*(5), 253–262. https://doi.org/10.1016/j.tibs.2013.02.003.

Rüdiger, S., Buchberger, A., & Bukau, B. (1997). Interaction of Hsp70 chaperones with substrates. *Nature Structural Biology, 4*(5), 342–349. Retrieved from http://www.ncbi.nlm.nih.gov/pubmed/9145101.

Salido, E., Pey, A. L., Rodriguez, R., & Lorenzo, V. (2012). Primary hyperoxalurias: Disorders of glyoxylate detoxification. *Biochimica et Biophysica Acta, 1822*(9), 1453–1464. https://doi.org/10.1016/j.bbadis.2012.03.004.

Santana, A., Salido, E., Torres, A., & Shapiro, L. J. (2003). Primary hyperoxaluria type 1 in the Canary Islands: A conformational disease due to I244T mutation in the P11L-containing alanine: Glyoxylate aminotransferase. *Proceedings of the National Academy of Sciences of the United States of America, 100*(12), 7277–7282. https://doi.org/10.1073/pnas.1131968100.

Scannevin, R. H. (2018). Therapeutic strategies for targeting neurodegenerative protein misfolding disorders. *Current Opinion in Chemical Biology, 44*, 66–74. https://doi.org/10.1016/j.cbpa.2018.05.018.

Schmid, D., Baici, A., Gehring, H., & Christen, P. (1994). Kinetics of molecular chaperone action. *Science (New York, NY), 263*(5149), 971–973. Retrieved from http://www.ncbi.nlm.nih.gov/pubmed/8310296.

Schopf, F. H., Biebl, M. M., & Buchner, J. (2017). The HSP90 chaperone machinery. *Nature Reviews. Molecular Cell Biology, 18*(6), 345–360. https://doi.org/10.1038/nrm.2017.20.

Schymkowitz, J., Borg, J., Stricher, F., Nys, R., Rousseau, F., & Serrano, L. (2005). The FoldX web server: An online force field. *Nucleic Acids Research, 33*, W382–W388. https://doi.org/10.1093/nar/gki387.

Shiau, A. K., Harris, S. F., Southworth, D. R., & Agard, D. A. (2006). Structural analysis of *E. coli* hsp90 reveals dramatic nucleotide-dependent conformational rearrangements. *Cell, 127*(2), 329–340. https://doi.org/10.1016/j.cell.2006.09.027.

Shorter, J. (2011). The mammalian disaggregase machinery: Hsp110 synergizes with Hsp70 and Hsp40 to catalyze protein disaggregation and reactivation in a cell-free system. *PLoS One, 6*(10), e26319. https://doi.org/10.1371/journal.pone.0026319.

Skjærven, L., Cuellar, J., Martinez, A., & Valpuesta, J. M. (2015). Dynamics, flexibility, and allostery in molecular chaperonins. *FEBS Letters*, *589*(19 Pt. A), 2522–2532. https://doi.org/10.1016/j.febslet.2015.06.019.

Skjaerven, L., Grant, B., Muga, A., Teigen, K., McCammon, J. A., Reuter, N., et al. (2011). Conformational sampling and nucleotide-dependent transitions of the GroEL subunit probed by unbiased molecular dynamics simulations. *PLoS Computational Biology*, 7(3), e1002004. https://doi.org/10.1371/journal.pcbi.1002004.

Sousa, R., & Lafer, E. M. (2015). The role of molecular chaperones in clathrin mediated vesicular trafficking. *Frontiers in Molecular Biosciences*, *2*, 26. https://doi.org/10.3389/fmolb.2015.00026.

Steel, G. J., Fullerton, D. M., Tyson, J. R., & Stirling, C. J. (2004). Coordinated activation of Hsp70 chaperones. *Science (New York, N.Y.)*, *303*(5654), 98–101. https://doi.org/10.1126/science.1092287.

Taipale, M., Krykbaeva, I., Koeva, M., Kayatekin, C., Westover, K. D., Karras, G. I., et al. (2012). Quantitative analysis of HSP90-client interactions reveals principles of substrate recognition. *Cell*, *150*(5), 987–1001. https://doi.org/10.1016/j.cell.2012.06.047.

Tao, Y.-X., & Conn, P. M. (2018). Pharmacoperones as novel therapeutics for diverse protein conformational diseases. *Physiological Reviews*, *98*(2), 697–725. https://doi.org/10.1152/physrev.00029.2016.

Tsutsumi, S., Mollapour, M., Prodromou, C., Lee, C.-T., Panaretou, B., Yoshida, S., et al. (2012). Charged linker sequence modulates eukaryotic heat shock protein 90 (Hsp90) chaperone activity. *Proceedings of the National Academy of Sciences of the United States of America*, *109*(8), 2937–2942. https://doi.org/10.1073/pnas.1114414109.

Tzankov, S., Wong, M. J. H., Shi, K., Nassif, C., & Young, J. C. (2008). Functional divergence between co-chaperones of Hsc70. *The Journal of Biological Chemistry*, *283*(40), 27100–27109. https://doi.org/10.1074/jbc.M803923200.

Vabulas, R. M., Raychaudhuri, S., Hayer-Hartl, M., & Hartl, F. U. (2010). Protein folding in the cytoplasm and the heat shock response. *Cold Spring Harbor Perspectives in Biology*, *2*(12), a004390. https://doi.org/10.1101/cshperspect.a004390.

van Woerden, C. S., Groothoff, J. W., Wijburg, F. A., Annink, C., Wanders, R. J. A., & Waterham, H. R. (2004). Clinical implications of mutation analysis in primary hyperoxaluria type 1. *Kidney International*, *66*(2), 746–752. https://doi.org/10.1111/j.1523-1755.2004.00796.x.

Verba, K. A., Wang, R. Y.-R., Arakawa, A., Liu, Y., Shirouzu, M., Yokoyama, S., et al. (2016). Atomic structure of Hsp90-Cdc37-Cdk4 reveals that Hsp90 traps and stabilizes an unfolded kinase. *Science (New York, N.Y.)*, *352*(6293), 1542–1547. https://doi.org/10.1126/science.aaf5023.

Vos, M. J., Hageman, J., Carra, S., & Kampinga, H. H. (2008). Structural and functional diversities between members of the human HSPB, HSPH, HSPA, and DNAJ chaperone families. *Biochemistry*, *47*(27), 7001–7011. https://doi.org/10.1021/bi800639z.

Walker, C. L., Pomatto, L. C. D., Tripathi, D. N., & Davies, K. J. A. (2018). Redox regulation of homeostasis and proteostasis in peroxisomes. *Physiological Reviews*, *98*(1), 89–115. https://doi.org/10.1152/physrev.00033.2016.

Walther, D. M., Kasturi, P., Zheng, M., Pinkert, S., Vecchi, G., Ciryam, P., et al. (2015). Widespread proteome remodeling and aggregation in aging C. elegans. *Cell*, *161*(4), 919–932. https://doi.org/10.1016/j.cell.2015.03.032.

Walton, P. A., Wendland, M., Subramani, S., Rachubinski, R. A., & Welch, W. J. (1994). Involvement of 70-kD heat-shock proteins in peroxisomal import. *The Journal of Cell Biology*, *125*(5), 1037–1046. Retrieved from http://www.ncbi.nlm.nih.gov/pubmed/8195287.

Wayne, N., & Bolon, D. N. (2007). Dimerization of Hsp90 is required for in vivo function. Design and analysis of monomers and dimers. *The Journal of Biological Chemistry*, *282*(48), 35386–35395. https://doi.org/10.1074/jbc.M703844200.

Whittaker, J. W. (2016). Intracellular trafficking of the pyridoxal cofactor. Implications for health and metabolic disease. *Archives of Biochemistry and Biophysics*, *592*, 20–26. https://doi.org/10.1016/j.abb.2015.11.031.

Wiedemann, N., & Pfanner, N. (2017). Mitochondrial machineries for protein import and assembly. *Annual Review of Biochemistry*, *86*, 685–714. https://doi.org/10.1146/annurev-biochem-060815-014352.

Xu, Z., Horwich, A. L., & Sigler, P. B. (1997). The crystal structure of the asymmetric GroEL-GroES-(ADP)7 chaperonin complex. *Nature*, *388*(6644), 741–750. https://doi.org/10.1038/41944.

Xu, X., Sarbeng, E. B., Vorvis, C., Kumar, D. P., Zhou, L., & Liu, Q. (2012). Unique peptide substrate binding properties of 110-kDa heat-shock protein (Hsp110) determine its distinct chaperone activity. *The Journal of Biological Chemistry*, *287*(8), 5661–5672. https://doi.org/10.1074/jbc.M111.275057.

Yam, A. Y., Xia, Y., Lin, H.-T. J., Burlingame, A., Gerstein, M., & Frydman, J. (2008). Defining the TRiC/CCT interactome links chaperonin function to stabilization of newly made proteins with complex topologies. *Nature Structural & Molecular Biology*, *15*(12), 1255–1262. https://doi.org/10.1038/nsmb.1515.

CHAPTER FIVE

Inflammatory response and its relation to sphingolipid metabolism proteins: Chaperones as potential indirect anti-inflammatory agents

Z. Begum Yagci[a], Elif Esvap[a], Hatice Asuman Ozkara[b], Kutlu O. Ulgen[a], Elif Ozkirimli Olmez[a,*]
[a]Department of Chemical Engineering, Boğaziçi University, Istanbul, Turkey
[b]Department of Medical Biochemistry, Hacettepe University, Faculty of Medicine, Ankara, Turkey
*Corresponding author: e-mail address: elif.ozkirimli@boun.edu.tr

Contents

Advances in Protein Chemistry and Structural Biology, Volume 114
ISSN 1876-1623
https://doi.org/10.1016/bs.apcsb.2018.09.004

Abstract

Lysosome is the organelle responsible for breaking down macromolecules to maintain homeostasis and to fight infection. The disruption of normal lysosomal function due to mutations in the sphingolipid metabolism proteins leads to a class of lysosomal storage diseases (LSDs). Defective autophagy and activation of inflammation are observed in most LSDs. The crosstalk between these key metabolic pathways suggests that therapeutic approaches used in the treatment of LSDs may provide anti-inflammatory therapies against chronic inflammatory diseases such as multiple sclerosis, rheumatoid arthritis, inflammatory bowel disease. Here, we review the role of sphingolipids in the inflammatory response and build a protein-protein interaction network for proteins related with sphingolipid metabolism and inflammation to identify key interaction partners for the crosstalk between sphingolipids and inflammation. In addition, we present an overview of LSDs in relation with sphingolipids and inflammation, and review the pharmacological chaperones identified for these diseases.

Abbreviations

AD Alzheimer's disease
C1P ceramide-1-phosphate
CERK ceramide kinase
CNS central nervous system
COX-2 cyclooxygenase-2
ER endoplasmic reticulum
ERT enzyme replacement therapy
HSCT hematopoietic stem cell transplantation
IBD inflammatory bowel disease
IFG isofagomine
IFN-γ interferon-γ
IL interleukin
LSD lysosomal storage disorder
MLD metachromatic leukodystrophy
MS multiple sclerosis
NF-κB nuclear factor-κB
NPC Niemann-Pick type C disease
NSAID nonsteroidal anti-inflammatory drug
PC pharmacological chaperone
PCT pharmacological chaperone therapy
PD Parkinson's disease
RA rheumatoid arthritis
S1P sphingosine-1-phosphate
SMase sphingomyelinase
SphK1 sphingosine kinase 1

SRT substrate reduction therapy
TLR Toll-like receptors
TNF tumor necrosis factor

1. Introduction

Inflammation is an essential immune response that is normally protective and maintains tissue homeostasis against infection, injury and tissue damage but it can be pathological when dysregulated. Inflammatory pathway consists of four universal components: inducers (infection, toxins and tissue damage), sensors that detect these toxic inducers (macrophages, epithelial cells, microglia and dendritic cells), inflammatory mediators that are produced by sensors (cytokines and chemokines) and target tissues or effectors that are acted on by mediators to induce changes in these tissues and re-establish homeostasis (Medzhitov, 2010; Okin & Medzhitov, 2012). There are two main modalities for inflammatory states; acute and chronic inflammatory responses, which differ in duration and mechanism. The acute inflammatory response is ceased when the system adapts itself to toxic conditions and re-establishes homeostasis. However, if the inflammatory trigger persists and is not eliminated with the acute inflammatory response, chronic inflammatory state can be induced (Calder, 2006; Medzhitov, 2010). High concentrations of the proinflammatory cytokines tumor necrosis factor (TNF)-α, interleukin (IL)-1β, and IL-6 have been implicated not only in the pathogenesis of the chronic inflammatory diseases such as rheumatoid arthritis (RA), asthma and inflammatory bowel disease (IBD) (Calder, 2006; Medzhitov, 2010) but also in type 2 diabetes (Huang et al., 2011; Mirza et al., 2012), obesity (Goyal, Faizy, Siddiqui, & Singhai, 2012; Popko et al., 2010), and neurodegenerative diseases (Chitnis & Weiner, 2017; Hong, Kim, & Im, 2016). Furthermore, patients with chronic inflammatory diseases are at increased risk of developing cancer; for example, IBD and RA can lead to colon cancer and lymphoma (Nakamura & Smyth, 2017). Targeting the inflammatory pathway in chronic inflammatory diseases (Tabas & Glass, 2013) as well as these complex diseases such as cancer (Nakamura & Smyth, 2017), diabetes (Agrawal & Kant, 2014) and neurodegenerative diseases (Chitnis & Weiner, 2017) is gaining interest.

Sphingolipids are the building blocks of eukaryotic cell membranes and they are important for cell signaling and several cellular functions including cell growth, cell cycle, senescence, autophagy, vascularization and apoptosis (Chiurchiù, Leuti, & Maccarrone, 2018; Espaillat, Kew, & Obeid, 2017; Hannun & Obeid, 2018; Maceyka & Spiegel, 2014). Sphingolipid metabolism is a complex network of enzymes that take part in sphingolipid biosynthesis and catabolism. Ceramide is the hub of the sphingolipid metabolism (Hannun & Obeid, 2008).

Sphingolipids also take part in immune and inflammatory processes as well as apoptosis and senescence (Hannun & Obeid, 2008) and they play crucial roles in the pathogenesis of several immune-dependent and vascular-related chronic inflammatory diseases including asthma, IBD, diabetes and obesity, multiple sclerosis (MS), RA, and cancer (Chiurchiù et al., 2018; Maceyka & Spiegel, 2014). Lysosomal dysfunction also leads to impaired autophagy (Aflaki et al., 2016; Convertino, Das, & Dokholyan, 2016; Rigante, Cipolla, Basile, Gulli, & Savastano, 2017). Excessive ceramide signaling was observed in case of adipose tissue inflammation and insulin resistance, which can result in diabetes and obesity (Chiurchiù et al., 2018; Gomez-Muñoz et al., 2016). Sphingosine-1-phosphate (S1P) plays important roles in cell survival and inflammation, and ceramide-1-phosphate (C1P) takes part in inflammation (Hannun & Obeid, 2008). The crosstalk between the sphingolipid metabolism proteins and inflammatory agents such as the cytokine TNF-α or the protein cyclooxygenase (COX)-2 (Nakamura & Murayama, 2014) suggests that targeting sphingolipid metabolic pathway might be an alternative therapy for the diseases with inflammatory component.

The defects in sphingolipid metabolism result in sphingolipidoses, which are a class of lysosomal storage disorders (LSDs), such as Gaucher disease, Fabry disease or Niemann-Pick disease (Hannun & Obeid, 2018; Platt, 2018). In the pathogenesis of LSDs, mutations in sphingolipid metabolism enzymes cause errors in protein folding (Shin & Lim, 2017) and trafficking from the endoplasmic reticulum (Ringe & Petsko, 2009). As a result, enzymatic activity is lower or absent, metabolic intermediates accumulate, hence the name is "storage disorder." LSDs are rare diseases (e.g., about 2:100,000 births for Gaucher (Nalysnyk, Rotella, Simeone, Hamed, & Weinreb, 2017) and for Fabry (van der Tol et al., 2014)) but they are more prevalent in some populations in which consanguineous marriages are common (Afroze & Brown, 2017; Ozkara & Topcu, 2004; Sheth et al., 2014). Even though sphingolipidoses are rare diseases, mutations in sphingolipid enzymes may be genetic risk factors for other complex but common diseases. For example, mutations in glucocerebrosidase (GBA) are risk factors for Parkinson's disease (PD) (Aflaki, Westbroek, & Sidransky, 2017; Goker-Alpan et al., 2004) and mutations in Niemann-Pick type C (NPC) are risk factors for Alzheimer's disease (AD) (Kresojević, Dobričić, Svetel, & Kostić, 2014). Furthermore, most sphingolipidoses cause neurodegeneration, exacerbated

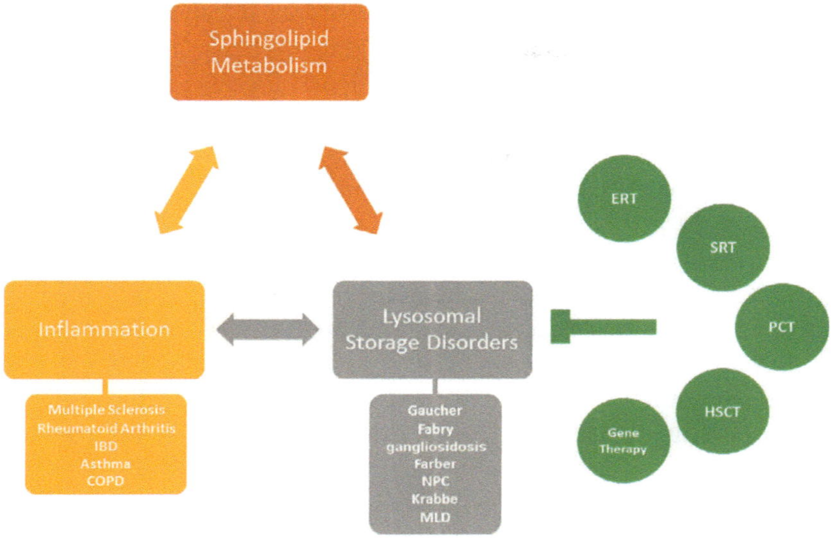

Fig. 1 Sphingolipid metabolism, inflammation and LSD interrelate with each other.

by inflammation (Bosch & Kielian, 2015; Mc Donald & Krainc, 2017; Plotegher & Duchen, 2017). The relation between sphingolipids and inflammation, and the role of these concepts in inflammatory diseases and sphingolipidoses suggest that inflammation, sphingolipids and LSDs are all interrelated with each other (Fig. 1).

The treatment of LSDs depends on decreasing the amount of metabolite that accumulates due to the absence or reduced activity of an enzyme. Most LSDs are currently treated by enzyme replacement therapy (ERT) and substrate reduction therapy (SRT); however, these approaches are limited by the requirement for the frequent intravenous administration of the enzyme due to its short half-life, poor tissue distribution and the unwanted immunologic responses in ERT and the side effects in SRT (Khanna et al., 2010; Porto et al., 2009; Syed Haneef & George Priya Doss, 2016; Xu et al., 2015). Pharmacological chaperones (PCs), also known as pharmacoperones, are small-molecule chaperones that are designed specifically to bind to and stabilize target proteins against proteolytic degradation, thermal denaturation or aggregation and they can be used to rescue the misfolded proteins in LSDs (Ringe & Petsko, 2009; Shin & Lim, 2017; Tao & Conn, 2018). Because of limitations in ERT and SRT, pharmacological chaperone therapy (PCT) is gaining interest as an alternative therapy in sphingolipid related disorders (Shin & Lim, 2017; Syed Haneef & George Priya Doss, 2016).

PCT can be relatively inexpensive, orally available and effective in all body tissue types (Ringe & Petsko, 2009). PCT is being tested on many sphingolipid related disorders such as Gaucher disease, Fabry disease and GM1 gangliosidosis (Convertino et al., 2016). In the last 5 years, an annual average of 1087 publications used the phrase "pharmacological chaperone" according to PubMed (PubMed; https://www.ncbi.nlm.nih.gov/pubmed/) (see Fig. 2).

Recent research shows that therapies developed against LSDs may find use in the treatment of neurodegenerative diseases, infectious diseases, and cancer because of the central role that sphingolipids play in multiple signaling pathways (Sidransky & Lopez, 2012). In this review, we first introduce inflammation and the sphingolipid metabolism. Then, we describe the three key bioactive sphingolipids: ceramide, S1P and C1P that have been reported to be involved in inflammation. We build a protein-protein interaction network for proteins involved in sphingolipid metabolism and inflammation and examine this network to identify the key interactions. The role of inflammation in relation with sphingolipids is then discussed for inflammatory diseases and sphingolipidoses. Finally, PCs identified for the sphingolipidoses; Gaucher, Fabry, GM1 and GM2 gangliosidosis, Farber, NPC, Krabbe diseases and MLD are reviewed. The focus of this chapter is to investigate the interrelationship between sphingolipids, inflammation and LSDs and highlight the potential of PCs for the treatment of the diseases with inflammatory components.

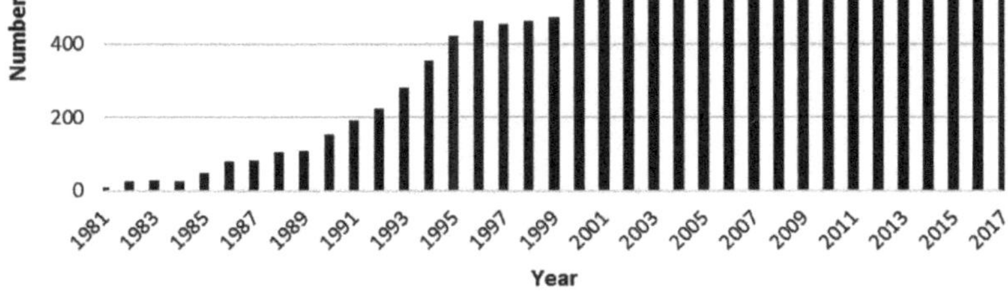

Fig. 2 Number of publications in the past years that use the phrase "pharmacological chaperone." *Data retrieved from PubMed.*

2. Inflammation

A physiological response to any injury or harmful stimuli, *inflammation* involves metabolic and signaling pathways that aim to heal and repair damaged tissue or to fight harmful stimuli such as microbial pathogens and toxic cellular components (Gomez-Muñoz et al., 2016; Maceyka & Spiegel, 2014; Medzhitov, 2010; Turner, Nedjai, Hurst, & Pennington, 2014). The nature of the inflammatory trigger affects the type of the inflammatory pathway stimulated (Medzhitov, 2008, 2010).

The inflammatory response is initiated by the inflammatory inducers: infection or tissue damage. These inducers are detected by essential signal transduction proteins and pattern recognition receptors such as Toll-like receptors (TLRs), which are expressed by mast cells, dendritic cells, macrophages in the bloodstream, and microglia in the brain (Bosch & Kielian, 2015; Lyman, Lloyd, Ji, Vizcaychipi, & Ma, 2014; Medzhitov, 2010; Tabas & Glass, 2013). These receptors recognize pathogen- and damage-associated molecular patterns (PAMPs and DAMPs) and induce signal transduction pathways that trigger the activation of the transcription factors such as nuclear factor κB (NF-κB), which regulates the expression of more than 150 different genes that encode cytokines and chemokines such as IL-1β, IL-6, or IL-8, monocyte chemotactic protein (MCP)-1, and also proinflammatory enzymes including COX-2 (Gomez-Muñoz et al., 2016; Nixon, 2009; Tabas & Glass, 2013). When these genes are expressed, the inflammatory mediators including chemokines, cytokines and bioactive lipids (prostaglandins) are produced. These mediators affect the target tissues by attracting the immune cells to the site of infection and eliciting a defense response to induce appropriate changes in the states of these tissues to restore homeostasis (Medzhitov, 2010; Tabas & Glass, 2013; Turner et al., 2014).

There are two types of immune systems: innate and adaptive (Iwasaki & Medzhitov, 2010). To protect the host and maintain homeostasis, both innate and adaptive immune cells (including neutrophils, macrophages and lymphocytes) mobilize during inflammation and travel to the region of infection or injury, and cytokine networks are activated (Gomez-Muñoz et al., 2016; Ketelhuth & Hansson, 2016; Maceyka & Spiegel, 2014; Medzhitov, 2010). The innate immune cells such as dendritic cells, macrophages (mature form of monocytes) and neutrophils detect infection through invariant receptors (Barton, 2008; Iwasaki & Medzhitov, 2010) and they initiate the inflammatory cascade, affect vascular permeability and activate adaptive immune response (Barton, 2008; Janeway, Travers, Walport, & Shlomchik, 2001). The adaptive immune system is highly specific and long-lasting in contrast to the innate immune system and includes two major types of lymphocytes (white blood cells): B- and T-cells, which are microbial-specific and have a memory to recognize different foreign antigens through their variant receptors (Barton, 2008; Iwasaki & Medzhitov, 2010; Janeway et al., 2001; Ketelhuth & Hansson, 2016). B-cells, which originate and mature in the bone marrow, produce antibodies when activated and differentiated into plasma cells, while T-cells, which originate in the bone marrow but mature in the thymus, kill the infected and tumor cells, or activate other cells such as macrophages and B-cells (Janeway et al., 2001). These cells are activated by antigens and once activated, they regulate immune responses and protect the host by producing antibodies and destroying the infected and tumor cells (Janeway et al., 2001; LaRosa & Orange, 2008).

The small cell signaling proteins, cytokines and chemokines produced by immune cells play crucial roles in both acute and chronic inflammation and regulate the communication between innate and adaptive immune systems (Getz, 2005; Turner et al., 2014; Zhang & An, 2007). Cytokines elicit a defense response during inflammation, and also take part in cell growth, survival and differentiation (Medzhitov, 2010; Ramesh, MacLean, & Philipp, 2013; Turner et al., 2014). Chemokines, which have chemotactic activities, recruit the immune cells to the target tissues (Medzhitov, 2010; Ramesh et al., 2013; Turner et al., 2014; Zhang & An, 2007). The chemokine MCP-1 is required for migration and infiltration of monocytes/macrophages and has both anti-inflammatory and proinflammatory effects (Alayoubi et al., 2013; Deshmane, Kremlev, Amini, & Sawaya, 2009).

Cytokines can be proinflammatory and anti-inflammatory (Lyman et al., 2014; Ng et al., 2003; Zhang & An, 2007). Proinflammatory cytokines initiate the inflammatory response against pathogens or tissue damage, upregulate inflammatory reactions, and have a role in pathological inflammation (Ng et al., 2003; Zhang & An, 2007). The key proinflammatory cytokines are IL-1, IL-6, IL-8 and TNF-α (Turner et al., 2014). Several sphingolipid metabolites are mediators of the TNF-α inflammatory response (discussed in Section 3). Anti-inflammatory cytokines such as IL-4 and IL-10 are immunoregulatory molecules that suppress the activity of proinflammatory cytokines and downregulate the inflammatory response when required, to maintain homeostasis (Akdis et al., 2011; Ng et al., 2003; Sultani, Stringer, Bowen, & Gibson, 2012; Zhang & An, 2007). The immune function can be suppressed if the anti-inflammatory mediators are overproduced (Akdis et al., 2011; Ng et al., 2003).

Eicosanoids such as leukotrienes, epoxides, and prostaglandins (PGD_2, PGE_2 and $PGF_{2\alpha}$) (Lone & Taskén, 2013) are inflammatory mediators produced by immune cells, and due to their roles in mediating inflammatory responses and physiological properties such as vascular permeability and platelet aggregation, they are associated with many inflammatory disorders (Hannun & Obeid, 2008; Lone & Taskén, 2013; Maceyka & Spiegel, 2014). Eicosanoids are lipid signaling molecules that are generated mainly through the metabolism of arachidonic acid by COX, lipoxygenase and P450 epoxygenase (Wang & DuBois, 2010). Arachidonic acid contributes to the fluidity and flexibility of the cell membrane, and it regulates the function of many ion channels and receptors (Leslie, 2015; Lone & Taskén, 2013; Tallima & El Ridi, 2018).

Inflammation elicits a protective adaptive response that can be beneficial; however, if not controlled, it can lead to chronic inflammation and result in various pathophysiological states such as edema, asthma, IBD, cancer, and autoimmune disorders such as MS and RA (Gomez-Muñoz et al., 2016; Maceyka & Spiegel, 2014; Medzhitov, 2008). Inflammation can also induce senescence and the clearance of senescent cells (Lasry & Ben-Neriah, 2015; Ren, Pan, Lu, Sun, & Han, 2009). The acute senescent cells release proinflammatory molecules to adapt to stress, but, when chronic, inflammation can contribute to the pathogenesis of several diseases including type 2 diabetes, AD and PD (Naylor, Baker, & Deursen, 2012; Trayssac, Hannun, & Obeid, 2018). The activation of senescence-associated secretory phenotype (SASP) is induced by the NF-κB pathway and the cellular concentration of proinflammatory cytokines (e.g., IL-1α, IL-1β, IL-6, and IL-8), chemokines and growth factors (Bernal et al., 2014; Lasry & Ben-Neriah, 2015; Ren et al., 2009; Trayssac et al., 2018) is increased. Interrupting the inflammatory response and removing the senescent cells were suggested as a therapeutic approach for aging-related diseases (Lasry & Ben-Neriah, 2015; Naylor et al., 2012).

The role of inflammation in chronic inflammatory diseases makes the inflammatory response a potential therapeutic target (Tabas & Glass, 2013). Anti-inflammatory therapy (e.g., nonsteroidal anti-inflammatory drugs (NSAIDs) and disease-modifying agents of rheumatoid diseases (DMARDs)) are used in the treatment of chronic inflammatory diseases. Biologic DMARDs used for the treatment of RA target inflammatory mediators or receptors such as TNF, IL-6, IL-1R and IL-1 (Quan, Thiele, Tian, & Wang, 2008; Tabas & Glass, 2013). TNF blockade by TNF inhibitors including infliximab, etanercept and adalimumab (Murdaca, Spanò, & Puppo, 2013; Quan et al., 2008), IL-1 blockade by

IL-1 inhibitors (e.g., anakinra; a recombinant IL-1R antagonist) (Dinarello, Simon, & van der Meer, 2012; Quan et al., 2008) and IL-6 blockade by IL-6 inhibitors such as tocilizumab, an antibody against IL-6 receptor (Lee, Bhang, Lee, Kim, & Hahn, 2017; Quan et al., 2008; Smolen, Avila, & Aletaha, 2012), are biologic DMARDs that were reported to relieve the symptoms associated with RA such as inflammation of the joints, anemia and fatigue. Although biologic DMARDs are successfully used against RA, they are costly, may produce some adverse effects such as increased susceptibility to infections and can have a short half-life as in the case of anakinra (Quan et al., 2008).

Long-term therapy with NSAIDs is used for the treatment of chronic inflammatory diseases such as osteoarthritis and RA (Crofford, 2013; Pereira-Leite, Nunes, Jamal, Cuccovia, & Reis, 2016; Quan et al., 2008). These drugs are competitive inhibitors that target and bind to the COX isoforms (COX-1 and COX-2) and inhibit the formation of prostaglandins which are produced by COX enzymes, especially by COX-2 during inflammation, and highly expressed in inflamed tissues (Crofford, 2013; Pereira-Leite et al., 2016; Ricciotti & FitzGerald, 2011). Even though the short-term therapy with NSAIDs is well tolerated, adverse effects in the gastrointestinal tract were observed because of long-term administration of NSAIDs, and these effects were suggested to be related to COX-1 inhibition (Pereira-Leite et al., 2016; Quan et al., 2008). Thus, to overcome these adverse effects, COX-2 selective inhibitors were developed, which, unfortunately, led to cardiovascular toxicity. Hence there are many ongoing studies to improve NSAIDs safety or to propose alternative mechanisms of action (Pereira-Leite et al., 2016).

Targeting inflammatory pathways is also used in the treatment of cancer and there are many ongoing clinical studies that test anti-inflammatory agents against cancer (Nakamura & Smyth, 2017). For example, NSAIDs improved disease-free survival in colon cancer. Anti-IL-1 therapy (IL-1 blockade) with anakinra, used for the treatment of rheumatic diseases, has been found to be effective against multiple myeloma (Lust et al., 2016; Nakamura & Smyth, 2017), and other cancer types including breast (NCT01802970 (ClinicalTrials.gov, 2013a)), metastatic colorectal (NCT02090101 (ClinicalTrials.gov, 2014)) and pancreas (NCT02021422 (ClinicalTrials.gov, 2013b)) cancer. Anti-TNF therapy is successfully utilized for the treatment of inflammatory diseases, while it exhibited limited antitumor efficacy in the cases of pancreas, breast and ovarian cancer. An anti-IL-6 receptor monoclonal antibody, tocilizumab, in combination with nivolumab is currently being tested against multiple myeloma, acute/chronic leukemia, lymphoma and myelodysplastic syndrome (NCT03588936 (ClinicalTrials.gov, 2018; Nakamura & Smyth, 2017)).

Even though anti-inflammatory therapy has been successfully used against chronic inflammatory diseases and cancer, there are some limitations encountered in targeting inflammation. Inflammation is a key defense mechanism necessary for cellular homeostasis and turning off the inflammatory response is not always favorable and may increase risk of infection for the patient. There are several molecules involved in inflammation and there is redundancy among them, hence targeting one or two of them may not be sufficient (Tabas & Glass, 2013). Furthermore, many feedback pathways participate in inflammation and downregulating one pathway may trigger inflammatory response in another pathway. It is also hard to pinpoint the precise identity of the inflammatory stimulus of the chronic inflammatory diseases (Tabas & Glass, 2013). Achieving a durable clinical response is still challenging (Nakamura & Smyth, 2017). Treating the root cause of the disease with high specificity and precision such that off-target effects are avoided would be an ideal therapeutic strategy given the complexity of the cellular inflammation pathways and their entanglement with most cellular processes. Hence understanding the underlying mechanism of the inflammatory response can help to develop effective therapeutic strategies against inflammatory diseases and cancer (Nakamura & Smyth, 2017; Tabas & Glass, 2013). One cause of inflammation has been shown to be lysosomal storage diseases associated with mutations in sphingolipid metabolizing enzymes and targeting these mutations is now recognized as an attractive therapeutic approach. For example, the sphingoid base analog, FTY720, is used for the treatment of neuroinflammation, MS and arthritis. Therapeutic approaches that target the sphingolipid metabolism with a special focus on PCT are discussed in the next sections.

3. Sphingolipid metabolism

3.1 Sphingolipids

Sphingolipids are a family of sphingosine containing phospholipids and glycolipids. They are ubiquitous and complex molecules of eukaryotic cell membranes and important components of lipid rafts. More than 60 different sphingolipids have been identified in cellular membranes of human tissue, but a specific function is identified only for some of them. They work as receptors for some proteins and have roles in inflammation (Kolter, 2011).

Sphingolipids have a polar head group and two hydrophobic tails. A fatty acid is attached in amide linkage to the sphingosine and yields a **ceramide**, which is the structural ancestor of all sphingolipids. Heterogeneity of the polar head groups and fatty-acid chain lengths of the ceramide moieties is responsible for the complexity of the sphingolipid structure (Park, Park, & Futerman, 2014; Platt, 2014).

Sphingomyelins, neutral (uncharged) glycolipids, and gangliosides are subgroups of sphingolipids. They are derivatives of ceramide, but they have different head groups. Sphingomyelins contain phosphocholine or phosphoethanolamine attached to the ceramide. They are present in the plasma membranes of animal cells and are essential components of myelin (Nelson & Cox, 2013). **Glycosphingolipids** have head groups with one or more sugars connected directly to the ceramide and are mostly located in the outer leaflet of plasma membranes. **Cerebrosides** have either a galactose linked to ceramide and characteristically found in the plasma membranes of cells in neural tissue, or a glucose linked to ceramide and found in the plasma membranes of cells in nonneural tissues. **Globosides** have two or more sugars such as D-glucose, D-galactose, or *N*-acetyl-D-galactosamine linked to ceramide. Cerebrosides and globosides are **neutral glycolipids**. **Gangliosides** are acidic and the most complex sphingolipids. Their head groups are oligosaccharides that contain *N*-acetylneuraminic acid (sialic acid) residue/residues at their ends. Sialic acid gives gangliosides the negative charge at pH 7 which distinguishes them from globosides. According to the number of the sialic acid residue/residues gangliosides are named in GM (M for mono-) series, GD (D for di-) series, and GT (T for three); GQ (Q for four) series.

3.1.1 Biosynthesis of sphingolipids

Sphingolipids can be synthesized by *de novo* biosynthesis or by salvage pathway (Fig. 3).

The *de novo* biosynthesis of sphingolipids starts with palmitoyl-CoA and serine in the endoplasmic reticulum (ER). It has four stages:

1. Synthesis of **sphinganine** from palmitoyl-CoA and serine by serine palmitoyltransferase. This is the rate-limiting step of the *de novo* biosynthesis.
2. Attachment of a fatty acid in amide linkage to sphinganine to yield ***N*-acylsphinganine**.
3. Desaturation of sphinganine moiety to form **ceramide** by six ceramide synthetases which result in heterogeneity of the fatty acid chain in the ceramide structure.

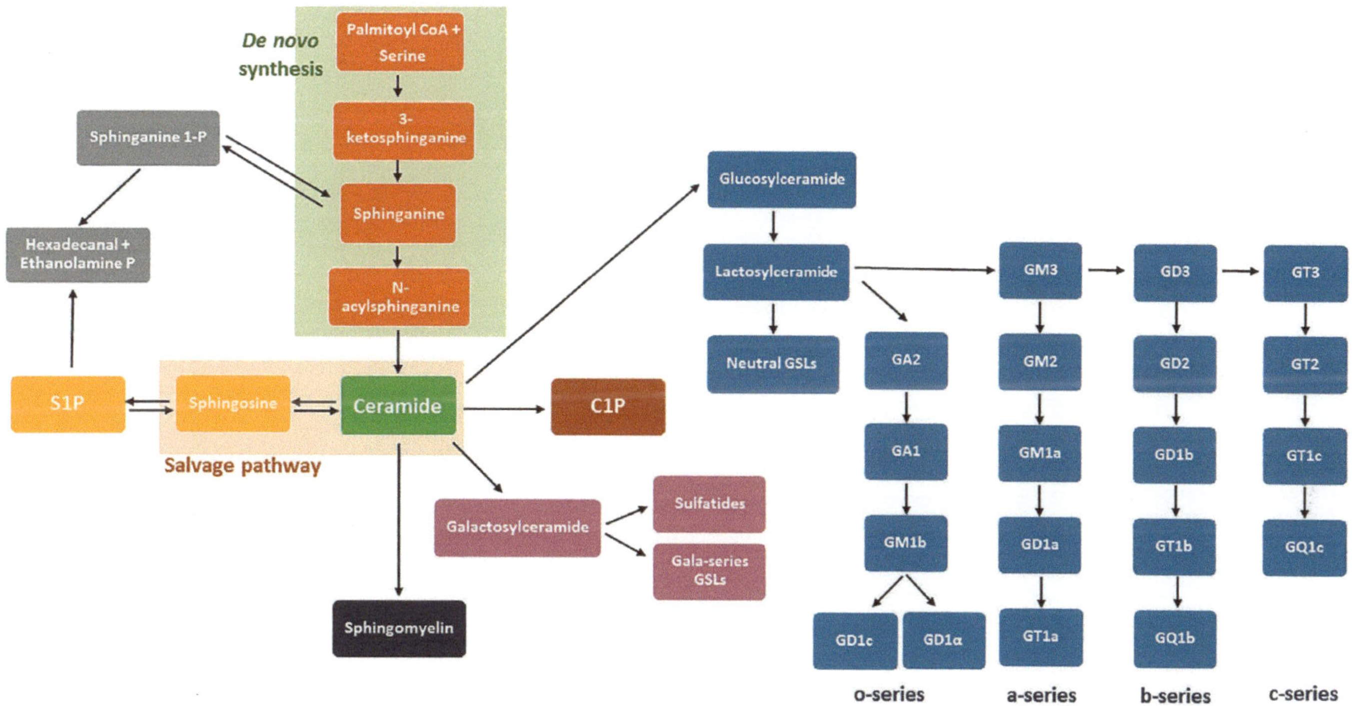

Fig. 3 *De novo* biosynthesis of sphingolipids and sphingolipid salvage pathway.

4. Attachment of a head group to produce a sphingolipid such as a cerebroside, ganglioside or sphingomyelin (Lahiri & Futerman, 2007; Nelson & Cox, 2013).

The first three steps yielding ceramide take place at the cytoplasmic leaflet of the ER membrane (Kolter, 2011; Mandon, Ehses, Rother, van Echten, & Sandhoff, 1992). The attachment of head group occurs in ER, in outer leaflet of the Golgi, in cytoplasmic face of Golgi and in the lumen of the Golgi. Ceramide transfer protein (CERT) transports ceramide from the ER to Golgi and four-phosphate adaptor protein 2 (FAPP2) transports glucosylceramide from trans-Golgi network to the cell surface (D'Angelo, Rega, & De Matteis, 2012; Hanada, 2014).

Salvage pathway uses catabolic building blocks of sphingolipids for new sphingolipid synthesis. Differentiated cells like neurons can prefer to use salvage pathway for sphingolipid synthesis (Kitatani, Idkowiak-Baldys, & Hannun, 2008; Kolter, 2011; Qin, Berdyshev, Goya, Natarajan, & Dawson, 2010; Tettamanti, Bassi, Viani, & Riboni, 2003).

3.1.2 Catabolism of sphingolipids

Sphingolipids are catabolized by lysosomal enzymes in endosomes and lysosomes. In some steps activator proteins are required for the activity of the enzymes (Kolter & Sandhoff, 2005) (Fig. 4). Sphingosine is produced only by the catabolic pathway. Sphingosine can then be used in salvage pathway, which is important for maintaining the sphingolipid rheostat (Maceyka & Spiegel, 2014) (Fig. 3). Ceramide and sphingomyelin are also catabolized by non-lysosomal enzymes. Cytoplasmic glucosylceramide is degraded by a cytoplasmic glucosylceramide cleaving enzyme (Kolter & Sandhoff, 2010).

Deficiency of a single step in sphingolipid degradation causes accumulation of the substrates in the endolysosomal compartment (Fig. 4) (Kolter, 2011). The mutations in genes encoding enzymes that take part in sphingolipid catabolism can lead to sphingolipidoses (Platt, 2018).

4. Sphingolipids involved in inflammation

Sphingolipids take part in the regulation of trafficking and functions of immune cells (Gomez-Muñoz et al., 2016; Maceyka & Spiegel, 2014). Sphingolipids mediate inflammatory signaling by the proinflammatory cytokine TNF. Sphingolipid metabolic enzymes such as sphingomyelinases, ceramidase and sphingosine kinase 1 (SphK1) are stimulated by TNF and

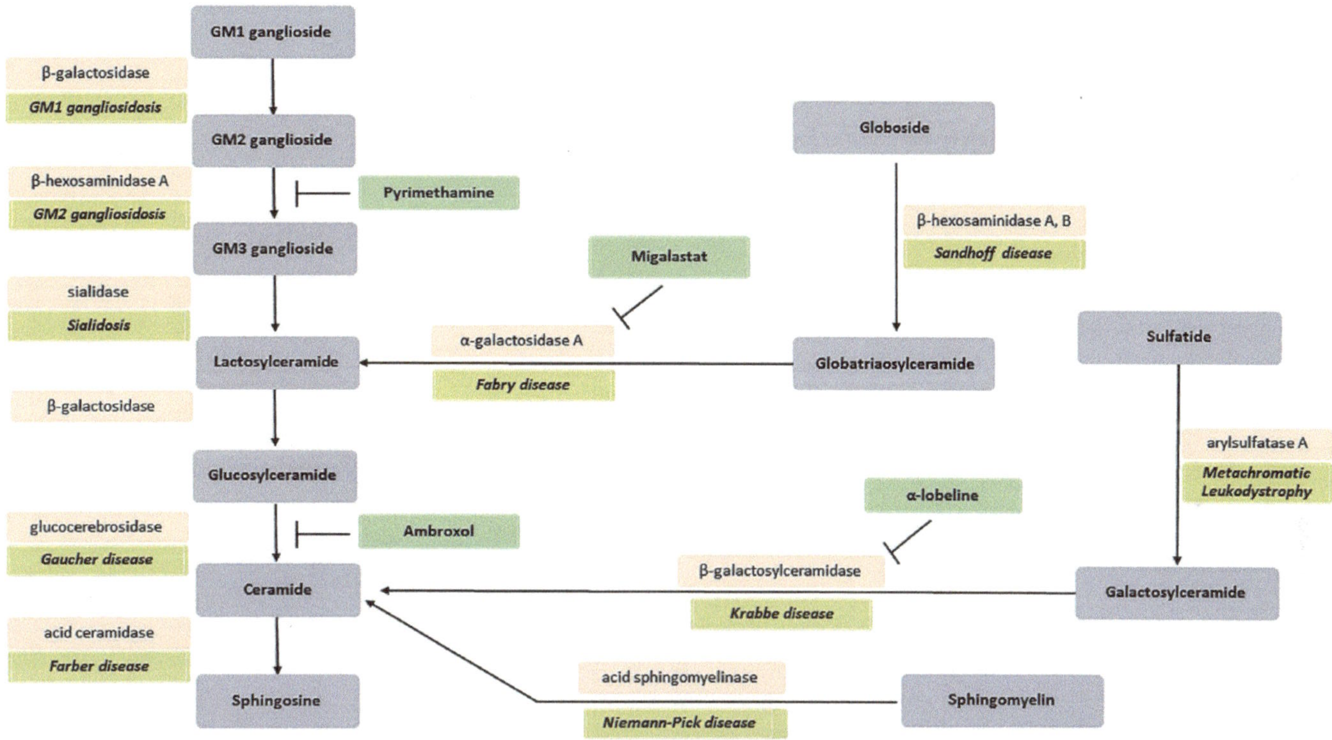

Fig. 4 Catabolism of sphingolipids and its diseases: sphingolipidoses. Sphingolipid metabolites (purple) and enzymes (pink) that catalyze these reactions are shown. Mutations in these enzymes can lead to sphingolipidoses (green). Pharmacological chaperones (blue) can be used to treat these diseases by preventing the misfolding of these proteins and stabilizing them.

sphingolipid metabolites are involved in TNF signaling and chemokine production (Maceyka & Spiegel, 2014).

Ceramide, S1P, and C1P take part in several cellular processes that are essential in immunity, inflammation and inflammatory disorders (Gomez-Muñoz et al., 2016; Maceyka & Spiegel, 2014) and maintaining their relative intracellular concentrations through control of the metabolic pathway is crucial for cell survival, cell growth, inflammation, senescence, and apoptosis (Chen, Chan, Stone, & Mandal, 2014). Ceramides promote apoptosis and they are potent proinflammatory agents while C1P and S1P promote cell survival, and based on the cell type where they are produced, they can exert pro- or anti-inflammatory effects (Gomez-Muñoz et al., 2016; Presa et al., 2016). Maintaining the balance of these sphingolipids via control of the metabolic pathway is critical for the cell. Glycosphingolipids such as lactosylceramide and GM3 are suggested to have a role in the control of cellular signaling in inflammatory processes (Kitatani, Iwabuchi, Snider, & Riboni, 2016). Here, we focus on three sphingolipids: ceramide, S1P and C1P (see Fig. 5A).

4.1 Ceramide

Ceramide levels in the cell are maintained by the *de novo* synthesis pathway, sphingomyelinase (SMase) pathway, and salvage pathway (Fig. 3). High plasma ceramide levels are observed in individuals with obesity and type 2 diabetes (Gomez-Muñoz et al., 2016), AD, and chronic obstructive pulmonary disease (COPD), which are also associated with aging (Gomez-Muñoz et al., 2016; Naylor et al., 2012; Trayssac et al., 2018). Ceramide also takes part in senescence, insulin sensitivity and inflammation in adipose tissue (Trayssac et al., 2018).

Ceramides are potent proinflammatory agents (Gomez-Muñoz et al., 2016) and proinflammatory ceramide signaling is regulated through S1P and C1P (Chiurchiù et al., 2018). SMase-derived ceramides have been associated with many inflammatory processes triggered by proinflammatory agonists, namely proinflammatory cytokines including TNF-α, interferon-γ (IFN- γ), ILs, or platelet-activating factor (PAF) (Gomez-Muñoz et al., 2016; Presa et al., 2016). Ceramides can also induce the activation of the proinflammatory NF-κB (Gomez-Muñoz et al., 2016; Nixon, 2009).

The relation between inflammation and diabetes is well known (Wellen & Hotamisligil, 2005). It has been suggested that inflammation, diabetes and obesity are interconnected by ceramide signaling (Gomez-Muñoz

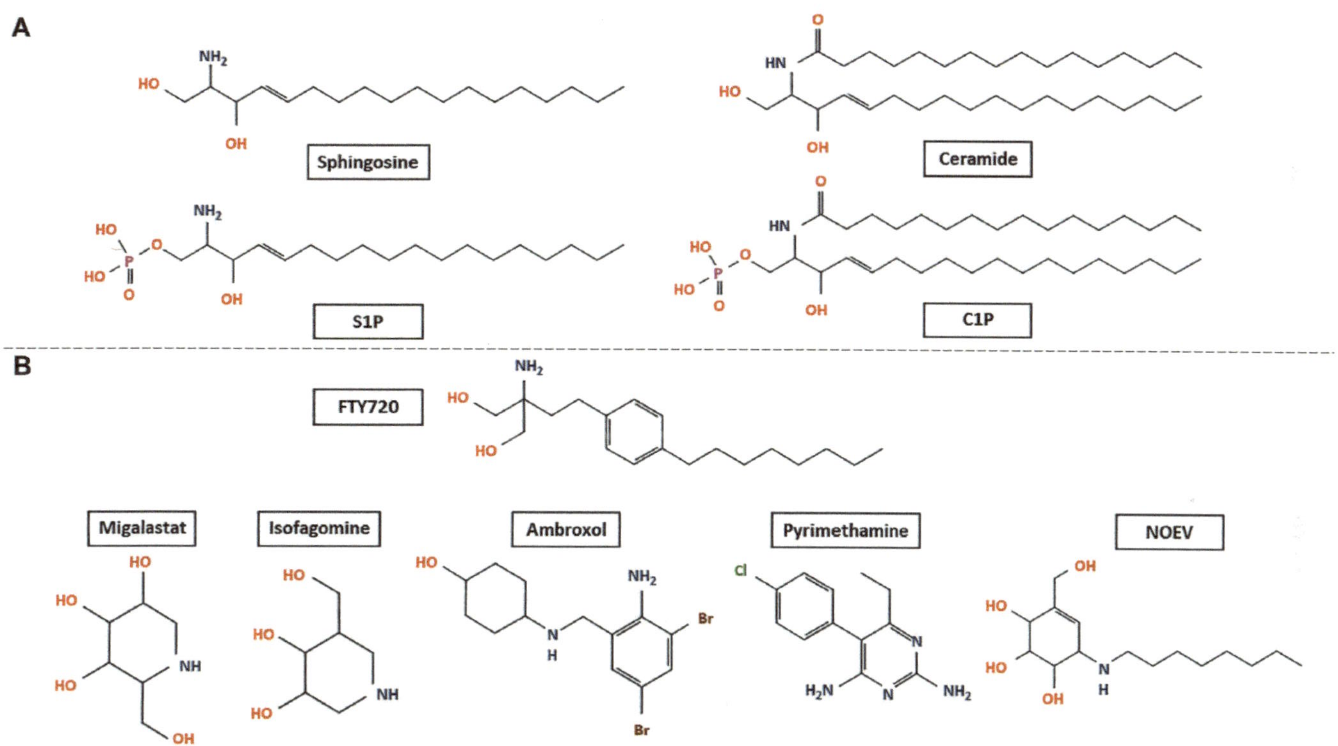

Fig. 5 Chemical structures of sphingosine, ceramide, S1P and C1P (A) and drug candidates for diseases with inflammatory component (B). 2D structures are created with Maestro (2015).

et al., 2016) and long chain ceramide synthesis was suggested as a potential therapeutic target for the treatment of these diseases (Gosejacob et al., 2016). Ceramides are linked to insulin resistance through inflammatory signaling pathways. Ceramide levels and plasma TNF-α levels were increased in obese type 2 diabetics while a decrease in ceramide levels and in plasma TNF-α levels was observed in severely obese patients after gastric bypass surgery (Huang et al., 2011). The reduction in C16:0 ceramide levels due to ceramide synthase 5 deficiency was suggested to reduce obesity and associated diseases (Gosejacob et al., 2016). C16:0 ceramide was suggested as a metabolic marker for obesity and associated diseases due to its implication in these diseases (Fucho, Casals, Serra, & Herrero, 2016).

Ceramides have also been related to some lung diseases, including asthma, COPD, pulmonary fibrosis, or pulmonary infections. Altered sphingolipid metabolism in the alveolar compartment is associated with the inflammatory reaction and ceramide increase (Ghidoni, Caretti, & Signorelli, 2015). Lysosomal acid sphingomyelinase (aSMase) was suggested to be involved in PAF-mediated pulmonary edema and cystic fibrosis (Gomez-Muñoz et al., 2016). Ceramide accumulation in the lungs of cystic fibrosis patients increased inflammation and cell death (Gomez-Muñoz et al., 2016; Grassmé, Carpinteiro, Edwards, Gulbins, & Becker, 2014) while inhibition of ceramide production by inhibiting aSMase reduced inflammation (Simonaro, 2016).

Excess ceramide levels induced by proinflammatory cytokines, oxidative stress and injurious insults can also cause endothelial barrier dysfunction and hence vascular dysfunctions such as emphysema, sepsis and acute respiratory distress syndrome (Maceyka & Spiegel, 2014). Cigarette smoke-induced ceramide generation, which can lead to apoptosis in pulmonary cells, was shown to be regulated by neutral sphingomyelinase (nSMase2) (Filosto et al., 2011). It was also found that nSMase2 was overexpressed in the lung tissues of emphysema patients (Filosto et al., 2011; Gomez-Muñoz et al., 2016).

Ceramide rich lipid rafts were shown to be critical for host defense against *Pseudomonas aeruginosa* (Grassmé et al., 2003) and ceramide accumulation was also linked to infection susceptibility in cystic fibrosis (Teichgräber et al., 2008). Reducing ceramide levels by inhibiting glucocerebrosidase by miglustat reduced the inflammatory response to infection (Loberto et al., 2014).

Ceramide takes part in the regulation of senescence in which inflammatory response plays an important role (Lasry & Ben-Neriah, 2015; Trayssac et al., 2018). Senescence and inflammation are both associated with aging and aging-related diseases (Lasry & Ben-Neriah, 2015; Naylor et al., 2012). Senescent cells generate higher amounts of potent proinflammatory agent, ceramide (Gomez-Muñoz et al., 2016; Trayssac et al., 2018; Venable, Webb-Froehlich, Sloan, & Thomley, 2006). nSMase2 was suggested to be the major senescence-associated ceramide generator (Venable et al., 2006). Ceramide also induces NF-κB that regulates the activation of SASP (Gomez-Muñoz et al., 2016; Lasry & Ben-Neriah, 2015; Nixon, 2009) highlighting the importance of sphingolipids for the relation between inflammation and senescence.

4.2 Sphingosine-1-phosphate

S1P modulates lymphocyte egress from the lymph nodes to the blood stream and its accumulation leads to TNF-α and IL-6 production (Suh & Saba, 2015). The intracellular concentration of S1P is regulated by phosphorylation by sphingosine kinase (SphK), dephosphorylation by sphingosine phosphatase and degradation by sphingosine 1-phosphate lyase (S1PL) (Serra & Saba, 2010). S1PL deficiency elicits proinflammatory responses (Allende et al., 2011; Gomez-Muñoz et al., 2016) while SphK1 and S1P have been implicated as mediators of the TNF-α inflammatory response (Hammad et al., 2008) necessary for the production of TNF-α-induced COX-2 and prostaglandin E2 production (Gomez-Muñoz et al., 2016). TNF-α plays a role in the translocation of SphK1 to the plasma membrane, where it generates S1P (Wollny et al., 2017). SphK/S1P signaling pathway and metabolism have been linked with many inflammatory-based diseases (Roviezzo et al., 2011; Wollny et al., 2017). High levels of S1P have been found in the cases of IBD, inflammatory arthritis, MS, and asthma which might be treated via pharmacological modulation of S1P/S1PR1 axis (Gomez-Muñoz et al., 2016). SphK/S1P pathway alters vascular permeability and promotes cell recruitment in acute inflammation (Gomez-Muñoz et al., 2016; Roviezzo et al., 2011). S1P has also been suggested to have an anti-inflammatory role by inducing a switch from proinflammatory M1 to anti-inflammatory M2 macrophage subtypes to affect the progress of diseases such as cancer (Gomez-Muñoz et al., 2016; Rivera, Proia, & Olivera, 2008). It is also suggested that sphingolipid metabolism and S1P receptors might be altered by inflammation (Maceyka & Spiegel, 2014).

S1P was also suggested to hinder senescence but the mechanism whereby S1P regulates senescence is not completely understood (Trayssac et al., 2018).

S1P was reported to induce human telomerase reverse transcriptase (hTERT) stability and prevention of this S1P-hTERT binding through SK inhibition increased senescence suggesting that S1P may delay senescence (Selvam et al., 2015; Trayssac et al., 2018).

S1P is a key mediator of lymphocyte trafficking. Increase of S1P gradient in the T-cell membrane results in T-cell delivery into the blood stream and causes MS. The S1P levels in the cell can be decreased at the receptor level or by modulating degradation by S1PL. Downregulation of S1PR1 by the S1P receptor modulator FTY720 (fingolimod) favors retention of T cells in the lymph nodes (Brinkmann et al., 2010). Our group has used a systematic virtual screening campaign to identify novel scaffolds that might be S1PL ligands (Deniz, Ozkirimli, & Ulgen, 2016).

4.3 Ceramide-1-phosphate

C1P is a bioactive sphingolipid metabolite that is involved in inflammation. The enzyme that catalyzes its production, ceramide kinase (CERK), has also been implicated in inflammatory response (Maceyka & Spiegel, 2014; Presa et al., 2016). Arachidonic acid is released in the rate-determining step in the biosynthesis of proinflammatory eicosanoids when cytosolic phospholipase A2 (cPLA2)α is activated by C1P (Maceyka & Spiegel, 2014; Simanshu et al., 2013). C1P produced intracellularly by CERK exerts proinflammatory actions. C1P was found to stimulate cell migration, which has a critical role in several chronic inflammatory diseases, such as asthma, MS, RA and IBD (Presa et al., 2016).

Wild-type mice expressed high levels of the proinflammatory cytokines IL-6 and TNF-α when they were fed a high fat diet but this increase in IL-6 and TNF-α levels was suppressed in CERK-null mice (Mitsutake et al., 2012; Presa et al., 2016). Furthermore, MCP-1/CCR2 signaling in macrophages infiltrating the adipose tissue was reported to be decreased due to CERK deficiency in CERK-null mice, eliminating inflammation in the adipocytes and hence suppressing obesity and diabetes (Mitsutake et al., 2012).

In addition to its proinflammatory properties, C1P has been proposed to exert anti-inflammatory actions in some cell types and tissues. C1P blocks ceramide induced NF-κB activation that leads to inflammation. It has also been demonstrated that C1P reduces airway inflammation and lung emphysema. Its anti-inflammatory actions in the lung suggest that synthetic C1P analogs may be used to treat lung inflammation (Presa et al., 2016). C1P prevented cigarette smoke-induced activation of NF-κB and nSMase and attenuated proinflammatory cytokine release in human airway epithelial cells and neutrophils (Baudiß et al., 2015).

C1P analogs PCERA-1 and ONO-SM-362 suppress the production of TNF-α and induce the release of anti-inflammatory IL-10 in macrophages suggesting that C1P analogs can be used to develop new treatment tools for inflammatory responses (Goldsmith et al., 2009; Presa et al., 2016).

5. Protein-protein interaction network for the crosstalk between inflammation and sphingolipid metabolism

To better understand the crosstalk between inflammation and sphingolipids, a protein-protein interaction (PPI) network of sphingolipid and inflammation pathways was constructed (Fig. 6). BioServices package (Cokelaer et al., 2013) of Python was used to acquire the proteins in relation to inflammation and sphingolipids. For inflammation (IN), the keywords used for acquisition are "inflammation," "inflammatory" and "GO:0006954" (Inflammatory Response). For sphingolipid (SL) related terms, the keywords "sphingolipid," "glycosphingolipid," "sphingomyelin," "GO:0006665" (Sphingolipid Metabolic Pathway) and "GO:0090520" (Sphingolipid Mediated Signaling Pathway), Sphingolipid Metabolism and Signaling Pathway terms from the KEGG database (Kanehisa, Sato, Kawashima, Furumichi, & Tanabe, 2016) were included. In total, 1382 proteins were collected. Experimental PPI data for humans were collected from BIOGRID (Chatr-Aryamontri et al., 2017), STRING (Szklarczyk et al., 2017), HPRD (Keshava Prasad et al., 2009) and IntAct (Orchard et al., 2014). The constructed network (SLIN) consists of 5690 interactions between 1117 proteins (194 SL, 891 IN and 32 SI) (Fig. 6). The proteins in the network were classified as SL for only sphingolipid related (yellow), IN for only inflammation related (blue) and SI for both sphingolipid and inflammation related (green) proteins. Proteins whose mutations cause sphingolipidoses are shown in red.

To reduce the network complexity and identify closely connected network members, we decomposed the big SLIN network into subnetworks using MCODE, a widely used clustering algorithm (Bader & Hogue, 2003). The first three clusters with largest number of members are shown in Fig. 7. As SI proteins are involved in both sphingolipid and inflammation processes, they were expected to be the hub and bottleneck nodes on main communication pathways. On the contrary, the sphingolipid related terms

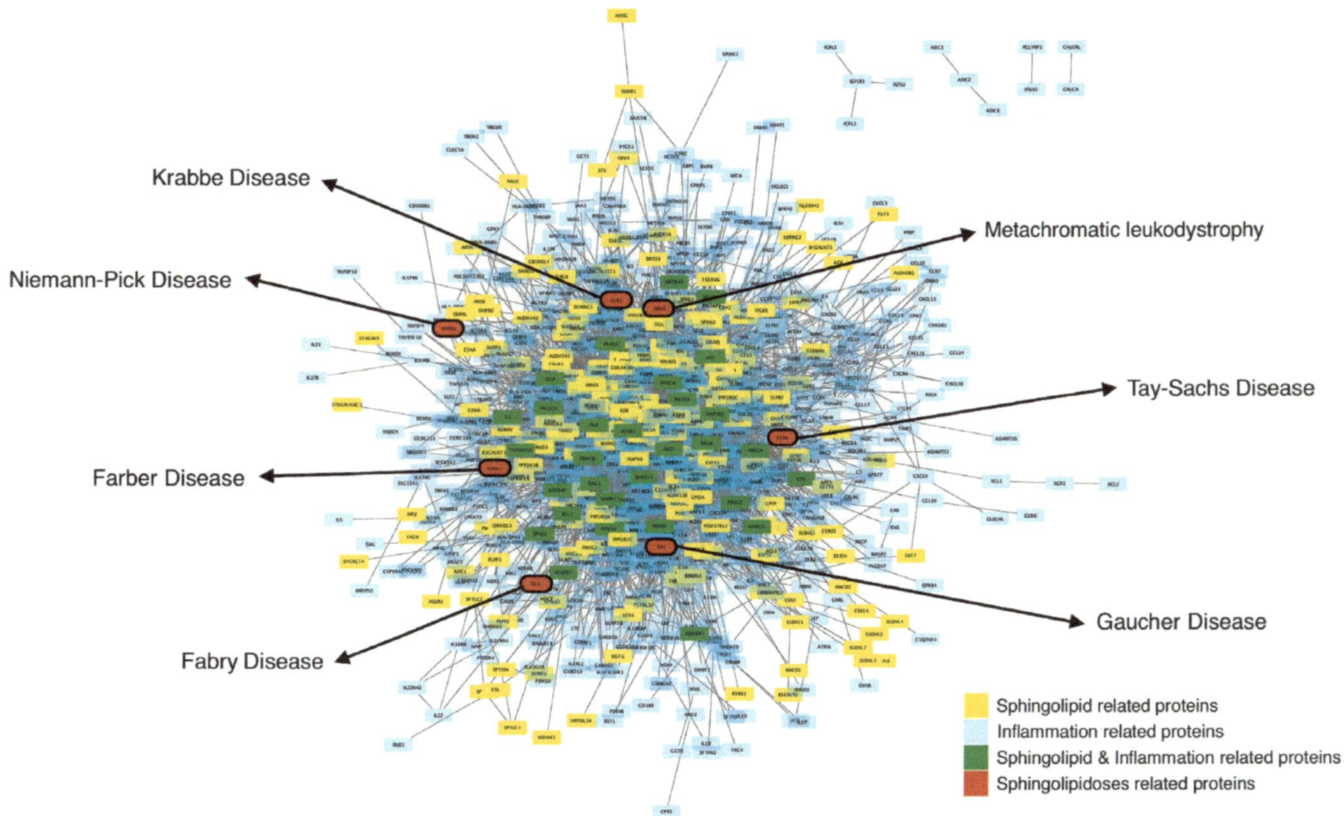

Fig. 6 PPI network of sphingolipids and inflammation.

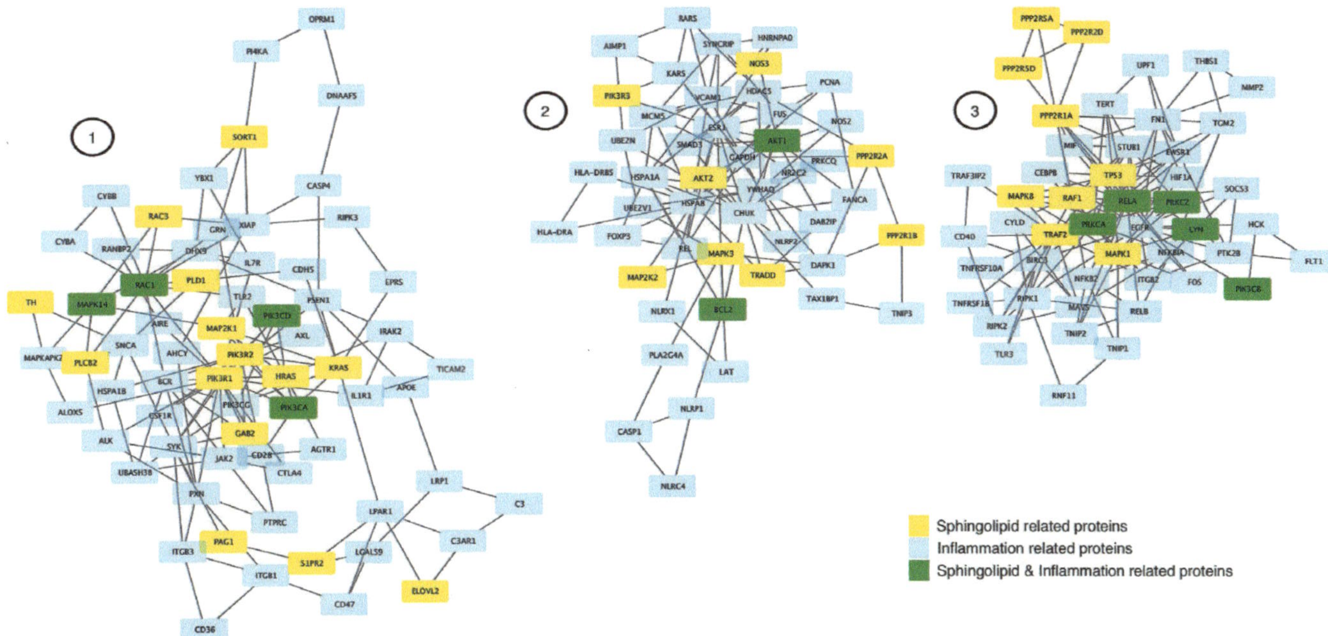

Fig. 7 First three clusters of PPI network.

(SL) were found to have the highest degree and betweenness centrality values in cluster 1, although the number of sphingolipid proteins is less than half of the number of inflammation proteins. The role of NF-κB and PI3K/AKT pathways in linking SL metabolism to the inflammatory response was highlighted by our analysis. In cluster 1, PIK3R1 (phosphatidylinositol 3-kinase regulatory subunit alpha) has the highest degree (21) and betweenness centrality (0.37) while the average degree and betweenness centrality values are 4.1 and 0.03, respectively, for the cluster. PIK3R1 is a member of the PI3K/AKT pathway, which is a key regulator of various cellular processes such as proliferation and apoptosis and is regulated by several sphingolipids (Oskouian & Saba, 2010). PIK3R1 expression plays a role in the initiation and propagation of inflammatory response in obese adipose tissue (McCurdy et al., 2012). Another protein with high betweenness centrality interacting with SL and IN proteins, RAC1 (Rac Family Small GTPase 1), is required for S1P-mediated cellular responses and also controls the initiation and progression of sepsis and inflammation (Gonzalez, Kou, & Michel, 2006; Qin, Liu, & Liu, 2015). In cluster 2, inflammation related proteins ESR1 (estrogen receptor 1) and CHUK (inhibitor of NF-κB kinase subunit alpha) provide the information flow between other members. Sphingolipids play a mediatory role in estrogen signaling and breast cancer, and as a member of the estrogen signaling pathway, ESR1 is a regulator of adipose tissue inflammation (Davis et al., 2013; Sukocheva, 2018). Inflammatory gene activation is limited by the deactivation of NF-κB by CHUK (Shembade, Pujari, Harhaj, Abbott, & Harhaj, 2011). Ceramide-S1P levels are also important for NF-κB activation or inactivation (Merrill et al., 1997; Patwardhan & Liu, 2011; Wątek et al., 2017). Another key node in cluster 2 is AKT1 (AKT Serine/Threonine Kinase 1), which regulates inflammation and apoptosis during carcinogenesis. AKT1 is mediated by ceramides and S1P through the PI3K/AKT1 pathway (Oskouian & Saba, 2010; Tang et al., 2016). EGFR (epidermal growth factor receptor) is the most central protein in cluster 3 and has a role in inflammation-related cell signaling and also takes part in progression of several carcinoma types (Berasain et al., 2009; Normanno et al., 2006). The transcription and translation of EGFR can also be induced by S1P via various signaling cascades (Hsieh et al., 2008). TRAF2 (TNF receptor-associated factor 2) is another key protein that promotes proinflammatory responses by regulating the TNF signaling pathway (Jin et al., 2015). TRAF2 is also a key member of the NF-κB signaling pathway and its activities are controlled by binding to S1P (Alvarez et al., 2010). The tumor suppressor protein 53 (p53) regulates many processes regarding cell cycle and also mediates cell senescence and its mutations are implicated in most cancer types (Rufini, Tucci, Celardo, & Melino, 2013). We note again that all the key proteins mentioned above interact with both the sphingolipid- and inflammatory proteins and act as a bridge between these two crosstalking pathways.

The dysregulation of sphingolipids due to the abnormalities in lysosomal function is known to lead to LSDs, called as sphingolipidoses, as well as to induction of inflammatory pathways. In Fig. 5, the nodes shown in red are the proteins, whose mutations cause the sphingolipidoses such as GBA, which is associated with Gaucher disease (Mistry et al., 2017), hexosaminidase (HEXA), which is associated with Tay-Sachs disease (Ozkara & Navon, 1998) and sphingomyelin phosphodiesterase or aSMase (SMPD1), which is associated with Niemann-Pick disease (Syed Haneef & George Priya Doss, 2016). The reconstructed SLIN protein-protein interaction network successfully captured this crosstalk between sphingolipid and inflammation related diseases that will be further discussed in Sections 6–8.

6. Neuroinflammation, inflammatory diseases and sphingolipids

6.1 Neuroinflammation

CNS injury, infection, or storage disease can lead to neuroinflammation, which is normally a protective response. The inflammatory response can protect the CNS; however, if it becomes chronic or heightened as seen in LSDs, the elevated levels of the inflammatory mediators can lead to neurodegeneration (Alroy, Garganta, & Wiederschain, 2014; Bosch & Kielian, 2015; Cox & Cachón-González, 2012). The effect of neuroinflammation on CNS homeostasis may depend on the duration and intensity of the inflammation (Bosch & Kielian, 2015). Therefore, besides the mutation of the lysosomal protein, another factor that affects the phenotype is its response to the accumulated sphingolipids in the endosomal-lysosomal system.

Microglial and astrocyte activation by TLRs, cytokines and chemokines are seen in many sphingolipidoses that affect the CNS. Microglia, the immune cells in the CNS, trigger the neuroinflammation reaction by producing cytokines and chemokines that attract and activate the peripheral immune cells to the site of infection, and elicit a defensive response in the CNS (Bosch & Kielian, 2015; Lyman et al., 2014). Microglia have a role in neural development, tissue repair and clearance of debris and they are implicated in the pathogenesis of several neurodegenerative disorders such as MS (Scott-Hewitt, Folts, & Noble, 2018). Astrocytes take part in the regulation of synaptic function, neural repair, CNS immunity and neuroinflammatory responses (Bosch & Kielian, 2015; Lee & MacLean, 2015). These cells have been observed to be activated in several cases of CNS injuries and diseases (Lee & MacLean, 2015).

Loss of control in the sphingolipid rheostat can lead to neuroinflammation (Bosch & Kielian, 2015). For example, aSMase activated by TNF-α induces ceramide production, and ceramide, in turn, activates the proinflammatory transcription factor, NF-κB. Cell death pathways and glia are activated when ceramide levels are increased with an accompanying a decrease in S1P (Bosch & Kielian, 2015; Nixon, 2009). FTY720, which is a structural analog of S1P and used for the treatment of the relapsing-remitting form of MS, has been observed to reduce glial activation, and neuroinflammation (Bosch & Kielian, 2015; Cipriani, Chara, Rodríguez-Antigüedad, & Matute, 2015).

Alzheimer's disease is related to chronic neuroinflammation. Microglia and astrocytes are activated by the amyloid peptides, and in turn these glial cells secrete proinflammatory cytokines including IL-1β and TNF-α (Dinkins, Dasgupta, Wang, Zhu, & Bieberich, 2014). The release of these cytokines induces nSMase2 activity, and hence, an increase in ceramide content which can lead to apoptosis of neurons and astrocytes (Dinkins et al., 2014, 2016). The pharmacological noncompetitive inhibitor of nSMase2, GW4869, has been suggested to prevent exosome secretion, which was upregulated during AD and reported to induce amyloid beta aggregation, by inhibiting nSMase2 (Dinkins et al., 2014). Hence developing therapeutic agents that target nSMase2 may block exosome formation which is upregulated during AD, reduce neuroinflammation and hinder apoptosis (Dinkins et al., 2014).

Parkinson's disease is characterized by the degeneration of dopaminergic neurons in the substantia nigra pars compacta, and chronic neuroinflammation has been suggested to play an important role in the pathogenesis of this disease (Tufekci, Meuwissen, Genc, & Genc, 2012; Wang, Liu, & Zhou, 2015). Activation of glial cells and increase in proinflammatory factor level have been observed in several studies conducted on PD brain. High levels of proinflammatory cytokines such as TNF-α, IL-1β and IFN-γ have been found in the midbrain of PD patients (Wang et al., 2015). Mutations of the glucocerebrosidase gene have been suggested as a significant risk factor of PD (Mc Donald & Krainc, 2017; Schapira, 2015). Isofagomine (IFG), which binds to glucocerebrosidase mutants as a pharmacological chaperone to increase their stability and improve their trafficking (Parenti, Andria, & Valenzano, 2015; Sardi, Cheng, & Shihabuddin, 2015; Weinreb et al., 2007), was found to improve motor performance and reduce microglial inflammation and α-synuclein immunoreactivity (Richter et al., 2014).

6.2 Multiple sclerosis

Multiple sclerosis (MS) is the most common autoimmune disorder of the CNS in which widespread myelin damage and inefficient remyelination are observed (Brinkmann et al., 2010; Scott-Hewitt et al., 2017). High levels of S1PR1 and S1PR3 on astrocytes are seen in MS (Van Doorn et al., 2010). S1P is a key mediator for lymphocyte trafficking, and the regulation of S1PR1 on lymphocytes in the lymph nodes is crucial since S1P-S1PR1-dependent egress of lymphocytes from these nodes might cause autoaggressive T lymphocytes to flow into the CNS and hence accelerate the pathogenesis of MS (Brinkmann et al., 2010; Wollny et al., 2017). FTY720, the S1PR1 functional antagonist mentioned previously, was found to reduce S1P-S1PR1-dependent egress of lymphocytes from the lymph nodes (Brinkmann et al., 2010) and prevent myelin damage (Chiurchiù et al., 2018). Th17 (IL-17-producing T helper) cells, considered as an important mediator of inflammation in MS, were reduced in MS patients treated with FTY720 (Mehling et al., 2010). The successful use of FTY720 suggests that novel MS treatment options can be designed via pharmacological modulation of the S1P signaling pathways (Podbielska, O'Keeffe, & Hogan, 2018). Identification of FTY720 analogs with improved selectivity has been widely studied; some of the agonists and antagonists on S1P receptors are reviewed in Sanllehí, Abad, Casas, and Delgado (2016).

Due to its role in the regulation of intracellular concentration of S1P and inflammatory responses (Allende et al., 2011; Gomez-Muñoz et al., 2016; Serra & Saba, 2010), S1PL was suggested as a promising therapeutic target against MS (Deniz et al., 2016). Our group proposed candidate active-site directed S1PL ligands with common scaffolds of benzimidazole and tetrahydrofuran, and dihydropyrimidine through a computational approach of virtual screening, docking and molecular dynamics simulations (Deniz et al., 2016). Among these scaffolds, dihydropyrimidine was reported as an anti-inflammatory agent (Deniz et al., 2016; Lauro et al., 2014).

6.3 Inflammatory bowel disease

Crohn's disease and ulcerative colitis are chronic inflammatory disorders of the gastrointestinal tract (Wollny et al., 2017). These disorders have high prevalence (>0.3%) and they are a heavy burden especially in western societies (Ng et al., 2017). Previously, antibody-based TNF inhibitors and integrin antagonists were used in the treatment of IBD, but these therapies were limited to a small group of patients and their efficacy was reduced with time (Nielsen, Li, Johansson-Lindbom, & Coskun, 2017). One of the most characteristic features of IBDs is the altered leukocyte recruitment into the intestine. Leukocyte traffic is regulated by S1P in gastrointestinal tract. S1P levels are elevated during colitis and SphK1 expression is increased in patients with ulcerative colitis (Maceyka & Spiegel, 2014; Snider et al., 2009). Mouse model studies have suggested that SphK1 and S1P might be a link between chronic intestinal inflammation and colitis-associated cancer (Maceyka & Spiegel, 2014). S1PR1 agonists that decrease lymphocyte egress to the gastrointestinal tract are being analyzed for the treatment of IBDs (Bamias, Clark, & Rivera-Nieves, 2013; Nielsen et al., 2017; Sandborn et al., 2016). For example, RPC1063 (ozanimod), an oral agonist of S1PR1 and S1PR5, is currently being tested clinically in patients with ulcerative colitis (Bamias et al., 2013; Sandborn et al., 2016). A novel sphingosine analog, LCL351, that selectively inhibits SphK1 was observed to ameliorate inflammation by reducing the expression of proinflammatory cytokines in dextran sulfate sodium-induced colitis murine model (Pulkoski-Gross et al., 2017).

TNF-α plays a crucial role in IBD and overproduction of TNF-α is observed in active IBDs (Abdel Hadi, Di Vito, & Riboni, 2016; Wollny et al., 2017). In a dextran sulfate sodium-induced colitis murine model, blood S1P levels were increased in wild-type mice but not in *Sphk1* knockout mice (Abdel Hadi et al., 2016; Gomez-Muñoz et al., 2016; Snider et al., 2009). Furthermore, colitis severity, cytokine production and systematic inflammation were reduced in *Sphk1* knockout mice. This study suggests that SphK1 inhibition might be an alternative treatment for systematic and local inflammation in IBD (Snider et al., 2009). Colitis-associated cancer has also been linked to STAT3 and NF-κB pathways which can trigger tumor formation and cause inflammation by increasing the expression of proinflammatory cytokines including TNF-α (Wollny et al., 2017). S1P-S1PR1 pathway plays a role in the persistent STAT3 activation in cancer cells (Lee, Deng, et al., 2010), and S1P and SphK1 were reported to be necessary for NF-κB activation as well as TNF-α signaling (Alvarez et al., 2010). Understanding SphK/S1P signaling can provide valuable clues toward developing effective therapies for IBD (Wollny et al., 2017).

6.4 Rheumatoid arthritis

RA is a chronic inflammatory disease that leads to the joint destruction due to severe synovial membrane inflammation (Lee et al., 2017; Murdaca et al.,

2013; Quan et al., 2008). Pain, stiffness and joint swelling are among the most common symptoms of RA (Quan et al., 2008).

The role of S1P/S1PR1 signaling in the pathogenesis of RA has been suggested in several studies. Significantly high levels of S1P were observed in the synovial fluid of RA patients compared to osteoarthritis patients (Kitano et al., 2006; Takeshita et al., 2012; Yu, Sun, & Argraves, 2015). Proinflammatory cytokines TNF-α and IL-1β were reported to play a central role in the pathogenesis of RA (Dinarello et al., 2012; Kitano et al., 2006; Murdaca et al., 2013; Quan et al., 2008), and elevated levels of these cytokines were observed in the synovial fluid and plasma of the patients suffering from this disease (Kitano et al., 2006). These proinflammatory cytokines, which were suggested to be enhanced by S1P/S1PR1 signaling, induced the production of COX-2 and prostaglandin E2 by RA synoviocytes (Kitano et al., 2006). The expression of proinflammatory receptor activator of NF-κB ligand (RANKL), a member of TNF, of RA synoviocytes (MH7A cells) and $CD4^+$ T cells were also enhanced by S1P/S1PR1 signaling, suggesting that S1P/S1PR1 signaling may be involved in RANKL-induced osteoclastogenesis of RA (Gomez-Muñoz et al., 2016; Takeshita et al., 2012). FTY720 was observed to inhibit arthritis in SGK mice, a murine RA model (Tsunemi et al., 2010). Taken together, novel therapeutic agents that regulate S1P/S1PR1 signaling might be used to treat RA (Kitano et al., 2006; Takeshita et al., 2012).

The role of SphK1 in RA pathogenesis is not clear (Maceyka & Spiegel, 2014). hTNF/SphK1$^{-/-}$ mice (transgenic TNF-α mice crossed with SphK1 knockout mice) displayed reduced synovial inflammation indicating that SphK1 plays a key role in TNF-α-induced chronic inflammatory arthritis (Baker, Barth, Chang, Obeid, & Gilkeson, 2010); however in collagen-induced arthritis model of RA, SphK1 knockout mice exhibited normal phenotype (Michaud, Kohno, Proia, & Hla, 2006).

7. Inherited disorders of lysosomal sphingolipid metabolism: Sphingolipidoses

7.1 Diseases of sphingolipid biosynthesis

In the ganglioside biosynthetic pathway two severe disorders, namely GM3 synthase and GM2 synthase deficiency, have been identified. The clinical finding is a severe epilepsy syndrome in GM3 synthase deficiency and it is spastic paraplegia in GM2 synthase deficiency. The biochemistry and pathogenesis of these diseases are not yet known (Boukhris et al., 2013; Harlalka et al., 2013; Simpson et al., 2004).

7.2 Diseases of sphingolipid catabolism: sphingolipidoses

Ten different human diseases, collectively known as sphingolipidoses, are known in the catabolic pathway of glycosphingolipids (Fig. 2). Those are GM1 gangliosidosis, Tay-Sachs disease, Sandhoff disease, Sialidosis, Gaucher disease, Fabry disease, MLD, Krabbe disease, Niemann-Pick disease type A and B and Farber disease. Sphingolipidoses occur due to deficiencies of lysosomal enzymes or nonenzyme lysosomal proteins (Kolter, 2011; Kolter & Sandhoff, 2006). In these diseases, undegraded substrate accumulates progressively within the lysosome. Endo-lysosomal accumulation of the substrate and other accompanying glycosphingolipids is the cellular hallmark of the diseases.

Sphingolipidoses are inherited in an autosomal recessive manner, except the X-linked Fabry disease. The incidence of the diseases is less than 1 in 20,000 live births (Meikle, Hopwood, Clague, & Carey, 1999; Ozkara & Topcu, 2004), making them rare diseases. The manifestation of the diseases can vary drastically between different sphingolipidoses depending on the type of mutations, accumulating substrate, tissue of the substrate accumulated and the age of onset. At least 100 mutations have been reported for each sphingolipid storage disease. Due to the prevalence of interfamily marriages, some mutations have been discovered in the Turkish population (Çeltikçi et al., 2012; Celtikci, Topcu, & Ozkara, 2011; Önder, Sinici, Sönmez, Topçu, & Özkara, 2009; Özkan & Özkara, 2016; Ozkara et al., 1995; Ozkara & Navon, 1998; Ozkara & Sandhoff, 2003a, 2003b; Ozkara, Topcu, & Renda, 1997). Most of them are missense mutations. But the other types of mutations like frameshift, splice site, deletion and insertion are also detected in the enzymes and nonenzyme protein genes (Brunetti-Pierri & Scaglia, 2008; Cesani et al., 2016; Graziano & Cardile, 2015; Mistry et al., 2017; Regier, Proia, D'Azzo, & Tifft, 2016; Schuchman & Desnick, 2017; Stirnemann et al., 2017; van Rappard, Boelens, & Wolf, 2015). The site of the mutation on the protein impacts the clinical phenotype. Severe changes in protein folding lead to the early onset type of diseases which present the most severe progression and are usually fatal. Juvenile form starts in the late childhood period while the adult form presents in young ages, respectively, and show mild clinical progress and longer longevity. Sphingolipidoses can present wide range of clinical signs and symptoms like organomegaly, mental and motor retardation, skin, eye, renal and cardiac involvement. Prenatal or early diagnosis, gene therapy and pharmacological chaperones are some of the current preventive and curative therapies for sphingolipidoses (Platt, 2018).

The characteristic feature of most of the sphingolipidoses is neurodegeneration including microglial and astrocyte activation (Baudry, Yao, Simmons, Liu, & Bi, 2003; Bosch & Kielian, 2015; Farfel-Becker et al., 2011; Wilkinson et al., 2012). Vitner et al. reviewed the innate immune responses in sphingolipid lysosomal storage diseases and described four possible mechanisms about the stored material and innate immune system activation: (i) accumulated sphingolipids activate microglia, (ii) after accumulation of undegraded substrates and secondary accumulated materials, danger signals are released from the cell and then microglial activation occurs, (iii) impaired phagocytosis of microglia leads to chronic inflammation, (iv) dying neurons release danger signals (Vitner, Futerman, & Platt, 2015). An inflammatory or immunological response to the metabolite stored in lysosomal storage diseases has been demonstrated in mucopolysaccharidoses (MPS), NPC, Farber, Gaucher, metachromatic leukodystrophy (MLD), Krabbe diseases, neuronal ceroid lipofuscinosis and Sandhoff diseases (Bosch & Kielian, 2015; Platt, 2018; Simonaro, 2016). In some cases, peripheral blood and lymphoid tissues of immune system cells migrate to the CNS.

Elucidating the role of sphingolipids in the innate immune response of the sphingolipidosis patient can provide valuable information for the treatment of these rare diseases, the inflammatory or neurodegenerative impact of sphingolipid accummulation as well the more common neurodegenerative disorders associated with inflammation.

7.3 Treatment options for lysosomal storage disorders

Lysosomal enzymes are synthesized in the ER at neutral pH and translocated to the lysosome through the Golgi apparatus. In lysosomes, the enzymes function at acidic pH (Convertino et al., 2016). Mutations in the genes encoding these lysosomal enzymes may lead to lower catalytic activity, less stable protein folding, ER retention and anomalies in lysosomal trafficking (Convertino et al., 2016; Khanna et al., 2010; Leidenheimer & Ryder, 2014). Protein misfolding is directly or indirectly attributed as the main cause of LSDs. In addition to causing loss of normal functions, misfolding can lead to toxicity because of accumulation of substrate that cannot be catalyzed. ERT, SRT, gene therapy, hematopoietic stem cell transplantation (HSCT) and PCT are the major therapeutic strategies to fight LSDs (Syed Haneef & George Priya Doss, 2016).

In ERT, missing enzyme is administered exogenously by repetitive intravenous infusion (Beutler, 2004; Parenti, Pignata, Vajro, & Salerno, 2013; Syed Haneef & George Priya Doss, 2016). The concentration of the required enzyme increases; however ERT is not a cure for the underlying genetic conditions (Syed Haneef & George Priya Doss, 2016). ERT has been used with some success in LSDs including Gaucher disease, Fabry disease and Pompe disease. The limitations encountered in ERT are instability and denaturation of the infused enzymes due to temperature and neutral pH levels, and limited distribution to the disease-relevant tissues (Khanna et al., 2010; Porto et al., 2009). The insufficient delivery of the infused enzymes might be one of the reasons for the high susceptibility of these exogenous enzymes to protein denaturation (Shen, Edwards, Hong, & Murray, 2008). ERT should be conducted frequently (weekly or biweekly) due to the short half-life of the infused enzymes (Syed Haneef & George Priya Doss, 2016). For instance, in Gaucher disease patients, response to ERT is often slow and the dose required is high (Shen et al., 2008). Bioavailability of recombinant enzymes, high cost of therapies and crossing the blood-brain barrier to reach the CNS are the other major challenges observed in ERT (Convertino et al., 2016; Parenti et al., 2013; Syed Haneef & George Priya Doss, 2016). ERT can also be immunogenic which may intervene with the efficacy of the treatment (Syed Haneef & George Priya Doss, 2016; Xu et al., 2015).

With SRT, the amount of substrate that accumulates in the LSD can be reduced using inhibitors that inhibit the enzyme that produces the substrate (Parenti et al., 2013; Syed Haneef & George Priya Doss, 2016). SRTs have been used in Gaucher disease but the patients suffer from gastrointestinal problems (Syed Haneef & George Priya Doss, 2016). In gene therapy, the activity of the defective enzyme is aimed to be restored or increased by introducing the normal copy of the defective gene to the recipient's cells (Parenti et al., 2013). The challenges of gene therapy are delivery to the correct cells and sustained expression (Syed Haneef & George Priya Doss, 2016). The aim of HSCT is to produce the missing enzyme from the transplanted hematopoietic stem cells (Hollak & Wijburg, 2014; Syed Haneef & George Priya Doss, 2016). HSCT is limited to few LSDs including MPS I, Krabbe disease and MLD (Parenti et al., 2013; van Gelder, Vollebregt, Plug, van der Ploeg, & Reuser, 2012). Some of the drawbacks for this treatment option are mortality related to the procedure, graft failure and lack of suitable donors (Parenti et al., 2013; Syed Haneef & George Priya Doss, 2016).

PCT is gaining interest as an alternative therapy in the treatment of the misfolding diseases (Parenti et al., 2015; Syed Haneef & George Priya Doss, 2016). Numerous noncovalent and covalent interactions contribute to the stability of the tertiary structure of a lysosomal protein. Due to mutations, some of these interactions might be lost, and the protein can become misfolded and labeled for proteolytic degradation (Leidenheimer & Ryder, 2014). PCs are small ligands that selectively bind mainly to the active site residues of these mutant enzymes through favorable interactions such as hydrogen bonding, electrostatic interactions and van der Waals interactions to stabilize and help in proper folding of these misfolded proteins, prevent them from proteolytic degradation, increase their cellular levels, and enhance lysosomal activity (Leidenheimer & Ryder, 2014; Parenti et al., 2015; Porto, Ferrara, et al., 2012; Porto, Pisani, et al., 2012; Ringe & Petsko, 2009; Syed Haneef & George Priya Doss, 2016). Correct folding of the misfolded proteins assisted by the chaperones also enables trafficking of these proteins from the ER to their designated location and hence prevents their degradation by the ER quality-control system (Pereira, Valentão, & Andrade, 2018; Syed Haneef & George Priya Doss, 2016). PCT is mutation specific and is not effective against certain mutations in which large insertions or deletions, frameshifts or premature stop codons are present (Sardi et al., 2015). PCT has better oral administration and bioavailability compared to other therapies (Parenti et al., 2015; Parenti, Moracci, Fecarotta, & Andria, 2014) and many pharmacological chaperones are in testing stage for LSDs (Syed Haneef & George Priya Doss, 2016). PCs have broad tissue distribution and can cross the blood-brain barrier to access the CNS due to its low molecular weight (Khanna et al., 2010; Xu et al., 2015). PCs were observed to stabilize several exogenous recombinant enzymes used for the treatment of LSDs against thermal denaturation *in vitro* and improve the activity and uptake of the enzymes (Khanna et al., 2010; Shen et al., 2008; Xu et al., 2015).

Glycomimetic-based PCs, which emulate functional carbohydrates, have been widely investigated for LSDs since many enzymes in the lysosomal pathway are glycosidases, and the accumulated substrates are oligosaccharide-, polysaccharide- or glycoconjugate-based (Pereira et al., 2018; Sánchez-Fernández, García Fernández, & Mellet, 2016). 1-Deoxynojirimycin-based, isofagomine-based and 1,5-dideoxy-1,5-imino-D-xylitol-based pyranose sugar mimics are widely observed among the PCs for LSDs (Pereira et al., 2018). Fluorinated iminosugars in formulation with cyclodextrin have been shown to effectively rescue the neuronopathic form of Gaucher disease (García-Moreno et al., 2017).

The crystal structures of various sphingolipid metabolizing enzymes in complex with a pharmacological chaperone provide insight about the molecular mechanism by which PCs facilitate correct folding and trafficking of the enzyme. These include Krabbe associated galactocerebrosidase with azasugar derivatives (PDB ID: 4UFH) (Hill et al., 2015), Fabry associated alpha-galactosidase in complex with iminosugar 1-deoxygalactonojirimycin (DGJ) (PDB ID: 3S5Y) (Guce, Clark, Rogich, & Garman, 2011) and 1-deoxygalactonojirimycin-arylthiourea (DGJ-ArT) (PDB ID: 4NSX) (Yu et al., 2014), the Gaucher associated acid beta glucosidase in complex with the iminosugar isofagomine (PDB ID: 2NSX), with NOEV (PDB ID: 3WEZ) or with the seven membered 3,4,5,6-tetrahydroxyazepane (PDB ID: 3RIK) (Orwig et al., 2011). These PCs are competitive active site inhibitors and enhance the thermal stability of the enzyme as shown by differential scanning fluorimetry measurements (Hill et al., 2015; Orwig et al., 2011). It is hypothesized that binding of the PC to the active site locks the enzyme in the activated conformation, enhancing its stability (Lieberman et al., 2006). However, to allow catalysis, the active site needs to be accessible to the substrate. This discrepancy has been explained by experiments that show that the interaction between the glucocerebrosidase enzyme and the ligand NB-DNJ is pH dependent, and the affinity is stronger at neutral pH and weaker at acidic pH suggesting that the PC can dissociate in the acidic environment of the lysosome (Abian et al., 2011). The stabilizing effect of the PC was increased at neutral pH compared to that at acidic pH supporting the hypothesis that the PC dissociates from the active site in the lysosome (Guce et al., 2011). Furthermore, the concentration of substrate in the lysosome will be higher than that of the PC both due to dosing limitations as well as reduced solubility in the acidic environment, allowing the substrate to compete for the active site (Parenti et al., 2015). On the other hand, the binding of IFG was not pH dependent (Abian et al., 2011) suggesting that the available literature does not give a thorough picture for the mechanism of chaperone action. Novel activator compounds that are not inhibitors have also been proposed as PCs (Patnaik et al., 2012).

Recently combined therapeutic approaches used against LSDs present more promising outcomes than monotherapy in which only one therapeutic approach is used. The limitations encountered with a single therapeutic approach can be overcome by combining two or more of these approaches to obtain a synergistic effect. ERT and PCT have shown synergy in the treatment and management of sphingolipidoses such that the PC improves the efficacy of the ERT (Benjamin et al., 2012; Porto et al., 2009; Porto, Pisani, et al., 2012; Xu et al., 2015). For example, using the PC migalastat in combination with ERT generated promising results in the case of Fabry disease (Xu et al., 2015).

8. Sphingolipidoses and inflammation

Abnormalities in lysosomal function, such in the case in sphingolipidoses, can cause dysregulation of the sphingolipid rheostat, especially the levels of ceramide, sphingosine, and S1P, and induction of inflammatory pathways. Lysosomes have an essential role in immunity and inflammation and abnormal autophagy, activation of inflammation, and decreased infection control are caused by defective lysosomes (Simonaro, 2016).

8.1 Gaucher disease

An example for the link between sphingolipidoses and inflammation is Gaucher disease (GD) in which glucocerebrosidase is deficient due to mutations in *GBA1* gene (Parenti et al., 2015; Simonaro, 2016). More than 300 mutations have been identified in *GBA1*, and N370S and L444P are the most prevalent missense mutations with prevalence about 53% and 18%, respectively (Kuech, Brogden, & Naim, 2016; Parenti et al., 2015; Pereira et al., 2018). The substrates glucosylceramide (GC) and glucosylsphingosine accumulate in macrophages of various organs including spleen, liver, lungs, bone marrow, and brain (Huang, Lee, & Chen, 2014; Simonaro, 2016). This, in turn, leads to macrophage activation, eliciting the release of cytokines, such as chitotriosidase, TNF-α, and IL-1β. The signaling mechanism is activated to recruit additional immune cells; however, this only worsens the condition of the patients with Gaucher disease since these cells are themselves dysfunctional (Simonaro, 2016).

Accumulation of GC in immune cells often leads to chronic inflammation (Pandey et al., 2014). Activation of the complement was suggested to be the driver of proinflammation in Gaucher disease (Pandey et al., 2017). Inflammation plays a role in the pathogenesis of Gaucher disease (Mucci et al., 2015). Type 2 and 3 Gaucher diseases are associated with neuropathology. Elevated proinflammatory cytokine (IL-1α, IL-1β, TNF-α, IL-6) levels were observed in Gaucher patient blood cells (Bosch & Kielian, 2015). Microglial activation has been observed in the Gaucher disease brain. Increased systemic oxidative stress is also observed in type 1 Gaucher patients (Simonaro, 2016).

Gaucher disease is principally considered as a skeletal disorder (Mucci et al., 2015, 2013). The balance between osteoblastic bone formation and osteoclastic bone resorption is crucial, and loss of this balance may lead to skeletal disorders. Hepatosplenomegaly, anemia, thrombocytopenia and skeletal alterations are seen in type 1 Gaucher disease (Mucci et al., 2015). The level of the proinflammatory $CD14^{+}CD16^{+}$ monocyte subset is increased, and monocytes and T cells produce inflammatory cytokines in Gaucher patients. $CD14^{+}CD16^{+}$ monocytes were demonstrated to promote expansion in the population of Th17 cells, which have been suggested as an osteoclastogenic T-cell subset. RANKL levels were observed to be high in T cells. Mucci et al. hypothesized that the interaction of $CD14^{+}CD16^{+}$ monocytes and T cells in circulation possibly mediates the pathological osteoclastogenesis in Gaucher disease (Mucci et al., 2015).

GD patients may be afflicted with polyclonal and monoclonal gammopathy, which is a risk factor for myeloid cancers observed with high prevalence among Gaucher patients. High levels of IL-6 observed in Gaucher patients are also linked with the expansion of myeloid cells. The mechanism by which the accumulation of glucosylceramide and glucosylsphingosine in GD patients triggers these events is not known but it has been suggested that accumulation of these precursors can increase ceramide and sphingosine levels, and hence trigger inflammatory pathways (Simonaro, 2016).

Elevated levels of IL-1β, TNF-α macrophage colony-stimulating factor, and transforming growth factor β observed in the brains of a neuropathic GD mouse model suggest that neuroinflammation and Gaucher disease are correlated. The small-molecule NCGC758, which acts as a glucocerebrosidase chaperone, reversed the adverse effects of inflammasome activation and elevated secretion of IL-1β, caused by impaired autophagy in Gaucher disease patients (Aflaki et al., 2016; Simonaro, 2016). Taken together, due to the role of inflammation in the pathology of Gaucher disease, targeting inflammatory pathway in Gaucher patients can be a potential therapeutic approach for this disease (Simonaro, 2016).

PCT has been suggested as a potential therapy for Gaucher disease as an alternative to ERT and SRT (Narita et al., 2016; Parenti et al., 2015; Sardi et al., 2015) because even a modest elevation in residual glucocerebrosidase is enough to improve the phenotype (Shawky & Elsayed, 2016). Carbohydrate mimetics (iminosugars, azasugars, and carbasugars) and noncarbohydrate compounds that increase glucocerebrosidase activity have been identified by high-throughput screening initiatives. Isofagomine (IFG, AT2101, afegostat tartrate) and Ambroxol were suggested as the most promising among these molecules (see Fig. 5B) (Narita et al., 2016; Parenti et al., 2015; Shen et al., 2008).

The first pharmacological chaperone for Gaucher disease to undergo clinical trials was IFG, which selectively binds and stabilizes β-glucocerebrosidase in the ER and increases trafficking of the enzyme to the lysosome, especially for the N370S variant (Parenti et al., 2015; Sardi et al., 2015; Weinreb et al., 2007). IFG administration also increases the levels of a number of missense mutant form of glucocerebrosidase such as in the case of L444P variant (Parenti et al., 2015). It was reported that small concentrations of IFG were enough to increase glucocerebrosidase activity (Sardi et al., 2015). The effects of IFG administration were studied in preclinical *in vivo* studies conducted on the transgenic mouse model of Gaucher disease homozygous for L444P, V394L, N370S, D409H, or D409V mutations. Glucocerebrosidase activity in liver, liver macrophages, spleen, lung, bone and brain increased with the administration of IFG in these models. Neuroinflammation was reduced in V394L model that was administered with IFG (Parenti et al., 2015). Although IFG was a promising PC targeting glucocerebrosidase, it failed during clinical trials (Convertino et al., 2016).

Preincubation of glucocerebrosidase with IFG stabilized glucocerebrosidase against protein denaturation due to heat and neutral pH *in vitro*. It also increased the enzymatic activity and uptake of the enzyme highlighting the synergistic effect between ERT and PCT (Shen et al., 2008).

Mutations of the *GBA* gene have also been suggested as a significant risk factor of the neurodegenerative Parkinson's disease (PD), and Gaucher patients have an increased risk of developing PD (Mc Donald & Krainc, 2017; Schapira, 2015). N370S and L444P mutations are highly encountered mutations in both Gaucher disease and PD (Schapira, 2015). IFG was also tested on the transgenic mouse models of PD, and it was seen that PD symptoms were relieved; motor performance was improved, and microglial inflammation and α-synuclein immunoreactivity were reduced suggesting that therapeutic approaches found for Gaucher disease might also be applicable to PD (Richter et al., 2014).

Ambroxol is a FDA-approved mucolytic agent which is a potential pharmacological GBA chaperone, which can stabilize wild-type glucocerebrosidase against thermal denaturation (Parenti et al., 2015; Sardi et al., 2015; Shawky & Elsayed, 2016). Glucocerebrosidase levels in spleen and liver were increased *in vivo* with the administration of Ambroxol for 14 days on N370S or L444P transgenic mice (Parenti et al., 2015). High-dose oral Ambroxol was found to be promising in treating neuronopathic Gaucher disease (Narita et al., 2016).

Bicyclic L-idonojirimycin derivative has been proposed as a potential PCT option for patients with homozygous L444P mutations (Alfonso et al., 2013). Mitochondrial dysfunction was associated with reduced glucocerebrosidase activity (Shawky & Elsayed, 2016). Enzyme folding and trafficking in fibroblasts were improved with the supplementation of Coenzyme Q10 (CoQ) along with PCT. In addition, mitochondrial function was improved (Shawky & Elsayed, 2016). Targeting both mitochondria function by CoQ and protein misfolding by pharmacological chaperones, L-idonojirimycin derivative, *N*-[*N*′-(4-adamantan-1-ylcarboxamidobutyl)thiocarbamoyl]-1,6-anhydro-L-idonojirimycin (NadBT-AIJ), was suggested to be a promising therapy for Gaucher disease (de la Mata et al., 2015).

A computational study conducted with docking and MD simulations has suggested *N*-methylmescaline and shihunine retrieved from traditional Chinese medicine (TCM) database as potential lead compounds to design pharmacological chaperone for glucocerebrosidase folding correctly. These two compounds were also suggested as a possible therapeutic approach to decrease the risk factor for PD (Huang et al., 2014).

A docking campaign using computational drug discovery methods, ligand-based and structure-based pharmacophore modeling, docking and molecular dynamics simulations, was utilized to identify pyrido-thieno-pyrimidine and dioxino quinolone scaffolds as potential leads for the treatment of Gaucher disease (Yilmazer et al., 2017).

8.2 Fabry disease

Fabry is one of the most common LSDs that is caused by the deficiency of the enzyme α-galactosidase A due to the mutations in the *GLA* gene (Kuech et al., 2016). Acroparesthesia, angiokeratoma, renal failure, stroke and gastrointestinal symptoms are observed in Fabry patients (Kuech et al., 2016). Inflammation and abnormal immune response also play an important role in the pathogenesis of Fabry. Deficiency of this enzyme leads to the accumulation of the glycosphingolipids, globotriaosylceramide (Gb3) and globotriaosylsphingosine (lyso-Gb3) (Hughes et al., 2017; Parenti et al., 2015; Simonaro, 2016; Xu et al., 2015). Glycosphingolipid accumulation (particularly Gb3) in Fabry patients caused dysregulation of iNKT distribution and Cd1 pathway (Simonaro, 2016).

Increased levels of inflammatory cytokines including IL-6, IL-1β, and TNF-α observed in Fabry disease show the proinflammatory side of this disease (Bosch & Kielian, 2015; Simonaro, 2016). This increase is associated with TLR4 pathway because cytokine release is reduced in Fabry cells when TLR4-blocking agents were used. This pathway is activated in many LSDs, and TLR4 has been reported to bind glycosaminoglycans that accumulate in

MPS, and glycosphingolipids including the gangliosides GM1 and GM2 (Simonaro, 2016). As a key receptor in proinflammatory signaling, TLR-4 induces the release of TNF-α and IL-1β when activated (Lyman et al., 2014).

ERT is the main treatment approach for Fabry disease in which the manufactured human α-galactosidase A is administered intravenously. Even though it is effective in alleviating symptoms and reducing plasma Gb3 and lyso-Gb3 levels, the intravenously administered enzymes tend to be unstable at physiological conditions (Benjamin et al., 2012; Xu et al., 2015). Hence alternative therapeutic approaches have been sought for Fabry disease.

1-Deoxygalactonojirimycin (DGJ, AT1001, migalastat), an iminosugar and an analog of the terminal galactose of Gb3 (Fig. 5B), has been suggested as a PC for α-galactosidase A which binds reversibly and selectively to the active site of α-galactosidase A and improves its folding (Hughes et al., 2017; Parenti et al., 2015; Xu et al., 2015). Based on a pharmacogenetic assay study, out of 600 α-galactosidase A mutations associated with Fabry disease, 268 mutant forms were reported to be responsive to migalastat (Benjamin et al., 2017), and it is estimated that about 35–50% of Fabry patients have the responsive mutations (Pereira et al., 2018). Recently, six mutant variants are identified for α-galactosidase A deficiency and concomitant Fabry disease, and N53K and I303F mutant forms were found to respond to migalastat (Ferri et al., 2018). Migalastat can elevate the cellular activity of α-galactosidase A mutant forms *in vitro* and facilitate normal lysosomal trafficking (Hughes et al., 2017; Parenti et al., 2015). *In vivo* studies conducted on transgenic mouse model that expresses human mutant α-galactosidase A demonstrated elevated activity of α-galactosidase A and decreased levels of Gb3 when administered with migalastat orally (ISHII, 2012; Parenti et al., 2015). Migalastat was reported to be safe, well-tolerated and a promising alternative to ERT for Fabry patients, and recently, it was approved as a treatment option for adult and adolescent Fabry patients over 16 years (Hughes et al., 2017; Kuech et al., 2016; Platt, 2018). Both co-administration and coformulation of migalastat with the recombinant human α-galactosidase A increased the stability of this enzyme *in vitro* and the uptake of the enzyme was improved in cultured fibroblasts of Fabry patients which in turn reduced the Gb3 levels in both cases (Benjamin et al., 2012; Porto, Pisani, et al., 2012; Xu et al., 2015). Hence synergistic effect between ERT and PCT may be more promising than the monotherapeutic approaches (Porto, Pisani, et al., 2012).

8.3 GM1 and GM2 gangliosidosis

GM1 gangliosidosis is observed when lysosomal acid β-galactosidase is deficient leading to accumulation of GM1 ganglioside (Kohyama et al., 2016). Deficiency of β-hexosaminidase can lead to two neurodegenerative GM2 gangliosidoses: Tay-Sachs disease and Sandhoff disease (Parenti et al., 2015). Mouse models of GM1 and GM2 gangliosidosis have demonstrated broad microglial activation in the brain prior to neuronal cell death. Inflammation and neuronal cell death were reduced with bone marrow transplantation, suggesting that there was a link between inflammation and neurodegeneration (Simonaro, 2016). Another study conducted on mouse models of GM1 and GM2 gangliosidosis has shown a progressive increase in local microglial activation and infiltration of inflammatory cells. The onset of the inflammatory process was postponed by slowing down the accumulation of glycosphingolipids in the CNS by applying substrate reduction therapy in Sandhoff mouse model. In this study, it was suggested that inflammation might have an essential role in the pathogenesis of the gangliosidoses (Jeyakumar et al., 2003). Neuroinflammation has been seen in the mouse models of Tay-Sachs and Sandhoff disease. Elevated levels of inflammatory mediators including MCP-1α, macrophage inflammatory protein (MIP)-1β, and tumor necrosis factor receptor 2 (TNFR2) have been observed in Tay-Sachs or infantile gangliosidosis (Bosch & Kielian, 2015).

An antimalarial drug pyrimethamine, which can cross the blood-brain barrier, was proposed as a potential chaperone of β-hexosaminidase in the treatment of GM2 gangliosidoses (Fig. 5B) (Chiricozzi et al., 2014; Parenti et al., 2015; Platt, 2018; Suzuki, 2014). Although pyrimethamine was reported to increase the activity of β-hexosaminidase A, significant side effects were observed in the subjects on higher doses of pyrimethamine (Chiricozzi et al., 2014; Clarke et al., 2011). *N*-Octyl-4-epi-β-valienamine (NOEV) and 5*N*,6*S*-(*N'*-butyliminomethylidene)-6-thio-1-deoxygalactonojirimycin (6S-NBI-DGJ) were proposed as chaperones for GM1 gangliosidosis and it was demonstrated that these molecules improved the activity of β-galactosidase both *in vitro* and in transgenic animal models of GM1 gangliosidosis (Higaki et al., 2011; Parenti et al., 2015; Takai et al., 2013). Treatment with NOEV, a galactose analog (Fig. 5B), at the early stage of the disease reduced disease progression and prolonged survival time (Parenti et al., 2015; Suzuki, 2014). It was reported that NOEV was delivered to the CNS of the GM1 gangliosidosis model mice through blood-brain barrier (Suzuki, 2014). A mutagenesis study demonstrated that

22 missense mutations out of 99 mutations were responsive to NOEV (Higaki et al., 2011), while 24 out of 88 β-galactosidase mutants were amenable to 6S-NBI-DGJ (Takai et al., 2013). Although both NOEV and 6S-NBI-DGJ stabilize β-galactosidase, the latter one formed a weaker complex with β-galactosidase (Takai et al., 2013). The combination of NOEV and the bicyclic DGJ derivative, MTD118, was suggested as a powerful therapeutic approach with a potential coverage of 60–70% of GM1 gangliosidosis patients. Galactose was also suggested as a potential chaperone for both Fabry disease and some GM1 gangliosidosis mutations (Suzuki, 2014). Furthermore, 4-epi-isofagomine and its derivatives were reported to increase the activity of mutant β-galactosidase highlighting the potential of these PCs for the treatment of GM1 gangliosidosis (Front et al., 2017; Ichikawa, Igarashi, Ichikawa, & Suhara, 1998; Thonhofer et al., 2016).

8.4 Farber disease

Farber disease is a very rare autosomal recessive LSD caused by acid ceramidase deficiency due to mutations in the lysosomal acid ceramidase (*ASAH1*) gene (Ehlert et al., 2007; Kim et al., 2016; Sands, 2013). An acid ceramidase residual activity below 10% was reported to lead to Farber disease (Zhou et al., 2012). Since acid ceramidase is crucial for the degradation of ceramide, a deficiency in this enzyme leads to a progressive accumulation of the proinflammatory and pro-apoptotic sphingolipid, ceramide in most tissues (Sands, 2013; Sólyom, Karabul, Hügle, Simonaro, & Schuchman, 2014). Painful joint deformity, hoarse voice, subcutaneous nodules, progressive neurological deterioration and premature death are some of the manifestations observed in Farber disease (Kim et al., 2016; Zhou et al., 2012). Most of the mutations identified for Farber disease are missense mutations including E138V, R254G, and P362R (Bashyam et al., 2013; Kim et al., 2016; Li et al., 1999). Inflammation has been suggested to play an important role in Farber disease (Simonaro, 2016). It was suggested that granulomatous inflammation seen in this disease is related to ceramide's role in triggering inflammatory response (Ehlert et al., 2007).

A murine model was constructed by Alayoubi et al. by introducing a single-nucleotide mutation observed in Farber disease. In this study, an increase in the chemokine MCP-1 in organs from $Asah1^{P361R/P361R}$ mice was observed (Alayoubi et al., 2013). It has been hypothesized that the therapies that decrease MCP-1 would be beneficial in patients with Farber disease (Simonaro, 2016). No PCT data for Farber disease are encountered in the literature but the implication of ceramide in the inflammatory response described previously suggests that pharmacological chaperones that recover acid ceramidase activity can be powerful treatment alternative against Farber disease and the related inflammation.

8.5 Niemann-Pick type C disease

NPC is a rare autosomal recessive LSD caused by *NPC1* or *NPC2* mutations (with *NPC1* mutations accounting for the 95% of the cases) affecting about 1:150,000 live births (Alobaidy, 2015; Fukuda et al., 2017; Kuech et al., 2016; Mc Donald & Krainc, 2017). Accumulation of cholesterol and sphingolipids, particularly glycosphingolipids and sphingomyelin, in the lysosomal and late endosomal compartments occurs as a result of a mutation in either *NPC1* or *NPC2* which may impair intracellular lipid transport and lead to neuronal death (Alobaidy, 2015; Bosch & Kielian, 2015; Fukuda et al., 2017). Since NPC affects visceral organs, e.g., liver, spleen and lungs, and central and peripheral nervous system, it is also defined as a neurovisceral disorder (McKay Bounford & Gissen, 2014). Ataxia, dystonia, dementia, seizures and hepatosplenomegaly are among the characteristic symptoms of NPC (Kuech et al., 2016). The most common disease-causing mutation in NPC disease is I1061T NPC1 mutant which is functional but degraded by ER because it is an unstable misfolded protein (Fukuda et al., 2017; Gelsthorpe et al., 2008; Kuech et al., 2016). Elevated levels of mutant NPC1 proteins led to improved cholesterol level indicating that PCT can be a therapeutic approach for NPC (Kuech et al., 2016).

In a study conducted by Peake et al., an increase in the number of microglia was observed in *Npc1*$^{-}$/$^{-}$ mouse cerebellum and cortex (Peake, Campenot, Vance, & Vance, 2011). High levels of TNF-α and IL-1β along with heightened apoptosis were seen in several brain regions of NPC^{-}/$^{-}$ mice. Delay in the disease progression with the treatment with NSAIDs has indicated that neuroinflammation plays an important role in NPC. Furthermore, a combination therapy that includes NSAIDs, miglustat and curcumin, which acts as an antioxidant and cell penetrating agent, reduces microglial activation and increases the survival of NPC^{-}/$^{-}$ mice (Bosch & Kielian, 2015).

Oxysterol derivatives and nonsteroidal compounds with an amide-bearing phenanthridine-6-one scaffold, identified through screening compound libraries and structure-relationship studies, have been suggested as candidate PCs for NPC1 (Fukuda et al., 2017; Ohgane, Karaki, Dodo, & Hashimoto, 2013; Ohgane, Karaki, Noguchi-Yachide, Dodo, & Hashimoto, 2014). Both sterol and nonsteroidal PCs were reported to bind and stabilize I1061T NPC1 mutant with the latter one having more drug-like characteristics, and correct the localization, stability and function defects of the mutant NPC1 (Fukuda et al., 2017; Ohgane et al., 2013).

8.6 Krabbe disease

Krabbe disease, or globoid cell leukodystrophy, is an autosomal recessive neurodegenerative disorder which is caused by a mutation in the lysosomal enzyme galactosylceramidase or galactocerebrosidase (GALC) (Deane et al., 2011; O'Sullivan & Dev, 2015). At least 40% of the GALC mutant forms are missense mutations, most of which cause misfolding in the GALC enzyme (Spratley & Deane, 2016). Accumulation of the cytotoxic metabolite galactosylsphingosine (psychosine) and, to a lesser extent, β-galactosylceramide occurs as a result of a deficiency in GALC leading to rapid demyelination and cell death (Deane et al., 2011; O'Sullivan & Dev, 2015; Spratley & Deane, 2016). White matter of the peripheral nervous system and CNS is damaged by this cytotoxic metabolite, psychosine (Pavuluri, Vadakedath, Gundu, Uppulety, & Kandi, 2017). Among its manifestations are neurosensory deficits, muscle atrophy and premature death if not treated (Marshall & Bongarzone, 2016). Krabbe disease was suggested to be possibly associated with MS in which demyelination occurs just like in the case of Krabbe disease since *GALC* variants were proposed as a risk factor for MS. However, the disease-relevance between these diseases is still not clear (Sahai, Baris, Kimonis, & Levy, 2005; Scott-Hewitt et al., 2017, 2018).

Inflammation plays an important role in the pathogenesis of Krabbe disease. Inflammatory molecules (e.g., AMP-activated protein kinase (AMPK), prostaglandin D) and proinflammatory cytokines have been involved in Krabbe disease. The inhibition of oligodendrocyte differentiation, widespread demyelination and concomitant aberrant cell signaling has been suggested to be caused by apoptotic process and abnormal inflammatory responses (O'Sullivan & Dev, 2015; Spratley & Deane, 2016). O'Sullivan and Dev have investigated the effect of the regulation of S1P signaling on psychosine-induced astrocyte dysfunction, proinflammatory cytokine release and demyelination in Krabbe disease. Psychosine-induced human astrocyte cell death was decreased by the S1P receptor agonist, pFTY720. In addition, psychosine-induced mitochondrial dysfunction was also restored by pFTY720. Psychosine was shown to enhance lipopolysaccharide (LPS)-induced production of proinflammatory cytokines (IL-6, TNF-α, IL-1β) in mouse astrocytes even though it alone had no effect on the levels of proinflammatory cytokines. These effects of psychosine were again attenuated by pFTY720 (O'Sullivan & Dev, 2015).

Although ERT can ameliorate the peripheral symptoms of Krabbe disease, it is not a suitable approach for this neurodegenerative disease since the infused enzymes cannot cross the blood-brain barrier (Hill et al., 2015). While HSCT is the current treatment option, the risk of mortality related to the procedure highlights the urgent need to find new therapies for Krabbe disease (Pereira et al., 2018). Hence PCT may be a promising approach to treat Krabbe disease.

α-Lobeline has been suggested as a possible allosteric pharmacological chaperone for Krabbe disease since it has been observed to increase intracellular GALC activity of the D528N mutant; however, other tested mutant forms were not amenable to this PC (Hill et al., 2015; Lee, Kang, et al., 2010; Mohamed, Al-Gazali, Al-Jasmi, & Ali, 2017; Valenzano et al., 2011). NOEV and 3′4′7-trihydroxyisoflavone have also been suggested as possible pharmacological chaperones for Krabbe disease (Berardi et al., 2014; Hossain et al., 2015; Spratley & Deane, 2016). As stated before, NOEV also improved the activity of β-galactosidase (Parenti et al., 2015); however, the specificity of NOEV toward GALC and β-galactosidase is still not compared (Spratley & Deane, 2016). The specificity of 3′4′7-trihydroxyisoflavone is also controversial since it has been reported to inhibit some cellular enzymes including cyclin-dependent kinases, macrophage inhibitory factor and xanthine oxidase (Spratley & Deane, 2016). Azasugar derivatives especially with iso-galacto-fagomine and aza-galacto-fagomine scaffolds were also postulated as possible active site-directed PCs for Krabbe disease and stabilizing agents for ERT (Hill et al., 2015).

8.7 Metachromatic leukodystrophy

MLD is a neurodegenerative LSD caused by the mutations in the gene encoding the enzyme arylsulfatase A (*ARSA*). As a result, the accumulation of sulfatides in the nervous system leads to demyelination and concomitant neurodegeneration (Özkan & Özkara, 2016; Patil & Maegawa, 2013; Syed Haneef & George Priya Doss, 2016).

Studies of MLD in mouse models of mild nondemyelinating and severe demyelinating variants have shown that in severe forms microgliosis, T-cell infiltration and cytokine-chemokine levels increased. When demyelination begins, there is an increase in MIP, MCP-1, and many interleukins. It has been suggested that factors such as MCP-1, IL-1Ra, IL-8, MIP-1b and

VEGF are increased in MLD and that they play a role in pathophysiology in MLD due to both sulfatide accumulation and damage in the neural tissue. When treated with an NSAID in the experimental setting, motor capacity increased and demyelination was reduced. This study suggests that some MLD types may be targets for anti-inflammatory therapy (Barrette, Nave, & Edgar, 2013; Stein, Stroobants, Gieselmann, D'Hooge, & Matzner, 2015). An interesting and important point is that this response also functions in the repair and remyelination of the tissue as well as the destruction. Therefore, how and how much it affects clinical findings may vary according to the mutation and the stage of the disease (Anlar, Waye, Eng, & Karli, 2006). There are no PCs proposed for ARSA to date but PCT has been proposed as a promising therapeutic approach for MLD (Patil & Maegawa, 2013).

Table 1 summarizes the relation between sphingolipidoses and inflammation, and the PCs suggested for these diseases.

9. Conclusions and future perspectives

The sphingolipid metabolism tightly regulates the sphingolipid rheostat in the cell. Loss of this regulation results in lysosomal storage diseases in which the defective sphingolipid metabolizing enzyme cannot turn over the substrate, the rheostat is not maintained, and the sphingolipid accumulates in the cell. These diseases are caused by monogenic mutations that are risk factors for chronic inflammatory diseases. Furthermore, there is a two-way communication between the sphingolipid metabolic pathway and the inflammatory response. Neuroinflammation has been shown to precede neurodegeneration in Sandhoff diseases suggesting that anti-inflammatory therapies can suppress the progression of LSDs (Wada, Tifft, & Proia, 2000) and act synergistically with LSD therapies (Jeyakumar et al., 2004). This two-way communication between the two pathways opens up a new avenue for the treatment of inflammatory diseases using LSD therapies. For example, rescuing the defective glucocerebrosidase enzyme implicated in Gaucher disease using the chaperone NCGC758 induced autophagy and reduced IL-1β secretion suggesting that LSD therapies can be effective against inflammatory diseases such as Parkinson's disease (Aflaki et al., 2016).

Due to their complex pathophysiology and rareness, LSDs were not the focus of pharmaceutical industry interest for a long time but recently there are many ongoing studies on these diseases to find new therapeutic approaches because they have been linked to more complex diseases such

Table 1 Relation between sphingolipidoses and inflammation, and the PCs suggested for these diseases.

Disease	Gene	Deficient enzyme	Substance accumulated	Inflammation	PCs
Gaucher	*GBA*	Glucocerebrosidase	Glucosylceramide, glucosylsphingosine	Microglial activation Increase in TNF-α, IL-1β, IL-1α, IL-6, chitotriosidase, RANKL	Ambroxol (Narita et al., 2016) Isofagomine (Shen et al., 2008)
Fabry	*GLA*	α-Galactosidase A	Glycosphingolipids, Gb3, lyso-Gb3	Increase in TNF-α, IL-1β, IL-6	Migalastat (Hughes et al., 2017)
GM1 gangliosidosis	*GLB1*	β-Galactosidase	GM1 ganglioside	Microglial activation, neuroinflammation	NOEV (Higaki et al., 2011) 6S-NBI-DGJ (Takai et al., 2013)
GM2 gangliosidosis	*HEXA* *HEXB*	β-Hexosaminidase A, B	GM2 ganglioside	Microglial activation, neuroinflammation Increase in MCP-1 and TNFR2	Pyrimethamine (Chiricozzi et al., 2014)
Farber	*ASAH1*	Acid ceramidase	Ceramide	Granulomatous inflammation Increase in MCP-1	
NPC	*NPC1* *NPC2*	NPC1 NPC2	Cholesterol, glycosphingolipids, sphingomyelin	Microglial activation, neuroinflammation Increase in TNF-α, IL-1β	Oxysterol derivatives (Ohgane et al., 2013) Phenanthridine-6-one derivatives (Fukuda et al., 2017)
Krabbe	*GALC*	Galactosylceramidase	Psychosine, β-galactosylceramide	Increase in proinflammatory cytokines, AMPK, PGD	α-Lobeline (Lee, Kang, et al., 2010) NOEV (Hossain et al., 2015) 3′4′7-trihydroxyisoflavon (Berardi et al., 2014)
MLD	*ARSA*	Arylsulfatase A	Sulfatide	T-Cell infiltration Increase in cytokine-chemokine levels, MIP, MCP-1, many ILs	

as inflammatory diseases, cancer and neurodegenerative diseases. Understanding the pathogenic mechanism of and developing new therapeutic approaches for LSDs can provide better understanding about more common diseases such as Parkinson disease, which has been linked to Gaucher disease through glucocerebrosidase enzyme (Aflaki et al., 2017; Platt, 2018). Some of the approaches, such as ERT for type 1 Gaucher disease, are successful in clinical trials (Platt, 2018; Suzuki, 2014). However, there are some limitations encountered in ERT such as the instability of the infused enzymes due to the changes in temperature and pH levels, short half-life of the infused enzymes, high cost and inability to cross the blood-brain barrier (Convertino et al., 2016; Khanna et al., 2010; Parenti et al., 2013; Porto et al., 2009; Syed Haneef & George Priya Doss, 2016). Hence novel, well-tolerated, and safe therapeutic approaches are extensively investigated.

Many pharmacological chaperones are being developed for protein misfolding diseases including LSDs and neurodegenerative diseases (Narita et al., 2016). PCs increase the cellular levels of the misfolded proteins by binding and stabilizing these proteins in the ER, and this, in turn, improves their lysosomal trafficking and activity (Kuech et al., 2016). It has been suggested that about 10% enzyme activity is enough to prevent the accumulation of the substrates. Hence PCT is a promising therapeutic approach that can increase the activity of a mutant protein and relieve LSD symptoms (Parenti, 2009). PCs are noninvasive small molecules that are orally administered (Suzuki, 2014). Furthermore, the ability to penetrate the blood-brain barrier makes PCT a promising therapy for LSDs since about 75% of the described LSDs are associated with the CNS (Spratley & Deane, 2016). Recently, the PC migalastat was approved as an alternative treatment option for Fabry patients (Hughes et al., 2017; Kuech et al., 2016; Parenti et al., 2015; Platt, 2018).

As seen in other therapeutic approaches, PCT also has some limitations. One of the biggest challenges when designing PCs is the ability to cross the blood-brain barrier and increase the activity of the target enzyme present in the brain (Mc Donald & Krainc, 2017). Another challenge is that PCT is a mutation-specific therapy (Suzuki, 2014). PCs can only be effective against some specific chaperone-responsive mutations; however, LSDs have complex phenotype and pathophysiology. For example, more than 600 mutations including missense and nonsense mutations and deletions were identified for Fabry disease with only about 30–50% of the missense mutations potentially being responsive to PCT (Iemolo et al., 2014; Parenti, 2009; Parenti et al., 2015). Correct diagnosis of the sphingolipidoses as well as identification of the mutation requires personalized medicine approaches, increasing the cost of the treatment (Parenti, 2009). Increase in the residual enzymatic activity with PCs may also be too low to relieve the symptoms of some LSDs and improve lysosomal trafficking (Parenti et al., 2015).

Most of the PCs identified so far are reversible weak inhibitors that bind to the active site and facilitate correct folding of the enzyme by weak interactions that increase the stability of the folded state (Leidenheimer & Ryder, 2014; Parenti et al., 2015). These PCs bind more tightly in the neutral environment of the ER and are released in the acidic environment of the lysosome (Leidenheimer & Ryder, 2014; Parenti et al., 2015; Sánchez-Fernández et al., 2016). However, because the interaction is via the active site, binding of substrate may also be affected (Parenti et al., 2015). Developing allosteric PCs might be an approach to overcome these limitations (Citro et al., 2016; Convertino et al., 2016; Leidenheimer & Ryder, 2014; Parenti et al., 2015). The substrates do not have to compete with allosteric PCs that can remain bound to the enzyme without interfering with the binding of the natural substrate (Convertino et al., 2016; Parenti et al., 2015). However, it is important to state that allosteric chaperones may cause off-target effects, and since they are not active site-directed molecules, these adverse effects may be difficult to monitor (Spratley & Deane, 2016).

To overcome the limitations encountered in PCT and ERT therapeutic approaches, combined ERT/PCT therapies are also being extensively investigated for sphingolipidoses. The synergistic effect between ERT and PCT for Fabry disease is promising enough to further analyze this combined therapeutic approach for other diseases (Benjamin et al., 2012; Kuech et al., 2016; Porto, Pisani, et al., 2012; Xu et al., 2015). The infused recombinant enzymes that are generally susceptible to the changes in temperature and pH level can be stabilized with the addition of the PCs and the cellular uptake of these infused enzymes and lysosomal trafficking can be improved (Kuech et al., 2016; Parenti, 2009). PCs act on the infused wild-type enzymes, and hence they are mutation independent (Parenti et al., 2015; Spratley & Deane, 2016). Taken together, the impact of ERT and PCT on LSDs is enhanced with the synergistic effect of this combination therapy (Parenti, 2009).

Despite the plethora of anti-inflammatories, successful treatment of chronic inflammatory diseases is still an unmet medical need. In this review, we have reviewed the relationship between the inflammatory response and sphingolipid metabolism and aimed to show that, at least for some inflammatory states, the underlying cause is a defective sphingolipid metabolizing enzyme. Sphingolipidoses are often monogenic diseases that can be cured by targeting single proteins. Not all inflammatory diseases are caused by sphingolipids, but the fact that inflammation and sphingolipid pathways are closely linked suggests that targeting sphingolipid metabolic pathway might be an alternative therapy for inflammatory diseases as well. The biggest challenge here is to diagnose the sphingolipidoses early and correctly. Given the correct diagnosis, the root cause of the disease will be known, and the resultant inflammation can be cured with the correct, personalized therapy. PCT and combined ERT/PCT are promising therapeutic approaches for sphingolipidoses and may eliminate disease-related inflammatory components. Furthermore, the extensive understanding of inflammatory pathways, sphingolipid metabolism and their roles in the pathophysiology of LSDs will enable the development of efficient therapeutic strategies for LSDs and the diseases associated with them such as PD.

Acknowledgments

This work was financially supported by the Scientific and Technological Research Council of Turkey (TUBITAK) under the grant number of 115S208.

References

Abdel Hadi, L., Di Vito, C., & Riboni, L. (2016). Fostering inflammatory bowel disease: Sphingolipid strategies to join forces. *Mediators of Inflammation, 2016*, 3827684. https://doi.org/10.1155/2016/3827684.

Abian, O., Alfonso, P., Velazquez-Campoy, A., Giraldo, P., Pocovi, M., & Sancho, J. (2011). Therapeutic strategies for Gaucher disease: Miglustat (NB-DNJ) as a pharmacological chaperone for glucocerebrosidase and the different thermostability of velaglucerase alfa and imiglucerase. *Molecular Pharmaceutics*, *8*(6), 2390–2397. https://doi.org/10.1021/mp200313e.

Aflaki, E., Moaven, N., Borger, D. K., Lopez, G., Westbroek, W., Chae, J. J., et al. (2016). Lysosomal storage and impaired autophagy lead to inflammasome activation in Gaucher macrophages. *Aging Cell*, *15*(1), 77–88. https://doi.org/10.1111/acel.12409.

Aflaki, E., Westbroek, W., & Sidransky, E. (2017). The complicated relationship between Gaucher disease and parkinsonism: Insights from a rare disease. *Neuron*, *93*(4), 737–746. https://doi.org/10.1016/j.neuron.2017.01.018.

Afroze, B., & Brown, N. (2017). Ethical issues in managing lysosomal storage disorders in children in low and middle income countries. *Pakistan Journal of Medical Sciences*, *33*(4), 1036–1041. https://doi.org/10.12669/pjms.334.12975.

Agrawal, N. K., & Kant, S. (2014). Targeting inflammation in diabetes: Newer therapeutic options. *World Journal of Diabetes*, *5*(5), 697–710. https://doi.org/10.4239/wjd.v5.i5.697.

Akdis, M., Burgler, S., Crameri, R., Eiwegger, T., Fujita, H., Gomez, E., et al. (2011). Interleukins, from 1 to 37, and interferon-γ: Receptors, functions, and roles in diseases. *Journal of Allergy and Clinical Immunology*, *127*(3), 701–721.e70. https://doi.org/10.1016/j.jaci.2010.11.050.

Alayoubi, A. M., Wang, J. C. M., Au, B. C. Y., Carpentier, S., Garcia, V., Dworski, S., et al. (2013). Systemic ceramide accumulation leads to severe and varied pathological consequences. *EMBO Molecular Medicine*, *5*(6), 827–842. https://doi.org/10.1002/emmm.201202301.

Alfonso, P., Andreu, V., Pino-Angeles, A., Moya-García, A. A., García-Moreno, M. I., Rodríguez-Rey, J. C., et al. (2013). Bicyclic derivatives of L-idonojirimycin as pharmacological chaperones for neuronopathic forms of Gaucher disease. *Chembiochem*, *14*(8), 943–949. https://doi.org/10.1002/cbic.201200708.

Allende, M. L., Bektas, M., Lee, B. G., Bonifacino, E., Kang, J., Tuymetova, G., et al. (2011). Sphingosine-1-phosphate lyase deficiency produces a pro-inflammatory response while impairing neutrophil trafficking. *Journal of Biological Chemistry*, *286*(9), 7348–7358. https://doi.org/10.1074/jbc.M110.171819.

Alobaidy, H. (2015). Recent advances in the diagnosis and treatment of Niemann-Pick disease type C in children: A guide to early diagnosis for the general pediatrician. *International Journal of Pediatrics*, *2015*, 816593. https://doi.org/10.1155/2015/816593.

Alroy, J., Garganta, C., & Wiederschain, G. (2014). Secondary biochemical and morphological consequences in lysosomal storage diseases. *Biochemistry (Moscow)*, *79*(7), 619–636. https://doi.org/10.1134/S0006297914070049.

Alvarez, S. E., Harikumar, K. B., Hait, N. C., Allegood, J., Strub, G. M., Kim, E. Y., et al. (2010). Sphingosine-1-phosphate: A missing cofactor for the E3 ubiquitin ligase TRAF2. *Nature*, *465*(7301), 1084–1088. https://doi.org/10.1038/nature09128.

Anlar, B., Waye, J. S., Eng, B., & Karli, K. (2006). Atypical clinical course in juvenile metachromatic leukodystrophy involving novel arylsulfatase A gene mutations. *Developmental Medicine & Child Neurology*, *48*(5), 383–387. https://doi.org/10.1017/S001216220600082X.

Bader, G. D., & Hogue, C. W. (2003). An automated method for finding molecular complexes in large protein interaction networks. *BMC Bioinformatics*, *4*(2). https://doi.org/10.1186/1471-2105-4-2.

Baker, D. A., Barth, J., Chang, R., Obeid, L. M., & Gilkeson, G. S. (2010). Genetic sphingosine kinase 1 deficiency significantly decreases synovial inflammation and joint erosions in murine TNF-α-induced arthritis. *The Journal of Immunology*, *185*(4), 2570–2579. https://doi.org/10.4049/jimmunol.1000644.

Bamias, G., Clark, D. J., & Rivera-Nieves, J. (2013). Leukocyte traffic blockade in inflammatory bowel disease. *Current Drug Targets*, *14*(12), 1490–1500.

Barrette, B., Nave, K.-A., & Edgar, J. (2013). Molecular triggers of neuroinflammation in mouse models of demyelinating diseases. *Biological Chemistry*, *394*(12), 1571–1581. https://doi.org/10.1515/hsz-2013-0219.

Barton, G. M. (2008). A calculated response: Control of inflammation by the innate immune system. *The Journal of Clinical Investigation*, *118*(2), 413–420. https://doi.org/10.1172/JCI34431.

Bashyam, M. D., Chaudhary, A. K., Kiran, M., Reddy, V., Nagarajaram, H. A., Dalal, A., et al. (2013). Molecular analyses of novel ASAH1 mutations causing Farber lipogranulomatosis: Analyses of exonic splicing enhancer inactivating mutation. *Clinical Genetics*, *86*(6), 530–538. https://doi.org/10.1111/cge.12316.

Baudiß, K., Ayata, C. K., Lazar, Z., Cicko, S., Beckert, J., Meyer, A., et al. (2015). Ceramide-1-phosphate inhibits cigarette smoke-induced airway inflammation. *European Respiratory Journal*, *45*(6), 1669–1680. https://doi.org/10.1183/09031936.00080014.

Baudry, M., Yao, Y., Simmons, D., Liu, J., & Bi, X. (2003). Postnatal development of inflammation in a murine model of Niemann–Pick type C disease: Immunohistochemical observations of microglia and astroglia. *Experimental Neurology*, *184*(2), 887–903. https://doi.org/10.1016/S0014-4886(03)00345-5.

Benjamin, E. R., Della Valle, M. C., Wu, X., Katz, E., Pruthi, F., Bond, S., et al. (2017). The validation of pharmacogenetics for the identification of Fabry patients to be treated with migalastat. *Genetics in Medicine, 19*(4), 430–438. https://doi.org/10.1038/gim.2016.122.

Benjamin, E. R., Khanna, R., Schilling, A., Flanagan, J. J., Pellegrino, L. J., Brignol, N., et al. (2012). Co-administration with the pharmacological chaperone AT1001 increases recombinant human α-galactosidase A tissue uptake and improves substrate reduction in Fabry mice. *Molecular Therapy, 20*(4), 717–726. https://doi.org/10.1038/mt.2011.271.

Berardi, A. S., Pannuzzo, G., Graziano, A., Costantino-Ceccarini, E., Piomboni, P., & Luddi, A. (2014). Pharmacological chaperones increase residual β-galactocerebrosidase activity in fibroblasts from Krabbe patients. *Molecular Genetics and Metabolism, 112*(4), 294–301. https://doi.org/10.1016/j.ymgme.2014.05.009.

Berasain, C., Perugorria, M. J., Latasa, M. U., Castillo, J., Goñi, S., Santamaría, M., et al. (2009). The epidermal growth factor receptor: A link between inflammation and liver cancer. *Experimental Biology and Medicine (Maywood, NJ), 234*(7), 713–725. https://doi.org/10.3181/0901-MR-12.

Bernal, G. M., Wahlstrom, J. S., Crawley, C. D., Cahill, K. E., Pytel, P., Liang, H., et al. (2014). Loss of Nfkb1 leads to early onset aging. *Aging (Albany, NY), 6*(11), 931–942. https://doi.org/10.18632/aging.100702.

Beutler, E. (2004). Enzyme replacement in Gaucher disease. *PLoS Medicine, 1*(2), e21. https://doi.org/10.1371/journal.pmed.0010021.

Bosch, M. E., & Kielian, T. (2015). Neuroinflammatory paradigms in lysosomal storage diseases. *Frontiers in Neuroscience, 9*(417). https://doi.org/10.3389/fnins.2015.00417.

Boukhris, A., Schule, R., Loureiro, J. L., Lourenço, C. M., Mundwiller, E., Gonzalez, M. A., et al. (2013). Alteration of ganglioside biosynthesis responsible for complex hereditary spastic paraplegia. *American Journal of Human Genetics, 93*(1), 118–123. https://doi.org/10.1016/j.ajhg.2013.05.006.

Brinkmann, V., Billich, A., Baumruker, T., Heining, P., Schmouder, R., Francis, G., et al. (2010). Fingolimod (FTY720): Discovery and development of an oral drug to treat multiple sclerosis. *Nature Reviews Drug Discovery, 9*(11), 883–897. https://doi.org/10.1038/nrd3248.

Brunetti-Pierri, N., & Scaglia, F. (2008). GM1 gangliosidosis: Review of clinical, molecular, and therapeutic aspects. *Molecular Genetics and Metabolism, 94*(4), 391–396. https://doi.org/10.1016/j.ymgme.2008.04.012.

Calder, P. C. (2006). n-3 Polyunsaturated fatty acids, inflammation, and inflammatory diseases. *The American Journal of Clinical Nutrition, 83*(6), 1505S–1519S. https://doi.org/10.1093/ajcn/83.6.1505S.

Çeltikçi, B., Aydın, H. İ., Sivri, S., Sönmez, M., Topçu, M., & Özkara, H. A. (2012). Four novel mutations in the β-galactosidase gene identified in infantile type of GM1 gangliosidosis. *Clinical Biochemistry, 45*(7–8), 571–574. https://doi.org/10.1016/j.clinbiochem.2011.12.019.

Celtikci, B., Topcu, M., & Ozkara, H. A. (2011). Two novel alpha-galactosidase A mutations causing Fabry disease: A missense mutation M11V in a heterozygote woman and a nonsense mutation R190X in a hemizygote man. *Clinical Biochemistry, 44*(10–11), 809–812. https://doi.org/10.1016/j.clinbiochem.2011.04.022.

Cesani, M., Lorioli, L., Grossi, S., Amico, G., Fumagalli, F., Spiga, I., et al. (2016). Mutation update of ARSA and PSAP genes causing metachromatic leukodystrophy. *Human Mutation, 37*(1), 16–27. https://doi.org/10.1002/humu.22919.

Chatr-Aryamontri, A., Oughtred, R., Boucher, L., Rust, J., Chang, C., Kolas, N. K., et al. (2017). The BioGRID interaction database: 2017 update. *Nucleic Acids Research, 45*(Database Issue), D369–D379. https://doi.org/10.1093/nar/gkw1102.

Chen, H., Chan, A. Y., Stone, D. U., & Mandal, N. A. (2014). Beyond the cherry-red spot: Ocular manifestations of sphingolipid-mediated neurodegenerative and inflammatory disorders. *Survey of Ophthalmology*, *59*(1), 64–76. https://doi.org/10.1016/j.survophthal.2013.02.005.

Chiricozzi, E., Niemir, N., Aureli, M., Magini, A., Loberto, N., Prinetti, A., et al. (2014). Chaperone therapy for GM2 gangliosidosis: Effects of pyrimethamine on β-hexosaminidase activity in Sandhoff fibroblasts. *Molecular Neurobiology*, *50*(1), 159–167. https://doi.org/10.1007/s12035-013-8605-5.

Chitnis, T., & Weiner, H. L. (2017). CNS inflammation and neurodegeneration. *The Journal of Clinical Investigation*, *127*(10), 3577–3587. https://doi.org/10.1172/JCI90609.

Chiurchiù, V., Leuti, A., & Maccarrone, M. (2018). Bioactive lipids and chronic inflammation: Managing the fire within. *Frontiers in Immunology*, *9*(38). https://doi.org/10.3389/fimmu.2018.00038.

Cipriani, R., Chara, J. C., Rodríguez-Antigüedad, A., & Matute, C. (2015). FTY720 attenuates excitotoxicity and neuroinflammation. *Journal of Neuroinflammation*, *12*(86). https://doi.org/10.1186/s12974-015-0308-6.

Citro, V., Peña-García, J., Den-Haan, H., Pérez-Sánchez, H., Del Prete, R., Liguori, L., et al. (2016). Identification of an allosteric binding site on human lysosomal alpha-galactosidase opens the way to new pharmacological chaperones for Fabry disease. *PLoS One*, *11*(10), e0165463. https://doi.org/10.1371/journal.pone.0165463.

Clarke, J. T. R., Mahuran, D. J., Sathe, S., Kolodny, E. H., Rigat, B. A., Raiman, J. A., et al. (2011). An open-label phase I/II clinical trial of pyrimethamine for the treatment of patients affected with chronic GM2 gangliosidosis (Tay–Sachs or Sandhoff variants). *Molecular Genetics and Metabolism*, *102*(1), 6–12. https://doi.org/10.1016/j.ymgme.2010.09.004.

ClinicalTrials.gov. (2013a). NCT01802970. Safety and blood immune cell study of Anakinra plus Physician's chemotherapy choice in metastatic breast cancer patients. Retrieved August 10, 2018, from https://clinicaltrials.gov/ct2/show/NCT01802970.

ClinicalTrials.gov. (2013b). NCT02021422. A pilot, prospective, non-randomized evaluation of the safety of Anakinra plus standard chemotherapy. Retrieved August 10, 2018, from https://clinicaltrials.gov/ct2/show/NCT02021422.

ClinicalTrials.gov. (2014). NCT02090101. Study evaluating the influence of LV5FU2 bevacizumab plus Anakinra association on metastatic colorectal Cancer (IRAFU). Retrieved August 10, 2018, from https://clinicaltrials.gov/ct2/show/NCT02090101.

ClinicalTrials.gov. (2018). NCT03588936. Nivolumab and tocilizumab for relapsed hematological malignancy post-allogeneic transplant. Retrieved August 22, 2018, from https://clinicaltrials.gov/ct2/show/NCT03588936.

Cokelaer, T., Pultz, D., Harder, L. M., Serra-Musach, J., Saez-Rodriguez, J., & Valencia, A. (2013). BioServices: A common Python package to access biological web services programmatically. *Bioinformatics*, *29*(24), 3241–3242. https://doi.org/10.1093/bioinformatics/btt547.

Convertino, M., Das, J., & Dokholyan, N. V. (2016). Pharmacological chaperones: Design and development of new therapeutic strategies for the treatment of conformational diseases. *ACS Chemical Biology*, *11*(6), 1471–1489. https://doi.org/10.1021/acschembio.6b00195.

Cox, T. M., & Cachón-González, M. B. (2012). The cellular pathology of lysosomal diseases. *The Journal of Pathology*, *226*(2), 241–254. https://doi.org/10.1002/path.3021.

Crofford, L. J. (2013). Use of NSAIDs in treating patients with arthritis. *Arthritis Research & Therapy*, *15*(Suppl. 3), S2. https://doi.org/10.1186/ar4174.

D'Angelo, G., Rega, L. R., & De Matteis, M. A. (2012). Connecting vesicular transport with lipid synthesis: FAPP2. *Biochimica et Biophysica Acta*, *1821*(8), 1089–1095. https://doi.org/10.1016/j.bbalip.2012.01.003.

Davis, K. E., Neinast, M. D., Sun, K., Skiles, W. M., Bills, J. D., Zehr, J. A., et al. (2013). The sexually dimorphic role of adipose and adipocyte estrogen receptors in modulating adipose tissue expansion, inflammation, and fibrosis. *Molecular Metabolism*, *2*(3), 227–242. https://doi.org/10.1016/j.molmet.2013.05.006.

Deane, J. E., Graham, S. C., Kim, N. N., Stein, P. E., McNair, R., Cachón-González, M. B., et al. (2011). Insights into Krabbe disease from structures of galactocerebrosidase. *Proceedings of the National Academy of Sciences of the United States of America*, *108*(37), 15169–15173. https://doi.org/10.1073/pnas.1105639108.

de la Mata, M., Cotán, D., Oropesa-Ávila, M., Garrido-Maraver, J., Cordero, M. D., Villanueva Paz, M., et al. (2015). Pharmacological chaperones and coenzyme Q10 treatment improves mutant β-glucocerebrosidase activity and mitochondrial function in neuronopathic forms of Gaucher disease. *Scientific Reports*, *5*(10903). https://doi.org/10.1038/srep10903.

Deniz, U., Ozkirimli, E., & Ulgen, K. O. (2016). A systematic methodology for large scale compound screening: A case study on the discovery of novel S1PL inhibitors. *Journal of Molecular Graphics and Modelling*, *63*, 110–124. https://doi.org/10.1016/j.jmgm.2015.11.004.

Deshmane, S. L., Kremlev, S., Amini, S., & Sawaya, B. E. (2009). Monocyte chemoattractant protein-1 (MCP-1): An overview. *Journal of Interferon & Cytokine Research*, *29*(6), 313–326. https://doi.org/10.1089/jir.2008.0027.

Dinarello, C. A., Simon, A., & van der Meer, J. W. M. (2012). Treating inflammation by blocking interleukin-1 in a broad spectrum of diseases. *Nature Reviews. Drug Discovery*, *11*(8), 633–652. https://doi.org/10.1038/nrd3800.

Dinkins, M. B., Dasgupta, S., Wang, G., Zhu, G., & Bieberich, E. (2014). Exosome reduction in vivo is associated with lower amyloid plaque load in the 5XFAD mouse model of Alzheimer's disease. *Neurobiology of Aging*, *35*(8), 1792–1800. https://doi.org/10.1016/j.neurobiolaging.2014.02.012.

Dinkins, M. B., Enasko, J., Hernandez, C., Wang, G., Kong, J., Helwa, I., et al. (2016). Neutral sphingomyelinase-2 deficiency ameliorates Alzheimer's disease pathology and improves cognition in the 5XFAD mouse. *The Journal of Neuroscience*, *36*(33), 8653–8667. https://doi.org/10.1523/JNEUROSCI.1429-16.2016.

Ehlert, K., Frosch, M., Fehse, N., Zander, A., Roth, J., & Vormoor, J. (2007). Farber disease: Clinical presentation, pathogenesis and a new approach to treatment. *Pediatric Rheumatology Online Journal*, *5*, 15. https://doi.org/10.1186/1546-0096-5-15.

Espaillat, M. P., Kew, R. R., & Obeid, L. M. (2017). Sphingolipids in neutrophil function and inflammatory responses: Mechanisms and implications for intestinal immunity and inflammation in ulcerative colitis. *Advances in Biological Regulation*, *63*, 140–155. https://doi.org/10.1016/j.jbior.2016.11.001.

Farfel-Becker, T., Vitner, E. B., Pressey, S. N. R., Eilam, R., Cooper, J. D., & Futerman, A. H. (2011). Spatial and temporal correlation between neuron loss and neuroinflammation in a mouse model of neuronopathic Gaucher disease. *Human Molecular Genetics*, *20*(7), 1375–1386. https://doi.org/10.1093/hmg/ddr019.

Ferri, L., Malesci, D., Fioravanti, A., Bagordo, G., Filippini, A., Ficcadenti, A., et al. (2018). Functional and pharmacological evaluation of novel GLA variants in Fabry disease identifies six (two de novo) causative mutations and two amenable variants to the chaperone DGJ. *Clinica Chimica Acta*, *481*, 25–33. https://doi.org/10.1016/j.cca.2018.02.021.

Filosto, S., Castillo, S., Danielson, A., Franzi, L., Khan, E., Kenyon, N., et al. (2011). Neutral sphingomyelinase 2: A novel target in cigarette smoke-induced apoptosis and lung injury. *American Journal of Respiratory Cell and Molecular Biology*, *44*(3), 350–360. https://doi.org/10.1165/rcmb.2009-0422OC.

Front, S., Biela-Banaś, A., Burda, P., Ballhausen, D., Higaki, K., Caciotti, A., et al. (2017). (5aR)-5a-C-Pentyl-4-epi-isofagomine: A powerful inhibitor of lysosomal β-galactosidase and a remarkable chaperone for mutations associated with GM1-gangliosidosis and

Morquio disease type B. *European Journal of Medicinal Chemistry*, *126*, 160–170. https://doi.org/10.1016/j.ejmech.2016.09.095.

Fucho, R., Casals, N., Serra, D., & Herrero, L. (2016). Ceramides and mitochondrial fatty acid oxidation in obesity. *The FASEB Journal*, *31*(4), 1263–1272. https://doi.org/10.1096/fj.201601156R.

Fukuda, H., Karaki, F., Dodo, K., Noguchi-Yachide, T., Ishikawa, M., Hashimoto, Y., et al. (2017). Phenanthridin-6-one derivatives as the first class of non-steroidal pharmacological chaperones for Niemann-Pick disease type C1 protein. *Bioorganic & Medicinal Chemistry Letters*, *27*(12), 2781–2787. https://doi.org/10.1016/j.bmcl.2017.04.062.

García-Moreno, M. I., de la Mata, M., Sánchez-Fernández, E. M., Benito, J. M., Díaz-Quintana, A., Fustero, S., et al. (2017). Fluorinated chaperone-β-cyclodextrin formulations for β-glucocerebrosidase activity enhancement in neuronopathic Gaucher disease. *Journal of Medicinal Chemistry*, *60*(5), 1829–1842. https://doi.org/10.1021/acs.jmedchem.6b01550.

Gelsthorpe, M. E., Baumann, N., Millard, E., Gale, S. E., Langmade, S. J., Schaffer, J. E., et al. (2008). Niemann-Pick type C1 I1061T mutant encodes a functional protein that is selected for endoplasmic reticulum-associated degradation due to protein misfolding. *Journal of Biological Chemistry*, *283*(13), 8229–8236. https://doi.org/10.1074/jbc.M708735200.

Getz, G. S. (2005). Bridging the innate and adaptive immune systems. *Journal of Lipid Research*, *46*(4), 619–622. https://doi.org/10.1194/jlr.E500002-JLR200.

Ghidoni, R., Caretti, A., & Signorelli, P. (2015). Role of sphingolipids in the pathobiology of lung inflammation. *Mediators of Inflammation*, *2015*, 487505 https://doi.org/10.1155/2015/487508.

Goker-Alpan, O., Schiffmann, R., LaMarca, M. E., Nussbaum, R. L., McInerney-Leo, A., & Sidransky, E. (2004). Parkinsonism among Gaucher disease carriers. *Journal of Medical Genetics*, *41*(12), 937–940. https://doi.org/10.1136/jmg.2004.024455.

Goldsmith, M., Avni, D., Levy-Rimler, G., Mashiach, R., Ernst, O., Levi, M., et al. (2009). A ceramide-1-phosphate analogue, PCERA-1, simultaneously suppresses tumour necrosis factor-α and induces interleukin-10 production in activated macrophages. *Immunology*, *127*(1), 103–115. https://doi.org/10.1111/j.1365-2567.2008.02928.x.

Gomez-Muñoz, A., Presa, N., Gomez-Larrauri, A., Rivera, I. G., Trueba, M., & Ordoñez, M. (2016). Control of inflammatory responses by ceramide, sphingosine 1-phosphate and ceramide 1-phosphate. *Progress in Lipid Research*, *61*, 51–62. https://doi.org/10.1016/j.plipres.2015.09.002.

Gonzalez, E., Kou, R., & Michel, T. (2006). Rac1 modulates sphingosine 1-phosphate-mediated activation of phosphoinositide 3-kinase/Akt signaling pathways in vascular endothelial cells. *Journal of Biological Chemistry*, *281*(6), 3210–3216. https://doi.org/10.1074/jbc.M510434200.

Gosejacob, D., Jäger, P. S., vom Dorp, K., Frejno, M., Carstensen, A. C., Köhnke, M., et al. (2016). Ceramide synthase 5 is essential to maintain C16:0-ceramide pools and contributes to the development of diet-induced obesity. *Journal of Biological Chemistry*, *291*(13), 6989–7003. https://doi.org/10.1074/jbc.M115.691212.

Goyal, R., Faizy, A. F., Siddiqui, S. S., & Singhai, M. (2012). Evaluation of TNF-α and IL-6 levels in obese and non-obese diabetics: Pre- and postinsulin effects. *North American Journal of Medical Sciences*, *4*(4), 180–184. https://doi.org/10.4103/1947-2714.94944.

Grassmé, H., Carpinteiro, A., Edwards, M. J., Gulbins, E., & Becker, K. A. (2014). Regulation of the inflammasome by ceramide in cystic fibrosis lungs. *Cellular Physiology and Biochemistry*, *34*(1), 45–55. https://doi.org/10.1159/000362983.

Grassmé, H., Jendrossek, V., Riehle, A., von Kürthy, G., Berger, J., Schwarz, H., et al. (2003). Host defense against Pseudomonas aeruginosa requires ceramide-rich membrane rafts. *Nature Medicine*, *9*, 322–330. https://doi.org/10.1038/nm823.

Graziano, A. C. E., & Cardile, V. (2015). History, genetic, and recent advances on Krabbe disease. *Gene, 555*(1), 2–13. https://doi.org/10.1016/j.gene.2014.09.046.

Guce, A. I., Clark, N. E., Rogich, J. J., & Garman, S. C. (2011). The molecular basis of pharmacological chaperoning in human α-galactosidase. *Chemistry & Biology, 18*(12), 1521–1526. https://doi.org/10.1016/j.chembiol.2011.10.012.

Hammad, S. M., Crellin, H. G., Wu, B. X., Melton, J., Anelli, V., & Obeid, L. M. (2008). Dual and distinct roles for sphingosine kinase 1 and sphingosine 1 phosphate in the response to inflammatory stimuli in RAW macrophages. *Prostaglandins & Other Lipid Mediators, 85*(3–4), 107–114. https://doi.org/10.1016/j.prostaglandins.2007.11.002.

Hanada, K. (2014). Co-evolution of sphingomyelin and the ceramide transport protein CERT. *Biochimica et Biophysica Acta, 1841*(5), 704–719. https://doi.org/10.1016/j.bbalip.2013.06.006.

Hannun, Y. A., & Obeid, L. M. (2008). Principles of bioactive lipid signalling: Lessons from sphingolipids. *Nature Reviews. Molecular Cell Biology, 9*(2), 139–150. https://doi.org/10.1038/nrm2329.

Hannun, Y. A., & Obeid, L. M. (2018). Sphingolipids and their metabolism in physiology and disease. *Nature Reviews Molecular Cell Biology, 19*(3), 175–191. https://doi.org/10.1038/nrm.2017.107.

Harlalka, G. V., Lehman, A., Chioza, B., Baple, E. L., Maroofian, R., Cross, H., et al. (2013). Mutations in B4GALNT1 (GM2 synthase) underlie a new disorder of ganglioside biosynthesis. *Brain, 136*(12), 3618–3624. https://doi.org/10.1093/brain/awt270.

Higaki, K., Li, L., Bahrudin, U., Okuzawa, S., Takamuram, A., Yamamoto, K., et al. (2011). Chemical chaperone therapy: Chaperone effect on mutant enzyme and cellular pathophysiology in β-galactosidase deficiency. *Human Mutation, 32*(7), 843–852. https://doi.org/10.1002/humu.21516.

Hill, C. H., Viuff, A. H., Spratley, S. J., Salamone, S., Christensen, S. H., Read, R. J., et al. (2015). Azasugar inhibitors as pharmacological chaperones for Krabbe disease. *Chemical Science, 6*(5), 3075–3086. https://doi.org/10.1039/c5sc00754b.

Hollak, C. E. M., & Wijburg, F. A. (2014). Treatment of lysosomal storage disorders: Successes and challenges. *Journal of Inherited Metabolic Disease, 37*(4), 587–598. https://doi.org/10.1007/s10545-014-9718-3.

Hong, H., Kim, B. S., & Im, H.-I. (2016). Pathophysiological role of neuroinflammation in neurodegenerative diseases and psychiatric disorders. *International Neurourology Journal, 20*(Suppl. 1), S2–S7. https://doi.org/10.5213/inj.1632604.302.

Hossain, M. A., Higaki, K., Saito, S., Ohno, K., Sakuraba, H., Nanba, E., et al. (2015). Chaperone therapy for Krabbe disease: Potential for late-onset GALC mutations. *Journal of Human Genetics, 60*(9), 539–545. https://doi.org/10.1038/jhg.2015.61.

Hsieh, H.-L., Sun, C.-C., Wu, C.-B., Wu, C.-Y., Tung, W.-H., Wang, H.-H., et al. (2008). Sphingosine 1-phosphate induces EGFR expression via Akt/NF-kappaB and ERK/AP-1 pathways in rat vascular smooth muscle cells. *Journal of Cellular Biochemistry, 103*(6), 1732–1746. https://doi.org/10.1002/jcb.21563.

Huang, H. J., Lee, C. C., & Chen, C. Y. C. (2014). Pharmacological chaperone design for reducing risk factor of Parkinson's disease from traditional Chinese medicine. *Evidence-Based Complementary and Alternative Medicine, 2014*, 830490. https://doi.org/10.1155/2014/830490.

Huang, H., Kasumov, T., Gatmaitan, P., Heneghan, H. M., Kashyap, S. R., Schauer, P. R., et al. (2011). Gastric bypass surgery reduces plasma ceramide subspecies and improves insulin sensitivity in severely obese patients. *Obesity (Silver Spring, Md.), 19*(11). https://doi.org/10.1038/oby.2011.107 https://doi.org/10.1038/oby.2011.107.

Hughes, D. A., Nicholls, K., Shankar, S. P., Sunder-Plassmann, G., Koeller, D., Nedd, K., et al. (2017). Oral pharmacological chaperone migalastat compared with enzyme replacement therapy in Fabry disease: 18-Month results from the randomised phase III ATTRACT study. *Journal of Medical Genetics*, *54*(4), 288–296. https://doi.org/10.1136/jmedgenet-2016-104178.

Ichikawa, Y., Igarashi, Y., Ichikawa, M., & Suhara, Y. (1998). 1-N-Iminosugars: Potent and selective inhibitors of β-glycosidases. *Journal of the American Chemical Society*, *120*(13), 3007–3018. https://doi.org/10.1021/ja973443k.

Iemolo, F., Pizzo, F., Albeggiani, G., Zizzo, C., Colomba, P., Scalia, S., et al. (2014). De novo mutation in a male patient with Fabry disease: A case report. *BMC Research Notes*, 7, 11. https://doi.org/10.1186/1756-0500-7-11.

ISHII, S. (2012). Pharmacological chaperone therapy for Fabry disease. *Proceedings of the Japan Academy. Series B, Physical and Biological Sciences*, *88*(1), 18–30. https://doi.org/10.2183/pjab.88.18.

Iwasaki, A., & Medzhitov, R. (2010). Regulation of adaptive immunity by the innate immune system. *Science (New York, N.Y.)*, *327*(5963), 291–295. https://doi.org/10.1126/science.1183021.

Janeway, C. A., Travers, P., Walport, M., & Shlomchik, M. (2001). The components of the immune system. In *Immunobiology: The immune system in health and disease*. (5th ed.). New York: Garland Science. Retrieved from https://www.ncbi.nlm.nih.gov/books/NBK27092/.

Jeyakumar, M., Smith, D. A., Williams, I. M., Borja, M. C., Neville, D. C. A., Butters, T. D., et al. (2004). NSAIDs increase survival in the Sandhoff disease mouse: Synergy with N-butyldeoxynojirimycin. *Annals of Neurology*, *56*(5), 642–649. https://doi.org/10.1002/ana.20242.

Jeyakumar, M., Thomas, R., Elliot-Smith, E., Smith, D. A., Van der Spoel, A. C., D'Azzo, A., et al. (2003). Central nervous system inflammation is a hallmark of pathogenesis in mouse models of GM1 and GM2 gangliosidosis. *Brain*, *126*(Pt 4), 974–987. https://doi.org/10.1093/brain/awg089.

Jin, J., Xiao, Y., Hu, H., Zou, Q., Li, Y., Gao, Y., et al. (2015). Proinflammatory TLR signalling is regulated by a TRAF2-dependent proteolysis mechanism in macrophages. *Nature Communications*, *6*, 5930 https://doi.org/10.1038/ncomms6930.

Kanehisa, M., Sato, Y., Kawashima, M., Furumichi, M., & Tanabe, M. (2016). KEGG as a reference resource for gene and protein annotation. *Nucleic Acids Research*, *44*(Database Issue), D457–D462. https://doi.org/10.1093/nar/gkv1070.

Keshava Prasad, T. S., Goel, R., Kandasamy, K., Keerthikumar, S., Kumar, S., Mathivanan, S., et al. (2009). Human protein reference database—2009 update. *Nucleic Acids Research*, *37*(Database issue), D767–D772. https://doi.org/10.1093/nar/gkn892.

Ketelhuth, D. F. J., & Hansson, G. K. (2016). Adaptive response of T and B cells in atherosclerosis. *Circulation Research*, *118*(4), 668–678. https://doi.org/10.1161/CIRCRESAHA.115.306427.

Khanna, R., Soska, R., Lun, Y., Feng, J., Frascella, M., Young, B., et al. (2010). The pharmacological chaperone 1-deoxygalactonojirimycin reduces tissue globotriaosylceramide levels in a mouse model of Fabry disease. *Molecular Therapy*, *18*(1), 23–33. https://doi.org/10.1038/mt.2009.220.

Kim, S. Y., Choi, S. A., Lee, S., Lee, J. S., Hong, C. R., Lim, B. C., et al. (2016). Atypical presentation of infantile-onset Farber disease with novel ASAH1 mutations. *American Journal of Medical Genetics Part A*, *170*(11), 3023–3027. https://doi.org/10.1002/ajmg.a.37846.

Kitano, M., Hla, T., Sekiguchi, M., Kawahito, Y., Yoshimura, R., Miyazawa, K., et al. (2006). Sphingosine 1-phosphate/sphingosine 1-phosphate receptor 1 signaling in rheumatoid synovium: Regulation of synovial proliferation and inflammatory gene expression. *Arthritis and Rheumatism*, *54*(3), 742–753. https://doi.org/10.1002/art.21668.

Kitatani, K., Idkowiak-Baldys, J., & Hannun, Y. A. (2008). The sphingolipid salvage pathway in ceramide metabolism and signaling. *Cellular Signalling*, *20*(6), 1010–1018. https://doi.org/10.1016/j.cellsig.2007.12.006.

Kitatani, K., Iwabuchi, K., Snider, A., & Riboni, L. (2016). Sphingolipids in inflammation: From bench to bedside. *Mediators of Inflammation*, *2016*, 7602526. https://doi.org/10.1155/2016/7602526.

Kohyama, M., Yabuki, A., Ochiai, K., Nakamoto, Y., Uchida, K., Hasegawa, D., et al. (2016). In situ detection of GM1 and GM2 gangliosides using immunohistochemical and immunofluorescent techniques for auxiliary diagnosis of canine and feline gangliosidoses. *BMC Veterinary Research*, *12*, 67. https://doi.org/10.1186/s12917-016-0691-y.

Kolter, T. (2011). A view on sphingolipids and disease. *Chemistry and Physics of Lipids*, *164*(6), 590–606. https://doi.org/10.1016/j.chemphyslip.2011.04.013.

Kolter, T., & Sandhoff, K. (2005). Principles of lysosomal membrane digestion: Stimulation of sphingolipid degradation by sphingolipid activator proteins and anionic lysosomal lipids. *Annual Review of Cell and Developmental Biology*, *21*, 81–103. https://doi.org/10.1146/annurev.cellbio.21.122303.120013.

Kolter, T., & Sandhoff, K. (2006). Sphingolipid metabolism diseases. *Biochimica et Biophysica Acta*, *1758*(12), 2057–2079. https://doi.org/10.1016/j.bbamem.2006.05.027.

Kolter, T., & Sandhoff, K. (2010). Lysosomal degradation of membrane lipids. *FEBS Letters*, *584*(9), 1700–1712. https://doi.org/10.1016/j.febslet.2009.10.021.

Kresojević, N., Dobričić, V., Svetel, M., & Kostić, V. (2014). Mutations in Niemann Pick type C gene are risk factor for Alzheimer's disease. *Medical Hypotheses*, *83*(5), 559–562. https://doi.org/10.1016/j.mehy.2014.08.025.

Kuech, E.-M., Brogden, G., & Naim, H. Y. (2016). Alterations in membrane trafficking and pathophysiological implications in lysosomal storage disorders. *Biochimie*, *130*, 152–162. https://doi.org/10.1016/j.biochi.2016.09.011.

Lahiri, S., & Futerman, A. H. (2007). The metabolism and function of sphingolipids and glycosphingolipids. *Cellular and Molecular Life Sciences*, *64*(17), 2270–2284. https://doi.org/10.1007/s00018-007-7076-0.

LaRosa, D. F., & Orange, J. S. (2008). 1. Lymphocytes. *Journal of Allergy and Clinical Immunology*, *121*(2, Supplement 2), S364–S369. https://doi.org/10.1016/j.jaci.2007.06.016.

Lasry, A., & Ben-Neriah, Y. (2015). Senescence-associated inflammatory responses: Aging and cancer perspectives. *Trends in Immunology*, *36*(4), 217–228. https://doi.org/10.1016/j.it.2015.02.009.

Lauro, G., Strocchia, M., Terracciano, S., Bruno, I., Fischer, K., Pergola, C., et al. (2014). Exploration of the dihydropyrimidine scaffold for the development of new potential anti-inflammatory agents blocking prostaglandin E2 synthase-1 enzyme (mPGES-1). *European Journal of Medicinal Chemistry*, *80*, 407–415. https://doi.org/10.1016/j.ejmech.2014.04.061.

Lee, H., Bhang, S. H., Lee, J. H., Kim, H., & Hahn, S. K. (2017). Tocilizumab–alendronate conjugate for treatment of rheumatoid arthritis. *Bioconjugate Chemistry*, *28*(4), 1084–1092. https://doi.org/10.1021/acs.bioconjchem.7b00008.

Lee, H., Deng, J., Kujawski, M., Yang, C., Liu, Y., Herrmann, A., et al. (2010). Stat3-induced S1PR1 expression is critical for persistent Stat3 activation in tumors. *Nature Medicine*, *16*(12), 1421–1428. https://doi.org/10.1038/nm.2250.

Lee, K. M., & MacLean, A. G. (2015). New advances on glial activation in health and disease. *World Journal of Virology*, *4*(2), 42–55. https://doi.org/10.5501/wjv.v4.i2.42.

Lee, W. C., Kang, D., Causevic, E., Herdt, A. R., Eckman, E. A., & Eckman, C. B. (2010). Molecular characterization of mutations that cause globoid cell leukodystrophy and pharmacological rescue using small molecule chemical chaperones. *The Journal of Neuroscience*, *30*(16), 5489–5497. https://doi.org/10.1523/JNEUROSCI.6383-09.2010.

Leidenheimer, N. J., & Ryder, K. G. (2014). Pharmacological chaperoning: A primer on mechanism and pharmacology. *Pharmacological Research*, *83*, 10–19. https://doi.org/10.1016/j.phrs.2014.01.005.

Leslie, C. C. (2015). Cytosolic phospholipase A2: Physiological function and role in disease. *Journal of Lipid Research*, *56*(8), 1386–1402. https://doi.org/10.1194/jlr.R057588.

Li, C.-M., Park, J.-H., He, X., Levy, B., Chen, F., Arai, K., et al. (1999). The human acid ceramidase gene (ASAH): Structure, chromosomal location, mutation analysis, and expression. *Genomics*, *62*(2), 223–231. https://doi.org/10.1006/geno.1999.5940.

Lieberman, R. L., Wustman, B. A., Huertas, P., Powe, A. C., Jr., Pine, C. W., Khanna, R., et al. (2006). Structure of acid β-glucosidase with pharmacological chaperone provides insight into Gaucher disease. *Nature Chemical Biology*, *3*(2), 101–107. https://doi.org/10.1038/nchembio850.

Loberto, N., Tebon, M., Lampronti, I., Marchetti, N., Aureli, M., Bassi, R., et al. (2014). GBA2-encoded β-glucosidase activity is involved in the inflammatory response to Pseudomonas aeruginosa. *PLoS One*, *9*(8), e104763. https://doi.org/10.1371/journal.pone.0104763.

Lone, A. M., & Taskén, K. (2013). Proinflammatory and immunoregulatory roles of eicosanoids in T cells. *Frontiers in Immunology*, *4*(130). https://doi.org/10.3389/fimmu.2013.00130.

Lust, J. A., Lacy, M. Q., Zeldenrust, S. R., Witzig, T. E., Moon-Tasson, L. L., Dinarello, C. A., et al. (2016). Reduction in C-reactive protein indicates successful targeting of the IL-1/IL-6 axis resulting in improved survival in early stage multiple myeloma. *American Journal of Hematology*, *91*(6), 571–574. https://doi.org/10.1002/ajh.24352.

Lyman, M., Lloyd, D. G., Ji, X., Vizcaychipi, M. P., & Ma, D. (2014). Neuroinflammation: The role and consequences. *Neuroscience Research*, *79*, 1–12. https://doi.org/10.1016/j.neures.2013.10.004.

Maceyka, M., & Spiegel, S. (2014). Sphingolipid metabolites in inflammatory disease. *Nature*, *510*(7503), 58–67. https://doi.org/10.1038/nature13475.

Maestro. (2015). *version 10.4*. New York, NY: Schrödinger, LLC.

Mandon, E. C., Ehses, I., Rother, J., van Echten, G., & Sandhoff, K. (1992). Subcellular localization and membrane topology of serine palmitoyltransferase, 3-dehydrosphinganine reductase, and sphinganine N-acyltransferase in mouse liver. *Journal of Biological Chemistry*, *267*(16), 11144–11148.

Marshall, M. S., & Bongarzone, E. R. (2016). Beyond Krabbe's disease: The potential contribution of GALC deficiency to neuronal vulnerability in late onset synucleinopathies. *Journal of Neuroscience Research*, *94*(11), 1328–1332. https://doi.org/10.1002/jnr.23751.

Mc Donald, J. M., & Krainc, D. (2017). Lysosomal proteins as a therapeutic target in neurodegeneration. *Annual Review of Medicine*, *68*, 445–458. https://doi.org/10.1146/annurev-med-050715-104432.

McCurdy, C. E., Schenk, S., Holliday, M. J., Philp, A., Houck, J. A., Patsouris, D., et al. (2012). Attenuated Pik3r1 expression prevents insulin resistance and adipose tissue macrophage accumulation in diet-induced obese mice. *Diabetes*, *61*(10), 2495–2505. https://doi.org/10.2337/db11-1433.

McKay Bounford, K., & Gissen, P. (2014). Genetic and laboratory diagnostic approach in Niemann Pick disease type C. *Journal of Neurology*, *261*(Suppl. 2), 569–575. https://doi.org/10.1007/s00415-014-7386-8.

Medzhitov, R. (2008). Origin and physiological roles of inflammation. *Nature*, *454*(7203), 428–435. https://doi.org/10.1038/nature07201.

Medzhitov, R. (2010). Inflammation 2010: New adventures of an old flame. *Cell*, *140*(6), 771–776. https://doi.org/10.1016/j.cell.2010.03.006.

Mehling, M., Lindberg, R., Raulf, F., Kuhle, J., Hess, C., Kappos, L., et al. (2010). Th17 central memory T cells are reduced by FTY720 in patients with multiple sclerosis. *Neurology*, *75*(5), 403–410. https://doi.org/10.1212/WNL.0b013e3181ebdd64.

Meikle, P. J., Hopwood, J. J., Clague, A. E., & Carey, W. F. (1999). Prevalence of lysosomal storage disorders. *JAMA*, *281*(3), 249–254. https://doi.org/10.1001/jama.281.3.249.

Merrill, A. H., Schmelz, E. M., Dillehay, D. L., Spiegel, S., Shayman, J. A., Schroeder, J. J., et al. (1997). Sphingolipids—The enigmatic lipid class: Biochemistry, physiology and pathophysiology. *Toxicology and Applied Pharmacology*, *142*(1), 208–225. https://doi.org/10.1006/taap.1996.8029.

Michaud, J., Kohno, M., Proia, R. L., & Hla, T. (2006). Normal acute and chronic inflammatory responses in sphingosine kinase 1 knockout mice. *FEBS Letters*, *580*(19), 4607–4612. https://doi.org/10.1016/j.febslet.2006.07.035.

Mirza, S., Hossain, M., Mathews, C., Martinez, P., Pino, P., Gay, J. L., et al. (2012). Type 2-diabetes is associated with elevated levels of TNF-alpha, IL-6 and adiponectin and low levels of leptin in a population of Mexican American: A cross-sectional study. *Cytokine*, *57*(1), 136–142. https://doi.org/10.1016/j.cyto.2011.09.029.

Mistry, P. K., Lopez, G., Schiffmann, R., Barton, N. W., Weinreb, N. J., & Sidransky, E. (2017). Gaucher disease: Progress and ongoing challenges. *Molecular Genetics and Metabolism*, *120*(1–2), 8–21. https://doi.org/10.1016/j.ymgme.2016.11.006.

Mitsutake, S., Date, T., Yokota, H., Sugiura, M., Kohama, T., & Igarashi, Y. (2012). Ceramide kinase deficiency improves diet-induced obesity and insulin resistance. *FEBS Letters*, *586*(9), 1300–1305. https://doi.org/10.1016/j.febslet.2012.03.032.

Mohamed, F. E., Al-Gazali, L., Al-Jasmi, F., & Ali, B. R. (2017). Pharmaceutical chaperones and proteostasis regulators in the therapy of lysosomal storage disorders: Current perspective and future promises. *Frontiers in Pharmacology*, *8*(448). https://doi.org/10.3389/fphar.2017.00448.

Mucci, J. M., Cuello, M. F., Kisinovsky, I., Larroude, M., Delpino, M. V., & Rozenfeld, P. A. (2015). Proinflammatory and proosteoclastogenic potential of peripheral blood mononuclear cells from Gaucher patients: Implication for bone pathology. *Blood Cells, Molecules, and Diseases*, *55*(2), 134–143. https://doi.org/10.1016/j.bcmd.2015.05.009.

Mucci, J. M., Suqueli García, F., de Francesco, P. N., Ceci, R., Di Genaro, S., Fossati, C. A., et al. (2013). Uncoupling of osteoblast–osteoclast regulation in a chemical murine model of Gaucher disease. *Gene*, *532*(2), 186–191. https://doi.org/10.1016/j.gene.2013.09.072.

Murdaca, G., Spanò, F., & Puppo, F. (2013). Long-term treatment of rheumatoid arthritis with adalimumab. *Open Access Rheumatology: Research and Reviews*, *5*, 43–49. https://doi.org/10.2147/OARRR.S32582.

Nakamura, H., & Murayama, T. (2014). Role of sphingolipids in arachidonic acid metabolism. *Journal of Pharmacological Sciences*, *124*(3), 307–312. https://doi.org/10.1254/jphs.13R18CP.

Nakamura, K., & Smyth, M. J. (2017). Targeting cancer-related inflammation in the era of immunotherapy. *Immunology and Cell Biology*, *95*(4), 325–332. https://doi.org/10.1038/icb.2016.126.

Nalysnyk, L., Rotella, P., Simeone, J. C., Hamed, A., & Weinreb, N. (2017). Gaucher disease epidemiology and natural history: A comprehensive review of the literature. *Hematology*, *22*(2), 65–73. https://doi.org/10.1080/10245332.2016.1240391.

Narita, A., Shirai, K., Itamura, S., Matsuda, A., Ishihara, A., Matsushita, K., et al. (2016). Ambroxol chaperone therapy for neuronopathic Gaucher disease: A pilot study. *Annals of Clinical and Translational Neurology*, *3*(3), 200–215. https://doi.org/10.1002/acn3.292.

Naylor, R. M., Baker, D. J., & Deursen, J. M. (2012). Senescent cells: A novel therapeutic target for aging and age-related diseases. *Clinical Pharmacology & Therapeutics*, *93*(1), 105–116. https://doi.org/10.1038/clpt.2012.193.

Nelson, D. L., & Cox, M. M. (2013). Lipid biosynthesis. In *Lehninger principles of biochemistry*. (6th ed.). New York: W.H. Freeman and Company.

Ng, P., Li, K., Wong, R., Chui, K., Wong, E., Li, G., et al. (2003). Proinflammatory and anti-inflammatory cytokine responses in preterm infants with systemic infections. *Archives of Disease in Childhood Fetal and Neonatal Edition*, *88*(3), F209–F213. https://doi.org/10.1136/fn.88.3.F209.

Ng, S. C., Shi, H. Y., Hamidi, N., Underwood, F. E., Tang, W., Benchimol, E. I., et al. (2017). Worldwide incidence and prevalence of inflammatory bowel disease in the 21st century: A systematic review of population-based studies. *The Lancet*, *390*(10114), 2769–2778. https://doi.org/10.1016/S0140-6736(17)32448-0.

Nielsen, O. H., Li, Y., Johansson-Lindbom, B., & Coskun, M. (2017). Sphingosine-1-phosphate signaling in inflammatory bowel disease. *Trends in Molecular Medicine*, *23*(4), 362–374. https://doi.org/10.1016/j.molmed.2017.02.002.

Nixon, G. F. (2009). Sphingolipids in inflammation: Pathological implications and potential therapeutic targets. *British Journal of Pharmacology*, *158*(4), 982–993. https://doi.org/10.1111/j.1476-5381.2009.00281.x.

Normanno, N., De Luca, A., Bianco, C., Strizzi, L., Mancino, M., Maiello, M. R., et al. (2006). Epidermal growth factor receptor (EGFR) signaling in cancer. *Gene*, *366*(1), 2–16. https://doi.org/10.1016/j.gene.2005.10.018.

Ohgane, K., Karaki, F., Dodo, K., & Hashimoto, Y. (2013). Discovery of oxysterol-derived pharmacological chaperones for NPC1: Implication for the existence of second sterol-binding site. *Chemistry & Biology*, *20*(3), 391–402. https://doi.org/10.1016/j.chembiol.2013.02.009.

Ohgane, K., Karaki, F., Noguchi-Yachide, T., Dodo, K., & Hashimoto, Y. (2014). Structure–activity relationships of oxysterol-derived pharmacological chaperones for Niemann–Pick type C1 protein. *Bioorganic & Medicinal Chemistry Letters*, *24*(15), 3480–3485. https://doi.org/10.1016/j.bmcl.2014.05.064.

Okin, D., & Medzhitov, R. (2012). Evolution of inflammatory diseases. *Current Biology: CB*, *22*(17), R733–R740. https://doi.org/10.1016/j.cub.2012.07.029.

Önder, E., Sinici, I., Sönmez, F. M., Topçu, M., & Özkara, H. A. (2009). Identification of two novel arylsulfatase A mutations with a polymorphism as a cause of metachromatic leukodystrophy. *Neurological Research*, *31*(1), 60–66. https://doi.org/10.1179/016164108X323762.

Orchard, S., Ammari, M., Aranda, B., Breuza, L., Briganti, L., Broackes-Carter, F., et al. (2014). The MIntAct project—IntAct as a common curation platform for 11 molecular interaction databases. *Nucleic Acids Research*, *42*(Database issue), D358–D363. https://doi.org/10.1093/nar/gkt1115.

Orwig, S. D., Tan, Y. L., Grimster, N. P., Yu, Z., Powers, E. T., Kelly, J. W., et al. (2011). Binding of 3,4,5,6-tetrahydroxyazepanes to the acid-β-glucosidase active site: Implications for pharmacological chaperone design for Gaucher disease. *Biochemistry*, *50*(49), 10647–10657. https://doi.org/10.1021/bi201619z.

Oskouian, B., & Saba, J. D. (2010). Cancer treatment strategies targeting sphingolipid metabolism. *Advances in Experimental Medicine and Biology*, *668*, 185–205. https://doi.org/10.1007/978-1-4419-6741-1_13.

O'Sullivan, C., & Dev, K. K. (2015). Galactosylsphingosine (psychosine)-induced demyelination is attenuated by sphingosine 1-phosphate signalling. *Journal of Cell Science*, *128*(21), 3878–3887. https://doi.org/10.1242/jcs.169342.

Özkan, A., & Özkara, H. A. (2016). Metachromatic leukodystrophy: Biochemical characterization of two (p.307Glu → Lys, p.318Trp → Cys) arylsulfatase A mutations. *Intractable & Rare Diseases Research*, *5*(4), 280–283. https://doi.org/10.5582/irdr.2016.01085.

Ozkara, H. A., Akerman, B., Ciliv, G., Topcu, M., Renda, Y., & Gravel, R. (1995). Donor splice site mutation in intron 5 of the HEXA gene in a Turkish infant with Tay-Sachs disease. *Human Mutation*, *5*(2), 186–187. https://doi.org/10.1002/humu.1380050216.

Ozkara, H. A., & Navon, R. (1998). At least six different mutations in HEXA gene cause Tay-Sachs disease among the Turkish population. *Molecular Genetics and Metabolism*, *65*(3), 250–253. https://doi.org/10.1006/mgme.1998.2742.

Ozkara, H. A., & Sandhoff, K. (2003a). A new point mutation (G412 to A) at the last nucleotide of exon 3 of hexosaminidase α-subunit gene affects splicing. *Brain and Development*, *25*(3), 203–206. https://doi.org/10.1016/S0387-7604(02)00219-X.

Ozkara, H. A., & Sandhoff, K. (2003b). Characterization of two Turkish β-hexosaminidase mutations causing Tay-Sachs disease. *Brain and Development*, *25*(3), 191–194. https://doi.org/10.1016/S0387-7604(02)00213-9.

Ozkara, H. A., & Topcu, M. (2004). Sphingolipidoses in Turkey. *Brain and Development*, *26*(6), 363–366. https://doi.org/10.1016/j.braindev.2003.09.006.

Ozkara, H. A., Topcu, M., & Renda, Y. (1997). Sandhoff disease in the Turkish population. *Brain and Development*, *19*(7), 469–472. https://doi.org/10.1016/S0387-7604(97)00061-2.

Pandey, M. K., Burrow, T. A., Rani, R., Martin, L. J., Witte, D., Setchell, K. D., et al. (2017). Complement drives glucosylceramide accumulation and tissue inflammation in Gaucher disease. *Nature*, *543*(7643), 108–112. https://doi.org/10.1038/nature21368.

Pandey, M. K., Jabre, N. A., Xu, Y.-H., Zhang, W., Setchell, K. D. R., & Grabowski, G. A. (2014). Gaucher disease: Chemotactic factors and immunological cell invasion in a mouse model. *Molecular Genetics and Metabolism*, *111*(2), 163–171. https://doi.org/10.1016/j.ymgme.2013.09.002.

Parenti, G. (2009). Treating lysosomal storage diseases with pharmacological chaperones: From concept to clinics. *EMBO Molecular Medicine*, *1*(5), 268–279. https://doi.org/10.1002/emmm.200900036.

Parenti, G., Andria, G., & Valenzano, K. J. (2015). Pharmacological chaperone therapy: Preclinical development, clinical translation, and prospects for the treatment of lysosomal storage disorders. *Molecular Therapy*, *23*(7), 1138–1148. https://doi.org/10.1038/mt.2015.62.

Parenti, G., Moracci, M., Fecarotta, S., & Andria, G. (2014). Pharmacological chaperone therapy for lysosomal storage diseases. *Future Medicinal Chemistry*, *6*(9), 1031–1045. https://doi.org/10.4155/fmc.14.40.

Parenti, G., Pignata, C., Vajro, P., & Salerno, M. (2013). New strategies for the treatment of lysosomal storage diseases (review). *International Journal of Molecular Medicine*, *31*(1), 11–20. https://doi.org/10.3892/ijmm.2012.1187.

Park, J.-W., Park, W.-J., & Futerman, A. H. (2014). Ceramide synthases as potential targets for therapeutic intervention in human diseases. *Biochimica et Biophysica Acta*, *1841*(5), 671–681. https://doi.org/10.1016/j.bbalip.2013.08.019.

Patil, S. A., & Maegawa, G. H. B. (2013). Developing therapeutic approaches for metachromatic leukodystrophy. *Drug Design, Development and Therapy*, 7, 729–745. https://doi.org/10.2147/DDDT.S15467.

Patnaik, S., Zheng, W., Choi, J. H., Motabar, O., Southall, N., Westbroek, W., et al. (2012). Discovery, structure–activity relationship, and biological evaluation of noninhibitory small molecule chaperones of glucocerebrosidase. *Journal of Medicinal Chemistry*, *55*(12), 5734–5748. https://doi.org/10.1021/jm300063b.

Patwardhan, G. A., & Liu, Y.-Y. (2011). Sphingolipids and expression regulation of genes in cancer. *Progress in Lipid Research*, *50*(1), 104–114. https://doi.org/10.1016/j.plipres.2010.10.003.

Pavuluri, P., Vadakedath, S., Gundu, R., Uppulety, S., & Kandi, V. (2017). Krabbe disease: Report of a rare lipid storage and neurodegenerative disorder. *Cureus*, *9*(1), e949. https://doi.org/10.7759/cureus.949.

Peake, K. B., Campenot, R. B., Vance, D. E., & Vance, J. E. (2011). Niemann-Pick type C1 deficiency in microglia does not cause neuron death in vitro. *Biochimica et Biophysica Acta*, *1812*(9), 1121–1129. https://doi.org/10.1016/j.bbadis.2011.06.003.

Pereira-Leite, C., Nunes, C., Jamal, S. K., Cuccovia, I. M., & Reis, S. (2016). Nonsteroidal anti-inflammatory therapy: A journey toward safety. *Medicinal Research Reviews, 37*(4), 802–859. https://doi.org/10.1002/med.21424.

Pereira, D. M., Valentão, P., & Andrade, P. B. (2018). Tuning protein folding in lysosomal storage diseases: The chemistry behind pharmacological chaperones. *Chemical Science, 9*(7), 1740–1752. https://doi.org/10.1039/C7SC04712F.

Platt, F. M. (2014). Sphingolipid lysosomal storage disorders. *Nature, 510*(7503), 68–75. https://doi.org/10.1038/nature13476.

Platt, F. M. (2018). Emptying the stores: Lysosomal diseases and therapeutic strategies. *Nature Reviews Drug Discovery, 17*(2), 133–150. https://doi.org/10.1038/nrd.2017.214.

Plotegher, N., & Duchen, M. R. (2017). Mitochondrial dysfunction and neurodegeneration in lysosomal storage disorders. *Trends in Molecular Medicine, 23*(2), 116–134. https://doi.org/10.1016/j.molmed.2016.12.003.

Podbielska, M., O'Keeffe, J., & Hogan, E. L. (2018). Autoimmunity in multiple sclerosis: Role of sphingolipids, invariant NKT cells and other immune elements in control of inflammation and neurodegeneration. *Journal of the Neurological Sciences, 385*, 198–214. https://doi.org/10.1016/j.jns.2017.12.022.

Popko, K., Gorska, E., Stelmaszczyk-Emmel, A., Plywaczewski, R., Stoklosa, A., Gorecka, D., et al. (2010). Proinflammatory cytokines IL-6 and TNF-α and the development of inflammation in obese subjects. *European Journal of Medical Research, 15*(Suppl. 2), 120–122. https://doi.org/10.1186/2047-783X-15-S2-120.

Porto, C., Cardone, M., Fontana, F., Rossi, B., Tuzzi, M. R., Tarallo, A., et al. (2009). The pharmacological chaperone N-butyldeoxynojirimycin enhances enzyme replacement therapy in Pompe disease fibroblasts. *Molecular Therapy, 17*(6), 964–971. https://doi.org/10.1038/mt.2009.53.

Porto, C., Ferrara, M. C., Meli, M., Acampora, E., Avolio, V., Rosa, M., et al. (2012). Pharmacological enhancement of α-glucosidase by the allosteric chaperone N-acetylcysteine. *Molecular Therapy, 20*(12), 2201–2211. https://doi.org/10.1038/mt.2012.152.

Porto, C., Pisani, A., Rosa, M., Acampora, E., Avolio, V., Tuzzi, M. R., et al. (2012). Synergy between the pharmacological chaperone 1-deoxygalactonojirimycin and the human recombinant alpha-galactosidase A in cultured fibroblasts from patients with Fabry disease. *Journal of Inherited Metabolic Disease, 35*(3), 513–520. https://doi.org/10.1007/s10545-011-9424-3.

Presa, N., Gomez-Larrauri, A., Rivera, I.-G., Ordoñez, M., Trueba, M., & Gomez-Muñoz, A. (2016). Regulation of cell migration and inflammation by ceramide 1-phosphate. *Biochimica et Biophysica Acta, 1861*(5), 402–409. https://doi.org/10.1016/j.bbalip.2016.02.007.

Pulkoski-Gross, M. J., Uys, J. D., Orr-Gandy, K. A., Coant, N., Bialkowska, A. B., Szulc, Z. M., et al. (2017). Novel sphingosine kinase-1 inhibitor, LCL351, reduces immune responses in murine DSS-induced colitis. *Prostaglandins & Other Lipid Mediators, 130*, 47–56. https://doi.org/10.1016/j.prostaglandins.2017.03.006.

Qin, C., Liu, R., & Liu, H. (2015). The conflicting role of Rac1 in inflammation. *Inflammation and Cell Signaling, 2*(3), 1–14. https://doi.org/10.14800/ics.922.

Qin, J., Berdyshev, E., Goya, J., Natarajan, V., & Dawson, G. (2010). Neurons and oligodendrocytes recycle sphingosine 1-phosphate to ceramide: Significance for apoptosis and multiple sclerosis. *The Journal of Biological Chemistry, 285*(19), 14134–14143. https://doi.org/10.1074/jbc.M109.076810.

Quan, L., Thiele, G. M., Tian, J., & Wang, D. (2008). The development of novel therapies for rheumatoid arthritis. *Expert Opinion on Therapeutic Patents, 18*(7), 723–738. https://doi.org/10.1517/13543776.18.7.723.

Ramesh, G., MacLean, A. G., & Philipp, M. T. (2013). Cytokines and chemokines at the crossroads of neuroinflammation, neurodegeneration, and neuropathic pain. *Mediators of Inflammation, 2013*, 480739. https://doi.org/10.1155/2013/480739.

Regier, D. S., Proia, R. L., D'Azzo, A., & Tifft, C. J. (2016). The GM1 and GM2 gangliosidoses: Natural history and progress toward therapy. *Pediatric Endocrinology Reviews*, *13*(Suppl. 1), 663–673.

Ren, J.-L., Pan, J.-S., Lu, Y.-P., Sun, P., & Han, J. (2009). Inflammatory signaling and cellular senescence. *Cellular Signalling*, *21*(3), 378–383. https://doi.org/10.1016/j.cellsig.2008.10.011.

Ricciotti, E., & FitzGerald, G. A. (2011). Prostaglandins and inflammation. *Arteriosclerosis, Thrombosis, and Vascular Biology*, *31*(5), 986–1000. https://doi.org/10.1161/ATVBAHA.110.207449.

Richter, F., Fleming, S. M., Watson, M., Lemesre, V., Pellegrino, L., Ranes, B., et al. (2014). A GCase chaperone improves motor function in a mouse model of synucleinopathy. *Neurotherapeutics*, *11*(4), 840–856. https://doi.org/10.1007/s13311-014-0294-x.

Rigante, D., Cipolla, C., Basile, U., Gulli, F., & Savastano, M. C. (2017). Overview of immune abnormalities in lysosomal storage disorders. *Immunology Letters*, *188*, 79–85. https://doi.org/10.1016/j.imlet.2017.07.004.

Ringe, D., & Petsko, G. A. (2009). Q&A: What are pharmacological chaperones and why are they interesting? *Journal of Biology*, *8*(9), 80. https://doi.org/10.1186/jbiol186.

Rivera, J., Proia, R. L., & Olivera, A. (2008). The alliance of sphingosine-1-phosphate and its receptors in immunity. *Nature Reviews Immunology*, *8*(10), 753–763. https://doi.org/10.1038/nri2400.

Roviezzo, F., Brancaleone, V., De Gruttola, L., Vellecco, V., Bucci, M., D'Agostino, B., et al. (2011). Sphingosine-1-phosphate modulates vascular permeability and cell recruitment in acute inflammation in vivo. *Journal of Pharmacology and Experimental Therapeutics*, *337*(3), 830–837. https://doi.org/10.1124/jpet.111.179168.

Rufini, A., Tucci, P., Celardo, I., & Melino, G. (2013). Senescence and aging: The critical roles of p53. *Oncogene*, *32*(43), 5129–5143. https://doi.org/10.1038/onc.2012.640.

Sahai, I., Baris, H., Kimonis, V., & Levy, H. L. (2005). Krabbe disease: Severe neonatal presentation with a family history of multiple sclerosis. *Journal of Child Neurology*, *20*(10), 826–828. https://doi.org/10.1177/08830738050200100901.

Sánchez-Fernández, E. M., García Fernández, J. M., & Mellet, C. O. (2016). Glycomimetic-based pharmacological chaperones for lysosomal storage disorders: Lessons from Gaucher, GM1-gangliosidosis and Fabry diseases. *Chemical Communications*, *52*(32), 5497–5515. https://doi.org/10.1039/C6CC01564F.

Sandborn, W. J., Feagan, B. G., Wolf, D. C., D'Haens, G., Vermeire, S., Hanauer, S. B., et al. (2016). Ozanimod induction and maintenance treatment for ulcerative colitis. *The New England Journal of Medicine*, *374*(18), 1754–1762. https://doi.org/10.1056/NEJMoa1513248.

Sands, M. S. (2013). Farber disease: Understanding a fatal childhood disorder and dissecting ceramide biology. *EMBO Molecular Medicine*, *5*(6), 799–801. https://doi.org/10.1002/emmm.201302781.

Sanllehí, P., Abad, J.-L., Casas, J., & Delgado, A. (2016). Inhibitors of sphingosine-1-phosphate metabolism (sphingosine kinases and sphingosine-1-phosphate lyase). *Chemistry and Physics of Lipids*, *197*, 69–81. https://doi.org/10.1016/j.chemphyslip.2015.07.007.

Sardi, S. P., Cheng, S. H., & Shihabuddin, L. S. (2015). Gaucher-related synucleinopathies: The examination of sporadic neurodegeneration from a rare (disease) angle. *Progress in Neurobiology*, *125*, 47–62. https://doi.org/10.1016/j.pneurobio.2014.12.001.

Schapira, A. H. V. (2015). Glucocerebrosidase and Parkinson disease: Recent advances. *Molecular and Cellular Neuroscience*, *66*(Part A), 37–42. https://doi.org/10.1016/j.mcn.2015.03.013.

Schuchman, E. H., & Desnick, R. J. (2017). Types A and B Niemann-Pick disease. *Molecular Genetics and Metabolism*, *120*(1–2), 27–33. https://doi.org/10.1016/j.ymgme.2016.12.008.

Scott-Hewitt, N., Folts, C., & Noble, M. (2018). Heterozygous carriers of galactocerebrosidase mutations that cause Krabbe disease have impaired microglial function and defective repair of myelin damage. *Neural Regeneration Research*, *13*(3), 393–401. https://doi.org/10.4103/1673-5374.228712.

Scott-Hewitt, N. J., Folts, C. J., Hogestyn, J. M., Piester, G., Mayer-Pröschel, M., & Noble, M. D. (2017). Heterozygote galactocerebrosidase (GALC) mutants have reduced remyelination and impaired myelin debris clearance following demyelinating injury. *Human Molecular Genetics*, *26*(15), 2825–2837. https://doi.org/10.1093/hmg/ddx153.

Selvam, S. P., De Palma, R. M., Oaks, J. J., Oleinik, N., Peterson, Y. K., Stahelin, R. V., et al. (2015). Binding of the sphingolipid S1P to hTERT stabilizes telomerase at the nuclear periphery by allosterically mimicking protein phosphorylation. *Science Signaling*, *8*(381) ra58. https://doi.org/10.1126/scisignal.aaa4998.

Serra, M., & Saba, J. D. (2010). Sphingosine 1-phosphate lyase, a key regulator of sphingosine 1-phosphate signaling and function. *Advances in Enzyme Regulation*, *50*(1), 349–362. https://doi.org/10.1016/j.advenzreg.2009.10.024.

Shawky, R. M., & Elsayed, S. M. (2016). Treatment options for patients with Gaucher disease. *Egyptian Journal of Medical Human Genetics*, *17*(3), 281–285. https://doi.org/10.1016/j.ejmhg.2016.02.001.

Shembade, N., Pujari, R., Harhaj, N. S., Abbott, D. W., & Harhaj, E. W. (2011). The kinase IKKα inhibits activation of the transcription factor NF-κB by phosphorylating the regulatory molecule TAX1BP1. *Nature Immunology*, *12*(9), 834–843. https://doi.org/10.1038/ni.2066.

Shen, J.-S., Edwards, N. J., Hong, Y. B., & Murray, G. J. (2008). Isofagomine increases lysosomal delivery of exogenous glucocerebrosidase. *Biochemical and Biophysical Research Communications*, *369*(4), 1071–1075. https://doi.org/10.1016/j.bbrc.2008.02.125.

Sheth, J., Mistri, M., Sheth, F., Shah, R., Bavdekar, A., Godbole, K., et al. (2014). Burden of lysosomal storage disorders in India: Experience of 387 affected children from a single diagnostic facility. *JIMD Reports*, *12*, 51–63. https://doi.org/10.1007/8904_2013_244.

Shin, M. H., & Lim, H.-S. (2017). Screening methods for identifying pharmacological chaperones. *Molecular BioSystems*, *13*(4), 638–647. https://doi.org/10.1039/C6MB00866F.

Sidransky, E., & Lopez, G. (2012). The link between the GBA gene and parkinsonism. *The Lancet. Neurology*, *11*(11), 986–998. https://doi.org/10.1016/S1474-4422(12)70190-4.

Simanshu, D. K., Kamlekar, R. K., Wijesinghe, D. S., Zou, X., Zhai, X., Mishra, S. K., et al. (2013). Non-vesicular trafficking by a ceramide-1-phosphate transfer protein regulates eicosanoids. *Nature*, *500*(7463), 463–467. https://doi.org/10.1038/nature12332.

Simonaro, C. M. (2016). Lysosomes, lysosomal storage diseases, and inflammation. *Journal of Inborn Errors of Metabolism and Screening*, *4*, 1–8. https://doi.org/10.1177/2326409816650465.

Simpson, M. A., Cross, H., Proukakis, C., Priestman, D. A., Neville, D. C. A., Reinkensmeier, G., et al. (2004). Infantile-onset symptomatic epilepsy syndrome caused by a homozygous loss-of-function mutation of GM3 synthase. *Nature Genetics*, *36*(11), 1225–1229. https://doi.org/10.1038/ng1460.

Smolen, J. S., Avila, J. C. M., & Aletaha, D. (2012). Tocilizumab inhibits progression of joint damage in rheumatoid arthritis irrespective of its anti-inflammatory effects: Disassociation of the link between inflammation and destruction. *Annals of the Rheumatic Diseases*, *71*(5), 687–693. https://doi.org/10.1136/annrheumdis-2011-200395.

Snider, A. J., Kawamori, T., Bradshaw, S. G., Orr, K. A., Gilkeson, G. S., Hannun, Y. A., et al. (2009). A role for sphingosine kinase 1 in dextran sulfate sodium-induced colitis. *The FASEB Journal*, *23*(1), 143–152. https://doi.org/10.1096/fj.08-118109.

Sólyom, A., Karabul, N., Hügle, B., Simonaro, C., & Schuchman, E. (2014). Polyarticular arthritis as presenting feature of Farber disease: A lysosomal storage disease involving

inflammation. *Pediatric Rheumatology Online Journal, 12*(Suppl. 1), P285. https://doi.org/10.1186/1546-0096-12-S1-P285.
Spratley, S. J., & Deane, J. E. (2016). New therapeutic approaches for Krabbe disease: The potential of pharmacological chaperones. *Journal of Neuroscience Research, 94*(11), 1203–1219. https://doi.org/10.1002/jnr.23762.
Stein, A., Stroobants, S., Gieselmann, V., D'Hooge, R., & Matzner, U. (2015). Anti-inflammatory therapy with simvastatin improves neuroinflammation and CNS function in a mouse model of metachromatic leukodystrophy. *Molecular Therapy, 23*(7), 1160–1168. https://doi.org/10.1038/mt.2015.69.
Stirnemann, J., Belmatoug, N., Camou, F., Serratrice, C., Froissart, R., Caillaud, C., et al. (2017). A review of Gaucher disease pathophysiology, clinical presentation and treatments. *International Journal of Molecular Sciences, 18*(2), 441. https://doi.org/10.3390/ijms18020441.
Suh, J. H., & Saba, J. D. (2015). Sphingosine-1-phosphate in inflammatory bowel disease and colitis-associated colon cancer: The fat's in the fire. *Translational Cancer Research, 4*(5), 469–483. https://doi.org/10.3978/j.issn.2218-676X.2015.10.06.
Sukocheva, O. A. (2018). Expansion of sphingosine kinase and sphingosine-1-phosphate receptor function in normal and cancer cells: From membrane restructuring to mediation of estrogen signaling and stem cell programming. *International Journal of Molecular Sciences, 19*(2), 420. https://doi.org/10.3390/ijms19020420.
Sultani, M., Stringer, A. M., Bowen, J. M., & Gibson, R. J. (2012). Anti-inflammatory cytokines: important immunoregulatory factors contributing to chemotherapy-induced gastrointestinal mucositis. *Chemotherapy Research and Practice, 2012*, 490804. https://doi.org/10.1155/2012/490804.
Suzuki, Y. (2014). Emerging novel concept of chaperone therapies for protein misfolding diseases. *Proceedings of the Japan Academy. Series B, Physical and Biological Sciences, 90*(5), 145–162. https://doi.org/10.2183/pjab.90.145.
Syed Haneef, S. A., & George Priya Doss, C. (2016). Personalized pharmacoperones for lysosomal storage disorder: Approach for next-generation treatment. In R. Donev (Ed.), *Advances in protein chemistry and structural biology: Vol. 102* (pp. 225–265). Academic Press. https://doi.org/10.1016/bs.apcsb.2015.10.001.
Szklarczyk, D., Morris, J. H., Cook, H., Kuhn, M., Wyder, S., Simonovic, M., et al. (2017). The STRING database in 2017: Quality-controlled protein-protein association networks, made broadly accessible. *Nucleic Acids Research, 45*(Database issue), D362–D368. https://doi.org/10.1093/nar/gkw937.
Tabas, I., & Glass, C. K. (2013). Anti-inflammatory therapy in chronic disease: Challenges and opportunities. *Science, 339*(6116), 166–172. https://doi.org/10.1126/science.1230720.
Takai, T., Higaki, K., Aguilar-Moncayo, M., Mena-Barragán, T., Hirano, Y., Yura, K., et al. (2013). A bicyclic 1-deoxygalactonojirimycin derivative as a novel pharmacological chaperone for GM1 gangliosidosis. *Molecular Therapy, 21*(3), 526–532. https://doi.org/10.1038/mt.2012.263.
Takeshita, H., Kitano, M., Iwasaki, T., Kitano, S., Tsunemi, S., Sato, C., et al. (2012). Sphingosine 1-phosphate (S1P)/S1P receptor 1 signaling regulates receptor activator of NF-κB ligand (RANKL) expression in rheumatoid arthritis. *Biochemical and Biophysical Research Communications, 419*(2). https://doi.org/10.1016/j.bbrc.2012.01.103.
Tallima, H., & El Ridi, R. (2018). Arachidonic acid: Physiological roles and potential health benefits—A review. *Journal of Advanced Research, 11*, 33–41. https://doi.org/10.1016/j.jare.2017.11.004.
Tang, B., Tang, F., Wang, Z., Qi, G., Liang, X., Li, B., et al. (2016). Upregulation of Akt/Nf-κB-regulated inflammation and Akt/Bad-related apoptosis signaling pathway involved in hepatic carcinoma process: Suppression by carnosic acid nanoparticle. *International Journal of Nanomedicine, 11*, 6401–6420. https://doi.org/10.2147/IJN.S101285.

Tao, Y.-X., & Conn, P. M. (2018). Pharmacoperones as novel therapeutics for diverse protein conformational diseases. *Physiological Reviews*, *98*(2), 697–725. https://doi.org/10.1152/physrev.00029.2016.

Teichgräber, V., Ulrich, M., Endlich, N., Riethmüller, J., Wilker, B., De Oliveira–Munding, C. C., et al. (2008). Ceramide accumulation mediates inflammation, cell death and infection susceptibility in cystic fibrosis. *Nature Medicine*, *14*(4), 382–391. https://doi.org/10.1038/nm1748.

Tettamanti, G., Bassi, R., Viani, P., & Riboni, L. (2003). Salvage pathways in glycosphingolipid metabolism. *Biochimie*, *85*(3–4), 423–437. https://doi.org/10.1016/S0300-9084(03)00047-6.

Thonhofer, M., Weber, P., Gonzalez Santana, A., Tysoe, C., Fischer, R., Pabst, B. M., et al. (2016). Synthesis of C-5a-substituted derivatives of 4-epi-isofagomine: Notable β-galactosidase inhibitors and activity promotors of GM1-gangliosidosis related human lysosomal β-galactosidase mutant R201C. *Carbohydrate Research*, *429*, 71–80. https://doi.org/10.1016/j.carres.2016.03.020.

Trayssac, M., Hannun, Y. A., & Obeid, L. M. (2018). Role of sphingolipids in senescence: Implication in aging and age-related diseases. *Journal of Clinical Investigation*, *128*(7), 2702–2712. https://doi.org/10.1172/JCI97949.

Tsunemi, S., Iwasaki, T., Kitano, S., Imado, T., Miyazawa, K., & Sano, H. (2010). Effects of the novel immunosuppressant FTY720 in a murine rheumatoid arthritis model. *Clinical Immunology*, *136*(2), 197–204. https://doi.org/10.1016/j.clim.2010.03.428.

Tufekci, K. U., Meuwissen, R., Genc, S., & Genc, K. (2012). Inflammation in Parkinson's disease. *Advances in Protein Chemistry and Structural Biology*, *88*, 69–132. https://doi.org/10.1016/B978-0-12-398314-5.00004-0.

Turner, M. D., Nedjai, B., Hurst, T., & Pennington, D. J. (2014). Cytokines and chemokines: At the crossroads of cell signalling and inflammatory disease. *Biochimica et Biophysica Acta*, *1843*(11), 2563–2582. https://doi.org/10.1016/j.bbamcr.2014.05.014.

Valenzano, K. J., Khanna, R., Powe, A. C., Boyd, R., Lee, G., Flanagan, J. J., et al. (2011). Identification and characterization of pharmacological chaperones to correct enzyme deficiencies in lysosomal storage disorders. *Assay and Drug Development Technologies*, *9*(3), 213–235. https://doi.org/10.1089/adt.2011.0370.

van der Tol, L., Smid, B. E., Poorthuis, B. J. H. M., Biegstraaten, M., Deprez, R. H. L., Linthorst, G. E., et al. (2014). A systematic review on screening for Fabry disease: Prevalence of individuals with genetic variants of unknown significance. *Journal of Medical Genetics*, *51*(1), 1–9. https://doi.org/10.1136/jmedgenet-2013-101857.

Van Doorn, R., Van Horssen, J., Verzijl, D., Witte, M., Ronken, E., Van Het Hof, B., et al. (2010). Sphingosine 1-phosphate receptor 1 and 3 are upregulated in multiple sclerosis lesions. *Glia*, *58*(12), 1465–1476. https://doi.org/10.1002/glia.21021.

van Gelder, C. M., Vollebregt, A. A. M., Plug, I., van der Ploeg, A. T., & Reuser, A. J. J. (2012). Treatment options for lysosomal storage disorders: Developing insights. *Expert Opinion on Pharmacotherapy*, *13*(16), 2281–2299. https://doi.org/10.1517/14656566.2012.729039.

van Rappard, D. F., Boelens, J. J., & Wolf, N. I. (2015). Metachromatic leukodystrophy: Disease spectrum and approaches for treatment. *Best Practice & Research Clinical Endocrinology & Metabolism*, *29*(2), 261–273. https://doi.org/10.1016/j.beem.2014.10.001.

Venable, M. E., Webb-Froehlich, L. M., Sloan, E. F., & Thomley, J. E. (2006). Shift in sphingolipid metabolism leads to an accumulation of ceramide in senescence. *Mechanisms of Ageing and Development*, *127*(5), 473–480. https://doi.org/10.1016/j.mad.2006.01.003.

Vitner, E., Futerman, A., & Platt, N. (2015). Innate immune responses in the brain of sphingolipid lysosomal storage diseases. *Biological Chemistry*, *396*(6–7), 659–667. https://doi.org/10.1515/hsz-2014-0301.

Wada, R., Tifft, C. J., & Proia, R. L. (2000). Microglial activation precedes acute neurodegeneration in Sandhoff disease and is suppressed by bone marrow transplantation. *Proceedings of the National Academy of Sciences of the United States of America, 97*(20), 10954–10959.

Wang, D., & DuBois, R. N. (2010). Eicosanoids and cancer. *Nature Reviews Cancer, 10*(3), 181–193. https://doi.org/10.1038/nrc2809.

Wang, Q., Liu, Y., & Zhou, J. (2015). Neuroinflammation in Parkinson's disease and its potential as therapeutic target. *Translational Neurodegeneration, 4*(1), 19. https://doi.org/10.1186/s40035-015-0042-0.

Wątek, M., Durnaś, B., Wollny, T., Pasiarski, M., Góźdź, S., Marzec, M., et al. (2017). Unexpected profile of sphingolipid contents in blood and bone marrow plasma collected from patients diagnosed with acute myeloid leukemia. *Lipids in Health and Disease, 16*(235). https://doi.org/10.1186/s12944-017-0624-1.

Weinreb, N. J., Grabowski, G. A., Pastores, G. M., Fernhoff, P. M., Kaplan, P. B., Palling, D., et al. (2007). Pharmacological chaperone therapy for the treatment of Gaucher disease: Results of phase 1 clinical trials and a clinical ex vivo response study with a survey of blood markers for 53 Gaucher patients. *Blood, 110*(11), 2404.

Wellen, K. E., & Hotamisligil, G. S. (2005). Inflammation, stress, and diabetes. *Journal of Clinical Investigation, 115*(5), 1111–1119. https://doi.org/10.1172/JCI200525102.

Wilkinson, F. L., Holley, R. J., Langford-Smith, K. J., Badrinath, S., Liao, A., Langford-Smith, A., et al. (2012). Neuropathology in mouse models of mucopolysaccharidosis type I, IIIA and IIIB. *PLoS One*, 7(4), e35787 https://doi.org/10.1371/journal.pone.0035787.

Wollny, T., Wątek, M., Durnaś, B., Niemirowicz, K., Piktel, E., Żendzian-Piotrowska, M., et al. (2017). Sphingosine-1-phosphate metabolism and its role in the development of inflammatory bowel disease. *International Journal of Molecular Sciences, 18*(4), 741. https://doi.org/10.3390/ijms18040741.

Xu, S., Lun, Y., Brignol, N., Hamler, R., Schilling, A., Frascella, M., et al. (2015). Coformulation of a novel human α-galactosidase A with the pharmacological chaperone AT1001 leads to improved substrate reduction in Fabry mice. *Molecular Therapy, 23*(7), 1169–1181. https://doi.org/10.1038/mt.2015.87.

Yilmazer, B., Yagci, Z. B., Bakar, E., Ozden, B., Ulgen, K., & Ozkirimli, E. (2017). Investigation of novel pharmacological chaperones for Gaucher disease. *Journal of Molecular Graphics and Modelling, 76*, 364–378. https://doi.org/10.1016/j.jmgm.2017.07.014.

Yu, H., Sun, C., & Argraves, K. M. (2015). Periodontal inflammation and alveolar bone loss induced by Aggregatibacter actinomycetemcomitans is attenuated in sphingosine kinase 1-deficient mice. *Journal of Periodontal Research, 51*(1), 38–49. https://doi.org/10.1111/jre.12276.

Yu, Y., Mena-Barragán, T., Higaki, K., Johnson, J. L., Drury, J. E., Lieberman, R. L., et al. (2014). Molecular basis of 1-deoxygalactonojirimycin arylthiourea binding to human α-galactosidase A: Pharmacological chaperoning efficacy on Fabry disease mutants. *ACS Chemical Biology, 9*(7), 1460–1469. https://doi.org/10.1021/cb500143h.

Zhang, J.-M., & An, J. (2007). Cytokines, inflammation and pain. *International Anesthesiology Clinics, 45*(2), 27–37. https://doi.org/10.1097/AIA.0b013e318034194e.

Zhou, J., Tawk, M., Tiziano, F. D., Veillet, J., Bayes, M., Nolent, F., et al. (2012). Spinal muscular atrophy associated with progressive myoclonic epilepsy is caused by mutations in ASAH1. *American Journal of Human Genetics, 91*(1), 5–14. https://doi.org/10.1016/j.ajhg.2012.05.001.

CHAPTER SIX

When safeguarding goes wrong: Impact of oxidative stress on protein homeostasis in health and neurodegenerative disorders

Ravit Mesika, Dana Reichmann*
Department of Biological Chemistry, The Alexander Silberman Institute of Life Sciences, The Hebrew University of Jerusalem, Jerusalem, Israel
*Corresponding author: e-mail address: danare@mail.huji.ac.il

Contents

Abstract

Cellular redox status is an established player in many different cellular functions. The buildup of oxidants within the cell is tightly regulated to maintain a balance between the positive and negative outcomes of cellular oxidants. Proteins are highly sensitive to oxidation, since modification can cause widespread unfolding and the formation of toxic aggregates. In response, cells have developed highly regulated systems that

Advances in Protein Chemistry and Structural Biology, Volume 114
ISSN 1876-1623
https://doi.org/10.1016/bs.apcsb.2018.11.001

contribute to the maintenance of both the global redox status and protein homeostasis at large. Changes to these systems have been found to correlate with aging and age-related disorders, such as neurodegenerative pathologies. This raises intriguing questions as to the source of the imbalance in the redox and protein homeostasis systems, their interconnectivity, and their role in disease progression.

Here we focus on the crosstalk between the redox and protein homeostasis systems in neurodegenerative diseases, specifically in Alzheimer's, Parkinson's, and ALS. We elaborate on some of the main players of the stress response systems, including the master regulators of oxidative stress and the heat shock response, Nrf2 and Hsf1, which are essential features of protein folding, and mediators of protein turnover. We illustrate the elegant mechanisms used by these components to provide an immediate response, including protein plasticity controlled by redox-sensing cysteines and the recruitment of naive proteins to the redox homeostasis array that act as chaperons in an ATP-independent manner.

1. Introduction

Cellular redox status is an established player in many different cellular functions and disorders. The buildup of oxidants (namely, reactive oxygen species, or ROS) has long been associated with the phenomenon referred to as "oxidative stress." Together with ROS, other reactive species generated from nitrogen (RNS) or chlorines (RCS) are produced by cells as a natural byproduct of oxygen-dependent cellular processes, and can either serve as signaling molecules in the downstream reactions or, in the undesirable scenario, lead to oxidative stress. Oxidative stress occurs when oxidants accumulate within the cell, beyond a certain physiological threshold, and cannot be efficiently detoxified. Oxidative stress has also been found to correlate with extensive physiological changes ranging from the morphological to the biochemical which are rooted in the damage of diverse macromolecules including proteins, nucleic acids, lipids, and others. Here we primarily focus on protein damage and ways the cell can protect its healthy proteome during elevated levels of ROS, RNS, and RCS. Even though oxidants can cause severe damage, they also play a pivotal role in beneficial regulation, both in terms of the oxidative stress response itself and through other pathways. Thus, cells must walk a fine line between counterbalancing the harmful effects of oxidant accumulation and utilizing these oxidants for signaling, immune responses, and biosynthesis. Perturbations in this balance can lead to diverse dysfunctions and disorders.

The key pathologies associated with oxidative stress and protein misfolding include neurodegenerative disorders such as Alzheimer's, amyotrophic lateral sclerosis (ALS), and Parkinson's disease, which are of particular interest due to their increased prevalence during aging. In this chapter, we discuss known cellular strategies to regulate protein and redox homeostasis and will discuss what happens when these safeguards go wrong, in particular as regards age-related neurodegenerative pathologies.

2. Cellular oxidants: Origin, targets, and benefits

2.1 What are the most common cellular oxidants and what are they good for?

Aerobic metabolism has certain advantages in terms of efficient energy production; however, oxygen is one of the most reactive of all molecules—to the extent that if suggested as a drug it most probably would not be approved for consumption. Despite this potentially harmful mode of existence our cells are unambiguously oxygen dependent, and accordingly rely on the fundamental cellular respiration chain process and other downstream oxygen-dependent metabolism pathways such oxidative folding in ER (Tu & Weissman, 2002), NADPH oxidases (e.g., NOX) (Alderton, Cooper, & Knowles, 2001; Babior, Lambeth, & Nauseef, 2002; Panday, Sahoo, Osorio, & Batra, 2015), lipid oxidation in the peroxisome (Schrader & Fahimi, 2006), mitochondrial enzymes such as cytochrome P450 (Hrycay & Bandiera, 2015), xanthine oxidase (Welles Kellogg & Fridovich, 1975), lipoxygenases and others (Brash, 1999; Holmström & Finkel, 2014). Among these processes, respiration is the main source tapped by reactive oxygen species (ROS) which are detoxified by cells or used for specific cellular processes including signaling, antimicrobial defense, and others (Devasagayam et al., 2004). ROS are also produced by external sources (e.g., environmental pollution, ultraviolet radiation, high temperatures, and other factors) (Lavranos, Balla, Tzortzopoulou, Syriou, & Angelopoulou, 2012).

In general, there are two main types of ROS: non-radical (e.g., hydrogen peroxide (H_2O_2), hypochlorous acid (HOCl), organic hydroperoxide (ROOH)) and radical (e.g., superoxide ($^{\bullet}O_2^-$), hydroxyl radical ($^{\bullet}OH$), and nitric oxide radical ($NO^{\bullet}$)). Free oxygen radicals are characterized by one or two unpaired electrons in the external orbital shell that makes these agents extremely reactive (Augusto & Sayuri, 2011; Ortiz et al., 2017). Non-radical oxygen species are less chemically reactive than radical species, but can easily become radical upon reaction with cellular metabolites and macromolecules (Gupta et al., 2012).

ROS reactiveness has dual implications for cellular functions. When oxidants are found in excess it causes a major set of detrimental conditions, known in general as oxidative stress, which can lead to extensive damage of macromolecules, including proteins, polynucleotides, polysaccharides, and lipids. Therefore, it is not surprising that cells have developed extensive and diverse antioxidant systems to balance redox homeostasis in different compartments.

Though devastating effects are associated with ROS, they actually play a key role in several cellular processes involved in the immune system, cell proliferation, signal transduction, and tissues regeneration defense (Gauron et al., 2013; Holmström & Finkel, 2014; Reichmann, Voth, & Jakob, 2018; Winterbourn & Kettle, 2013).

One of the ways to restock the cellular ROS arsenal is through a set of oxidase enzymes that catalase oxidation-reduction reactions by secreting different oxidants in a regulatory manner. The extensively studies oxidase are the family of NADPH oxidases, dubbed NOX, which were first identified in white blood cells and specifically neutrophils (Bedard & Krause, 2007; Panday et al., 2015). In last 10 years, other oxidases have been characterized and found to mediate a wide range of functions including gene expression regulation, host immunity, and regulation of cell signaling via ROS production. Upon activation of neutrophils, a signaling cascade activates the NOX enzymes to generate superoxide radicals. The superoxide is converted to hydrogen peroxide which is diffused in the extracellular space and kills the invading bacteria. Residues of this hydrogen peroxide diffuse back into the cell and activate downstream signaling cascades leading to phagocytosis and an anti-inflammation response.

Another ingredient activated by neutrophils is HOCl, a member of the reactive chlorine species (RCS), more famous as the active ingredient in household bleach (Winterbourn & Kettle, 2013; Winterbourn, Kettle, & Hampton, 2016). The mechanism and role of ROS in signaling in antibacterial defenses, cancer, and autophagy are extensively summarized in Gupta et al. (2012), Holmström and Finkel (2014), and Little et al. (2017). Most of the known redox-dependent signaling pathways rely on the reversible thiol oxidation of the target cysteines as described below.

Interestingly, ROS had also been suggested as a treatment for tumors in cancer due to its pro-apoptotic effect. However, ROS-based treatment is controversial, because of the double-edged sword effect of ROS: on the one hand ROS has a deleterious effect on cancer cells but on the other hand it contributes to cell proliferation and tumorigenesis (De Sá Junior et al., 2017; Wang & Yi, 2008).

2.2 Mitochondria are the main source of ROS and RNS

The mitochondria are pivotal eukaryote organelles that play a primary role in the regulation of cell survival and death, and specifically servers as the main energy producer supplying ATP molecules to the entire cell. Adaptation to an anaerobic environment resulted in the ability of cells to utilize di-oxygen to provide energy in the form of ATP during oxidative phosphorylation. This is one of the most important and well-studied cellular pathways and is comprised of a series of oxidation-reduction reactions generating transmembrane potentials by applying hydrogen pumping via mitochondrial complexes I, III, and IV of the electron transport chain (ETC) (Mailloux, 2015). During the respiration process, electrons leak to oxygen, mainly leading to secretions of the unstable superoxide anion radical, which is then propagated to hydrogen peroxide and hydroxyl radical (Cadenas & Davies, 2000; Mailloux, 2015; Murphy, 2009). In addition to ROS production during the electron transfer in ETC, the enzyme monoamine oxidase located in the outer mitochondrial membrane is another major source of hydrogen peroxide (Cadenas & Davies, 2000).

Thus, the mitochondria are considered the main source of endogenous ROS which needs be detoxified when found in large quantities to prevent cellular damage and death. Yet, cells rely on mitochondrial oxidants (e.g., hydrogen peroxide) to deliver signals through redox-dependent signaling processes in a well-controlled manner. Maintaining this delicate balance is not an easy task; therefore, it is highly regulated by enzymatic and small molecular scavengers which will be described in details below. The perturbation of this balance and the accumulation of oxidants in the mitochondria are associated with a diverse range of disorders including neurological diseases as well as metabolic disorders (e.g., obesity and type 2 diabetes). Here we focus on the mitochondrial dysfunction related to neurodegenerative diseases such as Parkinson's, Alzheimer's, ALS, and Huntington's diseases.

2.3 Effects of ROS on cellular macromolecules

Oxidants can produce different forms of damage affecting lipids, nucleic acids, proteins, and small molecules. Here we briefly describe some of the potential undesirable oxygen-mediated modifications that are specifically associated with neurogenerative disorders.

ROS target oxygen-sensitive lipids such as glycolipids, cholesterol, and phospholipids, which may undergo lipid peroxidation. Oxidation leads to formation of unsaturated bonds within the target lipid molecule, converting it to a highly reactive lipid radical (LOO·) and the subsequent generation of lipid peroxides (LOOH) (Butterfield, Castegna, Lauderback, & Drake, 2002). Some of these oxidation products, particularly highly electrophilic aldehydes, electrophile 4-hydroxynonenol (4-HNE), and malondialdehyde (MDA) can then further perpetuate the cellular damage when found in large quantities. Interestingly, these lipid peroxides are beneficial to cells in small quantities and are used for cell signaling by mediating ligand-receptor interactions using reversible thiol modifications. To serve as a signaling molecule, lipid peroxides (e.g., 4-HNE) are synthesized via enzymatic (by lipoxygenases) and non-enzymatic (by Fenton reaction and oxidants) regulated pathways (Higdon, Diers, Oh, Landar, & Darley-Usmar, 2012). When these lipid peroxides are generated in non-specific manner and accumulate in the membranes, this can lead to perturbation of the membrane bilayer structure, which affects its plasticity, permeability, and as result, the ion gradient in the cells. Another outcome is the formation of secondary toxic agents under various reactions. When found in elevated quantities 4-hydroxynonenal (4-HNE) and malondialdehyde (MDA) modify protein, nucleic acids, and antioxidant molecules, leading to cell death (Esterbauer, Schaur, & Zollner, 1991; Gaschler & Stockwell, 2017; Higdon et al., 2012).

Another target of non-specific oxygen modification is nucleic acids, which leads to distortion in the DNA structure and nucleotide modification. Specifically, the sugar base within the purines and pyrimidines is susceptible to attack by the hydroxyl radical ($OH^{\bullet}$), a process generating a wide range of labile oxidized derivatives (Halliwell & Aruoma, 1991). Consequently, DNA damage is often accompanied by mutagenesis and carcinogenesis (Devasagayam et al., 2004) during elevated ROS levels in cells. In addition, DNA was also found to form cross-link species with proteins in the surroundings though binding of guanine to lysine residues (Xu, Muller, Ye, & Burrows, 2008) in a non-specific manner which alters specific protein-DNA interactions and leads to protein dysfunction and aggregation.

Finally, one of the major cellular targets of ROS and RNS is protein. This is because proteins are comprised of a high number of oxidation prone residues, whose modification can lead to protein misfolding, loss of function, and toxic aggregation (Devasagayam et al., 2004; Stadtman & Levine, 2001). The major pathway was brilliantly demonstrated by Stadtman and Levine (2001). Briefly, protein oxidation initiates a set of chain reactions forming a toxic alkoxyl radical, which either directly leads to peptide bond cleavage or reacts with the protonated form of superoxide anion ($HO_2^{\bullet}$) giving rise to the generation of a hydroxyl derivative, thus damaging the protein's ternary and secondary structure. In addition, many of the chain reaction intermediates can individually convert adjacent amino acids into reactive radicals which may generate non-specific inter- and intra-protein covalent interactions, thus potentially disturbing the structure and function of the oxidized protein.

3. How do cells overcome oxidative stress and protect their proteome?

3.1 First line of defense: Detoxifying enzymes and small molecular antioxidants

ROS levels in the intracellular fraction need to be tightly regulated and balanced to ensure oxidant production for biogenesis and signaling purposes without damaging self-macromolecules. Desirable ROS levels are buffered by multiple, diverse members of the redox homeostasis system which is comprised of small molecular weight antioxidants (e.g., glutathione, vitamins A, C, and E) and detoxifying enzymes (e.g., thioredoxin, peroxiredoxin, and other oxidoreductases) (Uttara, Singh, Zamboni, & Mahajan, 2009). The diversity and high abundance of the members of the redox homeostasis system make both a specific and prompt response to restore the cellular balance possible. Fig. 1A summarizes the range of mechanisms of ROS and RNS production along with their detoxification strategies.

The antioxidant defense and protection of protein sulfhydryl groups rely on the coupling of protein donors and acceptors. For example, the oxidation of glutathione (GSH), which is considered to be the most abundant small molecule antioxidant, leads to the formation of a disulfate bond conjugating two GSH molecules into GSSG, which can be reduced back by glutathione reductase using NADH as a proton donor (Fig. 1A). In the majority of eukaryotes, including humans, glutathione is synthesized by cells usually from cysteine, glutamic acid, and glycine in a two-step enzymatic reaction using first glutamate cysteine ligase and then glutathione synthetase (Lu, 2009). Because of the high oxidative metabolism of the brain from mitochondrial respiration (Cadenas & Davies, 2000), the antioxidant function is highly important. Therefore, it is not surprising that the brain has one of the highest concentrations of GSH, which is mainly produced from cysteine by astrocytes and microglial cells (McBean, Aslan, Griffiths, & Torrão, 2015). The synthesis pathways of glutathione, the uptake of precursors in the brain and in neurons specifically has been discussed extensively in Angelova and Abramov (2018), McBean (2017), and McBean et al. (2015).

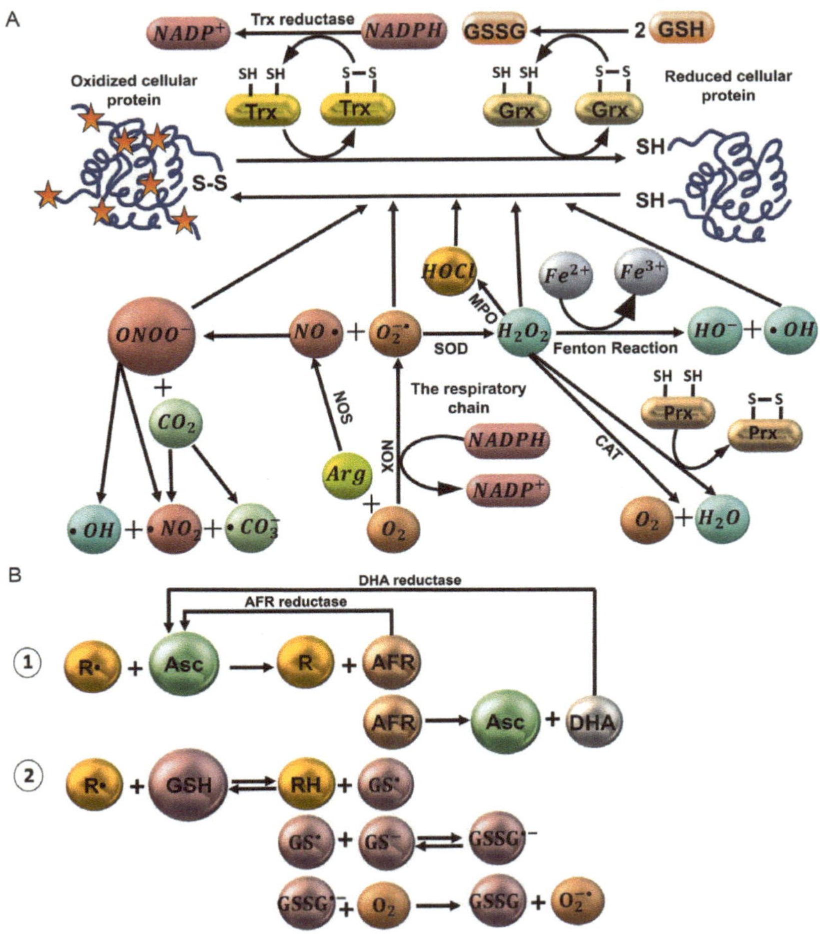

Fig. 1 Schematic representation of the cellular ROS and antioxidant pathways in cells. (A) Formation of ROS and RNS in the cell and its detoxification by the enzymatic antioxidant system. Oxygen is consumed during mitochondrial respiration and NADPH oxidase (NOX) activity producing the superoxide radical ($O_2^{-\bullet}$) (Babior et al., 2002; Cadenas & Davies, 2000). The SOD enzyme converts superoxide into hydrogen peroxide which is detoxified by catalase (Cat) and peroxiredoxins (Prx) (Holmström & Finkel, 2014). Peroxide is a reactive molecule that can be converted into other oxidants such as hypochlorous acid (HOCl) using myeloperoxidase (MPO) and the hydroxyl radical under Fenton reaction through the conversion of ferrous (Fe^{2+}) to ferric (Fe^{3+}). The nitric oxide radical ($NO^{\bullet}$) can further react with superoxide, resulting in the formation of the strong oxidant peroxinitrite ($ONOO^{-}$) (Alderton et al., 2001), producing other oxidant radicals such as the hydroxyl radical ($^{\bullet}OH$), the nitrogen dioxide radical ($^{\bullet}NO_2$) and carbonate radical anion ($^{\bullet}CO_3^{-}$) (Jakob & Reichmann, 2013; Winterbourn, 2008). The rapier system includes thiroedoxin (Trx) and glutaredoxin (Grx) which keep protein thiols in

In addition to direct ROS and RNS detoxification, glutathione is used to reduce protein disulfate bonds, protein sulfenic acids, as well as in the modification of protein thiols by glutathionylation in a non-enzymatic or in a regulated enzymatic manner via glutathione S-transferases, GSTs. Like other regulated post-translational modifications, glutathionylation is usually a reversible modification which has a wide range of consequences on protein structure and activity (Dalle-Donne et al., 2007; Ghezzi, 2013). Extensive research in recent years shows clearly that protein glutathionylation is not necessarily a byproduct of GSH oxidation during oxidative stress but rather is an important player in the regulation of cellular signaling during the cell cycle and stress response, inflammation, cytoskeleton organization, and protein quality control (Appenzeller-Herzog, 2011; Yu & Long, 2016). Therefore, it is not surprising that the protein glutathionylation is highly associated with numerous disorders and pathologies (Ghezzi, 2013; Stadtman, 2006; Suh et al., 2004). Proteomic studies have pointed to massive protein glutathionylation upon oxidative stress and pathologies (Checconi et al., 2015; Lind et al., 2002; Pajares, Cuadrado, & Rojo, 2017); however, at this stage it is challenging to discriminate between targeted glutathionylation mediating a specific protein function and the side effects of glutathionylation which might be a byproduct of increased levels of GSH and are not site and protein specific.

Another class of antioxidants are detoxifying enzymes, which either detoxify harmful oxygen and nitrogen radicals by converting them into more stable and less harmful forms, or reduce protein thiols, thus restoring their redox status and hence their function. To specifically treat a wide spectrum of oxidants, organisms have developed a large repertoire of enzymes targeting specific oxidants which are found in different compartments and organelles. For example, superoxide is converted into hydrogen peroxide by a highly diverse protein family of superoxide dismutase (SOD) enzymes

reduced state. (B) Ascorbate (Asc) and glutathione (GSH) are major small molecular weight scavengers in the cell. Ascorbate detoxifies a wide spectrum of radicals leading to the formation of the ascorbate free radical (AFR), which further converts to ascorbate and dehydroascorbate (DHA). Both AFR and DHA can subsequently be reversed by AFR reductase and DHA reductase, respectively (Rose & Bode, 1993; Winterbourn, 2013). GSH radical scavenging is accompanied by the formation of glutathionyl radical ($GS^{\bullet}$), which can further react with the thiolate anion (GS^{-}) to convert the oxidizing glutathionyl radical into the disulfide radical anion ($GSSG^{\bullet -}$), a highly reductant product. Finally, $GSSG^{\bullet -}$ can react with oxygen to oxidized glutathione (GSSG) and $O_2^{-\bullet}$ which can further be reduced by the antioxidant enzymes mentioned in (A).

(Fig. 1A). This way, SOD enzymes contribute to the hydrogen peroxide reservoir, which is utilized by the cells for signaling and antibacterial purposes (Holmström & Finkel, 2014). SOD enzymes form a large protein family of ubiquitous enzymes with a distinct structural shape, that are tissue and organelle specific, and utilize different metal ions, such as iron (Fe), manganese (Mn), zinc (Zn), and Copper (Cu) for their activity (Kawamata & Manfredi, 2010; Scandalios, 1993). Thus, SODs can be categorized into three main subgroups based on the metal co-factor: Fe-SOD, Mn-SOD, and Cu/Zn-SOD.

One of the most famous SOD enzymes is SOD1, which is found in the cytoplasm and in the intermembrane space in mitochondria. SOD1 is homodimer stabilized by a disulfide bridge between two highly conserved cysteines (Cys57 and Cys146), which is necessary for SOD1 activation (Kawamata & Manfredi, 2010). SOD1 activation during oxidative stress relies on a stepwise biogenesis pathway: the monomeric nascent polypeptide SOD1 binds Zn^{+2} ion, a process that underlies conformational changes in SOD1 that subsequently enables SOD1 interaction with the copper bound to its chaperone, CCS (Fig. 2). Binding to CCS-Cu triggers a disulfate bond formation, resulting in the active SOD1 enzyme (Furukawa et al., 2004). Since Cu binding to CCS is oxygen dependent, SOD1 activation is also tightly regulated by oxygen levels in the cells.

As mentioned above, the cell's main challenge is to preserve the balance between beneficial ROS production and detoxification of the residues; therefore, additional antioxidant enzymes developed to detoxify hydrogen peroxide into water and oxygen. This is done by another large group of enzymes known as catalases (CATs) and peroxidases (Prxs) (Fig. 1A), which are also tissue- and organelle-specific enzymes with diverse structures and co-factors. Catalase mainly uses Mn or Fe ions as co-factors (Chelikani, Fita, & Loewen, 2004; Glorieux & Calderon, 2017) to convert cellular hydrogen peroxides and hydroperoxides into water and oxygen, thus leveraging the desirable amount of peroxide species within the cellular organelles. Catalase is found in cytosol, peroxisome but is absent in the mitochondria, where glutathione peroxidases (GPX) play this role.

GPX are pivotal antioxidant enzymes since they play a key role in the detoxification of lipid and hydrogen peroxides. This detoxification is achieved through their conjunction with the selenium co-factor (Se), by converting hydrogen peroxides to water. GPX proteins are localized in the cytosol of various cell types. They are classified into four subspecies, GPx1–GPx4 (Espinoza et al., 2008), Gpx4, in contrast to its other family members,

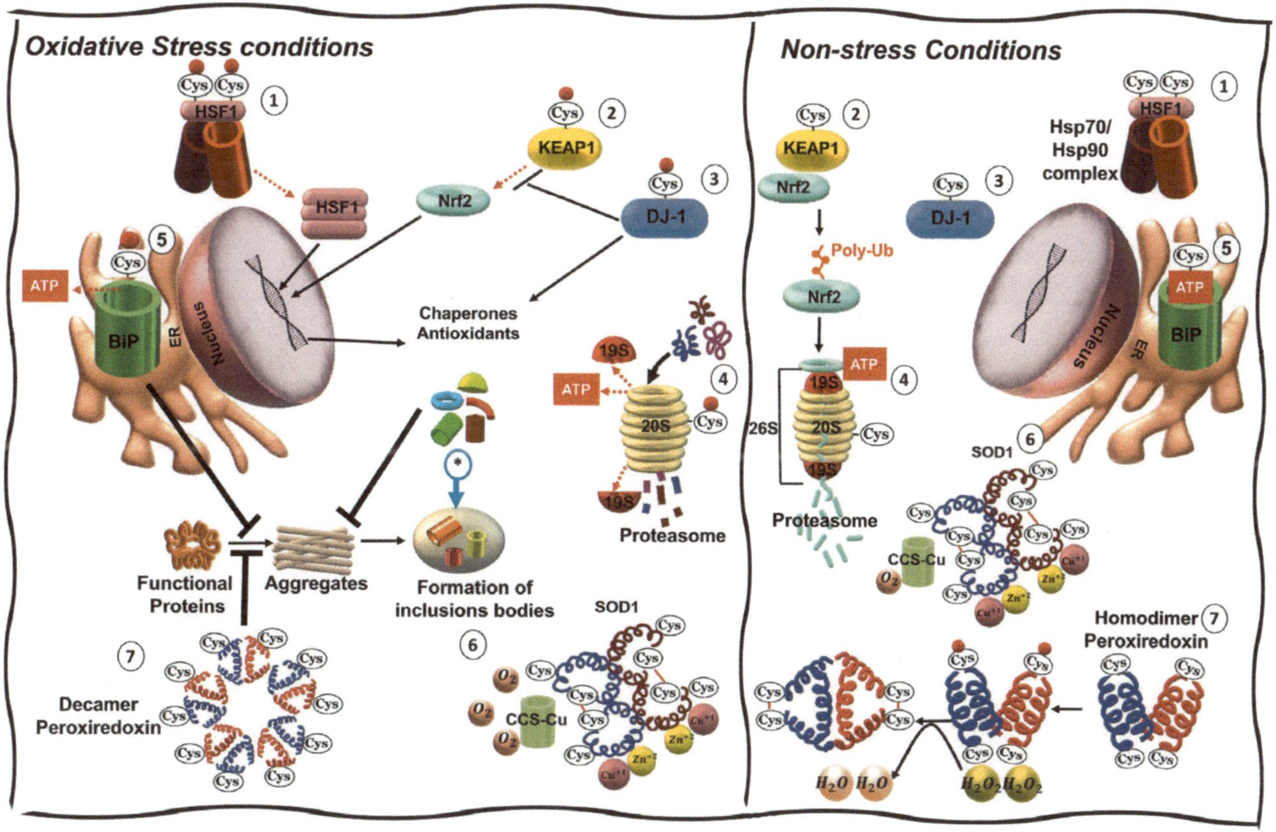

Fig. 2 See legend on next page.

is located in the mitochondria and detoxifies not only hydrogen peroxide but also lipid peroxides (Imai & Nakagawa, 2003). Other antioxidant enzymes that are considered to be the first line of defense are chelating metal ion proteins such as transferrin and ceruloplasmin.

In addition to protein antioxidants, there is a diverse spectrum of small molecular scavengers neutralizing free radicals by donating a self-electron and converting into a "new radical" which is either less active or neutralized by a dedicated enzyme. These groups of scavengers include ascorbic acid, uric acid, glutathione, and others (Fig. 1B).

Fig. 2 Schematic representation of cellular strategies to deal with oxidative stress. (1) Hsf1 is the global regulator of the stress response: during non-stress conditions (right panel) Hsf1 is protected by the Hsp70/Hsp90 complex from its trimerization and transport to the nucleus. Under oxidative stress conditions (left panel), the site-specific oxidation of Hsf1 leads to its dissociation from the Hsp70/Hsp90 complex, trimerization, accumulation in the nucleus, and activation of heat shock response genes (Ahn & Thiele, 2003). (2) Nrf2, the master regulator of the oxidative stress response: during non-stress conditions Nrf2 is constantly ubiquitinated by the E3-ligase KEAP and targeted to 26S proteasome-dependent degradation (Hayes & Dinkova-Kostova, 2014). When KEAP1 is oxidized in a site-specific manner, Nrf2 is released to the nucleus and mediates transcription of proteostasis components (Zhang, 2006). (3) DJ-1 is a pivotal modulator of the oxidative stress response: DJ-1 is activated under oxidation of the Cys106 residue, which leads to the onset of the oxidative stress response (Kiss et al., 2017). DJ-1 also facilitates the dissociation of Nrf2 from KEAP1 (Clements, McNally, Conti, Mak, & Ting, 2006; Milani, Ambrosi, Gammoh, Blandini, & Cereda, 2013). (4) Proteasome-depended protein degradation: during oxidative stress conditions the 20S proteasome has a higher load than 26S, and mediates the degradation of oxidized, non-ubiquitinated proteins by the ATP-independent manner (Reinheckel, Ullrich, Sitte, & Grune, 2000). (5) BiP (Grp48) is a redox-regulated chaperone in ER. Oxidation of BiP converts its canonical chaperone activity into a powerful ATP-independent holdase, preserving the limited ATP molecules, and preventing undesirable release of the folded proteins into the protein unfolding environments (Wang, Pareja, Kaiser, & Sevier, 2014). (6) An assembly of the antioxidant enzyme, superoxide dismutase 1 (SOD1), is redox dependent. During oxidative conditions SOD1 is post-translationally activated through its binding with the CCS-Cu chaperone and dimer formation which is stabilized by a disulfate bridge (Brown, Torres, Doan, & O'Halloran, 2004; Furukawa, Torres, & O'Halloran, 2004). (7) Peroxiredoxin is a protein with dual function. Under normal conditions peroxiredoxin functions as a homodimer detoxifying enzyme, converting peroxide into water molecules through the oxidation-reduction cycle of its catalytic cysteines (Georgiou & Masip, 2003; Rhee & Woo, 2011). Overoxidation of peroxiredoxins during oxidative stress leads to the formation of high oligomeric structures with ATP-independent holdase activity (Jang et al., 2004; Moon et al., 2005; Rhee & Woo, 2011).

3.2 The proteostasis network maintains a healthy proteome during oxidative stress

Oxidative stress has a ubiquitous influence and damages all the important macromolecules in the cell; therefore, cells activate multiple mechanisms to overcome this multilevel damage and restore existing conditions. One of the major mechanisms that is called upon to guarantee the health of the cell is the protein homeostasis (proteostasis) network. The proteostasis network is composed of three major systems: the chaperone system, which is considered to be the foundation of the proteostasis network; the protein degradation system, which works in conjunction with the proteasome; and the authophagic/lysosomal system, which comprises the major degradation pathway for damaged organelles, protein aggregates, and lipids.

Molecular chaperones play a key role in the protein quality control system because of their vast involvement in all aspects of this system including assistance in protein folding, translocation of proteins within cells, prevention of toxic aggregation formation, and mediation of the unfolded protein response (UPR) (Balchin, Hayer-Hartl, & Hartl, 2016; Hartl & Hayer-Hartl, 2002; Mayer & Bukau, 2005; Schopf, Biebl, & Buchner, 2017). Historically, upregulation of molecular chaperones was observed during heat shock conditions, which automatically classified them as heat shock proteins (HSPs). Over time, more proteins have been included in this class and their chaperone activity extended beyond heat shock conditions. By contrast, not all HSPs are heat-regulated chaperones or exhibit chaperone activity during heat shock conditions. Generally, HSPs are classified into subgroups according to their molecular weight and have self-explanatory names as such Hsp70, Hsp90, and others (Adachi et al., 2009).

When no longer repairable, oxidized damaged proteins are sent to degradation. Most proteins targeted for proteasomal degradation are "marked" by ubiquitin chains covalently attached to specific lysine residues. This signal is further recognized by the 19S cap of the proteasome complex, which mediates the unraveling of the marked protein to a thread-like structure which is followed by the internalization of the threaded protein to the 20S core proteasome unit, where the protein reacts with proteases on the surface of the cylinder core and subsequently is degraded into short peptides (Hochstrasser, 1996; Varshavsky, 1997).

The last pivotal system of proteostasis is autophagy. Autophagy is responsible for the clearance and recycling of impaired, non-functional, or unnecessary components of the cell via lysosomal degradation. Thus, autophagy plays a key role in repairing the severe damage caused by the oxidative stress when other proteostasis mechanisms can no longer overcome it. The autophagy system is classified into three pathways, according to the substrate delivery mechanism to the lysosome: macrophagy, microautophagy, and chaperone-mediated autophagy (Klionsky, 2008; Rubinsztein, Mariño, & Kroemer, 2011). The crosstalk between proteasomal and autophagy-mediated protein degradation is nicely summarized in Dikic (2017).

3.2.1 Effect of oxidative stress on canonical chaperones

Molecular chaperones maintain protein health from birth on the ribosome to transfer to the target organelle, assembly, and disassembly of protein complexes and finally protein death (degradation) (Balchin et al., 2016; Hartl & Hayer-Hartl, 2002; Mayer & Bukau, 2005; Reichmann et al., 2018). All these processes are highly orchestrated by an interactive network of different molecular chaperones and co-chaperones, which sense and react accordingly to any perturbation in cell conditions that might affect protein biogenesis. Failure of this network can lead to accumulation of toxic protein aggregates and cell death (Auluck, Chan, Trojanowski, Lee, & Bonini, 2002; Falsone, Kungl, Rek, Cappai, & Zangger, 2009; Mayer & Bukau, 2005).

The best-studied and well-defined chaperones are ATP-depended chaperones, which are also called canonical chaperones (e.g., Hsp70, Hsp90, Hsp60, and others) that require ATP hydrolysis for refolding and release of client proteins. It is clear today that the vast majority of protein unfolding stresses in cells (e.g., elevated temperature, osmotic shock, acidic stress) involve upregulation of the ATP-dependent chaperone. Dealing with oxidative protein unfolding requires additional strategies since some of the canonical chaperones are sensitive to oxidation, in addition to the fact that oxidative stress leads to ATP depletion which affects canonical chaperones (Khor, Fisher, & Schöneich, 2004; Winter, Linke, Jatzek, & Jakob, 2005).

One strategy is to convert the canonical chaperone into an efficient ATP-independent holding chaperone without refolding and release activities. Thus, the refolded protein is not released under unfavorable conditions which will lead to accidental misfolding and chaperone binding cycles that waste chaperone and ATP resources. One such example is the Hsp70 chaperone, BiP (or Grp78), located in ER, which serves as a redox-sensor through a conserved cysteine (Cys 63 in yeast) located in its ATP-binding domain (Fig. 2). It was shown in yeast that reversible oxidation or glutathionylation of Cys 63 abolishes the ATPase function and enhances BiP's binding affinity toward the misfolded client proteins (Wang et al., 2014;

Wang & Sevier, 2016). This new ATP-independent BiP acts as a holdase, preventing aggregation of client proteins in ER. Reduction of the BiP and restoring the refolding function is mediated, at least partially, by its co-chaperone Sli, which also serves as a nucleotide-exchange factor during normal conditions (Siegenthaler, Pareja, Wang, & Sevier, 2017). In mammalian cells, the redox-dependent activity of Grp78 is mediated by the reduction-oxidation cycles of its disulfate bond facilitated by the glutathione peroxidase proteins, NPGPs (Wei et al., 2012).

3.2.2 The dual function proteins: Peroxiredoxins

Peroxiredoxin (Prx) is an efficient antioxidant enzyme family, which either detoxify peroxides and peroxynitrite derivatives or directly oxidize specific proteins during single transduction pathways (Georgiou & Masip, 2003; Rhee & Woo, 2011). This is done through two highly conserved cysteines which undergo two oxidation stages: forming sulfenic acid (Cys-S-OH) intermediate and then intermolecular disulfate. Cellular thioredoxin enzymes restore the reduced state of the Prx cysteines enabling multiple catalytic cycles of Prx (Chae, Oubrahim, Park, Rhee, & Chock, 2012).

During severe oxidative stress some peroxides (human Prx1 and Prx2) form sulfinic acid intermediate (Cys-S-O_2^-) and are converted into oligomeric ATP-independent chaperone activity and a neglectable peroxidase activity (Jang et al., 2004; Moon et al., 2005; Rhee & Woo, 2011) (Fig. 2). It was shown that a repair enzyme, Sulfiredoxin, is specifically dedicated to re-reduce the overoxidized peroxiredoxins and convert them into the antioxidant enzyme without a chaperone activity (Lowther & Haynes, 2011). Thus, the dual function of peroxiredoxins is tightly regulated by redox environment of cells and is interchangeable.

3.2.3 Protein degradation in the ATP-independent manner

One of the major components of the proteostasis system is proteasome-dependent degradation of damaged or short-lived proteins (such as cell cycle regulators) which plays a major role in protein turnover in cell (Collins & Goldberg, 2017; Dikic, 2017).

During non-stress conditions, protein degradation is mainly facilitated by the 26S proteasome complex, composed of a multiprotein 20S barrel-shaped complex, and two cap 19S protein complexes containing ATPase domains closing the 20S cylinder. The 20S complex is comprised of seven different α or β subunits which are formed by two outer α rings and two inner β rings. The α rings serve as a selection gate to the proteolytic unit formed by the β rings harboring three different proteolytic activities (trypsin-like, chymotrypsin-like, caspase-like) (Collins & Goldberg, 2017; Yu & Matouschek, 2017). The 19S ATP-dependent regulatory particles server as an anchor, recognition, and de-ubiquitination unit which targets the proteins into 20S (Collins & Goldberg, 2017; Yu & Matouschek, 2017). Another role of the 19S particle is to unfold the targeted protein using ATP in order to ensure its passage via the α rings of the 20S core particle.

It was shown that upon high levels (>5 mM) of peroxide the 26S unit is not proteolytically active while the 20S particle by itself is completely functional (Reinheckel et al., 2000). It is in line with an observation that during oxidative stress there is a decrease in protein ubiquitination and that the oxidized proteins are targeted directly to the 20S proteasome without mediation of the 19S proteins. The 20S-mediated degradation can be a wonderful solution for saving ATP in the protein degradation and ubiquitination steps. The regulation of this process is not yet characterized well, however, it was suggested that oxidation of cysteine residues of the proteasomal proteins mediate accumulation of the free 20S proteasomes and ensures their availability for degradation of the oxidized proteins (Fig. 2) (Demasi et al., 2014; Livnat-Levanon et al., 2014).

In conclusion, developing reversible ATP-independent proteostasis network is an elegant adaption used by the cell to rapidly react and repair the damage caused by the oxidative stress.

4. Global transcription factors mediate the oxidative stress response

4.1 Nrf2 as the master regulator of the oxidative stress response

The nuclear factor (erythroid-derived 2)-like 2 (Nrf2) transcription factor is the master regulator of the antioxidant cell response. It regulates the expression of over 250 genes by binding to the antioxidant response element (ARE), an enhancer region shared by all Nrf2-regulated genes (Pajares et al., 2017). As a result multiple pathways are upregulated including ROS detoxification, glutathione metabolism, NADPH production, fatty acid oxidation, iron metabolism, as well as protein folding assistance (Hayes & Dinkova-Kostova, 2014).

Regulation of the Nrf2 activity relies on its redox-dependent turnover mediated by the ubiquitin member E3-ligase KEAP1 (Fig. 2). During non-stress conditions, newly synthesized Nrf2 is protected by KEAP1 which prevents upregulation of oxidative stress-related genes. KEAP1 sends the Nrf2 to degradation by mediating its ubiquitination, which is carried out by the CULLIN3/RBX1 protein complex (Hayes & Dinkova-Kostova, 2014). Under oxidative stress, KEAP1 undergoes a site-specific oxidation on redox-sensitive cysteine, resulting in the release of newly synthesized Nrf2 from KEAP1, the accumulation of Nrf2 in the nucleus, and activation of the downstream ARE-containing genes (Zhang, 2006) (Fig. 2).

4.2 The role of Hsf1 in the oxidative stress response

One of the Nrf2 activating genes is the heat shock factor 1 (Hsf1). Hsf1 is the primary modulator of the heat shock response (HSR) under stress conditions (Barna, Csermely, & Vellai, 2018; Pirkkala, 2001). It regulates the expression of a large array of heat shock proteins underlying the activation of many stress response pathways. Under normal conditions, Hsf1 exists as an inactive monomer which is suppressed by interactions between the Hsp90 and Hsp70 complex and the trimerization domain of Hsf1 (Fig. 2). Under diverse stress conditions such as heat shock, oxidation stress, and inflammation, HSF-1 is activated through oxidation of its redox-sensing cysteines (Cys35 and Cys105) into a disulfate, which leads to homotrimerization and its dissociation from the Hsp90/Hsp70 complex. The release of Hsf1 accumulates in the nucleus and initiates the global heat shock response (Ahn & Thiele, 2003).

In addition to inhibition of the Hsf1 trimer formation by the Hsp70-Hsp90 complex, cytoplasmic histone deacetylase (HDAC6), and VCP/p97 ATPase which target misfolded proteins to degradation mediate Hsf1 activity during stress conditions. In addition to the direct binding of negative regulators of Hsf1, a wide range of post-translational modifications including phosphorylation, acetylation, sumoylation, and others regulate the Hsf1 working cycle. These parallel regulation modes assure that Hsf1 translocation to the nucleus only takes place during protein damage stress.

4.3 DJ-1 is a pivotal multifunctional factor that modulates the oxidative stress response

DJ-1 is a relatively small protein (the monomer is 189 aa) which has attracted increasing attention in recent years due to its association with Parkinson's disease. Mutations in the autosomal recessive DJ-1 gene (termed PARK7) are found in Parkinson's as well as in other diseases such as amyotrophic lateral sclerosis (ALS), cancer, and others (Wilson, 2011).

DJ-1 is ubiquitously expressed in neurons and multiple functions, the majority of which are related to the oxidative stress response, including stabilization of the Nrf1 transcriptional factor, apoptosis attenuation by inhibiting apoptosis signal-regulating kinase 1 (ASK1), and mediating the ERK1/2 signal transduction pathway (Oh & Mouradian, 2018), antioxidant activity (Taira et al., 2004; Zhou & Freed, 2005), maintenance of the mitochondrial redox homeostasis (McCoy & Cookson, 2011), and chaperone activity (Shendelman, Jonason, Martinat, Leete, & Abeliovich, 2004).

A few studies have pointed to an intriguing interaction between DJ-1 and Nrf2. It was suggested that DJ-1 interferes with the interaction between Nrf2 to Keap1, and hence the releasing of Nrf2 to the nucleus (Fig. 2) (Clements et al., 2006; Milani et al., 2013).

Activation of DJ-1 depends on the redox status of its highly conserved cysteine residue, Cys106, which was reported to be in various oxidation states including reduced (Cys-SH), oxidized to sulfenic acid (Cys-SOH) or sulfinic acid (Cys-SO2), and overoxidized to sulfonic acid (Cys-SO3) (Kiss et al., 2017; Wilson, 2011). The oxidized form of DJ-1 was found to be active whereas overoxidation switches off DJ-1 activity.

The cellular localization of DJ-1 was shown to be regulated by oxidative conditions. Under basal conditions, DJ-1 is localized predominantly in the cytoplasm, and to a slight extent in the mitochondria and the nucleus. Elevated oxidant levels lead to DJ-1 accumulation in the mitochondria and in the nucleus (Junn, Jang, Zhao, Jeong, & Mouradian, 2009), supporting the DJ-1's role in preventing mitochondrial dysfunction during oxidative stress.

5. The crosstalk between proteostasis and redox homeostasis in age-associated neurodegenerative disorders (NDs)

Why do some people experience "healthy aging" whereas others develop age-associated diseases, such as Alzheimer's and Parkinson's? What goes wrong with aging that leads to the onset of such disorders and their progression? Research in the last 10 years has clearly established that age-associated neurological diseases are strongly associated with the accumulation of both oxidants (Cadenas & Davies, 2000; Lin & Beal, 2006; Zhang, Davies, & Forman, 2015) and "toxic" protein aggregates (Balchin et al., 2016; Rubinsztein et al., 2011), which are "byproducts" of cellular protein biogenesis. The link between these two processes is still not fully characterized and much remains unclear about their role in aging and disorder pathogenesis.

The accumulation of ROS with age was described as early as the 1950s in Harman's "Free Radical Theory of Aging" (Harman, 1956). This theory is compellingly based on the documented relationship between ROS and aging; however, its direct source remains poorly understood, in particular whether aging (as well as age-related disorders) leads to oxidative stress-associated damage or the reverse. Neuronal cells are particularly sensitive to oxidative stress in the presence of a high percentage of unsaturated fatty acid in the membranes and a relatively high consumption of oxygen, which challenges the redox homeostasis system (Liu, Zhou, Ziegler, Dimitrion, & Zuo, 2017; Rego & Oliveira, 2003).

Another hallmark of aging and age-associated neurogenerative diseases is the failure of the protein homeostasis system, which leads to the accumulation of misfolded proteins during aging and particularly disease-specific proteins (e.g., TAU and α/β-amyloids in Alzheimer's, α-synuclein in Parkinson's) into the cellular inclusion bodies followed by progressive death of specific neural cell populations (Ross & Poirier, 2005).

The cellular toxicity of protein aggregates was shown to be rooted in the perturbation of various pathways, for example damage to the mitochondrial respiration chain, leading to overproduction of ROS (Abramov, Berezhnov, Fedotova, Zinchenko, & Dolgacheva, 2017), alteration of calcium homeostasis (Abramov, Canevari, & Duchen, 2003, 2004), activation of the microglia triggering inflammatory pathways (Stewart, Kawas, Corrada, & Metter, 1997), impairment of synaptic signaling (Ingelsson, 2016; Kopeikina, Hyman, & Spires-Jones, 2012; Shankar et al., 2008), and undesirable interactions with other cellular proteins that affect their stability and functions (Olzscha et al., 2011; Sweeney et al., 2017). Hence, protein oxidation and misfolding are closely related and found in diverse pathologies, but more needs to be known about the contribution of each component individually, as well as their influence on each other.

5.1 Origin of oxidative stress in the neurogenerative diseases

5.1.1 Mitochondrial dysfunction is a central target of oxidative stress

One of the key features of neurogenerative diseases is mitochondrial dysfunction which is manifested in the accumulation of mtDNA mutations, and ROS overproduction that leads to neurotoxicity and apoptosis (Angelova & Abramov, 2018; Beal, 2005). However, the damage to the accumulated mutated mtDNA does not necessarily involve ROS overproduction (Lin & Beal, 2006). Therefore, additional factors are likely to play a more significant role in the accumulation of oxidants in mitochondria affecting ATP production.

The direct connection between mitochondrial dysfunction and ND progression is not surprising, since neurons relay completely on functional mitochondria (Abramov et al., 2017; Kann & Kovacs, 2006) (Fig. 3). Here are several reasons for this dependence: (1) neurons consume most of their ATP from the mitochondrial respiratory chain and not from glycolysis; (2) brain consumes 10-fold more oxygen than other tissues making neurons are more susceptible to elevated ROS production than other cells; (3) neurons are also highly dependent on the tight regulation of calcium levels, which is exclusively carried out by the mitochondria. Increased calcium transfer from the ER to the mitochondria lead to higher permeability of the mitochondrial transition pore, the secretion of cytochrome *c*, and to the subsequent activation of neuronal apoptosis mechanisms (Abramov et al., 2017; Hashimoto et al., 2003) (Fig. 3, pathway 1); (4) the mitochondria are also involved in neurotransmitter synthesis and degradation, and play a central role in synaptic transmission (Bak, Sickmann, Schousboe, & Waagepetersen, 2005; Kann & Kovacs, 2006; Rowley, Madsen, Schousboe, & Steve White, 2012; Waagepetersen, Sonnewald, Larsson, & Schousboe, 2000); and (5) because the regeneration of neurons is usually limited in the adult brain and lasts for many years (Bergmann, Spalding, & Frisén, 2015), neurons are more susceptible to fluctuations in mitochondrial homeostasis than other types of cells.

Interestingly, the misfolding and aggregation of different ND-associated proteins such as α-synuclein (in Parkinson), α/β-Amyloid and Tau (in Alzheimer's), and Huntington's disease were found to inhibit mitochondrial respiration through direct inhibition of specific mitochondrial complexes (Abramov et al., 2017).

5.1.2 Perturbation of metal homeostasis and its effect on elevated ROS formation

One major hypothesis put forward to account for the origin of oxidative stress is metal imbalance within the damaged neuronal tissues (Deibel, Ehmann, & Markesbery, 1996; Jomova, Vondrakova, Lawson, & Valko, 2010). Most studies have focused on iron metabolism during ND onset and progression, and have suggested that its dysregulation is a major cause of oxidative stress damage in ND-affected neurons (Mena et al., 2015). Perturbation of iron homeostasis results in the elevation of Fe^{2+} ions in the mitochondria, leading to increased generation of hydroxyl free radicals

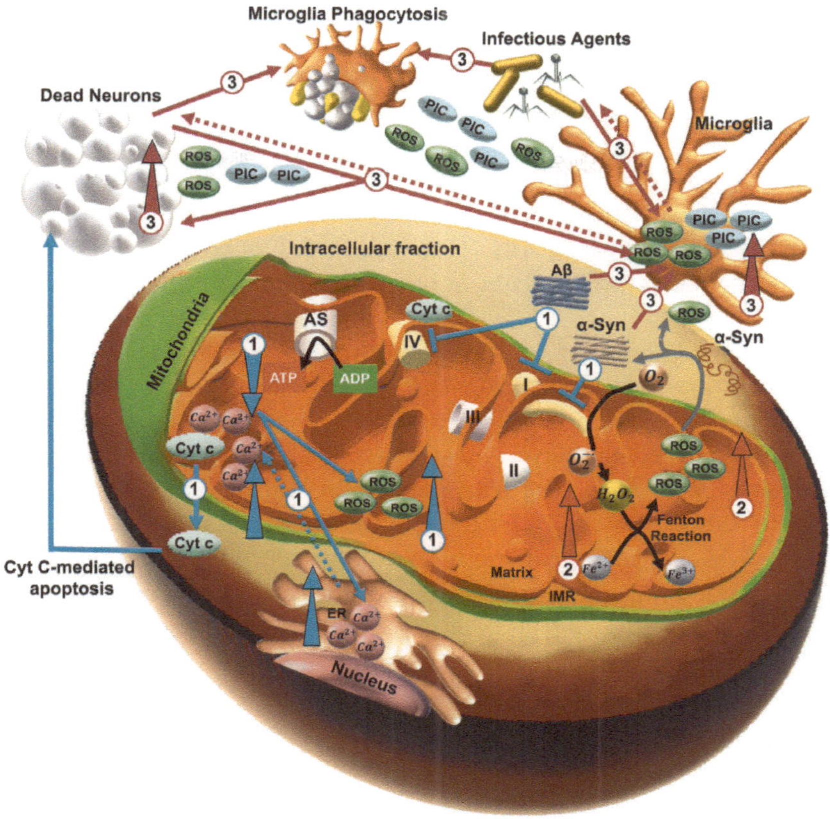

Fig. 3 The crosstalk between oxidative stress and protein aggregation in neurodegenerative diseases: (1) Mitochondrial dysfunction: The inhibition of the respiratory chain results in a decrease in ATP production, a rise in mitochondrial ROS levels, and the depletion of calcium ions from the ER with a corresponding increase of Ca(2+) concentrations in the mitochondria (Abramov et al., 2017; Calì, Ottolini, Negro, & Brini, 2012; Ferreiro, Oliveira, & Pereira, 2008; Resende, Ferreiro, Pereira, & De Oliveira, 2008). The increase in Ca(2+) levels results in greater mitochondrial membrane permeability and hence cytochrome *c* secretion from the mitochondria and apoptosis (Abramov et al., 2017; Hashimoto, Rockenstein, Crews, & Masliah, 2003). One of the reasons for these processes is the aggregation of the amyloid β-peptide (Aβ) and α-synuclein (α-Syn) which inhibits various complexes in the respiratory chain. (2) Iron imbalance: Elevated levels of Fe(2+) contribute to the increase in the ROS levels due to oxidation of peroxide in the Fenton reaction (Mena, Urrutia, Lourido, Carrasco, & Núñez, 2015). ROS excess facilitates the aggregation of alpha-synuclein proteins as well as other aggregation-prone proteins, thus exacerbating the lethal feedback resulting in neuronal death (Barnham, Masters, & Bush, 2004; Ward, Zucca, Duyn, Crichton, & Zecca, 2014). (3) Microglia overactivation induces apoptosis: Protein aggregates (plaques) were shown to induce chronic activation of microglia (Perry, 2004; Zhang et al., 2005) and lead to the production of ROS and cytokines PIC (Fischer & Maier, 2015; Hsieh & Yang, 2013). Although pathogenic infection is a short-term process, chronic brain injury, and ND lead to prolonged overactivation of the microglia and ROS-depended damage of neurons (Hsieh & Yang, 2013). Arrows pointing upwards or downwards indicate an increase or decrease of the related component, respectively. Colors indicate the corresponding pathways (blue, red, and purple linked to pathways 1, 2, and 3, respectively).

through the Fenton reaction, and causing neurotoxicity from irreversible damage to proteins, lipids, and nucleic acids (Ward et al., 2014) (Fig. 3, pathway 2).

Thus, iron dysregulation appears to be closely linked to protein aggregation in aging (Ashraf, Clark, & So, 2018) and diverse neurogenerative pathologies (Singh et al., 2014; Ward, Dexter, & Crichton, 2015), thus forming endless loop which results in massive neuronal death.

5.1.3 Neural inflammation during neurogenerative diseases contributes to ROS production

Regulation of central nervous system (CNS) homeostasis relies heavily on glia cells, which include astrocytes and microglia. Astrocytes are the primary suppliers of nutrients as well as serving as modulators of T cells and participate in the activation of inflammatory reactions (Amor et al., 2014; Fischer & Maier, 2015). Microglia are another type of CNS immune cells that function as macrophages engulfing extracellular debris including dead neurons and infectious agents. Thus astrocytes and microglia play a central role in tissue regeneration and in the neuroinflammation response (Biber, Owens, & Boddeke, 2014).

Activation of the microglia and astrocytes during a pathogenic bacterial invasion or the accumulation of dead neurons leads to the production of ROS and the secretion of proinflammatory cytokines (PIC) required for the activation of the immune response (Fischer & Maier, 2015; Hsieh & Yang, 2013) (Fig. 3, pathway 3). Neural inflammation is also found in different types of neurogenerative disorders (Amor et al., 2014), which contributes to the ROS arsenal in the damaged neuron cells of ND patients. It was suggested that protein aggregates or plaques associated with Parkinson's and Alzheimer's activate these inflammatory pathways (Perry, 2004; Zhang, 2005), leading to an increasingly oxidative environment.

5.2 Protein homeostasis challenges in neurodegenerative pathologies

Proteins are the most diverse and structurally complex macromolecules in the cell. The function of proteins is defined by their specific structure, making it a top priority for every cell to ensure that nascent polypeptide chains adopt the correct folded state and that mature proteins preserve their properly folded conformations. Cells constantly encounter misfolded proteins as a result of errors during biogenesis, changes in physiological conditions, potential mutations, and others. Thus, the protein quality control system is one of the most dynamic and complex intracellular communication pathways, which rapidly responds to any perturbation in the cellular environment, including changes in cellular oxidation associated with neurodegenerative pathologies.

Neurodegenerative diseases are associated with amyloid formations of disease-specific proteins, indicating that these proteins escaped the proteostatic system and could successfully be converted into thermodynamically favorable structures, i.e., amyloids. Amyloids are the stable fibril-like structures of non-functional proteins, whose assembly involves heterogeneous and highly toxic intermediate stages. There is ongoing debate as to the toxic origin of amyloidogenic proteins, although the consensual view is that pre-fibrillar species, rather than mature amyloids, are pathogenic. Extensive studies in the last 10 years have focused on understanding the sequence and structural features of amyloidogenic proteins, to better understand why a specific protein undergoes aggregation during ND pathogenesis while another is treated by the cellular protein quality control system (Knowles, Vendruscolo, & Dobson, 2014; Sweeney et al., 2017). There are ample evidences pointing out that amyloidogenesis is an intrinsic property of a wide range of proteins (Knowles et al., 2014), not only disease-related ones, which can be adopted freely in the absence of chaperones and the degradation system. This leads to the possibility that amyloid structures are energetically more stable than the native structure; however, this is still controversial. Regardless of their thermodynamic features, the core structure of amyloids is definitely resistant to denaturation conditions and proteases, thus constituting an enormous challenge to the protein quality control system. Sequence analysis of amyloidogenic proteins has given rise to numerous bioinformatic tools that can predict aggregation-prone regions (usually short patterns of hydrophobic-hydrophilic residues) that have a high probability of forming stable amyloidal structures (Belli, Ramazzotti, & Chiti, 2011; Chennamsetty, Voynov, Kayser, Helk, & Trout, 2010; Tartaglia et al., 2008; Tsolis, Papandreou, Iconomidou, & Hamodrakas, 2013). It was also suggested that during evolution such sequences were selected against by the insertion of aggregation-resistant residues to minimize protein instability (Knowles et al., 2014).

However, the expression of amyloidogenic proteins associated with neurodegenerative disorders is not tissue specific, since their accumulation in insoluble aggregates has only been found in CNS. Moreover, the accumulation of these proteins (either due to a somatic mutation or unknown factors) can be observed early in the lifetime of the individual whereas the onset of disease is only detected (at least clinically) at mid-age or later (Skovronsky, Lee, & Trojanowski, 2006). What underlies such a long lag and what defines the enormous difference between onset of the disease as well as its progression is a burning question in the field that has no clear answer to date.

Although age-related neurodegenerative diseases look very similar in terms of their main features, they differ not only as regards the affected proteins found in aggregates but also in their pathologies. Here we only discuss a few of them: Alzheimer's, Parkinson's, and ALS. There are many excellent reviews that provide more details about these and other neurodegenerative diseases (Andersen, 2004; Cox, Carver, & Ecroyd, 2014; Morimoto, 2008; Muchowski & Wacker, 2005; Robberecht & Philips, 2013; Ross & Poirier, 2005; Skovronsky et al., 2006; Xu & Pu, 2016).

5.2.1 Alzheimer's

Alzheimer's disease (AD) is the most common of all dementia disorders and involves memory loss and progressive deterioration of cognitive abilities. It is a late onset disease, affecting 4–8% of the elderly population, with a significant increase in incidence with age. The two major hallmarks of the disease are the presence of extracellular amyloid plaques, and intracellular neurofibrillary tangles. The amyloid plaques are formed as a result of accumulation of the 39–43 aa amyloid β-peptide (Aβ), whereas the neurofibrillary tangles are composed of hyperphosphorylated Tau (tubulin-associated unit) proteins (Kopeikina et al., 2012; Koren et al., 2009; Skovronsky et al., 2006; Sweeney et al., 2017; Walsh & Selkoe, 2007). The Aβ peptide is a product of proteolytic cleavage of a highly abundant membrane protein in neurons, APP, by β-secretase, followed by γ-secretase-mediated proteolysis (Zheng & Koo, 2006). When the APP protein is cleaved by α-secretase, the resulting peptide is not amyloidogenic. The Tau protein is also neuronal protein that promotes microtubule assembly and dynamics, as well as inhibits apoptosis and neuron development. The Tau protein has ~30 phosphorylation sites when it is hyperphosphorylated Tau's affinity for microtubules decreases, leading to its disassembly and aggregation (Rodríguez-Martín et al., 2013).

β-Amyloid aggregation was found to alter redox homeostasis in neurons in diverse ways. One mechanism through which Aβ aggregation causes toxicity is the inhibition of the respiratory chain by altering the activity of complexes I and IV (Casley et al., 2002) and increasing the superoxide secretion by NADPH oxidase (Shelat et al., 2008). Moreover, β-amyloid aggregation was also found to lead to calcium influx in the mitochondria of neurons which alters membrane depolarization and permeability, thus leading to neuronal death (Abramov, 2004; Abramov et al., 2017; Gunn et al., 2016). Along with decreased ATP production and ROS production, Aβ aggregation mediates lipid peroxidation by producing 4-hydroxynonenal (HNE) and 2-propenal (acrolein) reactive species that challenge the healthy proteome in neuron cells (Butterfield et al., 2002). Accordingly, it is not surprising that Aβ aggregation is associated with an activation of mitochondrial apoptotic pathways (Ferreiro et al., 2008).

As discussed above, redox and protein homeostasis are dependent on each other; thus, Aβ-mediated ROS production might contribute to elevated Aβ plaque formation and generates a neurotoxic positive feedback with the inevitable devastating outcome. Therefore, both antioxidant and proteostasis systems are engaged in attenuating Aβ aggregation. Analysis of the brains of AD patients and model animals has shown an upregulation of canonical chaperones and their co-factors (Lackie et al., 2017). Multiple studies by different groups have investigated the role of chaperones in AD progression (as well as its inhibition) and reported that Hsp70 and Hsp90 chaperones can successfully prevent β-amyloid self-assembly in vitro (Evans, Wisén, & Gestwicki, 2006) and in AD models (Lackie et al., 2017). Moreover, Cohen and colleagues showed that inhibition of the Hsf1 activity in *C. elegans* AD models resulted in high Aβ toxicity (Cohen, Bieschke, Perciavalle, Kelly, & Dillin, 2006), probably due to Aβ aggregation, which provides another link between AD and activation the stress response mechanisms.

In addition to Aβ containing plaques, some of AD brains show aggregations of hyperphosphorylated Tau proteins in brain neurons. Similar to Aβ-mediated ROS production, Tau aggregation has been linked to oxidative stress in diverse AD models through various mechanisms including copper-mediated peroxide production and the inhibition of GSH synthesis (Liu et al., 2015). As in the Aβ models, a link between chaperones and levels of Tau aggregates is very well established. Numerous studies have demonstrated that clearance of the misfolded Tau protein in different system and animal models is achieved by interaction with E3 ligase CHIP protein, targeting misfolded proteins from Hsp70 (or Hsp90) to proteasome for ubiquitin-dependent degradation (Koren et al., 2009; Petrucelli et al., 2004; Sahara et al., 2005).

5.2.2 Parkinson's

Parkinson's disease (PD) is one of the most common neurodegenerative movement disorders. It is characterized by the vulnerability of dopaminergic neurons in the substantia nigra pars compacta (SNpc), a midbrain region that has been linked to diverse motor functions. It is assumed that damage to the SN neurons results in movement disturbances in PD patients.

The major feature of PD is the formation of Lewy bodies in the affected neurons. These inclusions mainly consist of α-synuclein aggregates, therefore, PD is considered as one of the synucleinopathies diseases (Ingelsson, 2016). α-Synuclein is a small protein (140 aa) encoded by the SNAC gene, mutations (duplication or point mutations) of which lead to the autosomal dominant form of Parkinson's disease. Some PD-related mutations lead to α-synuclein misfolding and aggregation. The structural characterization of the α-synuclein protein is still unknown and has challenged structural biologists for years. Bacterial expressed α-synuclein was shown to be a monomer, but when purified from human erythrocyte cells it was found to be a stable tetramer with a-helical structures. NMR studies showed that α-synuclein can behave like an intrinsically disordered protein which undergoes a disorder-to-order transformation upon interaction with lipids. The same phenomenon was observed when α-synuclein was purified from brains (Burré et al., 2013). On the other hand, a recent NMR study has characterized a structure of the α-synuclein assembled into fibrils (Tuttle et al., 2016). For decades, α-synuclein has been considered as a model protein in the field of protein structural disorders and was suggested to have an ATP-independent chaperone activity of its own. Clearly, α-synuclein is a structurally dynamic protein which can be easily assembled into an amyloid upon mutation or in post-translational modification (Deas et al., 2016; Ingelsson, 2016; Kim, Kagedal, & Halliday, 2014; Xu & Pu, 2016).

Although the etiology of the disease is not yet fully understood, oxidative stress is considered to play a major role in disease pathogenesis, resulting in mitochondrial dysfunction and ER stress (Puspita, Chung, & Shim, 2017). It was also suggested that ER stress itself enhances α-synuclein aggregation by a still uncharacterized mechanism (Puspita et al., 2017).

Deficiency in mitochondrial complex I within the dopaminergic cells is considered the major cause of this neuronal degeneration, associated with reduced levels of ATP, and increased ROS formation (Beal, 2005). PD is linked to a loss of function of many redox homeostasis proteins, including mitochondrial or mitochondrial-associated Pink1, Parkin (Park2), DJ-1, and Lrrk2 genes, thus providing additional link to understanding PD pathogenesis.

Another major factor contributing to dopaminergic neurotoxicity is the presence of dopamine itself. Dopamine is an unstable molecule, which when oxidized is converted into reactive dopamine quinones (e.g., DOPAC quinones) accompanying ROS release (Gluck & Zeevalk, 2004). This effect is prevalent in the early stages of the disease when dopamine is produced in excess to compensate for dopaminergic neuron death (Scherman et al., 1989).

Another affected organelle in the PD progression is ER, which has dual function: protein maturation and maintaining calcium homeostasis. Failure of these functions results in accumulation of misfolded proteins and calcium leakage into both the cytosol and mitochondria, which triggers the unfolding protein response (UPR) pathways. Imaging of mitochondria and ER showed that they are closely associated, which leads to Ca^{2+} influx into mitochondria, causing increased ROS production, a UPR response mediated by Parkin and other proteins, autophagy upregulation, and inflammation of the affected neurons (Rodríguez-Arribas et al., 2017; Sala, Marinig, Arosio, & Ferrarese, 2016).

As mentioned above, an imbalance in redox homeostasis is interconnected with an imbalance in protein homeostasis. In PD, oligomerization of α-synuclein was shown to induce oxidative stress expressed by increases in lipid peroxidation and a reduction of GSH levels (Chinta & Andersen, 2008; Deas et al., 2016). To prevent α-synuclein oligomerization, the canonical chaperones as well as ATP-independent small Hsp (sHsp) proteins are upregulated to cope with the proteotoxic stress induced by oxidative stress and α-synuclein amyloid formation (Bruinsma et al., 2011; Cox et al., 2014; Klucken, Shin, Masliah, Hyman, & McLean, 2004).

5.2.3 Amyotrophic lateral sclerosis

Amyotrophic lateral sclerosis (ALS), also known as Lou Gehrig's disease, is a fatal neurodegenerative disorder characterized by the progressive loss of upper and lower motor neurons that act as a mediator between the spinal cord and the peripheral muscle fibers (Wijesekera & Leigh, 2009). Symptoms first appear in mid-life and include muscle weakness and spasticity; in most patients, paralysis and respiratory failure occur within 3–10 years of diagnosis (Turner & Swash, 2015). ALS is traditionally classified into two types: familial ALS (fALS) and sporadic ALS (sALS). fALS is the predominantly inherited form and is linked to autosomal dominant mutations in different genes, including SOD1 (Robberecht & Philips, 2013). Sporadic ALS is considered not to be inherited, although misclassification might occur due to incomplete family history records (Robberecht & Philips, 2013). Both forms of ALS are linked to mutations in heterogeneous genes, including RNA-binding and RNA-processing proteins such as Fus, TDP-43, TAF15, proteostasis proteins such as VCP, ubiquitin, and others (Robberecht & Philips, 2013). Therefore, ALS pathogenesis is rooted in the failure of proteasomal and autophagic processes, RNA metabolism, antioxidant activity resulting in protein clumping into inclusions, ER stress, as well as Golgi and mitochondrial disorders. The axonal architecture and function of ALS patients is damaged, causing denervation of neurons (e.g., lower and upper motor neurons) and muscles (Barber & Shaw, 2010; Robberecht & Philips, 2013; Saxena & Caroni, 2011).

As in the ND disorders described above, ALS is associated with a failure of the antioxidant system and mitochondrial dysfunction (Barber & Shaw, 2010). ALS patients undergo glutamate accumulation in synapses mediated by the redox-regulated proteins EAAC1 and NMDA (Foran & Trotti, 2009). When the equilibrium between uptake and release of glutamate is upset, glutamate accumulates in the synaptic clefts leading to excessive release of calcium into the intracellular space that overactivates these glutamate receptors, a process known as glutamate excitotoxicity. As previously described, the excess of calcium ions and its influx in mitochondria lead to mitochondrial dysfunction and a burst of oxidative stress in neurons.

One of the first links between ALS and oxidative stress resulted from the identification of autosomal dominant mutations in the mitochondrial superoxide dismutase 1 (SOD1) found in ~20% of familial ALS cases (Robberecht & Philips, 2013). Mutations in SOD1 lead to its inactivation and misfolding, forming stable aggregates. One of the genes affected by mutation in SOD1 is Nrf2, a central transcriptional factor of the antioxidant response as depicted in Fig. 2 (Kirby et al., 2005). Moreover, mutant SOD1 not only fails to dismutase the superoxide but also induces its production through NADPH oxidase within glial cells (Barber & Shaw, 2010). Therefore, NADPH oxidase deletion (Marden et al., 2007) or its inhibition (Harraz et al., 2008) results in a significant delay in disease progression and an increase in the lifespan of model mice. In addition to inactivation of the SOD1 function, mutations in SOD1 lead to its misfolding and aggregation which imposes another burden on the proteostasis system.

SOD1 is not the only ALS-linked aggregation-prone protein. TDP-43, a highly conserved nuclear RNA-binding protein, was found delocalized in cytoplasmic inclusions in somatic ALS and familial ALS patients. TDP-43 is a thermobile protein that is highly unstable and undergoes rapid aggregation upon oxidation or single mutations. Most probably, upon mild oxidative stress TDP-43 undergoes hyperphosphorylation and ubiquitination (which inactivate its function) followed by accumulation in the insoluble aggregates in the cytosol, which prevents translocation of the non-functional form of TDP-43 into the nucleus. Neumann et al. showed that short-length isoforms of TDP-43 were present in inclusions of the brains of ALS patients, suggesting another toxicity pathway for TDP-43 aggregation which involves TDP-43 cleavage in a yet undiscovered mechanism (Neumann et al., 2006). Recent biochemical studies of the aggregation properties of the full-length and short-length isoforms of TDP-43 defined potential differences in toxicity and structure of the oligomers. They demonstrated that the formation of TDP-43 oligomers is oxidation dependent and is triggered by first oxidation of its N-terminal cysteine residues and then completed by modification of other cysteines located in the RNA-recognition domain (Bozzo et al., 2016).

Similar to other neurodegenerative disorders, ALS is linked to altered chaperone functions combatting the proteotoxic stress induced by oxidative stress and aggregation of SOD1 or TDP-43 proteins, which in turn leads to misfolding of other functional proteins in the affected neuron cells and surrounded glia. In ALS the Hsf1-mediated pathway is upregulated, leading to clearance of the TDP-43 aggregates using DnaJ/Hsp70 and Hsp90 mechanisms (Chen et al., 2016; Webster, Smith, Shaw, & De Vos, 2017). In addition, both in vitro and in vivo studies in ALS models have observed the ubiquitination of SOD1 misfolded proteins in a Hsp70/CHIP-dependent manner, which facilitates its degradation by the proteasome. Moreover, it was shown that Hsp70/Hsp110 is involved in disaggregation of the SOD1 aggregates (Song et al., 2013).

Despite the commitment of the proteostasis network, SOD1 misfolded proteins escape the degradation system. In fact, SOD1 oligomers have a toxic effect on the degradation machinery, thus impairing proteasomal and autophagy degradation (Cozzolino, Ferri, & Carrì, 2008; Webster et al., 2017). ALS patients have inclusions of the ubiquitinated proteins spread throughout the degenerated neurons, and in related glia in different brain areas as well as in the spinal cord, suggesting failure of proteasomal degradation and protein turnover (Webster et al., 2017). The majority of these ubiquitinated aggregates contain either TDP-43 or SOD1 and FUS proteins.

In addition to the apparent effect of SOD1 and TDP-43 on protein turnover, some ALS disorders are associated with mutations of the proteins directly mediating protein degradation. For example, VCP (p97 or CDC48 in yeast) is an ATPase which is essential for targeting misfolded proteins from the ER, mitochondria, and nucleus to proteasomal degradation.

Mutations in VCP have been associated with 1–2% of all ALS cases. These VCP variants failed to promote mitochondrial and ER proteostasis, thus contributing to the increased accumulation of toxic aggregates in cells (Webster et al., 2017). Moreover, the correlation between mutations in VCP and a decrease in TDP-43 activity has also been nicely shown in different model systems (Custer, Neumann, Lu, Wright, & Taylor, 2010; Ritson et al., 2010), thus relating seemingly different ALS causes to one pathway.

6. Modulating protein homeostasis to treat neurodegenerative diseases

The redox and protein homeostasis systems are highly interconnected, and the collapse of one dramatically affects the other, leading to the progression of various neurodegenerative diseases. Thus, it is not surprising that the vast majority of drug development efforts have been centered on restoring proteostasis and redox homeostasis, specifically to increase the protein folding capacity, promote the removal of misfolded proteins, and boost the antioxidant capacity of cells to decrease cellular ROS levels.

Of the chaperones, Hsp90, Hsp70, and the Hsf1 are the most famous drug targets (Sweeney et al., 2017). Manipulation of their levels in ALS was shown to result in delays in pathogenesis, a reduction in protein aggregates and increase lifespan (Bruening et al., 1999; Krishnan et al., 2008; Sharp et al., 2008). However, in most cases there was only slight improvement and the success was limited to the early phase of the disease or to a specific ALS model. These Hsf1-targeted drugs include hydroxylamine derivatives (e.g., Arimoclomol) which stabilize Hsf1 in cells, leading to a prolonged stress response altering redox and protein homeostasis. Treatment with Arimoclomol was shown to delay SOD1 aggregation, increase lifespan and enhance motor activity (Kalmar et al., 2008).

Another promising target is Hsp90 since it was shown to promote amyloid formations of hyperphosphorylated Tau (Choi et al., 2015; Luo, Sun, Taldone, Rodina, & Chiosis, 2010) and α-synuclein (Jones, Moussaud, & Mclean, 2014). Various molecular drugs have been developed to inhibit Hsp90 singly or together with other chaperones (e.g., Hsp70) which have led to positive outcomes in tissue culture and animal models (Brandvold & Morimoto, 2015; Sweeney et al., 2017). One such molecule is dihydropyridine (DHP) which affects the heat shock response (Kasza et al., 2016) and stimulates a degradation of Tau (Dickey et al., 2007) as well as α-synuclein (Auluck & Bonini, 2002; Putcha et al., 2010) amyloids. Due to chaperone redundancy and interconnectivity, targeting more than one chaperone usually leads to better results than targeting a single one (Nollen, Brunsting, Roelofsen, Weber, & Kampinga, 1999; Patel, Payne Smith, De Belleroche, & Latchman, 2005; Wagstaff et al., 1999). Other strategies and cutting-edge drugs are discussed in the following recent reviews (Copple, Dinkova-Kostova, Kensler, Liby, & Wigley, 2017; Liu & Bitan, 2012; Sweeney et al., 2017).

In conclusion, we are in an era of the active research that aims to uncover common as well as disease-specific mechanisms underlying the onset and progression of neurodegenerative diseases. Close cooperation between pharmaceutical companies and academia has made it possible to discover multiple drugs that can potentially target different features of neurodegenerative disorders. The outcomes of these collaborations are currently being tested in clinical studies and hopefully will become successful therapeutic options for millions of patients around the world. However, the development of new drugs and new strategies targeting specific types of disorders are extremely challenging given the diversity of causes and drivers of these disorders which are still not fully understood. There is no doubt that, additional breakthroughs will make it possible to cope with the enormous heterogeneity of these pathologies, which should lead to more personalized medical strategies to prevent the onset and progression of these neurodegenerative diseases.

References

Abramov, A. Y. (2004). β-Amyloid peptides induce mitochondrial dysfunction and oxidative stress in astrocytes and death of neurons through activation of NADPH oxidase. *Journal of Neuroscience*, *24*, 565–575. https://doi.org/10.1523/JNEUROSCI.4042-03.2004.

Abramov, A. Y., Berezhnov, A. V., Fedotova, E. I., Zinchenko, V. P., & Dolgacheva, L. P. (2017). Interaction of misfolded proteins and mitochondria in neurodegenerative disorders. *Biochemical Society Transactions*, *45*(4), 1025–1033. https://doi.org/10.1042/BST20170024.

Abramov, A. Y., Canevari, L., & Duchen, M. R. (2003). Changes in intracellular calcium and glutathione in astrocytes as the primary mechanism of amyloid neurotoxicity. *Journal of Neuroscience*, *23*, 5088–5095. https://doi.org/10.1523/JNEUROSCI.4042-03.2004.

Abramov, A. Y., Canevari, L., & Duchen, M. R. (2004). Calcium signals induced by amyloid β peptide and their consequences in neurons and astrocytes in culture. *Biochimica et Biophysica Acta—Molecular Cell Research*, *1742*, 81–87. https://doi.org/10.1016/j.bbamcr.2004.09.006.

Adachi, H., Katsuno, M., Waza, M., Minamiyama, M., Tanaka, F., & Sobue, G. (2009). Heat shock proteins in neurodegenerative diseases: Pathogenic roles and therapeutic implications. *International Journal of Hyperthermia*, *25*, 647–684. https://doi.org/10.3109/02656730903315823.

Ahn, S. G., & Thiele, D. J. (2003). Redox regulation of mammalian heat shock factor 1 is essential for Hsp gene activation and protection from stress. *Genes and Development*, *17*, 516–528. https://doi.org/10.1101/gad.1044503.

Alderton, W. K., Cooper, C. E., & Knowles, R. G. (2001). Nitric oxide synthases: Structure, function and inhibition. *Biochemical Journal*, *357*, 593–615. https://doi.org/10.1042/0264-6021:3570593.

Amor, S., Peferoen, L. A. N., Vogel, D. Y. S., Breur, M., van der Valk, P., Baker, D., et al. (2014). Inflammation in neurodegenerative diseases—An update. *Immunology*, *142*, 151–166. https://doi.org/10.1111/imm.12233.

Andersen, J. K. (2004). Oxidative stress in neurodegeneration: Cause or consequence? *Nature Reviews Neuroscience*, *10*, S18–S25. https://doi.org/10.1038/nrn1434.

Angelova, P. R., & Abramov, A. Y. (2018). Role of mitochondrial ROS in the brain: From physiology to neurodegeneration. *FEBS Letters*, *592*, 692–702. https://doi.org/10.1002/1873-3468.12964.

Appenzeller-Herzog, C. (2011). Glutathione- and non-glutathione-based oxidant control in the endoplasmic reticulum. *Journal of Cell Science*, *124*(6), 847–855. Retrieved from: http://jcs.biologists.org/content/124/6/847.abstract.

Ashraf, A., Clark, M., & So, P.-W. (2018). The aging of iron man. *Frontiers in Aging Neuroscience*, *10*, 65. Retrieved from: https://www.frontiersin.org/article/10.3389/fnagi.2018.00065.

Augusto, O., & Sayuri, M. (2011). Oxygen radicals and related species. In K. Pantopoulos & H. M. Schipper (Eds.), *Vol. 1. Principles of Free Radical Biomedicine*. Nova Science Publishers, Inc. ISBN: 978-1-61209-773-2.

Auluck, P. K., & Bonini, N. M. (2002). Pharmacological prevention of Parkinson disease in Drosophila. *Nature Medicine*, *8*, 1185–1186. https://doi.org/10.1038/nm1102-1185.

Auluck, P. K., Chan, H. Y. E., Trojanowski, J. Q., Lee, V. M. Y., & Bonini, N. M. (2002). Chaperone suppression of α-synuclein toxicity in a Drosophila model for Parkinson's disease. *Science*, *295*, 865–868. https://doi.org/10.1126/science.1067389.

Babior, B. M., Lambeth, J. D., & Nauseef, W. (2002). The neutrophil NADPH oxidase. *Archives of Biochemistry and Biophysics*, *397*, 342–344. https://doi.org/10.1006/abbi.2001.2642.

Bak, L. K., Sickmann, H. M., Schousboe, A., & Waagepetersen, H. S. (2005). Activity of the lactate-alanine shuttle is independent of glutamate-glutamine cycle activity in cerebellar neuronal-astrocytic cultures. *Journal of Neuroscience Research*, *79*, 88–96. https://doi.org/10.1002/jnr.20319.

Balchin, D., Hayer-Hartl, M., & Hartl, F. U. (2016). In vivo aspects of protein folding and quality control. *Science*, *353*, aac4354. https://doi.org/10.1126/science.aac4354.

Barber, S. C., & Shaw, P. J. (2010). Oxidative stress in ALS: Key role in motor neuron injury and therapeutic target. *Free Radical Biology and Medicine*, *48*, 629–641. https://doi.org/10.1016/j.freeradbiomed.2009.11.018.

Barna, J., Csermely, P., & Vellai, T. (2018). Roles of heat shock factor 1 beyond the heat shock response. *Cellular and Molecular Life Sciences*, *75*(16), 2897–2916. https://doi.org/10.1007/s00018-018-2836-6.

Barnham, K. J., Masters, C. L., & Bush, A. I. (2004). Neurodegenerative diseases and oxidative stress. *Nature Reviews Drug Discovery*, *3*(3), 205.

Beal, M. F. (2005). Mitochondria take center stage in aging and neurodegeneration. *Annals of Neurology*, *58*, 495–505. https://doi.org/10.1002/ana.20624.

Bedard, K., & Krause, K.-H. (2007). The NOX family of ROS-generating NADPH oxidases: Physiology and pathophysiology. *Physiological Reviews*, *87*, 245–313. https://doi.org/10.1152/physrev.00044.2005.

Belli, M., Ramazzotti, M., & Chiti, F. (2011). Prediction of amyloid aggregation in vivo. *EMBO Reports*, *12*, 657–663. https://doi.org/10.1038/embor.2011.116.

Bergmann, O., Spalding, K. L., & Frisén, J. (2015). Adult neurogenesis in humans. *Cold Spring Harbor Perspectives in Biology*, *7*, a018994. https://doi.org/10.1101/cshperspect.a018994.

Biber, K., Owens, T., & Boddeke, E. (2014). What is microglia neurotoxicity (Not)? *Glia*, *62*, 841–854. https://doi.org/10.1002/glia.22654.

Bozzo, F., Salvatori, I., Iacovelli, F., Mirra, A., Rossi, S., Cozzolino, M., et al. (2016). Structural insights into the multi-determinant aggregation of TDP-43 in motor neuron-like cells. *Neurobiology of Disease*, *94*, 63–72. https://doi.org/10.1016/j.nbd.2016.06.006.

Brandvold, K. R., & Morimoto, R. I. (2015). The chemical biology of molecular chaperones—Implications for modulation of proteostasis. *Journal of Molecular Biology*, *427*, 2931–2947. https://doi.org/10.1016/j.jmb.2015.05.010.

Brash, A. R. (1999). Lipoxygenases: Occurrence, functions, catalysis, and acquisition of substrate. *Journal of Biological Chemistry*, *274*, 23679–23682. https://doi.org/10.1074/jbc.274.34.23679.

Brown, N. M., Torres, A. S., Doan, P. E., & O'Halloran, T. V. (2004). Oxygen and the copper chaperone CCS regulate posttranslational activation of Cu, Zn superoxide dismutase. *Proceedings of the National Academy of Sciences*, *101*(15), 5518–5523.

Bruening, W., Josée, R., Giasson, B., Figlewicz, D. A., Mushynski, W. E., & Durham, H. D. (1999). Up-regulation of protein chaperones preserves viability of cells expressing toxic Cu/Zn-superoxide dismutase mutants associated with amyotrophic lateral sclerosis. *Journal of Neurochemistry*, *72*, 693–699. https://doi.org/10.1046/j.1471-4159.1999.0720693.x.

Bruinsma, I. B., Bruggink, K. A., Kinast, K., Versleijen, A. A. M., Segers-Nolten, I. M. J., Subramaniam, V., et al. (2011). Inhibition of α-synuclein aggregation by small heat shock proteins. *Proteins: Structure, Function and Bioinformatics*, *79*, 2956–2967. https://doi.org/10.1002/prot.23152.

Burré, J., Vivona, S., Diao, J., Sharma, M., Brunger, A. T., & Südhof, T. C. (2013). Properties of native brain α-synuclein. *Nature*, *498*, E4–E6. https://doi.org/10.1038/nature12125.

Butterfield, D. A., Castegna, A., Lauderback, C. M., & Drake, J. (2002). Evidence that amyloid beta-peptide-induced lipid peroxidation and its sequelae in Alzheimer's disease brain contribute to neuronal death. *Neurobiology of Aging*, *23*, 655–664. https://doi.org/10.1016/S0197-4580(01)00340-2.

Cadenas, E., & Davies, K. J. A. (2000). Mitochondrial free radical generation, oxidative stress, and aging. *Free Radical Biology and Medicine*, *29*, 222–230. https://doi.org/10.1016/S0891-5849(00)00317-8.

Calì, T., Ottolini, D., Negro, A., & Brini, M. (2012). α-Synuclein controls mitochondrial calcium homeostasis by enhancing endoplasmic reticulum-mitochondria interactions. *Journal of Biological Chemistry*, *287*(22), 17914–17929.

Casley, C. S., Land, J. M., Sharpe, M. A., Clark, J. B., Duchen, M. R., & Canevari, L. (2002). β-Amyloid fragment 25-35 causes mitochondrial dysfunction in primary cortical neurons. *Neurobiology of Disease*, *10*, 258–267. https://doi.org/10.1006/nbdi.2002.0516.

Chae, H. Z., Oubrahim, H., Park, J. W., Rhee, S. G., & Chock, P. B. (2012). Protein glutathionylation in the regulation of peroxiredoxins: A family of thiol-specific peroxidases that function as antioxidants, molecular chaperones, and signal modulators. *Antioxidants & Redox Signaling*, *16*, 506–523. https://doi.org/10.1089/ars.2011.4260.

Checconi, P., Salzano, S., Bowler, L., Mullen, L., Mengozzi, M., Hanschmann, E.-M., et al. (2015). Redox proteomics of the inflammatory secretome identifies a common set of redoxins and other glutathionylated proteins released in inflammation, influenza virus infection and oxidative stress. *PLoS One*, *10*(5), e0127086. Retrieved from: https://doi.org/10.1371/journal.pone.0127086.

Chelikani, P., Fita, I., & Loewen, P. C. (2004). Diversity of structures and properties among catalases. *Cellular and Molecular Life Sciences*, *61*, 192–208. https://doi.org/10.1007/s00018-003-3206-5.

Chen, H. J., Mitchell, J. C., Novoselov, S., Miller, J., Nishimura, A. L., Scotter, E. L., et al. (2016). The heat shock response plays an important role in TDP-43 clearance: Evidence for dysfunction in amyotrophic lateral sclerosis. *Brain*, *139*, 1417–1432. https://doi.org/10.1093/brain/aww028.

Chennamsetty, N., Voynov, V., Kayser, V., Helk, B., & Trout, B. L. (2010). Prediction of aggregation prone regions of therapeutic proteins. *Journal of Physical Chemistry B*, *114*, 6614–6624. https://doi.org/10.1021/jp911706q.

Chinta, S. J., & Andersen, J. K. (2008). Redox imbalance in Parkinson's disease. *Biochimica et Biophysica Acta—General Subjects*, *1780*, 1362–1367. https://doi.org/10.1016/j.bbagen.2008.02.005.

Choi, S. H., Kim, Y. H., D'Avanzo, C., Aronson, J., Tanzi, R. E., & Kim, D. Y. (2015). Recapitulating amyloid ß and tau pathology in human neural cell culture models-clinical implications. *US Neurology*, *11*, 102–105. http://doi.org/10.17925/USN.2015.11.02.102.

Clements, C. M., McNally, R. S., Conti, B. J., Mak, T. W., & Ting, J. P.-Y. (2006). DJ-1, a cancer- and Parkinson's disease-associated protein, stabilizes the antioxidant transcriptional master regulator Nrf2. *Proceedings of the National Academy of Sciences*, *103*, 15091–15096. https://doi.org/10.1073/pnas.0607260103.

Cohen, E., Bieschke, J., Perciavalle, R. M., Kelly, J. W., & Dillin, A. (2006). Opposing activities protect against age-onset proteotoxicity. *Science*, *313*, 1604–1610. https://doi.org/10.1126/science.1124646.

Collins, G. A., & Goldberg, A. L. (2017). The logic of the 26S proteasome. *Cell*, *169*, 792–806. https://doi.org/10.1016/j.cell.2017.04.023.

Copple, I. M., Dinkova-Kostova, A. T., Kensler, T. W., Liby, K. T., & Wigley, W. C. (2017). NRF2 as an emerging therapeutic target. *Oxidative Medicine and Cellular Longevity*, *2017*, 8165458. https://doi.org/10.1155/2017/8165458.

Cox, D., Carver, J. A., & Ecroyd, H. (2014). Preventing α-synuclein aggregation: The role of the small heat-shock molecular chaperone proteins. *Biochimica et Biophysica Acta—Molecular Basis of Disease*, *1842*, 1830–1843. https://doi.org/10.1016/j.bbadis.2014.06.024.

Cozzolino, M., Ferri, A., & Carrì, M. T. (2008). Amyotrophic lateral sclerosis: From current developments in the laboratory to clinical implications. *Antioxidants & Redox Signaling*, *10*, 405–443. https://doi.org/10.1089/ars.2007.1760.

Custer, S. K., Neumann, M., Lu, H., Wright, A. C., & Taylor, J. P. (2010). Transgenic mice expressing mutant forms VCP/p97 recapitulate the full spectrum of IBMPFD including degeneration in muscle, brain and bone. *Human Molecular Genetics*, *19*(9), 1741–1755. Retrieved from: https://doi.org/10.1093/hmg/ddq050.

Dalle-Donne, I., Milzani, A., Gagliano, N., Colombo, R., Giustarini, D., & Rossi, R. (2007). Molecular mechanisms and potential clinical significance of S-glutathionylation. *Antioxidants & Redox Signaling*, *10*(3), 445–474. https://doi.org/10.1089/ars.2007.1716.

De Sá Junior, P. L., Câmara, D. A. D., Porcacchia, A. S., Fonseca, P. M. M., Jorge, S. D., Araldi, R. P., et al. (2017). The roles of ROS in cancer heterogeneity and therapy. *Oxidative Medicine and Cellular Longevity*, *2017*, 2467940. https://doi.org/10.1155/2017/2467940.

Deas, E., Cremades, N., Angelova, P. R., Ludtmann, M. H. R., Yao, Z., Chen, S., et al. (2016). Alpha-synuclein oligomers interact with metal ions to induce oxidative stress and neuronal death in Parkinson's disease. *Antioxidants & Redox Signaling*, *24*, 376–391. https://doi.org/10.1089/ars.2015.6343.

Deibel, M. A., Ehmann, W. D., & Markesbery, W. R. (1996). Copper, iron, and zinc imbalances in severely degenerated brain regions in Alzheimer's disease: Possible relation to oxidative stress. *Journal of the Neurological Sciences*, *143*, 137–142. https://doi.org/10.1016/S0022-510X(96)00203-1.

Demasi, M., Netto, L. E. S., Silva, G. M., Hand, A., de Oliveira, C. L. P., Bicev, R. N., et al. (2014). Redox regulation of the proteasome via S-glutathionylation. *Redox Biology*, *2*, 44–51. https://doi.org/10.1016/j.redox.2013.12.003.

Devasagayam, T. P. A., Tilak, J. C., Boloor, K. K., Sane, K. S., Ghaskadbi, S. S., & Lele, R. D. (2004). Free radicals and antioxidants in human health: Current status and future prospects. *Journal of the Association of Physicians of India*, *52*, 794–804. https://doi.org/10.1016/S0300-483X(03)00149-5.

Dickey, C. A., Kamal, A., Lundgren, K., Klosak, N., Bailey, R. M., Dunmore, J., et al. (2007). The high-affinity HSP90-CHIP complex recognizes and selectively degrades phosphorylated tau client proteins. *Journal of Clinical Investigation*, *117*, 648–658. https://doi.org/10.1172/JCI29715.

Dikic, I. (2017). Proteasomal and autophagic degradation systems. *Annual Review of Biochemistry*, *86*(1), 193–224. https://doi.org/10.1146/annurev-biochem-061516-044908.

Espinoza, S. E., Guo, H., Fedarko, N., DeZern, A., Fried, L. P., Xue, Q. L., et al. (2008). Glutathione peroxidase enzyme activity in aging. *Journals of Gerontology—Series A Biological Sciences and Medical Sciences*, *63*, 505–509. https://doi.org/10.1093/gerona/63.5.505.

Esterbauer, H., Schaur, R. J., & Zollner, H. (1991). Chemistry and biochemistry of 4-hydroxynonenal, malonaldehyde and related aldehydes. *Free Radical Biology and Medicine*, *11*, 81–128. https://doi.org/10.1016/0891-5849(91)90192-6.

Evans, C. G., Wisén, S., & Gestwicki, J. E. (2006). Heat shock proteins 70 and 90 inhibit early stages of amyloid β-(1-42) aggregation in vitro. *Journal of Biological Chemistry*, *281*, 33182–33191. https://doi.org/10.1074/jbc.M606192200.

Falsone, S. F., Kungl, A. J., Rek, A., Cappai, R., & Zangger, K. (2009). The molecular chaperone Hsp90 modulates intermediate steps of amyloid assembly of the Parkinson-related protein α-synuclein. *Journal of Biological Chemistry*, *284*, 31190–31199. https://doi.org/10.1074/jbc.M109.057240.

Ferreiro, E., Oliveira, C. R., & Pereira, C. M. F. (2008). The release of calcium from the endoplasmic reticulum induced by amyloid-beta and prion peptides activates the mitochondrial apoptotic pathway. *Neurobiology of Disease*, *30*, 331–342. https://doi.org/10.1016/j.nbd.2008.02.003.

Fischer, R., & Maier, O. (2015). Interrelation of oxidative stress and inflammation in neurodegenerative disease: Role of TNF. *Oxidative Medicine and Cellular Longevity*, *2015*, 610813. https://doi.org/10.1155/2015/610813.

Foran, E., & Trotti, D. (2009). Glutamate transporters and the excitotoxic path to motor neuron degeneration in amyotrophic lateral sclerosis. *Antioxidants & Redox Signaling*, *11*, 1587–1602. https://doi.org/10.1089/ars.2009.2444.

Furukawa, Y., Torres, A. S., & O'Halloran, T. V. (2004). Oxygen-induced maturation of SOD1: A key role for disulfide formation by the copper chaperone CCS. *The EMBO Journal*, *23*, 2872–2881. https://doi.org/10.1038/sj.emboj.7600276.

Gaschler, M. M., & Stockwell, B. R. (2017). Lipid peroxidation in cell death. *Biochemical and Biophysical Research Communications*, *482*, 419–425. https://doi.org/10.1016/j.bbrc.2016.10.086.

Gauron, C., Rampon, C., Bouzaffour, M., Ipendey, E., Teillon, J., Volovitch, M., et al. (2013). Sustained production of ROS triggers compensatory proliferation and is required for regeneration to proceed. *Scientific Reports*, *3*, 2084. https://doi.org/10.1038/srep02084.

Georgiou, G., & Masip, L. (2003). An overoxidation journey with a return ticket. *Science*, *300*, 592–594. https://doi.org/10.1126/science.1084976.
Ghezzi, P. (2013). Protein glutathionylation in health and disease. *Biochimica et Biophysica Acta—General Subjects*, *1830*, 3165–3172. https://doi.org/10.1016/j.bbagen.2013.02.009.
Glorieux, C., & Calderon, P. B. (2017). Catalase, a remarkable enzyme: Targeting the oldest antioxidant enzyme to find a new cancer treatment approach. *Biological Chemistry*, *398*, 1095–1108. https://doi.org/10.1515/hsz-2017-0131.
Gluck, M. R., & Zeevalk, G. D. (2004). Inhibition of brain mitochondrial respiration by dopamine and its metabolites: Implications for Parkinson's disease and catecholamine-associated diseases. *Journal of Neurochemistry*, *91*, 788–795. https://doi.org/10.1111/j.1471-4159.2004.02747.x.
Gunn, A. P., Wong, X. B. X., Johanssen, X. T., Griffith, X. J. C., Masters, C. L., Bush, A. I., et al. (2016). Amyloid- peptide A-3pE-42 induces lipid peroxidation, membrane permeabilization, and calcium influx in neurons. *Journal of Biological Chemistry*, *291*, 6134–6145. https://doi.org/10.1074/jbc.M115.655183.
Gupta, S. C., Hevia, D., Patchva, S., Park, B., Koh, W., & Aggarwal, B. B. (2012). Upsides and downsides of reactive oxygen species for cancer: The roles of reactive oxygen species in tumorigenesis, prevention, and therapy. *Antioxidants & Redox Signaling*, *16*, 1295–1322. https://doi.org/10.1089/ars.2011.4414.
Halliwell, B., & Aruoma, O. I. (1991). DNA damage by oxygen-derived species its mechanism and measurement in mammalian systems. *FEBS Letters*, *281*, 9–19. http://doi.org/10.1016/0014-5793(91)80347–6.
Harman, D. (1956). Aging: A theory based on free radical and radiation chemistry. *Journal of Gerontology*, *11*, 298–300. https://doi.org/10.1093/geronj/11.3.298.
Harraz, M. M., Marden, J. J., Zhou, W., Zhang, Y., Williams, A., Sharov, V. S., et al. (2008). SOD1 mutations disrupt redox-sensitive Rac regulation of NADPH oxidase in a familial ALS model. *Journal of Clinical Investigation*, *118*, 659–670. https://doi.org/10.1172/JCI34060.
Hartl, F. U., & Hayer-Hartl, M. (2002). Protein folding—Molecular chaperones in the cytosol: From nascent chain to folded protein. *Science*, *295*, 1852–1858. https://doi.org/10.1126/science.1068408.
Hashimoto, M., Rockenstein, E., Crews, L., & Masliah, E. (2003). Role of protein aggregation in mitochondrial dysfunction and neurodegeneration in Alzheimer's and Parkinson's diseases. *NeuroMolecular Medicine*, *4*, 21–36. https://doi.org/10.1385/NMM:4:1-2:21.
Hayes, J. D., & Dinkova-Kostova, A. T. (2014). The Nrf2 regulatory network provides an interface between redox and intermediary metabolism. *Trends in Biochemical Sciences*, *39*(4), 199–218. https://doi.org/10.1016/j.tibs.2014.02.002.
Higdon, A., Diers, A. R., Oh, J. Y., Landar, A., & Darley-Usmar, V. M. (2012). Cell signalling by reactive lipid species: New concepts and molecular mechanisms. *The Biochemical Journal*, *442*, 453–464. https://doi.org/10.1042/BJ20111752.
Hochstrasser, M. (1996). Ubiquitin-dependent protein degradation. *Annual Review of Genetics*, *30*(1), 405–439. https://doi.org/10.1146/annurev.genet.30.1.405.
Holmström, K. M., & Finkel, T. (2014). Cellular mechanisms and physiological consequences of redox-dependent signalling. *Nature Reviews Molecular Cell Biology*, *15*, 411–421. https://doi.org/10.1038/nrm3801.
Hrycay, E. G., & Bandiera, S. M. (2015). Involvement of cytochrome P450 in reactive oxygen species formation and cancer. *Advances in Pharmacology*, *74*, 35–84. https://doi.org/10.1016/bs.apha.2015.03.003.
Hsieh, H. L., & Yang, C. M. (2013). Role of redox signaling in neuroinflammation and neurodegenerative diseases. *BioMed Research International*, *2013*, 484613. https://doi.org/10.1155/2013/484613.

Imai, H., & Nakagawa, Y. (2003). Biological significance of phospholipid hydroperoxide glutathione peroxidase (PHGPx, GPx4) in mammalian cells. *Free Radical Biology and Medicine*, *34*, 145–169. https://doi.org/10.1016/S0891-5849(02)01197-8.

Ingelsson, M. (2016). Alpha-synuclein oligomers-neurotoxic molecules in Parkinson's disease and other Lewy body disorders. *Frontiers in Neuroscience*, *10*, 408. https://doi.org/10.3389/fnins.2016.00408.

Jakob, U., & Reichmann, D. (Eds.), (2013). *Oxidative Stress and Redox Regulation* (pp. 1–42). Dodrecht: Springer.

Jang, H. H., Lee, K. O., Chi, Y. H., Jung, B. G., Park, S. K., Park, J. H., et al. (2004). Two enzymes in one: Two yeast peroxiredoxins display oxidative stress-dependent switching from a peroxidase to a molecular chaperone function. *Cell*, *117*, 625–635. https://doi.org/10.1016/j.cell.2004.05.002.

Jomova, K., Vondrakova, D., Lawson, M., & Valko, M. (2010). Metals, oxidative stress and neurodegenerative disorders. *Molecular and Cellular Biochemistry*, *345*, 91–104. https://doi.org/10.1007/s11010-010-0563-x.

Jones, D. R., Moussaud, S., & Mclean, P. (2014). Targeting heat shock proteins to modulate α-synuclein toxicity. *Therapeutic Advances in Neurological Disorders*, *7*, 33–51. https://doi.org/10.1177/1756285613493469.

Junn, E., Jang, W. H., Zhao, X., Jeong, B. S., & Mouradian, M. M. (2009). Mitochondrial localization of DJ-1 leads to enhanced neuroprotection. *Journal of Neuroscience Research*, *87*, 123–129. https://doi.org/10.1002/jnr.21831.

Kalmar, B., Novoselov, S., Gray, A., Cheetham, M. E., Margulis, B., & Greensmith, L. (2008). Late stage treatment with arimoclomol delays disease progression and prevents protein aggregation in the SOD1G93Amouse model of ALS. *Journal of Neurochemistry*, *107*, 339–350. https://doi.org/10.1111/j.1471-4159.2008.05595.x.

Kann, O., & Kovacs, R. (2006). Mitochondria and neuronal activity. *American Journal of Physiology: Cell Physiology*, *292*, C641–C657. https://doi.org/10.1152/ajpcell.00222.2006.

Kasza, Á., Hunya, Á., Frank, Z., Fülöp, F., Török, Z., Balogh, G., et al. (2016). Dihydropyridine derivatives modulate heat shock responses and have a neuroprotective effect in a transgenic mouse model of Alzheimer's disease. *Journal of Alzheimer's Disease*, *53*, 557–571. https://doi.org/10.3233/JAD-150860.

Kawamata, H., & Manfredi, G. (2010). Import, maturation, and function of SOD1 and its copper chaperone CCS in the mitochondrial intermembrane space. *Antioxidants & Redox Signaling*, *13*, 1375–1384. https://doi.org/10.1089/ars.2010.3212.

Khor, H. K., Fisher, M. T., & Schöneich, C. (2004). Potential role of methionine sulfoxide in the inactivation of the chaperone GroEL by hypochlorous acid (HOCl) and peroxynitrite (ONOO-). *Journal of Biological Chemistry*, *279*, 19486–19493. https://doi.org/10.1074/jbc.M310045200.

Kim, W. S., Kagedal, K., & Halliday, G. M. (2014). Alpha-synuclein biology in Lewy body diseases. *Alzheimer's Research and Therapy*, *6*, 73. https://doi.org/10.1186/s13195-014-0073-2.

Kirby, J., Halligan, E., Baptista, M. J., Allen, S., Heath, P. R., Holden, H., et al. (2005). Mutant SOD1 alters the motor neuronal transcriptome: Implications for familial ALS. *Brain*, *128*, 1686–1706. https://doi.org/10.1093/brain/awh503.

Kiss, R., Zhu, M., Jójárt, B., Czajlik, A., Solti, K., Fórizs, B., et al. (2017). Structural features of human DJ-1 in distinct Cys106 oxidative states and their relevance to its loss of function in disease. *Biochimica et Biophysica Acta—General Subjects*, *1861*, 2619–2629. https://doi.org/10.1016/j.bbagen.2017.08.017.

Klionsky, D. J. (2008). Autophagy revisited: A conversation with Christian de Duve. *Autophagy*, *4*, 740–743. https://doi.org/10.4161/auto.6398.

Klucken, J., Shin, Y., Masliah, E., Hyman, B. T., & McLean, P. J. (2004). Hsp70 reduces α-synuclein aggregation and toxicity. *Journal of Biological Chemistry*, *279*, 25497–25502. https://doi.org/10.1074/jbc.M400255200.

Knowles, T. P. J., Vendruscolo, M., & Dobson, C. M. (2014). The amyloid state and its association with protein misfolding diseases. *Nature Reviews Molecular Cell Biology*, *15*, 384–396. https://doi.org/10.1038/nrm3810.

Kopeikina, K. J., Hyman, B. J., & Spires-Jones, T. L. (2012). Soluble forms of tau are toxic in Alzheimer's disease. *Translational Neuroscience*, *3*, 223–233. https://doi.org/10.2478/s13380-012-0032-y.

Koren, J., Jinwal, U. K., Lee, D. C., Jones, J. R., Shults, C. L., Johnson, A. G., et al. (2009). Chaperone signalling complexes in Alzheimer's disease. *Journal of Cellular and Molecular Medicine*, *13*, 619–630. https://doi.org/10.1111/j.1582-4934.2008.00557.x.

Krishnan, J., Vannuvel, K., Andries, M., Waelkens, E., Robberecht, W., & Van Den Bosch, L. (2008). Over-expression of Hsp27 does not influence disease in the mutant SOD1 G93A mouse model of amyotrophic lateral sclerosis. *Journal of Neurochemistry*, *106*, 2170–2183. https://doi.org/10.1111/j.1471-4159.2008.05545.x.

Lackie, R. E., Maciejewski, A., Ostapchenko, V. G., Marques-Lopes, J., Choy, W. Y., Duennwald, M. L., et al. (2017). The Hsp70/Hsp90 chaperone machinery in neurodegenerative diseases. *Frontiers in Neuroscience*, *11*, 254. https://doi.org/10.3389/fnins.2017.00254.

Lavranos, G., Balla, M., Tzortzopoulou, A., Syriou, V., & Angelopoulou, R. (2012). Investigating ROS sources in male infertility: A common end for numerous pathways. *Reproductive Toxicology*, *34*, 298–307. https://doi.org/10.1016/j.reprotox.2012.06.007.

Lin, M. T., & Beal, M. F. (2006). Mitochondrial dysfunction and oxidative stress in neurodegenerative diseases. *Nature*, *443*, 787–795. https://doi.org/10.1038/nature05292.

Lind, C., Gerdes, R., Hamnell, Y., Schuppe-Koistinen, I., von Löwenhielm, H. B., Holmgren, A., et al. (2002). Identification of S-glutathionylated cellular proteins during oxidative stress and constitutive metabolism by affinity purification and proteomic analysis. *Archives of Biochemistry and Biophysics*, *406*(2), 229–240. https://doi.org/10.1016/S0003-9861(02)00468-X.

Little, A. C., Sulovari, A., Danyal, K., Heppner, D. E., Seward, D. J., & van der Vliet, A. (2017). Paradoxical roles of dual oxidases in cancer biology. *Free Radical Biology and Medicine*, *110*, 117–132. https://doi.org/10.1016/j.freeradbiomed.2017.05.024.

Liu, T., & Bitan, G. (2012). Modulating self-assembly of amyloidogenic proteins as a therapeutic approach for neurodegenerative diseases: Strategies and mechanisms. *ChemMedChem*, *7*, 359–374. https://doi.org/10.1002/cmdc.201100585.

Liu, Z., Li, T., Li, P., Wei, N., Zhao, Z., Liang, H., et al. (2015). The ambiguous relationship of oxidative stress, tau hyperphosphorylation, and autophagy dysfunction in Alzheimer's disease. *Oxidative Medicine and Cellular Longevity*, *2015*, 352723. https://doi.org/10.1155/2015/352723.

Liu, Z., Zhou, T., Ziegler, A. C., Dimitrion, P., & Zuo, L. (2017). Oxidative stress in neurodegenerative diseases: From molecular mechanisms to clinical applications. *Oxidative Medicine and Cellular Longevity*, *2017*, 2525967. https://doi.org/10.1155/2017/2525967.

Livnat-Levanon, N., Kevei, É., Kleifeld, O., Krutauz, D., Segref, A., Rinaldi, T., et al. (2014). Reversible 26S proteasome disassembly upon mitochondrial stress. *Cell Reports*, *7*, 1371–1380. https://doi.org/10.1016/j.celrep.2014.04.030.

Lowther, W. T., & Haynes, A. C. (2011). Reduction of cysteine sulfinic acid in eukaryotic, typical 2-Cys peroxiredoxins by sulfiredoxin. *Antioxidants & Redox Signaling*, *15*, 99–109. https://doi.org/10.1089/ars.2010.3564.

Lu, S. C. (2009). Regulation of glutathione synthesis. *Molecular Aspects of Medicine*, *30*, 42–59. https://doi.org/10.1016/j.mam.2008.05.005.

Luo, W., Sun, W., Taldone, T., Rodina, A., & Chiosis, G. (2010). Heat shock protein 90 in neurodegenerative diseases. *Molecular Neurodegeneration*, *5*, 24. https://doi.org/10.1186/1750-1326-5-24.

Mailloux, R. J. (2015). Teaching the fundamentals of electron transfer reactions in mitochondria and the production and detection of reactive oxygen species. *Redox Biology*, *4*, 381–398. https://doi.org/10.1016/j.redox.2015.02.001.

Marden, J. J., Harraz, M. M., Williams, A. J., Nelson, K., Luo, M., Paulson, H., et al. (2007). Redox modifier genes in amyotrophic lateral sclerosis in mice. *Journal of Clinical Investigation*, *117*, 2913–2919. https://doi.org/10.1172/JCI31265.

Mayer, M. P., & Bukau, B. (2005). Hsp70 chaperones: Cellular functions and molecular mechanism. *Cellular and Molecular Life Sciences*, *62*, 670–684. https://doi.org/10.1007/s00018-004-4464-6.

McBean, G. (2017). Cysteine, glutathione, and thiol redox balance in astrocytes. *Antioxidants*, *6*, 62. https://doi.org/10.3390/antiox6030062.

McBean, G. J., Aslan, M., Griffiths, H. R., & Torrão, R. C. (2015). Thiol redox homeostasis in neurodegenerative disease. *Redox Biology*, *5*, 186–194. https://doi.org/10.1016/j.redox.2015.04.004.

McCoy, M. K., & Cookson, M. R. (2011). DJ-1 regulation of mitochondrial function and autophagy through oxidative stress. *Autophagy*, 7, 531–532. https://doi.org/10.4161/auto.7.5.14684.

Mena, N. P., Urrutia, P. J., Lourido, F., Carrasco, C. M., & Núñez, M. T. (2015). Mitochondrial iron homeostasis and its dysfunctions in neurodegenerative disorders. *Mitochondrion*, *21*, 92–105. https://doi.org/10.1016/j.mito.2015.02.001.

Milani, P., Ambrosi, G., Gammoh, O., Blandini, F., & Cereda, C. (2013). SOD1 and DJ-1 converge at Nrf2 pathway: A clue for antioxidant therapeutic potential in neurodegeneration. *Oxidative Medicine and Cellular Longevity*, *2013*, 836760. https://doi.org/10.1155/2013/836760.

Moon, J. C., Hah, Y. S., Kim, W. Y., Jung, B. G., Jang, H. H., Lee, J. R., et al. (2005). Oxidative stress-dependent structural and functional switching of a human 2-Cys peroxiredoxin isotype II that enhances HeLa cell resistance to H2O2-induced cell death. *Journal of Biological Chemistry*, *280*, 28775–28784. https://doi.org/10.1074/jbc.M505362200.

Morimoto, R. I. (2008). Proteotoxic stress and inducible chaperone networks in neurodegenerative disease and aging. *Genes and Development*, *22*, 1427–1438. https://doi.org/10.1101/gad.1657108.

Muchowski, P. J., & Wacker, J. L. (2005). Modulation of neurodegeneration by molecular chaperones. *Nature Reviews Neuroscience*, *6*, 11–22. https://doi.org/10.1038/nrn1587.

Murphy, M. P. (2009). How mitochondria produce reactive oxygen species. *The Biochemical Journal*, *417*, 1–13. https://doi.org/10.1042/BJ20081386.

Neumann, M., Sampathu, D. M., Kwong, L. K., Truax, A. C., Micsenyi, M. C., Chou, T. T., et al. (2006). Ubiquitinated TDP-43 in frontotemporal lobar degeneration and amyotrophic lateral sclerosis. *Science*, *314*, 130–133. https://doi.org/10.1126/science.1134108.

Nollen, E. A. A., Brunsting, J. F., Roelofsen, H., Weber, L. A., & Kampinga, H. H. (1999). In vivo chaperone activity of heat shock protein 70 and thermotolerance. *Molecular and Cellular Biology*, *19*, 2069–2079. https://doi.org/10.1128/MCB.19.3.2069.

Oh, S. E., & Mouradian, M. M. (2018). Cytoprotective mechanisms of DJ-1 against oxidative stress through modulating ERK1/2 and ASK1 signal transduction. *Redox Biology*, *14*, 211–217. https://doi.org/10.1016/j.redox.2017.09.008.

Olzscha, H., Schermann, S. M., Woerner, A. C., Pinkert, S., Hecht, M. H., Tartaglia, G. G., et al. (2011). Amyloid-like aggregates sequester numerous metastable proteins with essential cellular functions. *Cell*, *144*, 67–78. https://doi.org/10.1016/j.cell.2010.11.050.

Ortiz, G. G., Pacheco Moisés, F. P., Mireles-Ramírez, M., Flores-Alvarado, L. J., González-Usigli, H., Sánchez-González, V. J., et al. (2017). Oxidative stress: Love and hate history in central nervous system. *Advances in Protein Chemistry and Structural Biology*, *108*, 1–31. https://doi.org/10.1016/bs.apcsb.2017.01.003.

Pajares, M., Cuadrado, A., & Rojo, A. I. (2017). Modulation of proteostasis by transcription factor NRF2 and impact in neurodegenerative diseases. *Redox Biology*, *11*, 543–553. https://doi.org/10.1016/j.redox.2017.01.006.

Panday, A., Sahoo, M. K., Osorio, D., & Batra, S. (2015). NADPH oxidases: An overview from structure to innate immunity-associated pathologies. *Cellular and Molecular Immunology*, *12*, 5–23. https://doi.org/10.1038/cmi.2014.89.

Patel, Y. J. K., Payne Smith, M. D., De Belleroche, J., & Latchman, D. S. (2005). Hsp27 and Hsp70 administered in combination have a potent protective effect against FALS-associated SOD1-mutant-induced cell death in mammalian neuronal cells. *Brain Research. Molecular Brain Research*, *134*, 256–274. https://doi.org/10.1016/j.molbrainres.2004.10.028.

Perry, V. H. (2004). The influence of systemic inflammation on inflammation in the brain: Implications for chronic neurodegenerative disease. *Brain, Behavior, and Immunity*, *18*(5), 407–413. https://doi.org/10.1016/j.bbi.2004.01.004.

Petrucelli, L., Dickson, D., Kehoe, K., Taylor, J., Snyder, H., Grover, A., et al. (2004). CHIP and Hsp70 regulate tau ubiquitination, degradation and aggregation. *Human Molecular Genetics*, *13*, 703–714. https://doi.org/10.1093/hmg/ddh083.

Pirkkala, L. (2001). Roles of the heat shock transcription factors in regulation of the heat shock response and beyond. *The FASEB Journal*, *15*, 1118–1131. https://doi.org/10.1096/fj00-0294rev.

Puspita, L., Chung, S. Y., & Shim, J. W. (2017). Oxidative stress and cellular pathologies in Parkinson's disease. *Molecular Brain*, *10*, 53. https://doi.org/10.1186/s13041-017-0340-9.

Putcha, P., Danzer, K. M., Kranich, L. R., Scott, A., Silinski, M., Mabbett, S., et al. (2010). Brain-permeable small-molecule inhibitors of Hsp90 prevent α-synuclein oligomer formation and rescue α-synuclein-induced toxicity. *The Journal of Pharmacology and Experimental Therapeutics*, *332*, 849–857. https://doi.org/10.1124/jpet.109.158436.

Rego, A. C., & Oliveira, C. R. (2003). Mitochondrial dysfunction and reactive oxygen species in excitotoxicity and apoptosis: Implications for the pathogenesis of neurodegenerative diseases. *Neurochemical Research*, *28*, 1563–1574. https://doi.org/10.1023/A:1025682611389.

Reichmann, D., Voth, W., & Jakob, U. (2018). Maintaining a healthy proteome during oxidative stress. *Molecular Cell*, *69*, 203–213. https://doi.org/10.1016/j.molcel.2017.12.021.

Reinheckel, T., Ullrich, O., Sitte, N., & Grune, T. (2000). Differential impairment of 20S and 26S proteasome activities in human hematopoietic K562 cells during oxidative stress. *Archives of Biochemistry and Biophysics*, *377*, 65–68. https://doi.org/10.1006/abbi.2000.1717.

Resende, R., Ferreiro, E., Pereira, C., & De Oliveira, C. R. (2008). Neurotoxic effect of oligomeric and fibrillar species of amyloid-beta peptide 1–42: Involvement of endoplasmic reticulum calcium release in oligomer-induced cell death. *Neuroscience*, *155*(3), 725–737.

Rhee, S. G., & Woo, H. A. (2011). Multiple functions of peroxiredoxins: Peroxidases, sensors and regulators of the intracellular messenger H_2O_2, and protein chaperones. *Antioxidants & Redox Signaling*, *15*(3), 781–794. https://doi.org/10.1089/ars.2010.3393.

Ritson, G. P., Custer, S. K., Freibaum, B. D., Guinto, J. B., Geffel, D., Moore, J., et al. (2010). TDP-43 mediates degeneration in a novel Drosophila model of disease caused by mutations in VCP/p97. *The Journal of Neuroscience*, *30*(22), 7729–7739. Retrieved from: http://www.jneurosci.org/content/30/22/7729.abstract.

Robberecht, W., & Philips, T. (2013). The changing scene of amyotrophic lateral sclerosis. *Nature Reviews Neuroscience*, *14*, 248–264. https://doi.org/10.1038/nrn3430.

Rodríguez-Arribas, M., Yakhine-Diop, S. M. S., Pedro, J. M. B. S., Gómez-Suaga, P., Gómez-Sánchez, R., Martínez-Chacón, G., et al. (2017). Mitochondria-associated membranes (MAMs): Overview and its role in Parkinson's disease. *Molecular Neurobiology*, *54*, 6287–6303. https://doi.org/10.1007/s12035-016-0140-8.

Rodríguez-Martín, T., Cuchillo-Ibáñez, I., Noble, W., Nyenya, F., Anderton, B. H., & Hanger, D. P. (2013). Tau phosphorylation affects its axonal transport and degradation. *Neurobiology of Aging*, *34*, 2146–2157. https://doi.org/10.1016/j.neurobiolaging.2013.03.015.

Rose, R. C., & Bode, A. M. (1993). Biology of free radical scavengers: An evaluation of ascorbate. *The FASEB Journal*, 7(12), 1135–1142.

Ross, C. A., & Poirier, M. A. (2005). What is the role of protein aggregation in neurodegeneration? *Nature Reviews Molecular Cell Biology*, *6*, 891–898. https://doi.org/10.1038/nrm1742.

Rowley, N. M., Madsen, K. K., Schousboe, A., & Steve White, H. (2012). Glutamate and GABA synthesis, release, transport and metabolism as targets for seizure control. *Neurochemistry International*, *61*, 546–558. https://doi.org/10.1016/j.neuint.2012.02.013.

Rubinsztein, D. C., Mariño, G., & Kroemer, G. (2011). Autophagy and aging. *Cell*, *146*, 682–695. https://doi.org/10.1016/j.cell.2011.07.030.

Sahara, N., Murayama, M., Mizoroki, T., Urushitani, M., Imai, Y., Takahashi, R., et al. (2005). In vivo evidence of CHIP up-regulation attenuating tau aggregation. *Journal of Neurochemistry*, *94*, 1254–1263. https://doi.org/10.1111/j.1471-4159.2005.03272.x.

Sala, G., Marinig, D., Arosio, A., & Ferrarese, C. (2016). Role of chaperone-mediated autophagy dysfunctions in the pathogenesis of Parkinson's disease. *Frontiers in Molecular Neuroscience*, *9*, 157. https://doi.org/10.3389/fnmol.2016.00157.

Saxena, S., & Caroni, P. (2011). Selective neuronal vulnerability in neurodegenerative diseases: From stressor thresholds to degeneration. *Neuron*, *71*, 35–48. https://doi.org/10.1016/j.neuron.2011.06.031.

Scandalios, J. G. (1993). Oxygen stress and superoxide dismutases. *Plant Physiology*, *101*, 7–12. https://doi.org/10.1104/pp.101.1.7.

Scherman, D., Desnos, C., Darchen, F., Pollak, P., Javoy-Agid, F., & Agid, Y. (1989). Striatal dopamine deficiency in Parkinson's disease: Role of aging. *Annals of Neurology*, *26*, 551–557. https://doi.org/10.1002/ana.410260409.

Schopf, F. H., Biebl, M. M., & Buchner, J. (2017). The HSP90 chaperone machinery. *Nature Reviews Molecular Cell Biology*, *18*, 345–360. https://doi.org/10.1038/nrm.2017.20.

Schrader, M., & Fahimi, H. D. (2006). Peroxisomes and oxidative stress. *Biochimica et Biophysica Acta—Molecular Cell Research*, *1763*, 1755–1766. https://doi.org/10.1016/j.bbamcr.2006.09.006.

Shankar, G. M., Li, S., Mehta, T. H., Garcia-Munoz, A., Shepardson, N. E., Smith, I., et al. (2008). Amyloid-β protein dimers isolated directly from Alzheimer's brains impair synaptic plasticity and memory. *Nature Medicine*, *14*, 837–842. https://doi.org/10.1038/nm1782.

Sharp, P. S., Akbar, M. T., Bouri, S., Senda, A., Joshi, K., Chen, H. J., et al. (2008). Protective effects of heat shock protein 27 in a model of ALS occur in the early stages of disease progression. *Neurobiology of Disease*, *30*, 42–55. https://doi.org/10.1016/j.nbd.2007.12.002.

Shelat, P. B., Chalimoniuk, M., Wang, J. H., Strosznajder, J. B., Lee, J. C., Sun, A. Y., et al. (2008). Amyloid beta peptide and NMDA induce ROS from NADPH oxidase and AA release from cytosolic phospholipase A2in cortical neurons. *Journal of Neurochemistry*, *106*, 45–55. https://doi.org/10.1111/j.1471-4159.2008.05347.x.

Shendelman, S., Jonason, A., Martinat, C., Leete, T., & Abeliovich, A. (2004). DJ-1 is a redox-dependent molecular chaperone that inhibits α-synuclein aggregate formation. *PLoS Biology*, *2*, e362. https://doi.org/10.1371/journal.pbio.0020362.
Siegenthaler, K. D., Pareja, K. A., Wang, J., & Sevier, C. S. (2017). An unexpected role for the yeast nucleotide exchange factor Sil1 as a reductant acting on the molecular chaperone BiP. *eLife*, *6*. https://doi.org/10.7554/eLife.24141.
Singh, N., Haldar, S., Tripathi, A. K., McElwee, M. K., Horback, K., & Beserra, A. (2014). Iron in neurodegenerative disorders of protein misfolding: A case of prion disorders and Parkinson's disease. *Antioxidants & Redox Signaling*, *21*(3), 471–484. https://doi.org/10.1089/ars.2014.5874.
Skovronsky, D. M., Lee, V. M.-Y., & Trojanowski, J. Q. (2006). Neurodegenerative diseases: New concepts of pathogenesis and their therapeutic implications. *Annual Review of Pathology: Mechanisms of Disease*, *1*, 151–170. https://doi.org/10.1146/annurev.pathol.1.110304.100113.
Song, Y., Nagy, M., Ni, W., Tyagi, N. K., Fenton, W. A., Lopez-Giraldez, F., et al. (2013). Molecular chaperone Hsp110 rescues a vesicle transport defect produced by an ALS-associated mutant SOD1 protein in squid axoplasm. *Proceedings of the National Academy of Sciences of the United States of America*, *110*, 5428–5433. https://doi.org/10.1073/pnas.1303279110.
Stadtman, E. R. (2006). Protein oxidation and aging. *Free Radical Research*, *40*, 1250–1258. https://doi.org/10.1080/10715760600918142.
Stadtman, E. R., & Levine, R. L. (2001). Protein oxidation. *Annals of the New York Academy of Sciences*, *899*, 191–208. https://doi.org/10.1111/j.1749-6632.2000.tb06187.x.
Stewart, W. F., Kawas, C., Corrada, M., & Metter, E. J. (1997). Risk of Alzheimer's disease and duration of NSAID use. *Neurology*, *48*, 626–632. https://doi.org/10.1212/WNL.48.3.626.
Suh, J. H., Shenvi, S. V., Dixon, B. M., Liu, H., Jaiswal, A. K., Liu, R.-M., et al. (2004). Decline in transcriptional activity of Nrf2 causes age-related loss of glutathione synthesis, which is reversible with lipoic acid. *Proceedings of the National Academy of Sciences of the United States of America*, *101*, 3381–3386. https://doi.org/10.1073/pnas.0400282101.
Sweeney, P., Park, H., Baumann, M., Dunlop, J., Frydman, J., Kopito, R., et al. (2017). Protein misfolding in neurodegenerative diseases: Implications and strategies. *Translational Neurodegeneration*, *6*, 6. https://doi.org/10.1186/s40035-017-0077-5.
Taira, T., Saito, Y., Niki, T., Iguchi-Ariga, S. M. M., Takahashi, K., & Ariga, H. (2004). DJ-1 has a role in antioxidative stress to prevent cell death. *EMBO Reports*, *5*, 213–218. https://doi.org/10.1038/sj.embor.7400074.
Tartaglia, G. G., Pawar, A. P., Campioni, S., Dobson, C. M., Chiti, F., & Vendruscolo, M. (2008). Prediction of aggregation-prone regions in structured proteins. *Journal of Molecular Biology*, *380*, 425–436. https://doi.org/10.1016/j.jmb.2008.05.013.
Tsolis, A. C., Papandreou, N. C., Iconomidou, V. A., & Hamodrakas, S. J. (2013). A consensus method for the prediction of "aggregation-prone" peptides in globular proteins. *PLoS One*, *8*, e54175. https://doi.org/10.1371/journal.pone.0054175.
Tu, B. P., & Weissman, J. S. (2002). The FAD- and O2-dependent reaction cycle of Ero1-mediated oxidative protein folding in the endoplasmic reticulum. *Molecular Cell*, *10*, 983–994. https://doi.org/10.1016/S1097-2765(02)00696-2.
Turner, M. R., & Swash, M. (2015). The expanding syndrome of amyotrophic lateral sclerosis: A clinical and molecular odyssey. *Journal of Neurology, Neurosurgery and Psychiatry*, *86*, 667–673. https://doi.org/10.1136/jnnp-2014-308946.
Tuttle, M. D., Comellas, G., Nieuwkoop, A. J., Covell, D. J., Berthold, D. A., Kloepper, K. D., et al. (2016). Solid-state NMR structure of a pathogenic fibril of full-length human α-synuclein. *Nature Structural and Molecular Biology*, *23*, 409–415. https://doi.org/10.1038/nsmb.3194.

Uttara, B., Singh, A., Zamboni, P., & Mahajan, R. (2009). Oxidative stress and neurodegenerative diseases: A review of upstream and downstream antioxidant therapeutic options. *Current Neuropharmacology*, 7, 65–74. https://doi.org/10.2174/157015909787602823.

Varshavsky, A. (1997). The ubiquitin system. *Trends in Biochemical Sciences*, *22*(10), P383–P387. https://doi.org/10.1016/S0968-0004(97)01122-5.

Waagepetersen, H. S., Sonnewald, U., Larsson, O. M., & Schousboe, A. (2000). A possible role of alanine for ammonia transfer between astrocytes and glutamatergic neurons. *Journal of Neurochemistry*, *75*, 471–479. https://doi.org/10.1046/j.1471-4159.2000.0750471.x.

Wagstaff, M. J. D., Collaço-Moraes, Y., Smith, J., De Belleroche, J. S., Coffin, R. S., & Latchman, D. S. (1999). Protection of neuronal cells from apoptosis by Hsp27 delivered with a herpes simplex virus-based vector. *The Journal of Biological Chemistry*, *274*, 5061–5069. https://doi.org/10.1074/jbc.274.8.5061.

Walsh, D. M., & Selkoe, D. J. (2007). Aβ oligomers—A decade of discovery. *Journal of Neurochemistry*, *101*, 1172–1184. https://doi.org/10.1111/j.1471-4159.2006.04426.x.

Wang, J., Pareja, K. A., Kaiser, C. A., & Sevier, C. S. (2014). Redox signaling via the molecular chaperone BiP protects cells against endoplasmic reticulum-derived oxidative stress. *eLife*, *3*, e03496. https://doi.org/10.7554/eLife.03496.

Wang, J., & Sevier, C. S. (2016). Formation and reversibility of BiP protein cysteine oxidation facilitate cell survival during and post oxidative stress. *Journal of Biological Chemistry*, *291*, 7541–7557. https://doi.org/10.1074/jbc.M115.694810.

Wang, J., & Yi, J. (2008). Cancer cell killing via ROS: To increase or decrease, that is a question. *Cancer Biology and Therapy*, 7, 1875–1884. https://doi.org/10.4161/cbt.7.12.7067.

Ward, R. J., Dexter, D. T., & Crichton, R. R. (2015). Neurodegenerative diseases and therapeutic strategies using iron chelators. *Journal of Trace Elements in Medicine and Biology*, *31*, 267–273. https://doi.org/10.1016/j.jtemb.2014.12.012.

Ward, R. J., Zucca, F. A., Duyn, J. H., Crichton, R. R., & Zecca, L. (2014). The role of iron in brain ageing and neurodegenerative disorders. *Lancet Neurology*, *13*, 1045–1060. https://doi.org/10.1016/S1474-4422(14)70117-6.

Webster, C. P., Smith, E. F., Shaw, P. J., & De Vos, K. J. (2017). Protein homeostasis in amyotrophic lateral sclerosis: Therapeutic opportunities? *Frontiers in Molecular Neuroscience*, *10*, 123. https://doi.org/10.3389/fnmol.2017.00123.

Wei, P. C., Hsieh, Y. H., Su, M. I., Jiang, X., Hsu, P. H., Lo, W. T., et al. (2012). Loss of the oxidative stress sensor NPGPx compromises GRP78 chaperone activity and induces systemic disease. *Molecular Cell*, *45*, 747–756. https://doi.org/10.1016/j.molcel.2012.10.007.

Welles Kellogg, E., & Fridovich, I. (1975). Superoxide, hydrogen peroxide, and singlet oxygen in lipid peroxidation by a xanthine oxidase system. *Journal of Biological Chemistry*, *250*, 8812–8817. https://doi.org/10.1016/0022-4073(94)90028-0.

Wijesekera, L. C., & Leigh, P. N. (2009). Amyotrophic lateral sclerosis disease names. *Orphanet Journal of Rare Diseases*, *4*, 3. https://doi.org/10.1186/1750-1172-4-3.

Wilson, M. A. (2011). The role of cysteine oxidation in DJ-1 function and dysfunction. *Antioxidants & Redox Signaling*, *15*, 111–122. https://doi.org/10.1089/ars.2010.3481.

Winter, J., Linke, K., Jatzek, A., & Jakob, U. (2005). Severe oxidative stress causes inactivation of DnaK and activation of the redox-regulated chaperone Hsp33. *Molecular Cell*, *17*, 381–392. https://doi.org/10.1016/j.molcel.2004.12.027.

Winterbourn, C. C. (2008). Reconciling the chemistry and biology of reactive oxygen species. *Nature chemical biology*, *4*(5), 278.

Winterbourn, C. C. (2013). Radical scavenging by thiols and the fate of thiyl radicals. In *Oxidative Stress and Redox Regulation* (pp. 43–58). Dordrecht: Springer.

Winterbourn, C. C., & Kettle, A. J. (2013). Redox reactions and microbial killing in the neutrophil phagosome. *Antioxidants & Redox Signaling*, *18*, 642–660. https://doi.org/10.1089/ars.2012.4827.

Winterbourn, C. C., Kettle, A. J., & Hampton, M. B. (2016). Reactive oxygen species and neutrophil function. *Annual Review of Biochemistry*, *85*, 765–792. https://doi.org/10.1146/annurev-biochem-060815-014442.

Xu, X., Muller, J. G., Ye, Y., & Burrows, C. J. (2008). DNA-protein cross-links between guanine and lysine depend on the mechanism of oxidation for formation of C5 vs C8 guanosine adducts. *Journal of the American Chemical Society*, *130*, 703–709. https://doi.org/10.1021/ja077102a.

Xu, L., & Pu, J. (2016). Alpha-synuclein in Parkinson's disease: From pathogenetic dysfunction to potential clinical application. *Parkinson's Disease*, *2016*, 1720621. https://doi.org/10.1155/2016/1720621.

Yu, X., & Long, Y. C. (2016). Crosstalk between cystine and glutathione is critical for the regulation of amino acid signaling pathways and ferroptosis. *Scientific Reports*, *6*, 30033. Retrieved from: https://doi.org/10.1038/srep30033.

Yu, H., & Matouschek, A. (2017). Recognition of client proteins by the proteasome. *Annual Review of Biophysics*, *46*, 149–173. https://doi.org/10.1146/annurev-biophys-070816-033719.

Zhang, W. (2005). Aggregated α-synuclein activates microglia: A process leading to disease progression in Parkinson's disease. *The FASEB Journal*, *19*(6), 533–542. https://doi.org/10.1096/fj.04-2751com.

Zhang, D. D. (2006). Mechanistic studies of the Nrf2-Keap1 signaling pathway. *Drug Metabolism Reviews*, *38*(4), 769–789. https://doi.org/10.1080/03602530600971974.

Zhang, H., Davies, K. J. A., & Forman, H. J. (2015). Oxidative stress response and Nrf2 signaling in aging. *Free Radical Biology and Medicine*, *88*, 314–336. https://doi.org/10.1016/j.freeradbiomed.2015.05.036.

Zhang, W., Wang, T., Pei, Z., Miller, D. S., Wu, X., Block, M. L., et al. (2005). Aggregated α-synuclein activates microglia: A process leading to disease progression in Parkinson's disease. *The FASEB Journal*, *19*(6), 533–542.

Zheng, H., & Koo, E. H. (2006). The amyloid precursor protein: Beyond amyloid. *Molecular Neurodegeneration*, *1*, 5. https://doi.org/10.1186/1750-1326-1-5.

Zhou, W., & Freed, C. R. (2005). DJ-1 up-regulates glutathione synthesis during oxidative stress and inhibits A53T α-synuclein toxicity. *Journal of Biological Chemistry*, *280*, 43150–43158. https://doi.org/10.1074/jbc.M507124200.

CHAPTER SEVEN

The roles of cytosolic quality control proteins, SGTA and the BAG6 complex, in disease

Rashi Benarroch, Jennifer M. Austin, Fahmeda Ahmed, Rivka L. Isaacson*

Department of Chemistry, King's College London, London, United Kingdom
*Corresponding author: e-mail address: rivka.isaacson@kcl.ac.uk

Contents

Advances in Protein Chemistry and Structural Biology, Volume 114
ISSN 1876-1623
https://doi.org/10.1016/bs.apcsb.2018.11.002

Abstract

SGTA is a co-chaperone that, in collaboration with the complex of BAG6/UBL4A/TRC35, facilitates the biogenesis and quality control of hydrophobic proteins, protecting them from the aqueous cytosolic environment. This work includes targeting tail-anchored proteins to their resident membranes, sorting of membrane and secretory proteins that mislocalize to the cytoplasm and endoplasmic reticulum-associated degradation of misfolded proteins. Since these functions are all vital for the cell's continued proteostasis, their disruption poses a threat to the cell, with a particular risk of protein aggregation, a phenomenon that underpins many diseases. Although the specific disease implications of machinery involved in quality control of hydrophobic substrates are poorly understood, here we summarize much of the available information on this topic.

1. Introduction

The cytoplasm of cells is a bustling, crowded environment which requires layers of management to prevent descent into chaos. This chapter will address some of the machinery that maintains the cellular environment and what happens when it breaks down or gets out of control. In particular we will consider the case of stray hydrophobic proteins that become exposed to the aqueous cytosol. These pose an aggregation risk and, as relevant, must be sent to a membrane, refolded to tuck the hydrophobic parts back into the safety of the core or labeled for disposal with ubiquitin tags and then degraded by the proteasome. Since these processes are crucial for maintaining cellular proteostasis, it is inevitable that failures in them, which, among other things, can be caused by absence or disruption of any of their machine components, may result in disease. At present, the precise roles of these proteins in disease are not well understood. This review will consider some of the major known constituents of the mammalian quality control machinery for hydrophobic proteins exposed to the cytoplasm, SGTA, BAG6, TRC35 and UBL4A, and catalog some of the current disease connections that exist in today's literature. It is likely that these references barely scratch the surface of the disease relevance and therapeutic potential of these proteins and that this field will continue burgeoning in the coming years.

2. Hydrophobic parts of proteins exposed to the cytoplasm pose an aggregation risk

There are many scenarios in which hydrophobic areas of proteins become aberrantly exposed to the aqueous cell cytoplasm. Large proteins often reach their native folded state via a number of intermediate states. During these folding pathways there is the potential for protein misfolding, which can result in additional deleterious functions that can lead to a range of diseases (Hartl & Hayer-Hartl, 2009). The more complex the protein, the higher the chance it will misfold. During the folding pathway there has evolved a need for a quality control network. As proteins pass through intermediate states where they are incompletely folded, they can expose buried regions to the solvent which can lead to undesired interactions; the exposure of these regions can be an indicator of protein misfolding (Buchberger, Bukau, & Sommer, 2010). It's the role of molecular chaperones and co-chaperones of the quality control network to protect these exposed regions and thus aid either the refolding of the protein, further prevention of misfolding or transport of the protein to a desired safe site (Buchberger et al., 2010). While chaperones do not increase the rate of folding, they do increase its efficiency by reducing the chances of undesired reactions such as aggregation (Fernandez-Fernandez & Valpuesta, 2018). Hence molecular chaperones and co-chaperones play a role in the essential protein quality control network that deals with unfolded and misfolded proteins.

3. The specific case of tail-anchored proteins

Tail-anchored (TA) proteins represent a widespread but distinct case of hydrophobic regions requiring protection from the aqueous cytosol (Rabu, Schmid, Schwappach, & High, 2009). They are a class of membrane proteins that make up 3–5% of all membrane proteins including SNAREs (involved in vesicular trafficking); signaling proteins; endoplasmic reticulum translocon components and various enzymes that are spatially restricted (Borgese & Fasana, 2011). They therefore have critical roles in cell biology in membrane biogenesis, apoptosis, and protein degradation among many others. The specific localization of TA proteins within membranes is crucial to their task and, while their functions vary greatly, they are united in a common topology, based on the positioning of their single transmembrane helix (Kalbfleisch, Cambon, & Wattenberg, 2007). Newly synthesized TA proteins cannot be co-translationally directed to the endoplasmic reticulum (ER) for insertion, as is traditional for membrane proteins, as their single membrane-spanning helix is at the extreme C-terminus and therefore inaccessible to the signal recognition particle during translation (Borgese & Fasana, 2011). This is where quality control proteins come into play, they are nature's solution to the insertion of the TA proteins into the ER membrane and this process shares common sorting and targeting machinery with the MLPs described above (Shao, Rodrigo-Brenni, Kivlen, & Hegde, 2017).

4. Triage and fate of hydrophobic substrates in the cytoplasm

The details of these processes are still being established but many advances have been made in recent years including structure solution of some of the machinery for sorting and delivery (which will be discussed in the individual sections for each protein) and mechanistic understanding from cell biology and isolated systems (Shao et al., 2017). Our current understanding is summarized in Fig. 1. It is thought that SGTA can catch the TMDs of newly translated TA proteins or exposed hydrophobic patches on mislocalised membrane and secretory proteins (MLP), which bind to its C-terminal domain. In collaboration with the heterotrimeric BAG6 complex, which comprises BAG6, TRC35 and UBL4A, SGTA determines the fate of these proteins. Hydrophobic substrates bound to the BAG6

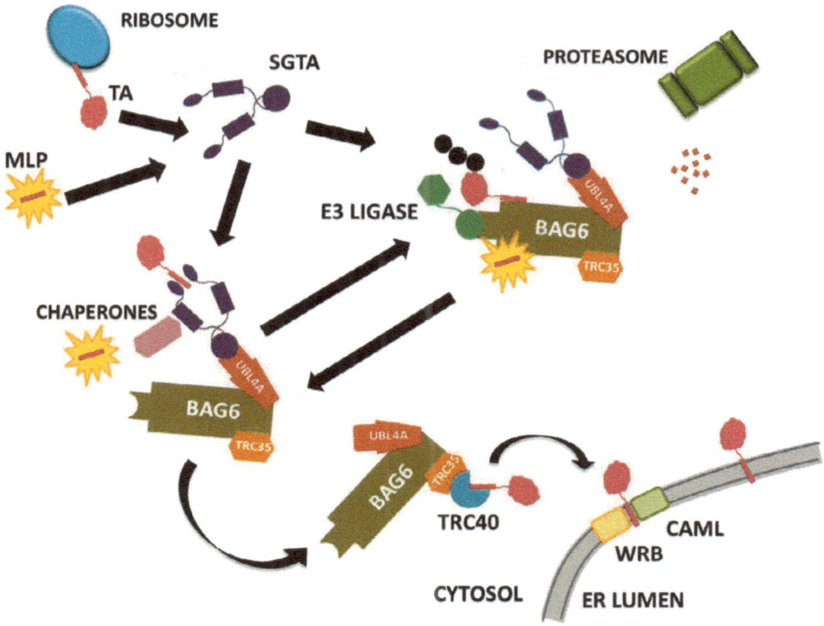

Fig. 1 Current ideas on quality control pathways for hydrophobic proteins exposed to the aqueous cytoplasm. SGTA can catch the TMDs of newly translated TA proteins or exposed hydrophobic patches on MLPs, which bind to its C-terminal domain. In collaboration with the heterotrimeric BAG6 complex, which comprises BAG6, TRC35 and UBL4A, SGTA determines the fate of these proteins. Hydrophobic substrates bound to the BAG6 complex can be ubiquitinated by the actions of the E3 ligase RNF126 and thus targeted for proteasomal degradation. SGTA can interact with the RPN13 subunit of the 19S regulatory particle of the proteasome through its TPR domain, which has led to the proposal of an SGTA/BAG6 cycle operating at the proteasome. SGTA hands tail-anchored (TA) proteins over to TRC40 facilitating their post-translational integration into the ER by means of the transmembrane proteins WRB and CAML. Furthermore, SGTA has been implicated in hormone receptor signaling and has been associated with viral lifecycles. SGTA's interactions with Hsp70/Hsp90 chaperones via its TPR domain likely provide substrate access to additional branches of the global cellular quality control network.

complex can be ubiquitinated by the actions of the E3 ligase RNF126 (Krysztofinska et al., 2016; Rodrigo-Brenni, Gutierrez, & Hegde, 2014) and thus targeted for proteasomal degradation. SGTA can interact with the RPN13 subunit of the 19S regulatory particle of the proteasome through its TPR domain (Leznicki et al., 2015; Thapaliya et al., 2016), which has led to the proposal of an SGTA/BAG6 cycle operating at the proteasome (Leznicki & High, 2012). SGTA hands tail-anchored (TA) proteins over to TRC40 facilitating their post-translational integration into the ER by means of the transmembrane proteins WRB and CAML (Vilardi, Stephan, Clancy, Janshoff, & Schwappach, 2014). Furthermore, SGTA has been implicated in hormone receptor signaling and has been associated with viral lifecycles (Philp et al., 2013). SGTA's interactions with Hsp70/Hsp90 chaperones via its TPR domain likely provide substrate access to additional branches of the global cellular quality control network (Yin et al., 2006).

5. SGTA

The focus of the first section of this chapter will be the molecular co-chaperone known as the small glutamine-rich tetratricopeptide repeat-containing protein alpha (SGTA) (Wunderley, Leznicki, Payapilly, & High, 2014). A major known function of SGTA is to aid in determining the fate of secretory and membrane proteins that have mislocalized to the cytosol (Leznicki et al., 2013). SGTA is also thought to be the first port of call for tail-anchored proteins upon termination of translation when the two ribosomal subunits separate (Philp et al., 2013). On many occasions SGTA works in concert with the BAG6 complex, a heterotrimeric complex comprising stoichiometric volumes of BAG6 (BCL2-associated athanogene 6, formerly known as BAT3 or Scythe), UBL4A (ubiquitin-like protein 4A, sometimes known as GDX) and TRC35 (transmembrane domain recognition complex 35), that is also involved in the protein quality control network (Xu, Cai, Yang, Huang, & Ye, 2012) and these three proteins will be addressed in detail later on in this chapter. It's been suggested that one of the roles of the SGTA/BAG6 combination is in facilitation of the ER-associated degradation pathway (ERAD) in which poorly folded proteins are retro-translocated through the ER membrane (Hampton & Sommer, 2012). SGTA has also been shown to rescue proteins from degradation by reversing the ubiquitination caused by BAG6 probably via recruitment of an, as yet unidentified, DUB or deubiquitinating enzyme (Leznicki & High, 2012). Disruption or incorrect use of these functions can lead to diseases such as cancer of the prostate, ovary, liver, polycystic ovary syndrome, Alzheimer's and many more (Philp et al., 2016). Here we will examine some of the current disease implications of SGTA.

6. SGTA structure

SGTA (shown in Fig. 2) is a homodimeric protein with each monomer composed of 313 amino acids encompassing 3 domains connected by

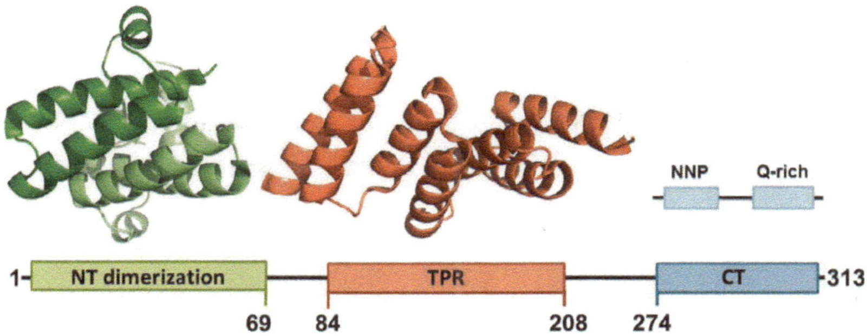

Fig. 2 Schematic representation of SGTA showing domain boundaries and solved structures. N-Terminal dimerization domain (monomer chains in two shades of green) residues 1–69 from PDB: 4CPG (Darby et al., 2014); TPR domain (red) residues 84–210 from PDB: 2VYI (Dutta & Tan, 2008). There is currently no solved structure for the C-terminal domain (blue) which encompasses NNP and Q-rich regions (Martinez-Lumbreras et al., 2018).

flexible linkers (Roberts, Thapaliya, Martinez-Lumbreras, Krysztofinska, & Isaacson, 2015). First is the N-terminal domain spanning the first 69 residues (Chartron, VanderVelde, & Clemons, 2012; Simon, Simpson, Goldstone, et al., 2013), and then following a 14 residue linker is the central tetratricopeptide repeat domain (TPR) (from which part of the name originates) which spans residues 86–208 (Dutta & Tan, 2008). Finally, from residues 211 to 313 is the C-terminal domain that contains the glutamine-rich region from residues 274 to 313 (Liou & Wang, 2005).

SGTA belongs to a family of co-chaperones that are categorized with a "*n*PR" nomenclature system. The "*n*" represents the varying number of residues in a single repeat, and thus SGTA belongs to the canonical 34PR motif (Thapaliya et al., 2016) as each repeat, made of 2 helices, consists of 34 amino acids; tetratrico means 34 in Greek.

SGTA is highly conserved throughout mammals and higher eukaryotes with the TPR domain sharing the greatest similarity across the different species (Martinez-Lumbreras et al., 2018). The N-terminal domain of SGTA forms a tight dimer with a negatively charged patch on its surface which has been shown to bind with positively charged UBL domains from both BAG6 and UBL4A domain through electrostatic interactions (Darby et al., 2014). The trade-off between binding ability of SGTA and these two UBLs is thought to be behind the sorting mechanism of the protein quality control network (Darby et al., 2014; Shao et al., 2017).

As previously discussed, it's the existence of the TPR domain that places SGTA in the 34PR family (Nguyen et al., 2017). Typically TPR domains are made up of between 3 and 16 tandem repeats—SGTA contains 3 repeats with each repeat containing 2 almost identically structured antiparallel folded alpha helices (Dutta & Tan, 2008; Krysztofinska et al., 2017). There is additionally a seventh alpha helix known as the C-terminal capping helix that aligns against the second helix of the third repeat, helix 6. The three repeats are structured in a parallel arrangement which creates a super helical structure containing a surface groove. This groove has been shown to play a role in intra-molecular interactions (Liou & Wang, 2005). The TPR domain is responsible for SGTA binding to Hsp70 and Hsp90, along with other varying receptors and proteins (Philp et al., 2013; Roberts et al., 2015). Due to this, the TPR domain has been linked with SGTA's role in diseases such as HIV (Dutta & Tan, 2008) and the parvovirus (Cziepluch et al., 1998).

Out of the three domains, the least is known about the C-terminal domain with no high resolution structure currently available. It has been shown that it contains an important glutamine-rich region between residues 274 and 313 where 13 glutamine residues are located. This region has been implicated as responsible for the C-terminal's ability to interact with hydrophobic amino acids within polypeptides (Liou & Wang, 2005). Removal of this region resulted in no interactions with hydrophobic substrates (Liou & Wang, 2005). This allows for SGTA to bind to and then shield improperly or incompletely folded proteins with exposed hydrophobic regions, and thus provide a vital role within the protein quality control network. Very recently another important motif has been identified in the C-terminus called the NNP motif consisting of three copies of asparagine-asparagine-proline (Martinez-Lumbreras et al., 2018). The C-terminal domain has also been found to transiently dimerize, probably around hydrophobic substrates in the manner of a pair of tweezers (Martinez-Lumbreras et al., 2018).

7. Regulation of cytosolic quality control by SGTA

As previously stated, exposed hydrophobic residues on proteins can form inappropriate interactions within the cytosol (Buchberger et al., 2010). Typically, when membrane proteins with exposed hydrophobic residues enter the secretory pathway, they are met and transported by signal recognition particle dependent delivery to their desired location. This process prevents interactions between the cytosol and the exposed hydrophobic residues (Cross, Sinning, Luirink, & High, 2009). However sometimes membrane and secretory proteins end up mislocalized which can lead to undesired interactions with the cytosol; it's the responsibility of the protein quality control network to deal with this (Ast, Cohen, & Schuldiner, 2013). SGTA, alongside the BAG6 complex, plays a role in responding to MLPs while encouraging the transportation of tail-anchored proteins into the ER membrane (Hessa et al., 2011). Studies have suggested that while BAG6 promotes the polyubiquitylation of mislocalized hydrophobic substrates, SGTA can antagonize this process by activating substrate deubiquitylation (Leznicki & High, 2012). Both complexes also play a role in the ERAD pathway for irregular proteins (Xu et al., 2012).

Studies have shown that SGTA binds directly to hydrophobic regions of MLPs both in vitro and in vivo, which highlights their importance in signaling for degradation. Results also suggest that SGTA promotes substrate deubiquitylation. Overexpression of SGTA is shown to inhibit proteasomal degradation of irregular proteins while also stabilizing MLPs; this links to undesirable results such as prolonging the lifespan of aberrant proteins, and several forms of cancer. Overall SGTA plays an important role in cytosolic quality control; at the ER it encourages the biogenesis of certain hydrophobic precursors while additionally promoting the removal of irregular and misfolded proteins (Shao et al., 2017). Without the presence of SGTA, not only would the lifespan of potentially dangerous proteins be extended, but undesirable interactions between proteins and the cytosol could lead to aggregation, a suggested cause of some neurodegenerative diseases (Chakrabarti & Hegde, 2009; Wunderley et al., 2014).

8. Role of SGTA in neurodegenerative diseases

Protein aggregation plays an important role in neurodegenerative diseases, such as Alzheimer's disease, Parkinson's disease, Huntington's disease, amyotrophic lateral sclerosis and prion diseases (Cox, Raeburn, Sui, & Hatters, 2018; Ross & Poirier, 2004). If proteins with exposed hydrophobic regions are allowed to interact with the cytosol they can form aggregates and inclusion bodies (Ross & Poirier, 2004). In Huntington's disorder, for example, inclusions have been found in regions of the brain that degenerate (Vonsattel et al., 1985). Another example is Alzheimer's disease, which is shown to be caused by two types of protein aggregates: extracellular and intracellular (Cox et al., 2018). Interestingly, both increased and decreased expression of SGTA can lead to the formation of aggregates which has been correlated with potentially fatal neurodegenerative diseases (Wunderley et al., 2014).

9. The androgen signaling pathway and SGTA

Steroid hormone receptors (SHRs) follow an incredibly complex and labile biogenesis which is aided by a number of different chaperones along the way (Echeverria & Picard, 2010). One such SHR, the androgen receptor (AR), is a testosterone-dependent, specific transcription factor (Cutress, Whitaker, Mills, Stewart, & Neal, 2008). Located in the cytoplasm, upon binding of the ligand to the AR, interactions take place within its N- and C-terminal domains that result in activation and transfer to the nucleus to activate transcription of specific genes (Cutress et al., 2008). ARs play an important role in locations such as the prostate gland; although their primary role in this organ is found during the growth of the gland, they still function after maturity (Brooke & Bevan, 2009).

SGTA comes into play as a chaperone that stabilizes the unliganded form of the AR within the cytoplasm by binding to the AR's hinge region with its TPR domain; after the ligand binds to AR, the receptor dissociates from the chaperone (Trotta et al., 2012). SGTA activity has been implicated in androgen regulation as a downregulator of the receptor by promoting retention within the cytoplasm (Buchanan et al., 2007). Studies suggest that this downregulation is caused by controlling the sensitivity to the AR ligand, which leads to retention within the cytoplasm (Trotta et al., 2013). However, one study showed that SGTA's role in this downregulation could be during the folding of the receptor, with SGTA affecting the ubiquitination and recycling of the receptor during the early folding stages (Paul et al., 2014). This links into its role in cytosolic quality control which supports this theory (Wunderley et al., 2014). The same study also suggests a link between SGTA and other SHRs, the glucocorticoid and progesterone receptor activity suggesting that SGTA's role in hormone signaling extends beyond the androgen receptors (Paul et al., 2014).

Improper regulation of AR signaling has been related to prostate cancer, polycystic ovary syndrome and breast cancer thus it's suggested that irregular activity of SGTA could be a factor in these diseases (Goodarzi et al., 2008; Trotta et al., 2013).

10. The role of SGTA in prostate cancer

Prostate cancer is one of the most common forms of cancer in males, and is considered one of the deadliest cancers. Usually, initial tumor growth is androgen dependent and driven by the AR activity within the cells (Brooke & Bevan, 2009). Androgen dependent prostate cancer involves the AR stimulating proliferation while inhibiting apoptosis of cells, thus ensuring the survival of cancerous cells (Brooke & Bevan, 2009). This form of the cancer is normally treated with androgen ablation and while most tumors respond to this treatment, some evolve into the androgen independent form which has been largely considered incurable (Cattrini et al., 2017). Androgen independent cancer evolves from genetic mutations within the cell that allow it to grow independent of AR activity; instead it follows pathways involving overexpression of BCL2 or inactivation of tumor suppressor genes among many others (Brooke & Bevan, 2009; Cattrini et al., 2017). These pathways have also been linked to other hormone independent cancers such as breast cancer (Bullock, 2016).

One such cause of androgen independent prostate cancer has been linked to SGTA; it has been suggested that SGTA is a negative regulator of AR transport into the nucleus by promoting AR retention within the cytoplasm, as previously stated (Kato et al., 2017). One theory as to the mechanism behind this involves Hsp70, which is suggested to provide the energy needed by chaperones that are responsible for association and dissociation of proteins. SGTA interactions with Hsp70 increase protein affinity and result in more efficient protein folding; in yeast cells that lack SGTA there is a decrease in Hsp70 activity, and thus it's suggested that SGTA can indirectly affect the ligand binding capacity of the AR via this mechanism (Buchanan et al., 2007). Since it has separately been suggested that SGTA directly binds to the AR's hinge region, it correlates that SGTA overexpression directly inhibits the AR by reducing its ability to bind with ligands (Trotta et al., 2012). Despite conflicting ideas behind the mechanism of which SGTA affects AR activity, there has been evidence to suggest that overexpression leads to decreased activity, while underexpression leads to increased activity (Buchanan et al., 2007). Thus its widely accepted SGTA plays a role in prostate cancer development; however the mechanism by which this occurs is still not fully understood.

11. The role of SGTA in polycystic ovary syndrome

Polycystic ovary syndrome (PCOS) is a common genetic disorder that affects between 6% and 8% of women of reproductive age. Common symptoms include infertility, insulin resistance, obesity and high cardiovascular risk factors (Azziz et al., 2004). While there is still much debate as to the cause of PCOS, with suggestions such as improper insulin signaling, genetic mutations and irregular androgen signaling; most patients present with hyperandrogenism (Goodarzi et al., 2008).

It has been suggested that SGTA's role in this disease is much the same as its role in prostate cancer, promoting retention of the AR within the cytoplasm and thus preventing further function (Buchanan et al., 2007). Overexpression of SGTA is implied to inhibit AR function by causing said retention in the cytoplasm, whereas knockdown of SGTA is said to result in increased AR activity and promiscuous activation of the receptor by ligands other than testosterone, such as progesterone (Buchanan et al., 2007; Goodarzi et al., 2008). Asides from affecting androgen signaling, SGTA has been linked to PCOS via its role in apoptosis, promoting cell death when both over- and under-expressed (Yin et al., 2006). A further role of SGTA within PCOS is insulin signaling. A study has shown that the SGTA gene is related to insulin resistance; however more research is needed to determine how and to what extent (Goodarzi et al., 2008). Overall it's shown that there is a probable role of SGTA in the cause of PCOS through multiple pathways, making it a possible target for therapy of this disease.

12. The role of SGTA in breast cancer

Another disease linked to androgen signaling is breast cancer, one of the most common cancers among women. A 2014 study (Zhu et al., 2014) first examined the relationship between SGTA and breast cancer cells and found that SGTA was overexpressed in breast cancer cells compared to normal tissue. They also found a positive correlation between SGTA concentration and breast cancer cell proliferation following SGTA knockout experiments. These results suggest that SGTA levels could be a potential indicator of breast cancer progression as concentration was found to increase at a higher histological grade. Additionally, SGTA was recognized as a potential prognosis indicator (Zhu et al., 2014). This leaves SGTA as an interesting potential treatment target for breast cancer and thus more research should be done into this relationship (Zhu et al., 2014).

13. SGTA in the cell cycle

SGTA is required for the cell to successfully complete the cell division pathway (Winnefeld, Rommelaere, & Cziepluch, 2004). Reduced levels of SGTA results in reduced proliferation through an arrest in mitosis, which is subsequently followed by cell death. Studies have shown that SGTA depleted cells mostly arrested in their metaphase stage, suggesting that SGTA was limiting the cells' exit from the metaphase; however some cells were able to progress through to the G1 stage before cell death took place (Winnefeld et al., 2004).

On the other hand, overexpression of SGTA seems to be just as disastrous for the cell lifespan. There is a correlation between SGTA and apoptosis with a proposal that the TPR domain was responsible for this function (Wang et al., 2005). The role of SGTA is also linked to that of Hsp90s (highly expressed protein chaperones that assist in protein degradation), suggesting that while they usually bind in the cytoplasm, during cell apoptosis SGTA dissociates and then accumulates in the nucleus, although much more research is needed to suggest how this further links to apoptosis (Yin et al., 2006). The importance of SGTA within the cell cycle is another likely contributor to its role in tumorigenesis.

14. The role of SGTA in lung cancer

Non-small-cell lung cancer (NSCLC) is the most common form of lung cancer, and a leading cause of cancer deaths worldwide. The prognosis is generally poor with 5 year survival rates less than 15%; this is mostly due to poor diagnostic rates and difficulty in treatment (Ji et al., 2011). There is a known relationship between NSCLC and SGTA expression; SGTA is positively correlated with cellular proliferation and was frequently highly expressed in NSCLC cells. A link is suggested between SGTA and the chemoresistance of NSCLC cells by SGTA suppressing cisplatin-induced apoptosis. This suggests SGTA as a potential target for NSCLC treatments especially due to its role in chemoresistance (Xue et al., 2013).

15. The role of SGTA in esophageal cancer

Like NSCLC, esophageal squamous cell carcinoma (ESCC) reports a poor prognosis. The etiology is complex and involves multiple genes; therefore important diagnostic indicators are needed to help with early stage diagnosis, which, like in NSCLC, is a reason for the poor prognosis (Talukdar et al., 2018). One study suggested SGTA contributed to the malignant progression of ESCC and thus could be used a prognostic biomarker (Yang et al., 2014). Much like the NSCLC study, overexpression of SGTA was shown to play a role in the development and progression of tumors, although more research is still needed to decipher the mechanisms behind these observations to further understand SGTA's role (Yang et al., 2014).

16. Early identifications of SGTA in viral infections

Early in the literature, SGTA was discovered in complex with the parvovirus H-1, specifically NS1, a non-structural protein of H-1 that is essential for DNA replication (Cziepluch et al., 1998). Another example of SGTA in complex with a virus involves Vpu, a viral core protein precursor from HIV-1 (Dutta & Tan, 2008). It was suggested that in the absence of Vpu, overexpression of SGTA leads to interactions with HIV-1 Gag, another HIV-1 protein, and the resulting complex inhibits HIV-1 release. However, in the presence of Vpu stable interactions form between SGTA and Vpu that prevent SGTA-Gag interactions (Waheed et al., 2016).

Another viral protein interacting with SGTA is protein 7a from the severe acute respiratory syndrome coronavirus (SARS-CoV), a member of the Coronaviridae family, the largest known RNA viruses (Fielding et al., 2006; Philp et al., 2013). SGTA is thought to play a role in virus assembly or release from the cell but more research is needed to fully understand the implications of SGTA's involvement with SARS-CoV and the mechanisms behind this (Fielding et al., 2006).

17. SGTA and the simian virus 40

A study into SGTA's relationship with the simian virus 40 (SV40) suggested that SGTA facilitates transport of the virus between the endoplasmic reticulum (ER) and the cytosol (Walczak, Ravindran, Inoue, & Tsai, 2014).

SV40 is a non-enveloped polyomavirus that has to travel through biological membranes to infect cells. As it lacks a surrounding lipid bilayer it cannot infect cells in the same way as enveloped viruses, which usually proceed via membrane fusion. Thus it needs to find a different path for entry to the cell; this is usually taken through virus-induced pore formation or disruption of the cell membrane's integrity (Wiethoff, Wodrich, Gerace, & Nemerow, 2005). Polyomaviruses are unique in that they travel past the endosomal system and reach the ER for membrane penetration; SV40 is known to infect cells via engaging the ganglioside receptor GM1 at the cell surface. Endocytosis then brings SV40 molecules complexed with lipid rafts into the cell where they proceed through endosomes before reaching the ER. After this SV40 has to exit through the ER membrane and cross the cytosol to reach the nucleus where it subsequently activates transcription and replication of the viral genome; thus cellular transformation is caused (Wiethoff et al., 2005). It is thought that SGTA associates with a B14 and B12 complex at the ER membrane and, during infection, SGTA binds directly to SV40 and thus promotes its ER membrane penetration. During this penetration, SGTA dissociates from the B14 and B12 complex and hence it is implied that SGTA is directly responsible for SV40's transport between the ER and the nucleus (Walczak et al., 2014).

A further study in 2017 showed that two of SGTA's domains are responsible for this function (Dupzyk, Williams, Bagchi, Inoue, & Tsai, 2017). First it linked the TPR domain's ability to interact with Hsc70; this aids penetration of the ER membrane and subsequent transport to the cytosol. Then it linked the N-terminal domain with the infection of SV40 into the nucleus; although the exact role the N-terminal plays is uncertain, results showed that the N-terminal was crucial for further infection after SV40 has entered the cytosol. The study suggested that this could be via the N-terminal being directly involved in transport to the nucleus, or the N-terminal could recruit cellular components that subsequently transport the virus to the nucleus. Although it has been seen in other studies that SGTA can localize to the nucleus, and this suggests the direct involvement of the N-terminal in SV40 infection, more research is needed to further understand the role of SGTA in virus infection, and how its domains affect each stage (Dupzyk et al., 2017).

18. BAG6

The human BAG family consists of six members, the largest being the BAG6 protein constituting of 1132 amino acids in the major isoform. It is situated in the gene cluster of the human major histocompatibility (MHC) class lll on chromosome 6 which contains many genes responsible for immune function (Banerji, Sands, Strominger, & Spies, 1990). The highly conserved BAG6 protein is flanked by a ubiquitin-like (UBL) domain on the N-terminus and a BAG domain on the C-terminus. The central region consists of a disordered proline-rich region, zinc-finger like and Nuclear

Localization Signal (NLS) domain. BAG6 was designated a member of the BAG family due to its observed sequence homology to the BAG domain and its collaboration with Heat Shock Protein 70 (Hsp70) family possessing chaperoning activity (Kuwabara et al., 2015).

BAG6 is an abundant cellular protein expressed ubiquitously in higher eukaryotes. Some tissues in multi-cellular organisms, such as testis and brain, express relatively higher levels of BAG6. This suggests a potential function in spermatogenesis and brain development respectively. Numerous other functions associated with BAG6 have been identified, including the regulation of apoptosis, T-cell response and protein homeostasis (Kuwabara et al., 2015; Lee & Ye, 2013). Due to the diversity of its functions, the significance of BAG6 in disease and therapy is of particular interest. As previously mentioned, a heterotrimeric complex (known as the BAG6 complex) is formed between BAG6, UBL4A and TRC35, which together determine the fate of hydrophobic nascent polypeptides. Hence, BAG6 has been identified as a multi-functional protein that plays a critical role in the quality control of hydrophobic polypeptides whereby it is at the interface of protein synthesis and degradation.

Fundamentally, the BAG6 complex can function in several cell conditions and carry out essential processes depending on individual cell requirements. Therefore, dysfunctions in the BAG6 complex cause various cellular abnormalities which lead to both physiological and pathological diseases (Lee & Ye, 2013). An overview of BAG6 functions and its association with disease are given in the following sections.

19. BAG6 structure

The primary sequence of BAG6 can be subdivided into three distinct domains: the N-terminal UBL-containing region spanning amino acids 1–474; a central proline-rich domain constituting amino acids 475–1029, a possible site for binding of hydrophobic substrates according to in vitro and in vivo studies (Leznicki et al., 2013); and a C-terminal NLS and designated unconventional BAG domain constituting amino acids 1030–1132. The small sections of BAG6 whose structures have been solved are shown in Fig. 3 but the vast majority of the protein still remains structurally undetermined. How BAG6 contributes to causing disease may be somewhat understood from understanding its primary sequence and the functions of the key domains illustrated in Fig. 3.

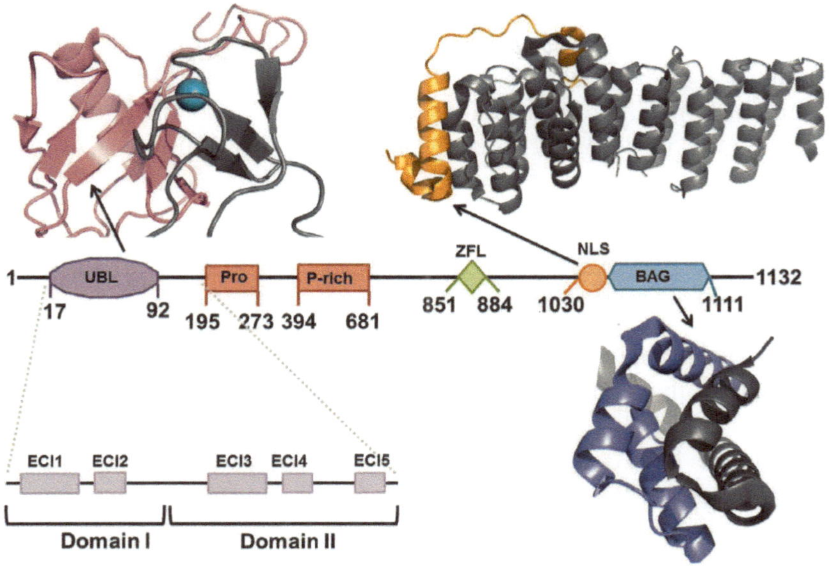

Fig. 3 Schematic representation of BAG6 showing domain boundaries and solved structures. UBL domain (residues 17–92, purple) in complex with E3 ligase RNF126 (residues 1–40, gray) from PDB: 2N9P (Krysztofinska et al., 2016); NLS (residues 1008–1050, orange) and TRC35 (residues 23–305, gray) from PDB: 6AU8 (Mock, Xu, Ye, & Clemons, 2017); BAG domain (blue) in complex with UBL4A (residues 95–147, gray), from PDB: 4X86 (Kuwabara et al., 2015).

20. BAG6 structure: N-Terminal UBL domain

Multiple alignments of the UBL-containing domain, responsible for recruiting ubiquitination machinery, revealed further subdivision of this region into two smaller domains: Domain I and Domain II, both containing designated Evolutionarily Conserved Island (ECI) regions (shown in Fig. 3). Hessa et al. (2011) discovered ECI1 to be the UBL domain which is directly linked to protein quality control, specifically ER substrates. This was experimentally found in in vitro experiments whereby absence of ECI1 failed to successfully ubiquitinate mislocalized proteins. Further experimental evidence showed an E3 ubiquitin ligase needed to interact with ECI1 to permit the degradation of ER substrates (Payapilly & High, 2014).

The binding of hydrophobic stretches of protein onto BAG6 is imperative for substrate discrimination. Domain I contains ECI1 and ECI2, both of which are capable of binding hydrophobic substrates, but cannot necessarily maintain adequate hydrophobic stabilizing interactions. Domain II contains further ECI with UBL sequence similarity, and therefore can potentially bind hydrophobic substrates. Therefore, collectively, the UBL-containing domain is capable of directly recognizing and binding hydrophobic substrates (Kawahara, Minami, & Yokota, 2013). Further structural understanding is required to gain a mechanistic understanding of substrate binding to the UBL domain.

21. BAG6 structure: C-Terminal BAG domain

A highly conserved BAG domain exists at the C-terminal region of BAG6. It is involved in direct interaction with the molecular chaperone family Hsp70/Hsc70 via the ATPase domain. Mechanistically, the interaction between Hsp70 and C-terminal BAG domain commences with Hsp70 in an ADP-bound state possessing a vacant nucleotide binding domain. This allows the binding of the BAG6 domain and Hsp70 via electrostatic and hydrophobic interactions. Thereafter, phosphorylation of ADP to ATP results in the simultaneous release of the BAG domain due to conformational changes to the nucleotide binding region. Regeneration of ADP-bound Hsp70 allows this cycle to continue and supports the recruitment of hydrophobic proteins by acting as a critical checkpoint in determining the fate of polypeptides.

Interestingly, multiple alignment of BAG domains from several species of the BAG family suggested that the BAG domain of BAG6 is atypical. In vivo binding assays have revealed the co-precipitation of the BAG6 and Hsp70, reinforcing that BAG6 is a Hsp70-binding motif. However, differences in the residual interaction between the BAG domain and Hsp70 across the six members of BAG proteins rationalizes *why* in vivo studies showed reduced affinity of BAG6 to Hsp70 (Thress, Henzel, Shillinglaw, & Kornbluth, 1998). Moreover, due to evolutionary pressure on the BAG domain, it has undergone phylogenetic divergence to accommodate for more complex demands in mammals. This indicates that on a functional and sequence level, the BAG6 C-terminal BAG domain is unique (Kuwabara et al., 2015; Mock et al., 2017).

Although Hsp70 participates in recognizing and binding hydrophobic polypeptides, BAG6 itself can also bind these nascent polypeptides independently. In cytosolic conditions, several chaperones are required to hinder undesired interactions leading to aggregate formation. This suggests that BAG6 exhibits critical chaperoning characteristics, supported by studies which shows preferential binding of hydrophobic Tail-Anchored proteins via the TMD which allows sufficient time frame for the fate of polypeptides to be decided (Slavotinek & Biesecker, 2001). BAG6 behaves as a special transporter which exploits its "holdase" activity to sort aggregation prone substrates depositing them into their desired functional system. Essentially, cofactors of BAG6 complex as well as the interplays between the BAG6 complex and nascent polypeptides determine its destination.

22. BAG6 structure: NLS domain

The significance of BAG6 protein lies in its position as a central hub mediating various cellular processes, involving but not limited to apoptosis, proteostasis and immunoregulation. A bipartite NLS exists in BAG6 adjacent to the N-terminal side of BAG domain in the primary sequence. The NLS signal in BAG6 is masked by TRC35-bound BAG6 promoting BAG6 nuclear translocation to the cytosol, indicative of a functional link between transmembrane protein biosynthesis and the NLS (Kawahara et al., 2013). BAG6 variants lacking the NLS due to alternative splicing have shown preferential localization at the cytosol. Although BAG6 is primarily localized in the cytosol, the NLS domain indicates that its presence in the nucleus is important too, and functions of BAG6 as a quality control site have been displayed in both localities (Mock et al., 2017). However, some cases show that increased nuclear levels of BAG6 can be associated with tumorigenesis, for example, in osteosarcoma tissue (Tsukahara et al., 2009).

Some functions of nuclear BAG6 have been proposed including DNA damage regulation via tempering of post-translational modification of nuclear factors such as histones. BAG6 regulates chromatin structure in testes as well as gene expression via employment of histone modifiers. During genotoxic stress in the testes, BAG6 transactivates tumor suppressor p53 target genes by enhancing p53 acetylation. This is aided by a direct stabilizing interaction between p53 gene and p300 acetyltransferase complex (Kawahara et al., 2013; Lee & Ye, 2013; Luce, Akpawu, Tucunduva, Mason, & Scott, 2016). Immunocytochemical analysis demonstrated that point mutations of the NLS domain led to the absence of BAG6 in the nucleus (Kawahara et al., 2013). This signifies the importance of the conserved NLS domain for BAG6 localization in the nucleus. Moreover, the studies emphasized the importance of regulated levels of nuclear localization of BAG6, which if overexpressed or underexpressed leads to disease.

23. BAG6 is an essential apoptotic regulator of nascent polypeptides

The earliest cellular function associated with BAG6 was its role as an apoptosis regulator. Reaper, an apoptosis-inducer protein in *Drosophila*, binds its C-terminal 50 amino acids to BAG6 (previously called Scythe) which inhibits apoptosis. This is possible due to Reaper having the ability to bind directly to ribosomes, thus inhibiting protein synthesis. This is the case for several cell systems including HeLa human cancerous cells, indicating possible health conditions that may be caused in humans due to unregulated apoptosis (Mock et al., 2015). In vitro assays revealed that Reaper induces rapid cytochrome *c* release during mitochondria-mediated apoptosis and therefore is BAG6 dependent (Thress et al., 1998). Mechanistically, Reaper works by disrupting the interaction between BAG6 and UBL4A which leads to BAG6 sequestering pro-apoptotic Reaper. Therefore, BAG6 serves as an anti-apoptotic regulator. An example of unregulated apoptosis causing health defects is in mice, whereby because of in vivo genetic ablation of BAG6, embryonic lethality is caused due to amplified levels of apoptosis and cell proliferation (Desmots, Russell, Lee, Boyd, & McKinnon, 2005). This concludes that functional BAG6 has a direct effect on regulating developmental defects. Although deficiency of BAG6 increases apoptosis which negatively impacts tissues, chemicals which affect calcium flow via ER stress are more resistant against cortical neurons deficient of BAG6 (Desmots et al., 2005; Kawahara et al., 2013; Tsukahara et al., 2009). This deduces the complex roles of BAG6 with the ability to adapt its consequences on its clients requirements.

Moreover, BAG6 is deemed compulsory for the operation of Reaper-induced apoptosis. This is deduced from cases of depleted endogenous BAG6 whereby the caspase activation of Reaper is hindered suggesting the need of functional BAG6 to act as an anti-apoptotic regulator (Thress et al., 1998). Surprisingly, BAG6 suppressed Reaper-induced apoptosis upon overexpression of wild-type BAG6 which is functionally beneficial for cells (Thress et al., 1998).

Despite reports of BAG6 being an apoptotic regulator, conflicting results show otherwise regarding BAG6 proteins role in apoptosis and viability. For that reason, during development, BAG6 is proposed to play ambiguous yet case-dependent roles. Several studies support this hypothesis whereby the absence of the gene encoding BAG6 in mice resulted in neonatal lethality due to extensive abnormal apoptosis.

Conversely, underexpression of BAG6 in mice with ICR background was viable. Strikingly, apoptosis was impaired in ICR mice upon deletion of BAG6 as a result of genotoxic stress. Other independent studies demonstrate that slight underexpression of BAG6 causes mitotic and G1/S phase arrest in mislocalized chromosomes and in mammalian in vitro cells respectively, as well as enhanced apoptotic activity in HCT116 cells, thus contributing negatively to health conditions (Kawahara et al., 2013). This portrays the differential roles of BAG6 in precise cellular contexts as well as its contribution to cells with varying cellular levels.

24. BAG6 involved in the triage system during protein quality control

As has already been discussed, BAG6 serves as a dynamic platform in determining the fate of post-translational polypeptides with exposed hydrophobicity. Importantly, BAG6 associates itself with nascent polypeptides and evidence supports this experimentally as, on co-treatment of eukaryotic cells with protein synthesis inhibitor, polyubiquitinated associated BAG6 proteins disappeared (Minami et al., 2010). There are three common pathways coupled to BAG6 to maintain quality control within the cytoplasm upon recognizing exposed hydrophobic regions of polypeptides.

25. BAG6 coupled degradation of defective nascent polypeptides

When membrane protein insertion fails, BAG6 triages uninserted substrates to the degradation pathway via the proteasome upon being ubiquitinated or attempts are made to correctly refold the protein and for reinsertion in the ER. Furthermore, during degradation deficiency, a ubiquitin-positive aggresome forms and BAG6 along with the hydrophobic substrates deposits into an aggregate (Kawahara et al., 2013).

One of the BAG6 quality control processes is the degradation of defective nascent polypeptides in the cytosol. The breakdown of BAG6 anatomy found the N-terminal UBL domain to be associated with the ubiquitin proteasome system (UPS). BAG6 associates with several protein binding partners when carrying out quality control functions. BAG6 also binds to polyubiquitinated substrates showing that it is interacting multivalently with the proteasome for degradation (Lee & Ye, 2013; Payapilly & High, 2014). This is also the case for Defective Ribosome Products (DRiP) which behave using a similar mechanism.

BAG6 can detect nascent defective polypeptides and subsequently target it to polyubiquitination by E3 ubiquitin ligase. This targets the substrates to the proteasome where it undergoes degradation (Lee & Ye, 2013). It was experimentally deduced that BAG6 heterotrimeric complex is involved in the direct interaction of the proteasome-bound substrate whereby BAG6 plays a key role in the interaction. Therefore, depletion of BAG6 prevented turnover as degradation at the proteasome was inhibited by the absence of BAG6. The turnover of DRiPs is proposed to create MHC class I antigenic peptides, which, in the case of BAG6 depletion, have limited cell surface presentation (Minami et al., 2010). Antigen production results due to degradation of misfolded and mislocalized proteins. The importance of BAG6 here lies in its underexpression causing limited suppression of cell surface presentation in MHC class I which may be a basis of peptide ligands. Alongside this, BAG6 is a possible participant in transmembrane protein assembly, vital for antigen presentation (Kawahara et al., 2013).

Initial studies proposed BAG6 to be a protein associated with the proteasome. However, newer research via MS analysis elucidated extremely high sequence similarity between the 26S proteasome subunits and BAG6 immunoprecipitates as well as BAG6's collaboration with interferon-γ (IFN-γ)-induced immunoproteasomes, which are typically favored due to their ability to produce peptides satisfactory to MHC class I molecules (Minami et al., 2010). Therefore, in controlling antigen presentation, IFN-γ exposure increased BAG6 expression which can be regulated in health conditions. These results conclude that BAG6 is a functional collaborator of the proteasome.

BAG6 also assists mislocalized protein degradation by BAG6 heterotrimeric complex. Nascent polypeptides existing the ribosomal tunnel are targeted by BAG6 and "sorted" to its appropriate destination. Issues associated with MLPs are that their hydrophobic region is exposed to the aqueous cytosolic environment; thus they are inclined to form aggregates. This poses challenges for cells experiencing ER-stress as membrane integration efficiency is further reduced. Ideally, TA proteins captured by BAG6 are transferred to TRC40 for ER membrane targeting. However, this pathway is not always efficient which can contribute to the build-up of mislocalized proteins in the cytosol. For some mislocalized transmembrane proteins, the TMD is targeted to polyubiquitination by E3 ligase and thus undergoes degradation at the proteasome (Amm, Sommer, & Wolf, 2014). Interestingly, BAG6 may behave in a similar manner to the ER lumen timer which controls the degradation of misfolded protein whereby only in the case of mislocalized proteins will BAG6 triage them to degradation at the proteasome (Lee & Ye, 2013).

A third major evolutionary conserved quality control process is ERAD of hydrophobic proteins which fail to be successfully inserted into the ER lumen or fail to obtain its native conformation. Typically, the efficiency of ER-targeting is relatively low depending on the strength of the ER signal. However, to enhance ERAD efficiency and, in turn, maintain homeostasis of both transmembrane and secretory proteins which are disassembled or misfolded, sophisticated chaperones are used to retro-translocate them to the proteasome for degradation (Payapilly & High, 2014). Sequentially, membrane-associated E3 ubiquitin ligase including gp78 targets retro-translocated proteins for polyubiquitination. An ATPase associated multiprotein complex then extracts the ubiquitinated protein from the membrane and BAG6 is recruited. This suggests preferential chaperoning activity of BAG6 as opposed to the assisting multiprotein complex which shuttles the substrates to the proteasome for degradation. However, whether BAG6 directly interacts with the proteasome or is facilitated by an adaptor remains elusive (Lee & Ye, 2013; Payapilly & High, 2014).

At a structural level, interaction between BAG6 and ERAD de-glycosylated intermediates was observed as well as the association of the BAG6 heterotrimeric complex with retro-translocated membrane targeted complex involving gp78 and p97, as revealed by affinity chromatography (Wang, Whynot, Tung, & Denic, 2011). This implies that BAG6 works as a p97-mediated downstream retro-translocation of ERAD. Additionally, soluble and membrane proteins are stabilized in the case of BAG6 depletion. Notably, some polyubiquitin conjugates remain which can potentially form aggregates, and thus become unaffected by proteasome degradation as a result of its inhibitory effect (Wang, Whynot, et al., 2011). This concludes that the absence of BAG6 prevents its detrimental role in maintaining client ERAD protein solubility and escorting of substrates to the proteasome for degradation.

The heterotrimeric BAG6 complex elevates the quality control of nascent polypeptides, particularly ERAD substrates. This has been observed in studies where underexpression of BAG6 impaired the degradation of ERAD substrate T-cell receptor, TRCα, a single-spanning transmembrane helix. On the other hand, non-ERAD substrates such as cytosolic Ub-V-GFP protein remained unaffected in degradation (Wang, Whynot, et al., 2011). This exemplifies the strong inhibitory effect of absent BAG6 on ERAD degradation. Although BAG6 can independently maintain retro-translocated substrates in a soluble, unfolded state, underexpression of its heterotrimeric complex counterparts TRC35 and UBL4A allowed ERAD substrates to remain stabilized. This is because BAG6 is the main chaperone protein involved in ERAD but requires its counterparts for increased ERAD efficiency (Wang, Whynot, et al., 2011).

26. BAG6 complex—Structural interaction with UBL4A

To successfully achieve protein homeostasis through ER-associated degradation and mislocalized protein pathways, the formation of the BAG6 complex with TRC35 and UBL4A is vital. Notably, the BAG domain of BAG6 is necessary and can bind UBL4A self-sufficiently which plays a pivotal role in the correct assembly of TA proteins (Kuwabara et al., 2015). The precise role of UBL4A is obscure; however it is proposed that it recruits additional chaperone machinery and ubiquitination enzymes through the UBL domains to module protein quality control. Thus, the essential interaction between the two binding partners was explored in depth. GST pull-down analysis on *Escherichia coli* presented the co-precipitation of both GST-BAG domain and UBL4A, indicating that the domains interact directly. Interestingly, the BAG domain of BAG6 preferentially binds with UBL4A relative to other members of the BAG family, supporting that its domain is distinct from the other canonical BAG family members and direct binding with UBL4A is optimized (Kuwabara et al., 2015).

To explore the key residues involved in the sufficient binding of BAG6 and UBL4a, the interactions between the BAG domain and evolutionarily conserved C-terminal half of UBL4A were closely examined. The secondary structure reveals overlapping of three α-helices from each domain. Across the dimer, several key hydrophobic and electrostatic interactions were observed playing a vital role in its strong binding character.

Characterization of the binding interface by point mutation allowed the most critical binding residues to be identified. Constructed combination of BAG6 mutations highlighted the significance of hydrophobic interactions. On single substitution of a BAG6 residue, no apparent effect was observed on the dimer interaction. However, double and triple substitution abolished the interaction altogether, particularly simultaneous Val-1068 and Leu-1086, establishing these residues as the most critical for the BAG6-UBL4A interaction. Fascinatingly, out of the four interactions illustrated, point mutations on UBL4A His-106 and Phe-107 significantly lowered the binding affinity of the BAG6-UBL4A complex indicating the importance of both hydrophobic and electrostatic interactions in stabilizing the dimer. However, point mutations of UBL4a Val-02 and Tyr-123 had a lesser effect on the binding stabilization as the former residues are located on the linker region which may be critical for complex formation (Kuwabara et al., 2015).

Importantly, the heterotrimeric complex of BAG6 effectively shields long hydrophobic exposed regions on the client proteins, thus preventing aggregates forming. Essentially, undesired interactions with client proteins are avoided before being triaged to the correct destination. Within mislocalized proteins, BAG6 can bind to broader specific hydrophobic stretches including GPI-anchors and signal sequences. An example of the necessary requirement of BAG6 in ERAD is demonstrated by interaction with Derlin2, an ER dislocon protein. BAG6-Derlin2 interaction facilitates the recruitment of BAG6 to the mislocalized protein site as well as retro-translocation of substrates back to the cytosol. Importantly, BAG6 is also essential during the ER exit of substrates and in the prevention of hydrophobic domains forming a blockade (Payapilly & High, 2014; Wang, Whynot, et al., 2011). Therefore, the heterotrimeric complex with BAG6 plays a crucial quality control role by functioning at the interface of protein biosynthesis and degradation. This regulatory mechanism needs the use of special "holdase" enzyme activity which sorts and channels hydrophobic polypeptides to its required destination (Lee & Ye, 2013).

27. BAG6 associations with disease

It is a logical corollary that dysfunctions in the BAG6 system can cause severe cellular abnormalities that may be associated with pathological conditions. Due to the diverse functions of the BAG6 complex in mammals, it is not surprising that deficiency in BAG6 may cause several developmental consequences and human diseases. Several roles of BAG6 in disease have already been identified and will be discussed below. However, it is likely that this review will barely scratch the surface of future disease implications of BAG6 that remain to be discovered and understood.

28. The role of BAG6 in male infertility

Consistent with the view that BAG6 binds to the family of heat shock proteins, detailed studies on the interaction between BAG6 and Hsp2A, a testis-enriched member of the Hsp70 family, have revealed a crucial regulatory role of BAG6 in human spermatozoa (Bromfield, Aitken, & Nixon, 2015). Extensive studies into the primary role of BAG6 localized in human testes have been implicated to protect Hsp2A from polyubiquitination and subsequent degradation, thereby maintaining its stability in germ cells. It has been established that hydrophobic and electrostatic interactions govern the binding between BAG6 and Hsp2A. The release of BAG6 from Hsp2A is promoted by these intermolecular interactions.

Recent studies have revealed Hsp2A as a regulator of zona pellucida-receptor complex whereby in infertile male subjects, this complex formation is perturbed due to a lack of Hsp2A expression due to a knock-on effect from the lack of BAG6. Ultimately, BAG6 plays a role in male infertility by stabilizing Hsp2A by a similar manner of Hsp70-BAG6 interaction mechanism. Therefore, BAG6 may be a crucial candidate in future understanding of male idiopathic infertility due to its suggested underexpression in zona pellucida-spermatozoa binding deficiency (Bromfield et al., 2015).

29. The role of BAG6 in lung cancer

As mentioned earlier, the major known roles of BAG6 in disease are associated with underexpression or overexpression of BAG6; however single nucleotide polymorphisms (SNPs) are also a factor. The genotype and alleles of BAG6 are prognostic for non-small cell lung cancer in Norwegian-Croatian subjects, as implicated by logistic regression analyses. It was observed that the C allele of BAG6 rs3117582 SNP posed a gene-dosage increased risk of lung cancer whereby subjects carrying a single C allele had 1.7 fold increased risk and an even greater 7-fold risk for subjects carrying two C alleles (Etokebe et al., 2015; Rosenberger et al., 2017).

Typically, mutations in the promoter-region of genes have an impact on expression and when the specific rs3117582 SNP is situated in the promoter-region of BAG6, it may perturb BAG6 expression. As a result, normal functions of BAG6 in regulating cellular processes are disturbed. The significantly increased susceptibility to lung cancer with SNPs in the BAG6 gene leads to consideration of these SNPs as potential therapeutic targets (Etokebe et al., 2015; Zhao, Wang, Hu, & Jin, 2014).

30. The role of BAG6 in osteoarthritis

Additionally, the influence of BAG6 expression as a result of the BAG6 rs3117582 SNP located in the promoter region effects the risk of osteoarthritis (Mock et al., 2017). The normal major allele homozygote was associated with reduced risk to osteoarthritis while the major-minor allele heterozygote contributed to the increased susceptibility, irrespective of anatomical site and gender in the Croatian population. Results from the study indicate that variations in alleles and genotype of BAG6 affect the etiology of osteoarthritis due to dysfunctions in BAG6 protein regulatory roles (Wang et al., 2016).

31. The role of BAG6 in autoimmune disease

BAG6 polymorphism is associated with increased incidence of several autoimmune diseases such as Kawasaki syndrome and type I diabetes (Hsieh et al., 2011; Piras et al., 2014). As BAG6 influences numerous substrate proteins essential to immune regulations, it suggests a functional link between BAG6 and autoimmune disorders. This is demonstrated by examination of single-spanning transmembrane protein T-cell immunoglobulin and mucin domain-containing molecule-3 (Tim-3), an inhibitory type I membrane receptor. In cases of tumor-bearing subjects, exhausted CD4+ T-cells show increased expression of Tim-3 which results in negative-regulation of immune functions as the T-cell response is repressed. Consequently, exhausted T-cells diminish effector functions including cytokine secretion and cytotoxicity in response to antigen stimulation. However, BAG6 plays a role in binding the tail of Tim-3 and galectin-9 facilitates the disruption of the binding by mediating phosphorylation at the Tim-3 tail on residues Tyr-256 and Tyr-263. Subsequently, BAG6 is released which permits the transduction of inhibitory signals (Binici et al., 2013). Therefore, when BAG6 is overexpressed, levels of Tim-3 is reduced which results in earlier onset of autoimmune development by promoting helper T-cell response. This is supported experimentally whereby increased levels of Tim-3 and low frequency of IFN-γ-production were observed in BAG6-deficient T-cells. This is suggestive of BAG6's behavior as a repressor of Tim-3 activity which functions by protecting helper T-cells from galectin-9-mediated death, thereby promoting pro-inflammatory cytokine production and proliferation (Binici & Koch, 2014; Kawahara et al., 2013; Rangachari et al., 2012). Therefore, manipulations to the BAG6/Tim-3 interaction may be a potential molecular target in treatment of several chronic infections and autoimmune disorders.

32. BAG6: future prospects

As outlined in this section, BAG6 is a crucial multi-functional protein contributing to several diseases. It is involved in several unrelated cellular pathways and plays a pivotal role by functioning at the interface of protein synthesis and quality control of nascent polypeptides. This is governed by the unique interaction of BAG6 with the two-opposing cellular machinery, the ribosome and proteasome. As BAG6 is involved in many key cellular processes in healthy mammalian cells, it makes sense that defects in such processes will lead to unwanted cellular functions, leading to disease.

The role of BAG6 in immune response regulation allows the multi-functional protein to behave as a promising diagnostic, more so a prognostic marker protein, whereby this unravels promising new perspectives for conventional therapies, for example, in autoimmune disease. However, to fully understand the functions of BAG6 in such diseases, a range of patient cohorts need to be investigated, not limiting to certain geographical regions to understand the full potential of BAG6 as a target, thus allowing elucidation of specific diseases that may benefit from BAG6 counteractions.

Further exploring BAG6 mechanisms into how it distinguishes and targets different substrates to its correct fate remains unsolved. It has been postulated that hydrophobic interactions between BAG6 and substrates govern substrate recognition. Furthermore, the BAG6 cofactors contribute to recognizing and determining the fate of hydrophobic substrates but the mechanism it follows is yet to be fully understood. This requires full mapping and structural characterization of substrate-BAG6 interactions and upon discovery, manipulations can be made helping combat disease.

A number of answers regarding BAG6's role in health and disease are yet to be discovered. Although BAG6 binds its substrates with high affinity, it is unknown how BAG6 is able to release individual substrates upon reaching their final destination. Additionally, how BAG6 interacts with accessory factors helps determine the fate of polypeptides. Understanding these questions would allow the function of BAG6 to be exploited and manipulated for treatments in disease.

33. Other members of the BAG6 complex and machinery for TA protein insertion

Understanding the targeting mechanism of the TA proteins is expected to have far reaching conjectures on molecular biology, and much work has been carried out to identify the components of the insertion system (Simpson, Schwappach, Dohlman, & Isaacson, 2010; Yamamoto & Sakisaka, 2012). Initial biochemical analyses between 1995 and 2005 (mostly yeast based studies) suggested that TA proteins were inserted via an energy dependent process, and an ATP requirement was confirmed. The process was defined in yeast systems as the guided entry of TA proteins (GET) pathway (Denic, 2012) or transmembrane domain recognition complex (TRC) pathway in mammalian systems (Johnson, Powis, & High, 2013). The studies made evident that TA protein targeting and insertion can be regulated, particularly in mammalian systems.

The challenges involved in identifying the components in the TRC pathway were likely a result of heterogeneity of insertion assays measuring binding of TA proteins (Stefanovic & Hegde, 2007). Using a new protease protection assay a key component of the pathway essential for membrane insertion was found: the ATPase TRC40 (also known as ASNA1, yeast homolog termed Get3), trailblazing the proximate identification of the other components in further studies using genetic and biochemical approaches.

Get3 forms a complex with Get1 and Get2, and these proteins provide a route for insertion of the TA proteins into the ER membrane by Get3 (TRC40 in mammalian constructs) (Schuldiner et al., 2008). WRB (Tryptophan-rich basic protein) is now thought to be the Get1 mammalian analogue (Vilardi, Lorenz, & Dobberstein, 2011), and CAML (calcium-modulating cyclophilin ligand) the Get2 mammalian analogue (Vilardi et al., 2014; Yamamoto & Sakisaka, 2012). Using the endogenous sensor of yeast cells (unfolded protein response), Jonikas et al. (2009) identified two new components of the TA protein biogenesis machinery: Get4 and Get5, these were expected to form a complex and capture nascent proteins to be delivered to Get3 (Jonikas et al., 2009). The mammalian analogues were discovered by Mariappan et al. (2010) to be TRC35 and UBL4A, respectively, forming a complex with an additional protein BAG6 (Bat3) (Mariappan et al., 2010). In large scale yeast genetic interaction profiles Sgt2 was found as another essential factor in protein targeting by virtue of its strong functional relationship with the GET pathway (Chang et al., 2010). The human homolog is known to be SGTA and has been found to interact with BAG6 (Darby et al., 2014).

Since discovery a complex model of the TA protein membrane insertion has been formulated with the most up-to-date configuration simulated by a fully constructed mammalian system proposed by Shao et al. (2017). They have conglomerated the investigations of the well-defined GET pathway in yeast and the evolutionarily conserved mammalian homologues to suggest the chaperones involved act as a molecular triage system. BAG6, SGTA and TRC40 are the three proteins that can recognize and shield the transmembrane domain (TMD) of a TA protein (Leznicki et al., 2013). The C-terminal 110 residues of BAG6 form a complex with two tightly associated subunits UBL4A and TRC35 (collectively known as the cBag6 complex) (Mock et al., 2015), while the N-terminal part of BAG6 contains a UBL domain (nBag6) (Tanaka et al., 2016). As mentioned earlier, SGTA is the most competitive binder to exposed hydrophobic stretches of proteins, and in the first step the proteins undergo binding and shielding by SGTA, following which the proteins have one of two fates. The first involves the targeting module which consists of SGTA and TRC40 bridged by the cBag6 complex, where the TA protein binds to the TRC40 and is delivered to the ER for insertion. The second involves the quality control module exclusive to mammalian systems comprised of SGTA, nBag6 and a newly defined protein RNF126 (Krysztofinska et al., 2016) where the TA protein binds to nBag6 and is targeted for ubiquitination. These events are depicted in Fig. 1.

This quality control pathway is well-balanced allowing nascent proteins to mature but forcing degradation of poorly folded proteins over time. Accurate triage of degradation versus biosynthesis is essential for maintaining the desired quantity of TA proteins in the membrane, and ensuring the quality of proteins meets the requirements. Upregulation or inhibition of these proteins is postulated to lead to cancer or disease. Important reviews (Denic, Dotsch, & Sinning, 2013; Hegde & Keenan, 2011; Shao & Hegde, 2016; Simpson et al., 2010) have focused much on the GET system and on the known structures and interactions of the protein chaperones.

34. Formation of the Get4/5 and cBag6 complex

The year 2010 was revolutionary to the understanding of much of the GET pathway, and the structure of Get4 and Get5 was determined and their interactions with the other proteins in the pathway better understood. Chang et al. (2010) obtained the first crystal structure of Get4 in a complex with a fragment of Get5 obtained by co-expression in *E. coli*. Almost simultaneously Bozkurt et al. (2010) published the structure of Get4, and Chartron, Suloway, Zaslaver, and Clemons (2010) further defined the structural elements of the Get4/5 complex. The first structural analyses of Get5 were the crystal structure of just the Get5-N domain in complex with Get4, followed by determination of the C-terminal Get5 (Get5-C) structure by X-ray and SAXS by Chartron et al. (2012). The full Get5 protein structure was only later published in complexes with Get3 and Get4 by Gristick et al. (2014). While the interactions of the mammalian analogues of Get4 and Get5 were well studied in relation to the TA protein insertion, their molecular structures were not elucidated until more recently. The UBL4A crystal structure was elucidated by Kuwabara et al. (2015), and the TRC35 structure was recently published by Mock et al. (2017).

The standalone Get4 structure had significant distortions compared to the structure of Get4 in complex with Get5-N, suggestive of a tightly bound complex of Get4/5. Based on immunoprecipitation results where both Get4 and Get5 were completely immunoprecipitated from yeast cytosol by anti-Get5 this tight Get4/5 complex was verified. It was reported that the full length Get4/5 complex exists as a dimer, mediated by the C-terminal homodimerization domain of Get5, resulting in a heterotetramer. Prior to the structure elucidation, sequencing alignment experiments suggested that UBL4A in higher eukaryotes lacks the Get5-N domain, and TRC35 does not have a β-loop, both of which interact in yeast to form the Get4/5 complex. BAG6 was found to bridge UBL4A and TRC35 and form a similar construct (cBag6). Knockdown experiments using siRNA demonstrated that knocking down UBL4A or TRC35 alone has negligible effect on BAG6 presence; however simultaneous knockdown results in a threefold reduction of BAG6. When BAG6 was knocked down the levels of UBL4A and TRC35 were greatly reduced, but PCR analysis demonstrated that the mRNA levels remain almost constant, and BAG6 is needed to maintain TRC35 and UBL4A at the protein level. This experiment further proved that Bag6 can functionally substitute the Get5-N domain.

35. Structures and binding domains of proteins in the Get4/5 and the cBag6 complexes

Get4 was found to contain 15 α-helices, composed of several helix-turn-helix motifs in an α-2-solenoid fold as well as an antiparallel β-sheet inserted between helices α11 and α12 only allowing packing of α13-α 14 to the side leading to an empty helix binding site. The N-terminal domain (NTD) consists of the first seven helices, they do not contain an obvious internal consensus however they do exhibit some curvature. The C-terminal domain (CTD) continues with right handed helical coils, more varied in length. Bar the absent β-hairpin, TRC35 was found by Mock et al. (2017) to have the same overall conserved construct as Get4, with the first 14 α-helices forming the α-solenoid fold comprised of the NTD (RMSD = 1.38 Å) and CTD (RMSD = 3.43 Å). As a result of the absent β-hairpin α11, α13 and α15 are shifted compared to Get4 resulting in the greater RMSD of the CTD. Get4 was found to physically interact with Get3 as well as Get5, the mammalian homolog TRC35 interacts with TRC40 as well as BAG6 (mediator of UBL4A interaction), and the residues of TRC35 at both the TRC40 and BAG6 binding interfaces are well conserved from the yeast homologs. The convex surface of Get4 has conserved regions in the C-domain as well as helices α1-α4, suggesting a usefulness of the convex surface for protein interactions. A cleft between α2 and α4 with conserved positive charges is proposed to be the N-terminal binding site to Get3 and similarly in TRC35 the binding site of TRC40.

The first Get4-Get5 binding interface is between α12 and α13, the cleft formed by empty helix binding site with conserved hydrophobic interactions, in which the Get5-N α1-helix lies (Bozkurt et al., 2010; Chang et al., 2010; Chartron et al., 2010). Similarly α1 and α2 of BAG6 (W1004, V1008 and W1012) bind within the cleft formed by α12 and α13 of TRC35 (V254, V257, F242, L258 and Y262) (Mock et al., 2017). An extended, highly ordered loop 1 following the α1-helix of Get5-N lies along the helices α9, α10, α11 of Get4 which form a well-conserved pronounced hydrophobic cleft, a result of an insertion in the loop connecting α6 and α7 and subsequent deviation of the α7 and α8 from the helical arrangement (Chang et al., 2010). This interface is also conserved in mammalian systems and the BA G6 α3 (L1032, Y1036 and M1040) interacts along the helices α9, α10, α11 of TRC35 (F195 and M271) (Mock et al., 2017). The loop 2 of Get5-N fits into the concavity on the surface of Get4 comprised of helices α8, α9, α10 and α14 via backbone contacts; however the BAG6 extended domain (second loop) wraps around the TRC35 at the concave side of the α-solenoid scaffold (Bozkurt et al., 2010; Chartron et al., 2010).

The binding interface between the C-terminal heterodimerization domains of BAG6 and UBL4A (BAG6-C and UBL4A-C) was structurally solved by Mock et al. (2015) and Kuwabara et al. (2015) verifying that BAG6 mediates the indirect interaction between TRC35 and UBL4A. UBL4A-C domain was found to contain three helices, helices 1 and 2 form the dimer interface with conserved hydrophobic residues, and helix 3 wraps around BAG6-C (Kuwabara et al., 2015). The UBL4A heterodimerization domain is identical to the Get5-C homodimerization domain, and the hydrophobic residues (W179, I182, L186, F190, V200, L204, W208) of the Get5-C core are conserved in UBL4A (W96, I99, L103, F107, V115, L119, Y123) (Mock et al., 2015). The central ubiquitin-like domain of Get5 (Get5-UBL) domain was found to form a complex with Sgt2, and structural information was elucidated by Chartron, Gonzalez, and Clemons (2011) via small angle X-ray scattering, then by Simon, Simpson, Hawthorne, et al. (2013) via NMR. Crystal structures of the Get5-UBL domain were published shortly thereafter by Chartron et al. (2012), Simon, Simpson, Goldstone, et al. (2013) and Tung, Li, Lin, and Hsiao (2013), and consist of an α-helix: two sets of antiparallel β-strands and two 3/10 (η) helices that form a UBL beta-grasp fold. Compared to the solution NMR structure of the UBL domain of UBL4A (UBL4A-UBL; PDB ID: 2DZI), the Get5-UBL domain is very similar with an average RMSD of 1.23 Å, the few differences due to small sequence insertions, and the β-sheet surface elements are conserved between the two proteins (Chartron et al., 2012). Using NMR, Xu et al. (2012) determined that the UBL site on UBL4A is a non-conventional type IIb class, unrecognizable by UBD and CUE. This Get5/UBL4a-UBL domain has been demonstrated to be unique due to a charged polar residue near the binding site, and is clearly distinguished from other UBLs resulting in specific and favored interactions (Xu et al., 2012).

36. Protein-protein interactions of the Get4/5 (cBag6) complex in the GET/TRC pathway

The cBag6 complex in mammals acts as a pre-targeting step for insertion of TA proteins, and it is a crucial part of the TRC pathway mediating the transfer of proteins from SGTA (Sgt2 in yeast) to TRC40 (Get3 in yeast). The interaction of the complex with these two proteins provides insight into how priority and time are encoded within the sorting system for different hydrophobic substrates (Shao et al., 2017).

37. Interaction of Get4/5 (cBag6) complex with Sgt2 (SGTA)

Sgt2 is a small glutamine-rich protein, and like its mammalian homolog, SGTA, it comprises an N-terminal homodimerization domain; a tetratrico-repeat (TPR) domain; and a glutamine-rich C-terminal domain. Early links were found between Sgt2, Get5, Get4 and Get3 in large scale yeast two-hybrid yeast two-hybrid and tandem-affinity assays, as well as mass spectrometry analyses of protein complexes (Gavin et al., 2006; Ito, Chiba, & Yoshida, 2001;

Uetz et al., 2000). In a large scale genetic yeast study using small molecules, Get4 and Get5 were found to have highest correlation scores as interactors of Sgt2 based on homozygous co-sensitivity for yeast deletion strains (Hillenmeyer et al., 2008). In vitro immunoprecipitation results showed Sgt2 in varying (almost invariably lower) concentrations compared to the identical amounts of Get4 and Get5, suggestive of a transient association of Sgt2 with the Get4/5 complex, therefore a dynamic stoichiometry (Chang et al., 2010). It was postulated following these studies that Sgt2 mediates the transfer to Get3 via the Get4/5 complex especially as Sgt2 was found to be the first point of contact for TA membrane proteins in yeast following release from the ribosome (Brodsky, 2010). The mammalian homologs SGTA, BAG6, UBL4A, TRC35 and TRC40 were found via proteomic analyses to interact with one another, indicative of similar TA protein transfer between the chaperones (Sowa, Bennett, Gygi, & Harper, 2009).

Liou, Cheng, and Wang (2007) used recombinant protein assays to identify the N-terminal region of Sgt2 (Sgt2-NT) to mediate the physical interaction between Get4/5 and Sgt2. Later structural studies and proteolysis assays indicated that it interacts with Get5-UBL forming a canonical binding interface facilitating the indirect interaction with Get4 (Chang et al., 2010; Chartron et al., 2011). NMR chemical shift experiments indicated that the UBL4A-UBL is sequentially and structurally highly conserved, as is its interaction surface with SGTA's N-terminal domain (SGTA-NT) (Simon, Simpson, Goldstone, et al., 2013). Xu et al. (2012) confirmed the direct interaction with mass spectrometry; co-immunoprecipitation and immunoblotting experiments. The results indicated that UBL4A promotes an interaction between SGTA and cBag6 complex.

The MALLS molecular weight of a complex of Sgt2 and Get5 purified by SEC was consistent with a 1:1 binding ratio, confirmed by SAXS, indicating that the Sgt2 dimer binds to just one Get5 (Chartron et al., 2012). This was confirmed in computer docking, ITC and NMR relaxation experiments (Chartron et al., 2012; Simon, Simpson, Goldstone, et al., 2013). ITC and microscale thermophoresis (MST) were used by Darby et al. (2014) to demonstrate a corresponding 1:1 stoichiometry between the mammalian homologs SGTA and UBL4A. Simon, Simpson, Goldstone, et al. (2013) analyzed the interface between constructs covering Sgt2-NT (residues 1–78) and that make up Get5-UBL (residues 70–152). Following X-ray crystallographic structure identification, reciprocal chemical shift perturbation and site-directed mutagenesis experiments were used to map the binding interactions. A well conserved helix in the Sgt2-NT helical dimer interface with an adjoining negatively charged surface (D28, D31, D38, E42) forms strong electrostatic interactions with the highly positively charged binding patch in the Get5-UBL fragment (K85, K79, K122, K124). The two pairs of antiparallel β-sheets of Get5-UBL provide the main interactions, specifically β1 and β4 and the loops connecting β1-β2 and β3-β 4 (Simon, Simpson, Goldstone, et al., 2013). Exposed hydrophobic residues of Sgt2-NT (V35, C39) were also found to be crucial for the binding interface, and interact with compatible residues of Get5-UBL (I81, L120, V125, G143, M147) (Chartron et al., 2012). A salt concentration dependency in the interaction of SGTA-NT with UBL4A-UBL was observed by Leznicki et al. (2013), indicating that the strong electrostatic binding is conserved in mammalian analogues. Darby et al. (2014) biochemically characterized the interaction using NMR perturbation shifts followed by computer docking analysis and confirmed the electrostatic binding. The negatively charged residues of SGTA-NT at the binding interface (D27, E30, E33, E40) are highly conserved, interacting with conserved positively charged residues of UBL4A-UBL (K6, K66, K46, K48). The greatest difference between the interactions of mammalian and yeast complexes is the 45° rotation of the mammalian system due to the shift in the charge distribution of key residues on the surface of each interacting partner.

38. Interaction of Get4/5 (cBag6) complex with Get3 (TRC40)

Get3 binding by a hydrophobic substrate is thought to be the point at which it is committed to the TA protein insertion route. Get3 is an ATPase, it forms a "closed" configuration when bound to ATP (and in some cases ADP); in this form a hydrophobic nucleotide-binding groove is available, thereby increasing its affinity for TA proteins (Mateja et al., 2009). Early analyses demonstrated that the Get4/5 complex functions upstream of Get3 in the GET targeting pathway enabling transfer of TA proteins from the ribosome to Get3, a direct, but transient, physical interaction was expected (Hu, Li, Qian, Denic, & Sha, 2009; Jonikas et al., 2009). Crystal structure analysis confirmed this reaction, and computational docking experiments performed found that the conserved patch of positive residues at the Get4 N-terminal face acts as the binding interface suggestive of complementary electrostatic interactions with Get3 justifying the salt-sensitivity of the interaction observed in size-exclusion chromatography (SEC)

(Chang et al., 2010). A 1:1 stoichiometry was confirmed by SAXS, wherein the Get3 exists as a dimer binding to two Get4-N termini of the Get4/5 dimer complex (Chang et al., 2010).

Rome, Rao, Clemons, and Shan (2013) challenged the model of the assumed passive bridging complex Get4/5, and suggested that it drives the transfer of TA proteins to Get3 via an active process. Using ATPase assays they found that binding of Get4/5 to Get3 reduces the ATPase activity of Get3 10-fold, and inhibits Get3 tetramerization indicating that Get4/5 induces tighter ATP binding by Get3, this demonstrates that ATP-bound "closed" Get3 binds with greater affinity than the apo-Get3 to Get4/5 (Rome et al., 2013). Gel filtration chromatography further established this relationship since a complex was only formed between Get4/5 and apo-Get3 with saturating ATP present. SEC, pull-down and ITC experiments carried out by Gristick et al. (2014) confirmed a preference of Get4/5 for the "closed" ATP-bound Get3 form. Get4/5 was verified to be actively involved in TA protein loading rather than just acting as convenient bridge bringing Get3 in closer proximity with the ribosome. It regulates the Get3 protein by induced ATP-hydrolysis delay, stabilizing it in a conformation favored for efficient substrate capture.

Mariappan et al. (2010) found that the Get4/5 mammalian analog, the cBag6 complex, regulates the nucleotide cycle. They observed that absence of cBag6 slows down the activity of TRC40, but in purified systems TRC40 can still operate suggesting that it is not essential, but increases efficiency and fidelity. It has since been proposed that it actively regulates the conformation of TRC40 to facilitate substrate loading, and perhaps as with Get3 preferentially binds open conformations.

In vitro isothermal titration calorimetry experiments (ITC) with purified wild type and mutant Get4/5-N and Get3 components together with the crystal structures of the Get3-Get4/5 complex found two functionally distinct binding interfaces between Get3 and Get4 (Chang et al., 2010; Chartron et al., 2010; Gristick et al., 2014). The first is the anchoring interface responsible for binding specificity, and it involves α2 (Y30, R37, R42) of Get4 forming hydrophobic and electrostatic contacts within the groove created by α10 and α11 of Get 3 (F246, Y250, E253, Q257, E258, E304, D308). α1 of Get4 containing conserved charged residues (K15, R19) is tilted toward α10 of Get3. The second interface is responsible for ATPase regulation, located on the opposite monomer, and involves conserved complementary charged residue interactions, and the C-terminal end of α4 (D74, E1) of Get4 lies in the post-α3 loop (K69, K72, R75) of Get3.

The regulatory interface was found by mutating residues on the expected regulatory surface: K69D on Get3 and D74K on Get4; Mock et al. (2017) demonstrated that the corresponding regulatory mutations: K86D in Trc40 and D84K in TRC35 resulted in a similar response observing a loss if ATPase inhibition and reduction of TA insertion. Gristick et al. (2014) refined the model for Get3-Get4/5 complex formation, using pre-steady state kinetics. They found that an intermediate is formed by initial electrostatic interactions with a single Get4/5 heterotetramer bound to one monomer of a Get3 dimer; this then rearranges in a second step to a stable final complex dominated by hydrophobic interactions. Upon binding of the TA protein ATP hydrolysis is promoted and Get3 is released from the Get4/5 complex. Mock et al. demonstrate that as expected when histidine tagged SGTA (hSGTA) was incubated with TRC40 in the presence of the cBag6 complex a resultant ATP-dependent increase in the rate of transfer of a TA protein was observed compared to its absence (Mock et al., 2015). In yeast the increased rate in transfer is not just a result of proximity, rather a regulatory role carried out by the Get4/5 complex; Mock et al. (2015) have illustrated that the cBag6 complex performs a similar regulatory role to the Get4/5 complex as originally postulated.

39. Protein interactions in the TRC(GET) pathway

In yeast systems the Get4/5 complex is associated with the ribosome via the interaction of Get5 with Sgt2, and Get4/5 interacts with Get3 targeting protein by the Get4. Get4/5 has been indisputably proven to use a dual mechanism to aid TA protein insertion, a first passive mode involves binding Get3 and tethering it near TA proteins held by Sgt2, acting simply as a bridge. The second mechanism involves the assembly of the Get3 membrane-targeting complex, and binding of Get3 to ATP shifts the protein to the closed conformation, the only form recognized by the Get4/5 heterotetramer. Get4/5 binding promotes inhibition of ATPase activity, enabling the Get3 in the tetrameric complex to bind a TA substrate transferred from Sgt2 (Wang, Brown, Mak, Zhuang, & Denic, 2010).

This has been related to mammalian systems as the TRC pathway, the picture that has emerged is of initial hydrophobic protein capture at the ribosome, then assembly into highly dynamic sorting complex containing many potential binding partners. The pathway protects against high levels of aggregation, and it allows a rapid and private bridging mechanism of transfer from SGTA to TRC40 using the Get4/5 complex equivalent cBag6 (Wang et al., 2010). These specific interactions: of SGTA with UBL4A and of TRC40 with TRC35 confirm the role of the cBag6 complex as an active tethering device, it brings TRC40 into proximity with its substrates and activates it for TMD recognition serving as more than just an adapter function or bridge (Hegde & Keenan, 2011). If any proteins in Get4/5 or cBag6 complexes are absent, then bridging is impaired and so is the capture of the protein by TRC40/Get3. It is clear therefore that the protein chaperones involved in bridging are crucial for delivery of TA proteins, and their up/downregulation will likely affect subsequent processes and lead to disease.

40. Disease related to this quality control system

Hu, Potthoff, Hollenberg, and Ramezani-Rad (2006) demonstrated that cells lacking Get4 or Get5 show identical phenotypes as expected in co-fitness chemical genomic assays, deletion of Get4 and Get5 showed reduction in mating efficiency in yeast cells. Get4 and Get5 were found to be closely related in a physical and functional sense, allowing the transfer of TA proteins from Sgt2 at the ribosome to the targeting Get3 (Chang et al., 2010). Mutagenesis experiments suggest that loss of regulation of Get3 by Get4 leads to growth defects in vivo, and knock out yeast strains of Get4 and Get5 show defects in TA protein biogenesis (Bozkurt et al., 2010; Gristick et al., 2014).

Following these observed phenotypes in yeast Get4 and Get5 deletion assays it appeared likely that the deletion of UBL4A and TRC35 which are closely related in sequence and function will have similar effects in mammalian constructs. As yet, there is no published report as to the effects of deletion of these crucial bridging proteins in relation to the delivery of TA proteins, and the consequences of reduced delivery. However, as will be discussed at length below, the BAG6 complex is found to be a multiprotein "holdase" complex having crucial other roles in cells; in particular the protein quality control processes that are the subject of this chapter. The holdase activity has been found to play a role in retro-translocation of ERAD (endoplasmic reticulum assisted degradation) substrates to the proteasome; mediation of DNA damage response signaling (DDR); degradation of mislocalized membrane proteins, as well as the above discussed chaperoning of nascent TA proteins to the ER. The effect of TRC35 and UBL4A in the role of some of these activities has been studied and it is likely the observations can be correlated in future experiments to their role in TA protein membrane insertion.

In the study by Wang, Song, Brancati, and Segatori (2011) to understand ERAD and how misfolded proteins are directed for proteasomal degradation, the BAG6 complex was found to act as a "holdase." They experimentally identified cBag6 complex as a key mediator in the chaperoning step in ERAD. The cBag6 complex maintains the solubility of retro-translocated polypeptides, facilitating movement of the misfolded proteins across the ER membrane for targeting to the cytosolic proteasome before aggregation. This action promotes cell survival under stress conditions. BAG6 is a nuclear protein, but was found to be present in the cytosol and in an ER-containing membrane fraction. For engagement in ERAD there must be a sufficient presence of BAG6 in cytosol; TRC35 maintains a cytosolic pool of BAG6 (Xu et al., 2012). TRC35 is also involved in the holdase activity of BAG6 in the cell (Mock et al., 2017). UBL4A function is still unclear. Although knockout experiments of UBL4A greatly reduced ERAD, in TA protein transfer UBL4A regulates substrate handoff to TRC40, so it may have a similar function in substrate transfer to the proteasome.

Krenciute et al. (2013) demonstrated that cBag6 complex mediates DNA damage response (DDR) signaling and damage-induced cell death. All three components of the cBag6 complex are essential for DDR signaling since they regulate the recruitment of BRCA1 (breast cancer protein) to sites of DNA damage, and depletion of the complex compromises the role of BRCA1 in maintaining genome integrity. Knockdown experiments using siRNA demonstrated that knocking down UBL4A or TRC40 alone has negligible effect on Bag6 presence and therefore cell death, knockdown of both simultaneously results in a threefold reduction of bag6 and cells are more resistant to death. UBL4A knockdown largely reduces BAG6 phosphorylation, while TRC40 had no obvious effect, suggesting that UBL4A may play a more direct role in DDR (Krenciute et al., 2013).

These effects indicate that cBag6 complex regulates cellular protein quality control in the cytoplasm as well as the nucleus by targeting multiple substrates for a variety of cellular functions. The cBag6 complex plays a similar "holdase" role in all its seemingly unrelated functions, by shielding hydrophobic or transmembrane domain (TMD) sequences in a variety of substrates, thereby maintaining solubility and preventing aggregation. The model proposed by Krenciute et al. (2013) suggests that UBL4A and TRC35 can interact independently of one another with BAG6, and both can act individually as a stabilizing partner; therefore only knockdown of both will lead to total resistance to cell death. It is interesting to note that in ERAD TRC35 plays the major role while in DDR UBL4A plays the major role. This may be linked to their roles in the GET/TRC pathways in which specifically Get4/TRC35 enables Get3/TRC40 binding, and Get5/UBL4A enables Sgt2/SGTA binding.

41. Disease seemingly unrelated to the chaperone system

UBL4A deficiency has been found to play roles in disease beyond the cBag6 complex. It was found by Zhao et al. (2015) to be critical for insulin-induced plasma membrane translocation, it interacts via Arp2/3 promoting Arp2/3 mediated actin branching, and therefore it was predicted that it regulates other cellular activities related to actin cytoskeletal reorganization dependent on Arp2/3. Recently new results published by Zhao et al. (2017) found that indeed UBL4A is crucial for other cellular activities that depend on Arp2/3, and found it critical for migration of fibroblasts and macrophages. Mice studies found that neutrophils lacking UBL4A showed significant directionality defects during chemotaxis, and loss of UBL4A delayed wound closure by fibroblasts.

STAT3 activation occurs via phosphorylation of tyrosine reside Y705; its activated form is widely accepted as a hallmark in a wide range of cancers such as leukemia or solid tumors. Failure of the balance of STAT3 activation results in tumorigenesis. Wang et al. (2014) revealed that UBL4A bridges an association between TC45 and STAT3, mediating efficient phosphorylation by TC45, and blocking p-STAT3-dependent cancer cell growth. The authors found that tumor growth rate and tumor size were both greatly reduced in B16 murine melanoma cells in vivo when transfected with UBL4A, and it does so by targeting p-STAT3. The results suggest that UBL4A may function as a tumor suppressor, further implied when the deletion of UBL4A facilitated MEF cell proliferation, potentially providing a new strategy for tumor therapy.

Liang et al. (2018) recently identified UBL4A as being a novel regulator in bone development. *GdX* is the UBL4A gene; *GdX* knockout mice indicate a critical role of UBL4A in skeletal biology; at the embryonic stage they have decreased bone mineral density, and later demonstrate a blended spine and reduced growth. UBL4A mutations may be a potential cause for skeletal dysplasias and other bone disorders via repressed osteoblast formation and increased chrondrocyte differentiation; however this is yet to be proven in human systems.

42. Conclusion

This chapter has described some of the core machinery involved in quality control of hydrophobic proteins exposed to the aqueous cytoplasm of cells. This field is still rapidly evolving and there are many outstanding questions, in particular with regard to understanding the roles of this machinery in disease states. It is likely that one or more of the proteins discussed here is already under investigation as a drug target and that many of the interaction partners for these proteins have yet to be identified. In the future these ambiguities will become clearer but for now, we have outlined the currently suggested connections between SGTA, BAG6, UBL4A and TRC35 and a range of diseases.

References

Amm, I., Sommer, T., & Wolf, D. H. (2014). Protein quality control and elimination of protein waste: The role of the ubiquitin-proteasome system. *Biochimica et Biophysica Acta*, *1843*(1), 182–196. https://doi.org/10.1016/j.bbamcr.2013.06.031.

Ast, T., Cohen, G., & Schuldiner, M. (2013). A network of cytosolic factors targets SRP-independent proteins to the endoplasmic reticulum. *Cell*, *152*(5), 1134–1145. https://doi.org/10.1016/j.cell.2013.02.003.

Azziz, R., Woods, K. S., Reyna, R., Key, T. J., Knochenhauer, E. S., & Yildiz, B. O. (2004). The prevalence and features of the polycystic ovary syndrome in an unselected population. *The Journal of Clinical Endocrinology and Metabolism*, *89*(6), 2745–2749. https://doi.org/10.1210/jc.2003-032046.

Banerji, J., Sands, J., Strominger, J. L., & Spies, T. (1990). A gene pair from the human major histocompatibility complex encodes large proline-rich proteins with multiple repeated motifs and a single ubiquitin-like domain. *Proceedings of the National Academy of Sciences of the United States of America*, *87*(6), 2374–2378.

Binici, J., Hartmann, J., Herrmann, J., Schreiber, C., Beyer, S., Guler, G., et al. (2013). A soluble fragment of the tumor antigen BCL2-associated athanogene 6 (BAG-6) is essential and sufficient for inhibition of NKp30 receptor-dependent cytotoxicity of natural killer cells. *The Journal of Biological Chemistry*, *288*(48), 34295–34303. https://doi.org/10.1074/jbc.M113.483602.

Binici, J., & Koch, J. (2014). BAG-6, a jack of all trades in health and disease. *Cellular and Molecular Life Sciences*, *71*(10), 1829–1837. https://doi.org/10.1007/s00018-013-1522-y.

Borgese, N., & Fasana, E. (2011). Targeting pathways of C-tail-anchored proteins. *Biochimica et Biophysica Acta*, *1808*(3), 937–946. https://doi.org/10.1016/j.bbamem.2010.07.010.

Bozkurt, G., Wild, K., Amlacher, S., Hurt, E., Dobberstein, B., & Sinning, I. (2010). The structure of Get4 reveals an alpha-solenoid fold adapted for multiple interactions in tail-anchored protein biogenesis. *FEBS Letters*, *584*(8), 1509–1514. https://doi.org/10.1016/j.febslet.2010.02.070.

Brodsky, J. L. (2010). The special delivery of a tail-anchored protein: Why it pays to use a dedicated courier. *Molecular Cell*, *40*(1), 5–7. https://doi.org/10.1016/j.molcel.2010.09.016.

Bromfield, E., Aitken, R. J., & Nixon, B. (2015). Novel characterization of the HSPA2-stabilizing protein BAG6 in human spermatozoa. *Molecular Human Reproduction*, *21*(10), 755–769. https://doi.org/10.1093/molehr/gav041.

Brooke, G. N., & Bevan, C. L. (2009). The role of androgen receptor mutations in prostate cancer progression. *Current Genomics*, *10*(1), 18–25. https://doi.org/10.2174/138920209787581307.

Buchanan, G., Ricciardelli, C., Harris, J. M., Prescott, J., Yu, Z. C., Jia, L., et al. (2007). Control of androgen receptor signaling in prostate cancer by the cochaperone small glutamine rich tetratricopeptide repeat containing protein alpha. *Cancer Research*, *67*(20), 10087–10096. https://doi.org/10.1158/0008-5472.CAN-07-1646.

Buchberger, A., Bukau, B., & Sommer, T. (2010). Protein quality control in the cytosol and the endoplasmic reticulum: Brothers in arms. *Molecular Cell*, *40*(2), 238–252. https://doi.org/10.1016/j.molcel.2010.10.001.

Bullock, M. (2016). FOXO factors and breast cancer: Outfoxing endocrine resistance. *Endocrine-Related Cancer*, *23*(2), R113–R130. https://doi.org/10.1530/ERC-15-0461.

Cattrini, C., Zanardi, E., Vallome, G., Cavo, A., Cerbone, L., Di Meglio, A., et al. (2017). Targeting androgen-independent pathways: New chances for patients with prostate cancer? *Critical Reviews in Oncology/Hematology*, *118*, 42–53. https://doi.org/10.1016/j.critrevonc.2017.08.009.

Chakrabarti, O., & Hegde, R. S. (2009). Functional depletion of mahogunin by cytosolically exposed prion protein contributes to neurodegeneration. *Cell*, *137*(6), 1136–1147. https://doi.org/10.1016/j.cell.2009.03.042.

Chang, Y. W., Chuang, Y. C., Ho, Y. C., Cheng, M. Y., Sun, Y. J., Hsiao, C. D., et al. (2010). Crystal structure of Get4-Get5 complex and its interactions with Sgt2, Get3, and Ydj1. *The Journal of Biological Chemistry*, *285*(13), 9962–9970. https://doi.org/10.1074/jbc.M109.087098.

Chartron, J. W., Gonzalez, G. M., & Clemons, W. M., Jr. (2011). A structural model of the Sgt2 protein and its interactions with chaperones and the Get4/Get5 complex. *The Journal of Biological Chemistry*, *286*(39), 34325–34334. https://doi.org/10.1074/jbc.M111.277798.

Chartron, J. W., Suloway, C. J., Zaslaver, M., & Clemons, W. M., Jr. (2010). Structural characterization of the Get4/Get5 complex and its interaction with Get3. *Proceedings of the National Academy of Sciences of the United States of America*, *107*(27), 12127–12132. https://doi.org/10.1073/pnas.1006036107.

Chartron, J. W., VanderVelde, D. G., & Clemons, W. M., Jr. (2012). Structures of the Sgt2/SGTA dimerization domain with the Get5/UBL4A UBL domain reveal an interaction that forms a conserved dynamic interface. *Cell Reports*, *2*(6), 1620–1632. https://doi.org/10.1016/j.celrep.2012.10.010.

Cox, D., Raeburn, C., Sui, X., & Hatters, D. M. (2018). Protein aggregation in cell biology: An aggregomics perspective of health and disease. *Seminars in Cell & Developmental Biology* https://doi.org/10.1016/j.semcdb.2018.05.003.

Cross, B. C., Sinning, I., Luirink, J., & High, S. (2009). Delivering proteins for export from the cytosol. *Nature Reviews Molecular Cell Biology*, *10*(4), 255–264. https://doi.org/10.1038/nrm2657.

Cutress, M. L., Whitaker, H. C., Mills, I. G., Stewart, M., & Neal, D. E. (2008). Structural basis for the nuclear import of the human androgen receptor. *Journal of Cell Science*, *121*(Pt. 7), 957–968. https://doi.org/10.1242/jcs.022103.

Ciepluch, C., Kordes, E., Poirey, R., Grewenig, A., Rommelaere, J., & Jauniaux, J. C. (1998). Identification of a novel cellular TPR-containing protein, SGT, that interacts with the nonstructural protein NS1 of parvovirus H-1. *Journal of Virology*, *72*(5), 4149–4156.

Darby, J. F., Krysztofinska, E. M., Simpson, P. J., Simon, A. C., Leznicki, P., Sriskandarajah, N., et al. (2014). Solution structure of the SGTA dimerisation domain and investigation of its interactions with the ubiquitin-like domains of BAG6 and UBL4A. *PLoS One*, *9*(11), e113281. https://doi.org/10.1371/journal.pone.0113281.

Denic, V. (2012). A portrait of the GET pathway as a surprisingly complicated young man. *Trends in Biochemical Sciences*, *37*(10), 411–417. https://doi.org/10.1016/j.tibs.2012.07.004.

Denic, V., Dotsch, V., & Sinning, I. (2013). Endoplasmic reticulum targeting and insertion of tail-anchored membrane proteins by the GET pathway. *Cold Spring Harbor Perspectives in Biology*, *5*(8), a013334. https://doi.org/10.1101/cshperspect.a013334.

Desmots, F., Russell, H. R., Lee, Y., Boyd, K., & McKinnon, P. J. (2005). The reaper-binding protein scythe modulates apoptosis and proliferation during mammalian development. *Molecular and Cellular Biology*, *25*(23), 10329–10337. https://doi.org/10.1128/MCB.25.23.10329-10337.2005.

Dupzyk, A., Williams, J. M., Bagchi, P., Inoue, T., & Tsai, B. (2017). SGTA-dependent regulation of Hsc70 promotes cytosol entry of simian virus 40 from the endoplasmic reticulum. *Journal of Virology*, *91*(12). https://doi.org/10.1128/JVI.00232-17, e00232-17.

Dutta, S., & Tan, Y. J. (2008). Structural and functional characterization of human SGT and its interaction with Vpu of the human immunodeficiency virus type 1. *Biochemistry*, *47*(38), 10123–10131. https://doi.org/10.1021/bi800758a.

Echeverria, P. C., & Picard, D. (2010). Molecular chaperones, essential partners of steroid hormone receptors for activity and mobility. *Biochimica et Biophysica Acta*, *1803*(6), 641–649. https://doi.org/10.1016/j.bbamcr.2009.11.012.

Etokebe, G. E., Zienolddiny, S., Kupanovac, Z., Enersen, M., Balen, S., Flego, V., et al. (2015). Association of the FAM46A gene VNTRs and BAG6 rs3117582 SNP with non small cell lung cancer (NSCLC) in Croatian and Norwegian populations. *PLoS One*, *10*(4), e0122651. https://doi.org/10.1371/journal.pone.0122651.

Fernandez-Fernandez, M. R., & Valpuesta, J. M. (2018). Hsp70 chaperone: A master player in protein homeostasis. *F1000Research*, 7, 1497. https://doi.org/10.12688/f1000research.15528.1, [version 1; referees: 3 approved] (F1000 Faculty Rev).

Fielding, B. C., Gunalan, V., Tan, T. H., Chou, C. F., Shen, S., Khan, S., et al. (2006). Severe acute respiratory syndrome coronavirus protein 7a interacts with hSGT. *Biochemical and Biophysical Research Communications*, *343*(4), 1201–1208. https://doi.org/10.1016/j.bbrc.2006.03.091.

Gavin, A. C., Aloy, P., Grandi, P., Krause, R., Boesche, M., Marzioch, M., et al. (2006). Proteome survey reveals modularity of the yeast cell machinery. *Nature*, *440*(7084), 631–636. https://doi.org/10.1038/nature04532.

Goodarzi, M. O., Xu, N., Cui, J., Guo, X., Chen, Y. I., & Azziz, R. (2008). Small glutamine-rich tetratricopeptide repeat-containing protein alpha (SGTA), a candidate gene for polycystic ovary syndrome. *Human Reproduction*, *23*(5), 1214–1219. https://doi.org/10.1093/humrep/den065.

Gristick, H. B., Rao, M., Chartron, J. W., Rome, M. E., Shan, S. O., & Clemons, W. M., Jr. (2014). Crystal structure of ATP-bound Get3-Get4-Get5 complex reveals regulation of Get3 by Get4. *Nature Structural & Molecular Biology*, *21*, 437–442. https://doi.org/10.1038/nsmb.2813.

Hampton, R. Y., & Sommer, T. (2012). Finding the will and the way of ERAD substrate retrotranslocation. *Current Opinion in Cell Biology*, *24*(4), 460–466. https://doi.org/10.1016/j.ceb.2012.05.010.

Hartl, F. U., & Hayer-Hartl, M. (2009). Converging concepts of protein folding in vitro and in vivo. *Nature Structural & Molecular Biology*, *16*(6), 574–581. https://doi.org/10.1038/nsmb.1591.

Hegde, R. S., & Keenan, R. J. (2011). Tail-anchored membrane protein insertion into the endoplasmic reticulum. *Nature Reviews. Molecular Cell Biology*, *12*(12), 787–798. https://doi.org/10.1038/nrm3226.

Hessa, T., Sharma, A., Mariappan, M., Eshleman, H. D., Gutierrez, E., & Hegde, R. S. (2011). Protein targeting and degradation are coupled for elimination of mislocalized proteins. *Nature*, *475*(7356), 394–397. https://doi.org/10.1038/nature10181.

Hillenmeyer, M. E., Fung, E., Wildenhain, J., Pierce, S. E., Hoon, S., Lee, W., et al. (2008). The chemical genomic portrait of yeast: Uncovering a phenotype for all genes. *Science*, *320*(5874), 362–365. https://doi.org/10.1126/science.1150021.

Hsieh, Y. Y., Chang, C. C., Hsu, C. M., Chen, S. Y., Lin, W. H., & Tsai, F. J. (2011). Major histocompatibility complex class I chain-related gene polymorphisms: Associated with susceptibility to Kawasaki disease and coronary artery aneurysms. *Genetic Testing and Molecular Biomarkers*, *15*(11), 755–763. https://doi.org/10.1089/gtmb.2011.0001.

Hu, J., Li, J., Qian, X., Denic, V., & Sha, B. (2009). The crystal structures of yeast Get3 suggest a mechanism for tail-anchored protein membrane insertion. *PLoS One*, *4*(11), e8061. https://doi.org/10.1371/journal.pone.0008061.

Hu, Z., Potthoff, B., Hollenberg, C. P., & Ramezani-Rad, M. (2006). Mdy2, a ubiquitin-like (UBL)-domain protein, is required for efficient mating in *Saccharomyces cerevisiae*. *Journal of Cell Science*, *119*(Pt. 2), 326–338. https://doi.org/10.1242/jcs.02754.

Ito, T., Chiba, T., & Yoshida, M. (2001). Exploring the protein interactome using comprehensive two-hybrid projects. *Trends in Biotechnology*, *19*(10 Suppl), S23–S27. https://doi.org/10.1016/S0167-7799(01)01790-5.

Ji, X. D., Li, G., Feng, Y. X., Zhao, J. S., Li, J. J., Sun, Z. J., et al. (2011). EphB3 is overexpressed in non-small-cell lung cancer and promotes tumor metastasis by enhancing cell survival and migration. *Cancer Research*, *71*(3), 1156–1166. https://doi.org/10.1158/0008-5472.CAN-10-0717.

Johnson, N., Powis, K., & High, S. (2013). Post-translational translocation into the endoplasmic reticulum. *Biochimica et Biophysica Acta*, *1833*(11), 2403–2409. https://doi.org/10.1016/j.bbamcr.2012.12.008.

Jonikas, M. C., Collins, S. R., Denic, V., Oh, E., Quan, E. M., Schmid, V., et al. (2009). Comprehensive characterization of genes required for protein folding in the endoplasmic reticulum. *Science*, *323*(5922), 1693–1697. https://doi.org/10.1126/science.1167983.

Kalbfleisch, T., Cambon, A., & Wattenberg, B. W. (2007). A bioinformatics approach to identifying tail-anchored proteins in the human genome. *Traffic*, *8*(12), 1687–1694. https://doi.org/10.1111/j.1600-0854.2007.00661.x.

Kato, Y., Ochiai, K., Kawakami, S., Nakao, N., Azakami, D., Bonkobara, M., et al. (2017). Canine REIC/Dkk-3 interacts with SGTA and restores androgen receptor signalling in androgen-independent prostate cancer cell lines. *BMC Veterinary Research*, *13*(1), 170. https://doi.org/10.1186/s12917-017-1094-4.

Kawahara, H., Minami, R., & Yokota, N. (2013). BAG6/BAT3: Emerging roles in quality control for nascent polypeptides. *Journal of Biochemistry*, *153*(2), 147–160. https://doi.org/10.1093/jb/mvs149.

Krenciute, G., Liu, S., Yucer, N., Shi, Y., Ortiz, P., Liu, Q., et al. (2013). Nuclear BAG6-UBL4A-GET4 complex mediates DNA damage signaling and cell death. *The Journal of Biological Chemistry*, *288*(28), 20547–20557. https://doi.org/10.1074/jbc.M112.443416.

Krysztofinska, E. M., Evans, N. J., Thapaliya, A., Murray, J. W., Morgan, R. M. L., Martinez-Lumbreras, S., et al. (2017). Structure and interactions of the TPR domain of Sgt2 with yeast chaperones and Ybr137wp. *Frontiers in Molecular Biosciences*, *4*, 68. https://doi.org/10.3389/fmolb.2017.00068.

Krysztofinska, E. M., Martinez-Lumbreras, S., Thapaliya, A., Evans, N. J., High, S., & Isaacson, R. L. (2016). Structural and functional insights into the E3 ligase, RNF126. *Scientific Reports*, *6*. https://doi.org/10.1038/srep26433.

Kuwabara, N., Minami, R., Yokota, N., Matsumoto, H., Senda, T., Kawahara, H., et al. (2015). Structure of a BAG6 (Bcl-2-associated athanogene 6)-UBL4a (ubiquitin-like protein 4a) complex reveals a novel binding interface that functions in tail-anchored protein biogenesis. *The Journal of Biological Chemistry*, *290*(15), 9387–9398. https://doi.org/10.1074/jbc.M114.631804.

Lee, J. G., & Ye, Y. (2013). Bag6/Bat3/Scythe: A novel chaperone activity with diverse regulatory functions in protein biogenesis and degradation. *BioEssays*, *35*(4), 377–385. https://doi.org/10.1002/bies.201200159.

Leznicki, P., & High, S. (2012). SGTA antagonizes BAG6-mediated protein triage. *Proceedings of the National Academy of Sciences of the United States of America*, *109*(47), 19214–19219. https://doi.org/10.1073/pnas.1209997109.

Leznicki, P., Korac-Prlic, J., Kliza, K., Husnjak, K., Nyathi, Y., Dikic, I., et al. (2015). Binding of SGTA to Rpn13 selectively modulates protein quality control. *Journal of Cell Science*, *128*(17), 3187–3196. https://doi.org/10.1242/jcs.165209.

Leznicki, P., Roebuck, Q. P., Wunderley, L., Clancy, A., Krysztofinska, E. M., Isaacson, R. L., et al. (2013). The association of BAG6 with SGTA and tail-anchored proteins. *PLoS One*, *8*(3), e59590. https://doi.org/10.1371/journal.pone.0059590.

Liang, J., Li, J., Fu, Y., Ren, F., Xu, J., Zhou, M., et al. (2018). GdX/UBL4A null mice exhibit mild kyphosis and scoliosis accompanied by dysregulation of osteoblastogenesis and chondrogenesis. *Cell Biochemistry and Function*, *36*(3), 129–136. https://doi.org/10.1002/cbf.3324.

Liou, S. T., Cheng, M. Y., & Wang, C. (2007). SGT2 and MDY2 interact with molecular chaperone YDJ1 in *Saccharomyces cerevisiae*. *Cell Stress & Chaperones*, *12*(1), 59–70.

Liou, S. T., & Wang, C. (2005). Small glutamine-rich tetratricopeptide repeat-containing protein is composed of three structural units with distinct functions. *Archives of Biochemistry and Biophysics*, *435*(2), 253–263. https://doi.org/10.1016/j.abb.2004.12.020.

Luce, M. J., Akpawu, A. A., Tucunduva, D. C., Mason, S., & Scott, M. S. (2016). Extent of pre-translational regulation for the control of nucleocytoplasmic protein localization. *BMC Genomics*, *17*, 472. https://doi.org/10.1186/s12864-016-2854-4.

Mariappan, M., Li, X., Stefanovic, S., Sharma, A., Mateja, A., Keenan, R. J., et al. (2010). A ribosome-associating factor chaperones tail-anchored membrane proteins. *Nature*, *466*(7310), 1120–1124. https://doi.org/10.1038/nature09296.

Martinez-Lumbreras, S., Alfano, C., Evans, N. J., Collins, K. M., Flanagan, K. A., Atkinson, R. A., et al. (2018). Structural and functional insights into *Bacillus subtilis* sigma factor inhibitor, CsfB. *Structure*, *26*(4), 640–648.e5. https://doi.org/10.1016/j.str.2018.02.007.

Mateja, A., Szlachcic, A., Downing, M. E., Dobosz, M., Mariappan, M., Hegde, R. S., et al. (2009). The structural basis of tail-anchored membrane protein recognition by Get3. *Nature*, *461*(7262), 361–366. https://doi.org/10.1038/nature08319.

Minami, R., Hayakawa, A., Kagawa, H., Yanagi, Y., Yokosawa, H., & Kawahara, H. (2010). BAG-6 is essential for selective elimination of defective proteasomal substrates. *The Journal of Cell Biology*, *190*(4), 637–650. https://doi.org/10.1083/jcb.200908092.

Mock, J. Y., Chartron, J. W., Zaslaver, M., Xu, Y., Ye, Y., & Clemons, W. M., Jr. (2015). Bag6 complex contains a minimal tail-anchor-targeting module and a mock BAG domain. *Proceedings of the National Academy of Sciences of the United States of America*, *112*(1), 106–111. https://doi.org/10.1073/pnas.1402745112.

Mock, J. Y., Xu, Y., Ye, Y., & Clemons, W. M., Jr. (2017). Structural basis for regulation of the nucleo-cytoplasmic distribution of Bag6 by TRC35. *Proceedings of the National Academy of Sciences of the United States of America*, *114*(44), 11679–11684. https://doi.org/10.1073/pnas.1702940114.

Nguyen, T. Q., Chenon, M., Vilela, F., Velours, C., Aumont-Nicaise, M., Andreani, J., et al. (2017). Structural plasticity of the N-terminal capping helix of the TPR domain of kinesin light chain. *PLoS One*, *12*(10), e0186354. https://doi.org/10.1371/journal.pone.0186354.

Paul, A., Garcia, Y. A., Zierer, B., Patwardhan, C., Gutierrez, O., Hildenbrand, Z., et al. (2014). The cochaperone SGTA (small glutamine-rich tetratricopeptide repeat-containing protein alpha) demonstrates regulatory specificity for the androgen, glucocorticoid, and progesterone receptors. *The Journal of Biological Chemistry*, *289*(22), 15297–15308. https://doi.org/10.1074/jbc.M113.535229.

Payapilly, A., & High, S. (2014). BAG6 regulates the quality control of a polytopic ERAD substrate. *Journal of Cell Science*, *127*(Pt. 13), 2898–2909. https://doi.org/10.1242/jcs.145565.

Philp, L. K., Butler, M. S., Hickey, T. E., Butler, L. M., Tilley, W. D., & Day, T. K. (2013). SGTA: A new player in the molecular co-chaperone game. *Hormones & Cancer*, *4*(6), 343–357. https://doi.org/10.1007/s12672-013-0151-0.

Philp, L. K., Day, T. K., Butler, M. S., Laven-Law, G., Jindal, S., Hickey, T. E., et al. (2016). Small glutamine-rich tetratricopeptide repeat-containing protein alpha (SGTA) ablation limits offspring viability and growth in mice. *Scientific Reports*, *6*, 28950. https://doi.org/10.1038/srep28950.

Piras, I. S., Angius, A., Andreani, M., Testi, M., Lucarelli, G., Floris, M., et al. (2014). BAT2 and BAT3 polymorphisms as novel genetic risk factors for rejection after HLA-related SCT. *Bone Marrow Transplantation*, *49*(11), 1400–1404. https://doi.org/10.1038/bmt.2014.177.

Rabu, C., Schmid, V., Schwappach, B., & High, S. (2009). Biogenesis of tail-anchored proteins: The beginning for the end? *Journal of Cell Science*, *122*(Pt. 20), 3605–3612. https://doi.org/10.1242/jcs.041210.

Rangachari, M., Zhu, C., Sakuishi, K., Xiao, S., Karman, J., Chen, A., et al. (2012). Bat3 promotes T cell responses and autoimmunity by repressing Tim-3-mediated cell death and exhaustion. *Nature Medicine*, *18*(9), 1394–1400. https://doi.org/10.1038/nm.2871.

Roberts, J. D., Thapaliya, A., Martinez-Lumbreras, S., Krysztofinska, E. M., & Isaacson, R. L. (2015). Structural and functional insights into small, glutamine-rich, tetratricopeptide repeat protein alpha. *Frontiers in Molecular Biosciences*, *2*, 71. https://doi.org/10.3389/fmolb.2015.00071.

Rodrigo-Brenni, M. C., Gutierrez, E., & Hegde, R. S. (2014). Cytosolic quality control of mislocalized proteins requires RNF126 recruitment to Bag6. *Molecular Cell*, *55*(2), 227–237. https://doi.org/10.1016/j.molcel.2014.05.025.

Rome, M. E., Rao, M., Clemons, W. M., & Shan, S. O. (2013). Precise timing of ATPase activation drives targeting of tail-anchored proteins. *Proceedings of the National Academy of Sciences of the United States of America*, *110*(19), 7666–7671. https://doi.org/10.1073/pnas.1222054110.

Rosenberger, A., Sohns, M., Friedrichs, S., Hung, R. J., Fehringer, G., McLaughlin, J., et al. (2017). Gene-set meta-analysis of lung cancer identifies pathway related to systemic lupus erythematosus. *PLoS One*, *12*(3), e0173339. https://doi.org/10.1371/journal.pone.0173339.

Ross, C. A., & Poirier, M. A. (2004). Protein aggregation and neurodegenerative disease. *Nature Medicine*, *10*(Suppl), S10–S17. https://doi.org/10.1038/nm1066.

Schuldiner, M., Metz, J., Schmid, V., Denic, V., Rakwalska, M., Schmitt, H. D., et al. (2008). The GET complex mediates insertion of tail-anchored proteins into the ER membrane. *Cell*, *134*(4), 634–645. https://doi.org/10.1016/j.cell.2008.06.025.

Shao, S., & Hegde, R. S. (2016). Target selection during protein quality control. *Trends in Biochemical Sciences*, *41*(2), 124–137. https://doi.org/10.1016/j.tibs.2015.10.007.

Shao, S., Rodrigo-Brenni, M. C., Kivlen, M. H., & Hegde, R. S. (2017). Mechanistic basis for a molecular triage reaction. *Science, 355*(6322), 298–302. https://doi.org/10.1126/science.aah6130.

Simon, A. C., Simpson, P. J., Goldstone, R. M., Krysztofinska, E. M., Murray, J. W., High, S., et al. (2013). Structure of the Sgt2/Get5 complex provides insights into GET-mediated targeting of tail-anchored membrane proteins. *Proceedings of the National Academy of Sciences of the United States of America, 110*(4), 1327–1332. https://doi.org/10.1073/pnas.1207518110.

Simon, A. C., Simpson, P. J., Hawthorne, W., Hale, L. R., Goldstone, R. M., & Isaacson, R. L. (2013). 1H, 13C and 15N assignments of Sgt2 N-terminal dimerisation domain and its binding partner, Get5 ubiquitin-like domain. *Biomolecular NMR Assignments*, 7(2), 271–274. https://doi.org/10.1007/s12104-012-9425-7.

Simpson, P. J., Schwappach, B., Dohlman, H. G., & Isaacson, R. L. (2010). Structures of Get3, Get4, and Get5 provide new models for TA membrane protein targeting. *Structure, 18*(8), 897–902. https://doi.org/10.1016/j.str.2010.07.003.

Slavotinek, A. M., & Biesecker, L. G. (2001). Unfolding the role of chaperones and chaperonins in human disease. *Trends in Genetics, 17*(9), 528–535.

Sowa, M. E., Bennett, E. J., Gygi, S. P., & Harper, J. W. (2009). Defining the human deubiquitinating enzyme interaction landscape. *Cell, 138*(2), 389–403. https://doi.org/10.1016/j.cell.2009.04.042.

Stefanovic, S., & Hegde, R. S. (2007). Identification of a targeting factor for posttranslational membrane protein insertion into the ER. *Cell, 128*(6), 1147–1159. https://doi.org/10.1016/j.cell.2007.01.036.

Talukdar, F. R., di Pietro, M., Secrier, M., Moehler, M., Goepfert, K., Lima, S. S. C., et al. (2018). Molecular landscape of esophageal cancer: Implications for early detection and personalized therapy. *Annals of the New York Academy of Sciences*. https://doi.org/10.1111/nyas.13876, (Epub ahead of print).

Tanaka, H., Takahashi, T., Xie, Y., Minami, R., Yanagi, Y., Hayashishita, M., et al. (2016). A conserved island of BAG6/Scythe is related to ubiquitin domains and participates in short hydrophobicity recognition. *The FEBS Journal, 283*(4), 662–677. https://doi.org/10.1111/febs.13618.

Thapaliya, A., Nyathi, Y., Martinez-Lumbreras, S., Krysztofinska, E. M., Evans, N. J., Terry, I. L., et al. (2016). SGTA interacts with the proteasomal ubiquitin receptor Rpn13 via a carboxylate clamp mechanism. *Scientific Reports, 6*, 36622. https://doi.org/10.1038/srep36622.

Thress, K., Henzel, W., Shillinglaw, W., & Kornbluth, S. (1998). Scythe: A novel reaper-binding apoptotic regulator. *The EMBO Journal, 17*(21), 6135–6143. https://doi.org/10.1093/emboj/17.21.6135.

Trotta, A. P., Need, E. F., Butler, L. M., Selth, L. A., O'Loughlin, M. A., Coetzee, G. A., et al. (2012). Subdomain structure of the co-chaperone SGTA and activity of its androgen receptor client. *Journal of Molecular Endocrinology, 49*(2), 57–68. https://doi.org/10.1530/JME-11-0152.

Trotta, A. P., Need, E. F., Selth, L. A., Chopra, S., Pinnock, C. B., Leach, D. A., et al. (2013). Knockdown of the cochaperone SGTA results in the suppression of androgen and PI3K/Akt signaling and inhibition of prostate cancer cell proliferation. *International Journal of Cancer, 133*(12), 2812–2823. https://doi.org/10.1002/ijc.28310.

Tsukahara, T., Kimura, S., Ichimiya, S., Torigoe, T., Kawaguchi, S., Wada, T., et al. (2009). Scythe/BAT3 regulates apoptotic cell death induced by papillomavirus binding factor in human osteosarcoma. *Cancer Science, 100*(1), 47–53. https://doi.org/10.1111/j.1349-7006.2008.00991.x.

Tung, J. Y., Li, Y. C., Lin, T. W., & Hsiao, C. D. (2013). Structure of the Sgt2 dimerization domain complexed with the Get5 UBL domain involved in the targeting of tail-anchored membrane proteins to the endoplasmic reticulum. *Acta Crystallographica. Section D, Biological Crystallography, 69*(Pt. 10), 2081–2090. https://doi.org/10.1107/S0907444913019379.

Uetz, P., Giot, L., Cagney, G., Mansfield, T. A., Judson, R. S., Knight, J. R., et al. (2000). A comprehensive analysis of protein-protein interactions in *Saccharomyces cerevisiae. Nature, 403*(6770), 623–627. https://doi.org/10.1038/35001009.

Vilardi, F., Lorenz, H., & Dobberstein, B. (2011). WRB is the receptor for TRC40/Asna1-mediated insertion of tail-anchored proteins into the ER membrane. *Journal of Cell Science, 124*(Pt. 8), 1301–1307. https://doi.org/10.1242/jcs.084277.

Vilardi, F., Stephan, M., Clancy, A., Janshoff, A., & Schwappach, B. (2014). WRB and CAML are necessary and sufficient to mediate tail-anchored protein targeting to the ER membrane. *PLoS One, 9*(1), e85033. https://doi.org/10.1371/journal.pone.0085033.

Vonsattel, J. P., Myers, R. H., Stevens, T. J., Ferrante, R. J., Bird, E. D., & Richardson, E. P., Jr. (1985). Neuropathological classification of Huntington's disease. *Journal of Neuropathology and Experimental Neurology, 44*(6), 559–577.

Waheed, A. A., MacDonald, S., Khan, M., Mounts, M., Swiderski, M., Xu, Y., et al. (2016). The Vpu-interacting protein SGTA regulates expression of a non-glycosylated tetherin species. *Scientific Reports, 6*, 24934. https://doi.org/10.1038/srep24934.

Walczak, C. P., Ravindran, M. S., Inoue, T., & Tsai, B. (2014). A cytosolic chaperone complexes with dynamic membrane J-proteins and mobilizes a nonenveloped virus out of the endoplasmic reticulum. *PLoS Pathogens, 10*(3), e1004007. https://doi.org/10.1371/journal.ppat.1004007.

Wang, F., Brown, E. C., Mak, G., Zhuang, J., & Denic, V. (2010). A chaperone cascade sorts proteins for posttranslational membrane insertion into the endoplasmic reticulum. *Molecular Cell, 40*(1), 159–171. https://doi.org/10.1016/j.molcel.2010.08.038.

Wang, T., Liang, Y., Li, H., Li, H., He, Q., Xue, Y., et al. (2016). Single nucleotide polymorphisms and osteoarthritis: An overview and a meta-analysis. *Medicine (Baltimore), 95*(7), e2811. https://doi.org/10.1097/MD.0000000000002811.

Wang, Y., Ning, H., Ren, F., Zhang, Y., Rong, Y., Wang, Y., et al. (2014). GdX/UBL4A specifically stabilizes the TC45/STAT3 association and promotes dephosphorylation of STAT3 to repress tumorigenesis. *Molecular Cell, 53*(5), 752–765. https://doi.org/10.1016/j.molcel.2014.01.020.

Wang, H., Shen, H., Wang, Y., Li, Z., Yin, H., Zong, H., et al. (2005). Overexpression of small glutamine-rich TPR-containing protein promotes apoptosis in 7721 cells. *FEBS Letters, 579*(5), 1279–1284. https://doi.org/10.1016/j.febslet.2004.12.092.

Wang, F., Song, W., Brancati, G., & Segatori, L. (2011). Inhibition of endoplasmic reticulum-associated degradation rescues native folding in loss of function protein misfolding diseases. *The Journal of Biological Chemistry, 286*(50), 43454–43464. https://doi.org/10.1074/jbc.M111.274332.

Wang, F., Whynot, A., Tung, M., & Denic, V. (2011). The mechanism of tail-anchored protein insertion into the ER membrane. *Molecular Cell, 43*(5), 738–750. https://doi.org/10.1016/j.molcel.2011.07.020.

Wiethoff, C. M., Wodrich, H., Gerace, L., & Nemerow, G. R. (2005). Adenovirus protein VI mediates membrane disruption following capsid disassembly. *Journal of Virology, 79*(4), 1992–2000. https://doi.org/10.1128/JVI.79.4.1992-2000.2005.

Winnefeld, M., Rommelaere, J., & Cziepluch, C. (2004). The human small glutamine-rich TPR-containing protein is required for progress through cell division. *Experimental Cell Research, 293*(1), 43–57.

Wunderley, L., Leznicki, P., Payapilly, A., & High, S. (2014). SGTA regulates the cytosolic quality control of hydrophobic substrates. *Journal of Cell Science, 127*(Pt. 21), 4728–4739. https://doi.org/10.1242/jcs.155648.

Xu, Y., Cai, M., Yang, Y., Huang, L., & Ye, Y. (2012). SGTA recognizes a noncanonical ubiquitin-like domain in the Bag6-UBL4A-Trc35 complex to promote endoplasmic reticulum-associated degradation. *Cell Reports*, *2*(6), 1633–1644. https://doi.org/10.1016/j.celrep.2012.11.010.

Xue, Q., Lv, L., Wan, C., Chen, B., Li, M., Ni, T., et al. (2013). Expression and clinical role of small glutamine-rich tetratricopeptide repeat (TPR)-containing protein alpha (SGTA) as a novel cell cycle protein in NSCLC. *Journal of Cancer Research and Clinical Oncology*, *139*(9), 1539–1549. https://doi.org/10.1007/s00432-013-1474-5.

Yamamoto, Y., & Sakisaka, T. (2012). Molecular machinery for insertion of tail-anchored membrane proteins into the endoplasmic reticulum membrane in mammalian cells. *Molecular Cell*, *48*(3), 387–397. https://doi.org/10.1016/j.molcel.2012.08.028.

Yang, X., Cheng, L., Li, M., Shi, H., Ren, H., Ding, Z., et al. (2014). High expression of SGTA in esophageal squamous cell carcinoma correlates with proliferation and poor prognosis. *Journal of Cellular Biochemistry*, *115*(1), 141–150. https://doi.org/10.1002/jcb.24641.

Yin, H., Wang, H., Zong, H., Chen, X., Wang, Y., Yun, X., et al. (2006). SGT, a Hsp90beta binding partner, is accumulated in the nucleus during cell apoptosis. *Biochemical and Biophysical Research Communications*, *343*(4), 1153–1158. https://doi.org/10.1016/j.bbrc.2006.03.090.

Zhao, Y., Lin, Y., Zhang, H., Manas, A., Tang, W., Zhang, Y., et al. (2015). UBL4A is required for insulin-induced Akt plasma membrane translocation through promotion of Arp2/3-dependent actin branching. *Proceedings of the National Academy of Sciences of the United States of America*, *112*(31), 9644–9649. https://doi.org/10.1073/pnas.1508647112.

Zhao, J., Wang, H., Hu, W., & Jin, Y. (2014). Effect of HLA-B-associated transcript 3 polymorphisms on lung cancer risk: A meta-analysis. *Medical Science Monitor*, *20*, 2461–2465. https://doi.org/10.12659/MSM.891141.

Zhao, Y., Zhang, H., Affonso, C., Bonomo, R., Manas, A., & Xiang, J. (2017). Deficiency in ubiquitin-like protein UBL4A impairs migration of fibroblasts and macrophages. *Biochemical and Biophysical Research Communications*, *483*(1), 617–623. https://doi.org/10.1016/j.bbrc.2016.12.094.

Zhu, T., Ji, Z., Xu, C., Peng, Z., Gu, L., Zhang, R., et al. (2014). Expression and prognostic role of SGTA in human breast carcinoma correlates with tumor cell proliferation. *Journal of Molecular Histology*, *45*(6), 665–677. https://doi.org/10.1007/s10735-014-9586-z.

CHAPTER EIGHT

A comparative computational approach toward pharmacological chaperones (NN-DNJ and ambroxol) on N370S and L444P mutations causing Gaucher's disease

D. Thirumal Kumar[a], Sharada Iyer[a], J. Priyadharshini Christy[a], R. Siva[a], Iftikhar Aslam Tayubi[b], C. George Priya Doss[a,*], Hatem Zayed[c,*]

[a]Department of Integrative Biology, School of Bio Sciences and Technology, VIT, Vellore, Tamil Nadu, India
[b]Faculty of Computing and Information Technology, King Abdulaziz University, Rabigh, Saudi Arabia
[c]Department of Biomedical Sciences, College of Health and Sciences, Qatar University, Doha, Qatar
*Corresponding authors: e-mail addresses: georgepriyadoss@vit.ac.in; hatem.zayed@qu.edu.qa

Contents

Abstract

Gaucher's disease (GD) is the most commonly known lysosomal disorder that occurs due to mutations in the β-glucocerebrosidase (GBA) protein. Our previous findings (Thirumal Kumar, Eldous, Mahgoub, George Priya Doss, Zayed, 2018) and other reports concluded

Advances in Protein Chemistry and Structural Biology, Volume 114
ISSN 1876-1623
https://doi.org/10.1016/bs.apcsb.2018.10.002

that the mutations N370S and L444P are the most significant mutations that could cause disruptions in protein stability and structure. These disruptions lead to protein misfolding and result in a diseased condition. Enzyme Replacement Therapy (ERT) and Pharmacological chaperone therapy (PCT) are currently used to treat GD caused by mutations in the GBA protein. The extreme disparity in cost between ERT and chaperone therapy, shifted the attention toward chaperone therapy. The most common chaperones in the market and trial phases to treat GD are Isofagomine, Miglustat, Eliglustat, NN-DNJ, and Ambroxol. In the era of personalized medicine, it is often necessary to understand the drug likeliness of each chaperone. In this context, the present study utilized molecular docking analysis to understand the interaction behavior of the chaperone toward the native and the two mutants N370S and L444P. The molecular dynamics simulation analyses performed on chaperones (NN-DNJ and Ambroxol) interaction showed that the chaperone NN-DNJ possesses better affinity toward the protein with N370S mutation whereas chaperone Ambroxol showed better activity against both the significant mutations (N370S and L444P). This study is expected to serve as a platform for drug repurposing.

1. Introduction

Gaucher disease (GD; *606463) is the most common lysosomal storage disorder (LSD) caused by an autosomal recessive inherited deficiency of a β-glucocerebrosidase enzyme (GBA; EC 3.2.1.45). This further leads to the accumulation of glucosylceramide, particularly in cells of the macrophage lineage (Mehta et al., 2010). The enzyme acid-β-glucosidase cleaves the glycolipid glucocerebrosidase (GlcCer) into glucose and ceramide (Brady, Kanfer, & Shapiro, 1965). Gaucher's disease is caused by mutations in the glucocerebrosidase (*GBA*) gene resulting in a non-functional enzyme. GD is a multisystem disorder that affects high-end organs like the brain, heart and kidneys (Ries et al., 2006). GD occurs in 1:57,000–1:75,000 births in the general population but is much more prevalent in the Ashkenazi Jewish population, where its incidence is estimated to be as high as 1:800 (Bohra & Nair, 2011). Based on the age and neurological features, GD is classified into three subtypes GD Type I (OMIM#230800), GD Type II (OMIM#230900) and GD Type III (OMIM#2301000). Adult-onset, GD Type I is considered to be a non-neuronopathic form and most common type among the GD subtypes. Symptoms of type I include hepatosplenomegaly, anemia, thrombocytopenia, lung disease, skeletal weakness and abnormalities of the bone (Neudorfer et al., 1996). Type I is highly prevalent in Ashkenazi Jews. GD Type II (OMIM#230900) is the acute neuronopathic type and also known as acute infantile neuropathic Gaucher's disease with a prevalence rate of 1 in 150,000. This form manifests soon after birth and the patient usually dies before or by 2 years of age (Sidransky, 1997). GD

Type III (OMIM#2301000) is also a chronic neuronopathic type, but has slower progression and a later onset. Since there are phenotypic variations in type III, it was further classified into type 3A, 3B, and 3C.

The *GBA* gene is located on the chromosome 1q22 and comprises of 11 exons and 10 introns spanning sequence length of 7.6 kb. The mature GBA enzyme contains 497 amino acids (Horowitz et al., 1989). Till date, 400 distinct mutations were reported in association with GD majority comprising of missense mutations resulting in decreased catalytic activity and/or stability of the protein (Hruska, LaMarca, Scott, & Sidransky, 2008). The most prevalent missense mutations associated with GD are N370S (1226A → G) and L444P (1448T → C) (Emre et al., 2008; Montfort, Chabas, Vilageliu, & Grinberg, 2004; Sidransky, Bottler, Stubblefield, & Ginns, 1994; Sidransky & Lopez, 2012). GBA mutations tend to develop misfolded glucocerebrosidase (GCase). This misfolded GCase affects the normal functions of the endoplasmic reticulum and increases the levels of the ubiquitin-proteasome system (UPS) and endoplasmic reticulum stress (ER) (Sawkar et al., 2005; Velayati, Yu, & Sidransky, 2010). Misfolded proteins due to missense mutations or frameshift or nonsense mutations are not transported out of the endoplasmic reticulum to the Golgi apparatus and are degraded by the degradation machinery in the endoplasmic reticulum leading to decreased enzymatic activity (Ishii, 2012). In many cases, the misfolding in the protein is relatively minor and will not have a hugely deleterious effect on the protein. They are functionally competent but less stable which leads to inefficient trafficking from the endoplasmic reticulum. Such proteins undergo endoplasmic reticulum-associated degradation and are destroyed prematurely (Bernier et al., 2006). However, to a certain extent, the use of pharmacological chaperones to transport misfolded proteins out the ER has been put to use. The chaperones are known to bind allosterically or competitively to the misfolded enzyme and facilitate its transport. Molecular chaperones are present in vivo and help native proteins fold into their proper configuration and their transport to its respective target. Chaperones are considered to oversee proteostasis, i.e., quality control of proteins. These macromolecules do not have a specific substrate but act on a variety of protein families by facilitating their folding, reconfiguration of the misfolded structures, transport and play a role in deciding the fate of the mutated proteins beyond their control (Saibil, 2013). The pharmacological chaperones are widely being used as a suitable therapy for various LSDs, specifically GD (Boyd et al., 2013; Compain et al., 2006). Enzyme replacement therapies (ERT) were understood to be the more viable option for the treatment of LSD as it simply was being a substitute for the misfolded or mutant protein.

However, there were multiple drawbacks associated with the purification of enzymes from their source, hypersensitivity and immune reactions, $<1\%$ of patients showing resistance to the therapy, requirements for intravenous infusions, accessibility and cost (Coutinho, Santos, & Alves, 2016; Cox, 2010). To overcome the disadvantages of ERT, the hypothesis of pharmacological chaperone therapy had been introduced. The effectiveness of chaperones had been demonstrated in LSDs characterized by the deficiency of glycosidases that had a role in the catabolism of glycosphingolipids in the lysosome (Compain et al., 2006).

One of the first drugs to be used was Miglustat (N-butyldeoxynojirimycin), developed as Zavesca under Oxford Glycosciences and is now marketed under orphan drug legislation by Actelion company (Cox, 2010). It was approved to be used for the treatment of GD Type I. Another approved drug as a chaperone in GD Type I is Eliglustat (Mistry, Lukina, Ben Turkia, et al., 2015). Isofagomine has been used in multiple studies to demonstrate the effects of chaperones with the protein (Sun, Liou, Xu, et al., 2012). Another N-alkylated imino sugar, N-nonyldeoxynojirimycin (NN-DNJ) was also used to study the chaperoning effect for GD with a comparison to Ambroxol (Castilla et al., 2012). Ambroxol, although a mucolytic agent for respiratory illnesses, has shown positive response as a chaperone and is under trials (Narita et al., 2016). In this study, we assessed the structural and functional impact of the reported mutations at N370 and L444 positions using combinatorial computational approach. Our study validated the efficacy of the available PCT against the selected mutations, using molecular docking and molecular simulation approaches. This study provides a possible application for bioinformatics tool in understanding the differential efficacy of action of various drugs.

2. Materials and methods

2.1 Protein and ligand modeling

Mutations N370S and L444P of the GBA protein were known to be the crucial mutations involved in the misfolding of protein. The action of these mutations were reported in our previous study (Thirumal Kumar, Eldous, Mahgoub, George Priya Doss, Zayed, 2018). The native 3D structure of a protein containing the structure length of 40–536 amino acids with the 2.40 Å resolution was obtained from the PDB (PDB ID-3GXI) (Lieberman, D'aquino, Ringe, & Petsko, 2009). The water molecules were eliminated from the protein file, and the mutation was introduced using SwissPDB Viewer (Guex & Peitsch, 1997). After introducing the individual mutations (N370S and L444P), the corresponding proteins were subjected to energy minimization to eliminate improper molecular geometries of the atoms. These mutant structures were taken for our further analysis. The canonical SMILES of ligands Isofagomine (CID 447607), Miglustat (CID 51634), Eliglustat (CID 23652731), NN-DNJ (CID 501640), Ambroxol (CID 2132) were obtained from PubChem database and converted into 3D coordinates using online Corina server.

2.2 Chaperone ADME analysis

Druglikeness of the chaperones Isofagomine, Miglustat, Eliglustat, NN-DNJ, and Ambroxol was studied using the SwissADME server (Daina, Michielin, & Zoete, 2017). The SMILES strings obtained from the PubChem database were provided as the input for the prediction.

2.3 Molecular docking

Molecular docking was performed with AutoDock v4.2 (Morris et al., 2009). The protein molecules were processed by adding all hydrogen, merging non-polar hydrogen atoms using AutoDock Tools. The charges were assigned using the Gasteiger method. The torsions were fixed for the ligand. The grid box of $60 \times 60 \times 60$ with 3.75 Å was set around the active sites (ASP-127, TRP-179, ASN-234, GLU-235, TYR-313, GLU-340, TRP-381, ASN-396) where the ligand can interact. The rigid grid box was attained using Autogrid. It was followed by AutoDock with Lamarckian Genetic Algorithm to obtain the best docking conformation. The docking was performed in two trials, and average binding energy was calculated. The pose with best binding affinity was visualized using Schrödinger Maestro.

2.4 Molecular dynamics simulations

Molecular dynamics simulations for the native-chaperone complexes and mutant-chaperone complexes were performed using the GROMACS software (Abraham et al., 2015). The topologies for the proteins were generated using the GROMOS96 53a6 force field, and the topologies for the ligands were created using ATB server (Malde et al., 2011; Oostenbrink, Villa, Mark, & van Gunsteren, 2004). The molecules were solvated in a box size of 10 Å with an explicit water box under periodic boundary conditions. 12Na + ions were replaced with the SOL molecules to maintain the neutrality of the system. The neutralized system was then subjected to equilibration using NVT and NPT ensemble for 50,000 steps each. Finally, molecular dynamics simulations were performed for 10 ns each. The trajectories were obtained as the output and files were analyzed to understand the protein behavior. The graphs were plotted using the Ubuntu version of Grace graph plotting tool.

3. Results

3.1 Chaperone ADME analysis

Absorption, Digestion, Metabolism, and Excretion (ADME) properties of the chaperones were predicted using the online SwissADME server. Various druglikeness algorithms like Lipinski, Ghose, Veber, Egan, and Muegge were used to understand the druglikeness of the chaperone. All the five chaperones had shown zero violation for the Lipinski and Egan rules. Isofagomine and Miglustat showed three and one violations for Ghose algorithm, respectively. Eliglustat showed one violation for the Veber algorithm and Isofagomine showed one violation for the Muegge algorithm. The detailed ADME results are tabulated in Table 1. Further, the brand name, 2D-Structure, FDA approval status, and phase trial identifiers are classified in Table 2.

3.2 Molecular docking

Molecular docking was performed with the aid of AutoDock standalone tool. The average binding energies between the native GBA protein with Isofagomine, Miglustat, Eliglustat, NN-DNJ, and Ambroxol were found to be −7.11, −6.84, −8.07, −7.09, and −8.66 kcal/mol, respectively. Protein with the N370S mutation was found to possess −7.08, −7.02, −7.37, −7.62, and −8.68 kcal/mol with Isofagomine, Miglustat, Eliglustat, NN-DNJ, and Ambroxol, respectively. Finally, a protein with the L444P mutation was found to possess −7.1, −6.87, −7.22, −7.74, and −8.65 kcal/mol with Isofagomine, Miglustat, Eliglustat, NN-DNJ, and Ambroxol, respectively. The residues involved in the binding are visualized in (Figs. 1–3).

3.3 Molecular dynamics simulations

Molecular dynamics simulations were performed for 10 ns using GROMACS simulation software. The obtained trajectories were analyzed to understand the Root Mean Square Deviation (RMSD), Intermolecular hydrogen bonds, and Radius of Gyration (Rg). RMSD plots were examined to understand the convergence of the molecules over the simulation period. All the three complexes: native-NN-DNJ, N370S-NN-DNJ, and L444P-NN-DNJ showed convergence at the end of the simulation period. Native with NN-DNJ showed higher deviation at ~8 ns (Fig. 4A). On the other

Table 1 Physicochemical descriptors of Isofagomine, Miglustat, Eliglustat, NN-DNJ, and Ambroxol calculated using SwissADME server.

Molecule	Isofagomine	Miglustat	Eliglustat	NN-DNJ	Ambroxol
Canonical SMILES	OCC1CNCC(C1O)O	CCCCN1CC(O)C(C(C1CO)O)O	CCCCCCCC(=O)NC(C(c1ccc2c(c1)OCCO2)O)CN1CCCC1	CCCCCCCCCN1CC(O)C(C(C1CO)O)O	OC1CCC(CC1)NCc1cc(Br)cc(c1N)Br
Formula	C6H13NO3	C10H21NO4	C23H36N2O4	C15H31NO4	C13H18Br2N2O
MW	147.17	219.28	404.54	289.41	378.1
#Heavy atoms	10	15	29	20	18
#Aromatic heavy atoms	0	0	6	0	6
Fraction Csp3	1	1	0.7	1	0.54
#Rotatable bonds	1	4	12	9	3
#H-bond acceptors	4	5	5	5	2
#H-bond donors	4	4	2	4	3
MR	39.04	59.53	118.44	83.56	81.9
TPSA	72.72	84.16	71.03	84.16	58.28
iLOGP	1.08	1.92	4.24	2.94	2.98
XLOGP3	−1.54	−0.63	3.69	2.08	2.64
WLOGP	−2.46	−1.84	2.73	0.12	3.04
MLOGP	−1.52	−1.04	1.86	0.32	2.71
Silicos-IT Log P	−0.88	−0.86	4.4	1.09	2.87

Continued

Table 1 Physicochemical descriptors of Isofagomine, Miglustat, Eliglustat, NN-DNJ, and Ambroxol calculated using SwissADME server.—cont'd

Molecule	**Isofagomine**	**Miglustat**	**Eliglustat**	**NN-DNJ**	**Ambroxol**
Consensus Log P	−1.06	−0.49	3.38	1.31	2.85
ESOL Log S	0.28	−0.54	−4.03	−2.35	−3.9
ESOL Solubility (mg/mL)	2.83E + 02	6.34E + 01	3.74E-02	1.29E + 00	4.80E − 02
ESOL Solubility (mol/L)	1.92E + 00	2.89E − 01	9.25E − 05	4.46E − 03	1.27E − 04
ESOL Class	Highly soluble	Very soluble	Moderately soluble	Soluble	Soluble
Ali Log S	0.52	−0.66	−4.87	−3.48	−3.51
Ali Solubility (mg/mL)	4.87E + 02	4.74E + 01	5.43E − 03	9.65E − 02	1.16E − 01
Ali Solubility (mol/L)	3.31E + 00	2.16E − 01	1.34E − 05	3.33E − 04	3.06E − 04
Ali Class	Highly soluble	Very soluble	Moderately soluble	Soluble	Soluble
Silicos-IT LogSw	0.71	0.6	−5.54	−1.41	−4.77
Silicos-IT Solubility (mg/mL)	7.60E + 02	8.79E + 02	1.18E − 03	1.13E + 01	6.42E − 03
Silicos-IT Solubility (mol/L)	5.16E + 00	4.01E + 00	2.91E − 06	3.92E − 02	1.70E − 05
Silicos-IT class	Soluble	Soluble	Moderately soluble	Soluble	Moderately soluble
GI absorption	Low	High	High	High	High
BBB permeant	No	No	Yes	No	Yes
Pgp substrate	No	Yes	No	Yes	Yes
CYP1A2 inhibitor	No	No	No	No	No

CYP2C19 inhibitor	No	No	No	No	No
CYP2C9 inhibitor	No	No	No	No	No
CYP2D6 inhibitor	No	No	Yes	No	Yes
CYP3A4 inhibitor	No	No	No	No	No
log Kp (cm/s)	−8.29	−8.08	−6.15	−6.59	−6.73
Lipinski #violations	0	0	0	0	0
Ghose #violations	3	1	0	0	0
Veber #violations	0	0	1	0	0
Egan #violations	0	0	0	0	0
Muegge #violations	1	0	0	0	0
Bioavailability Score	0.55	0.55	0.55	0.55	0.55
PAINS #alerts	0	0	0	0	0
Brenk #alerts	0	0	0	0	1
Leadlikeness #violations	1	1	3	1	1
Synthetic Accessibility	2.64	3.35	4.24	3.93	2.69

Table 2 General information on the chaperones Isofagomine, Miglustat, Eliglustat, NN-DNJ, and Ambroxol.

Drug name	Brand name	2D-Structure	FDA approval	Phase trial done
Isofagomine	Afegostat	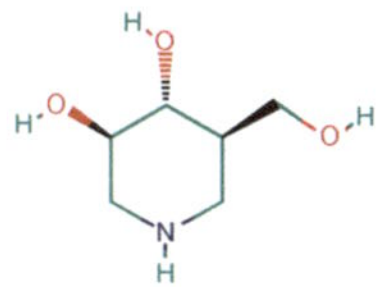	No	2 (ClinicalTrials.gov Identifier: NCT00813865)
Miglustat	Zavesca	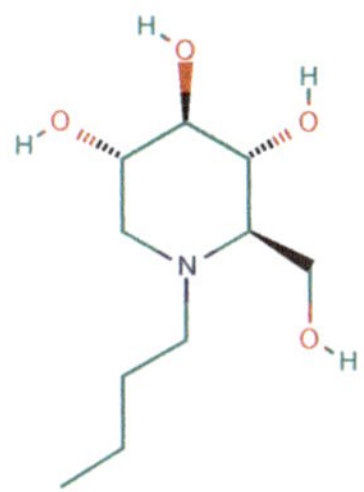	Yes	3 (ClinicalTrials.gov Identifier: NCT00319046)
Eliglustat	Cerdelga	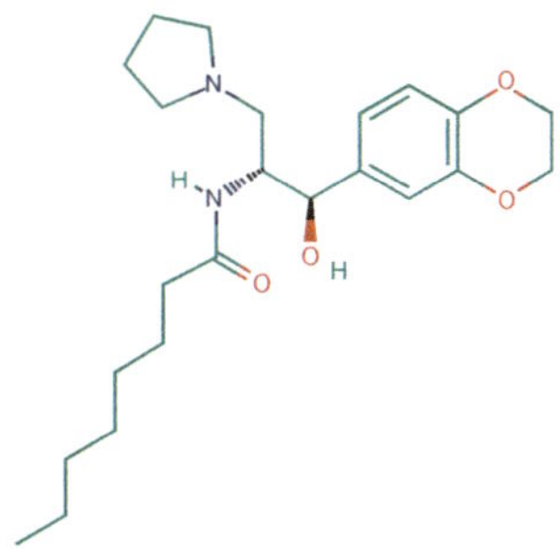	Yes	3 (ClinicalTrials.gov Identifier: NCT00891202)

NN-DNJ	–	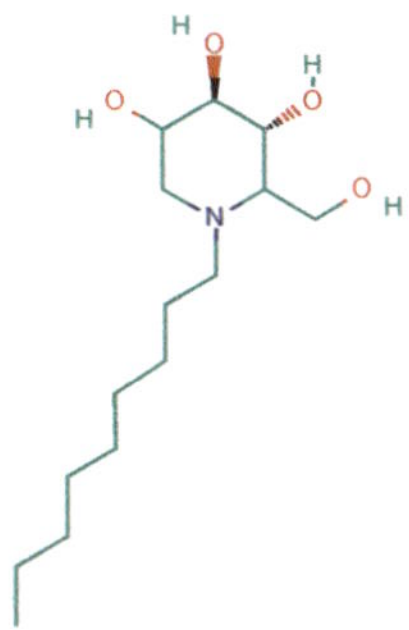	–	–
Ambroxol	Mucosolvon	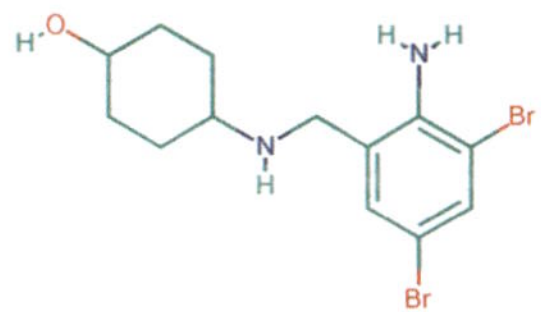	–	Ongoing (ClinicalTrials.gov Identifier: NCT01463215)

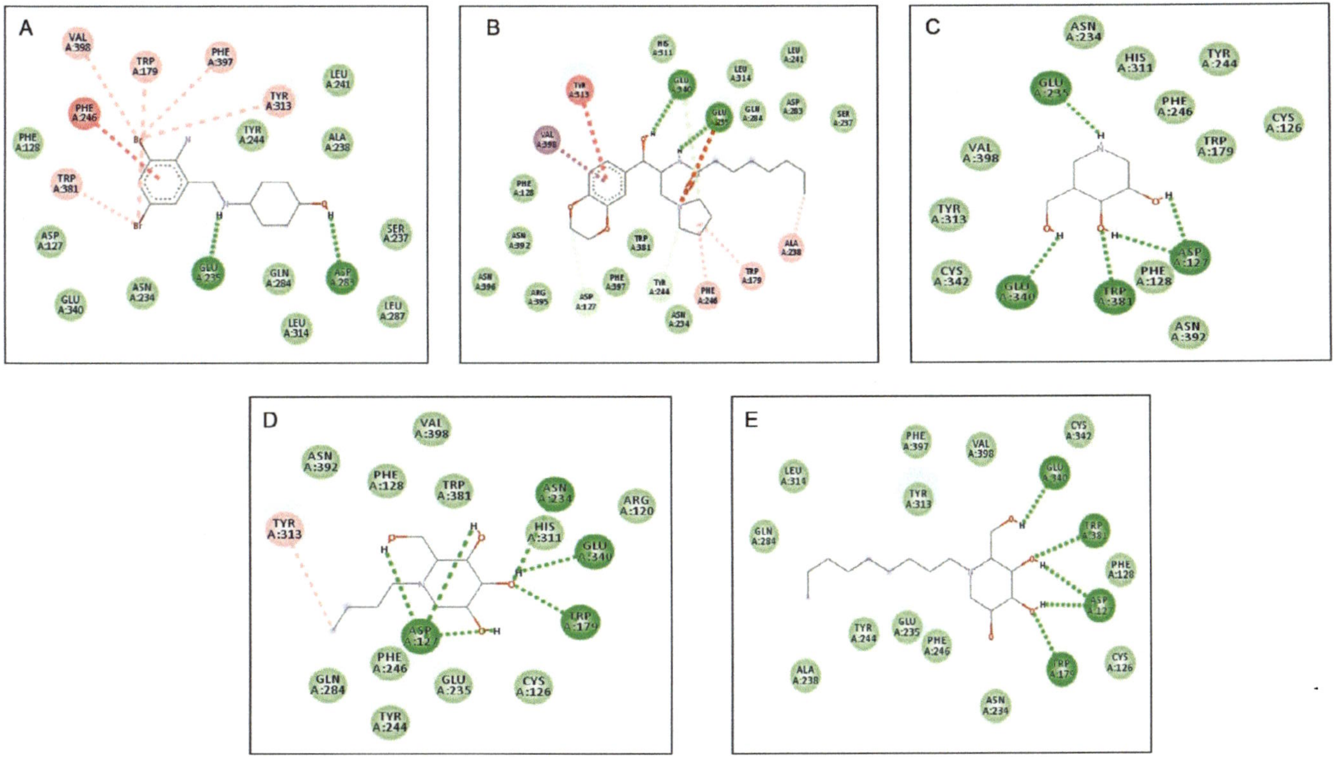

Fig. 1 Visualization of binding amino acids of the native protein to the chaperones. (A) Ambroxol, (B) Eliglustat, (C) Isofagomine, (D) Miglustat, and (E) NN-DNJ.

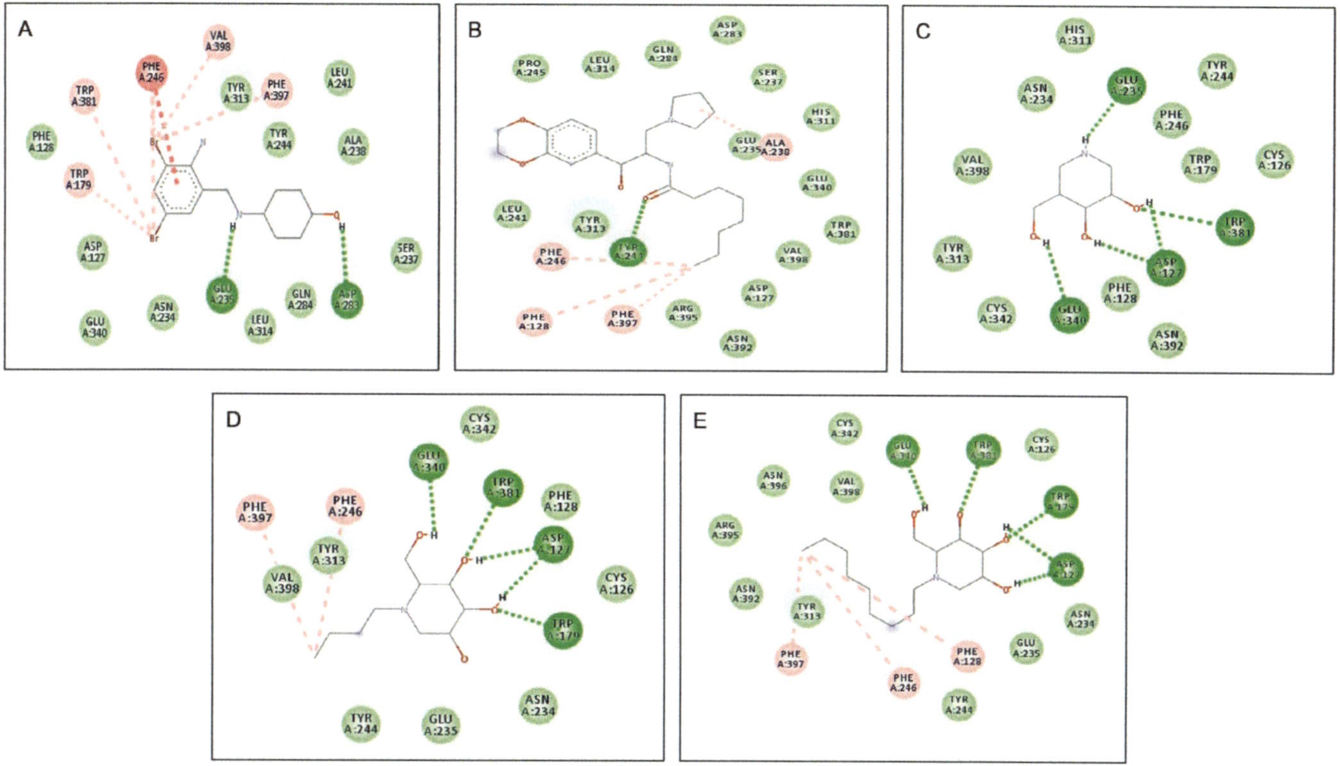

Fig. 2 Visualization of binding amino acids of the protein with L444P mutation to the chaperones. (A) Ambroxol, (B) Eliglustat, (C) Isofagomine, (D) Miglustat, and (E) NN-DNJ.

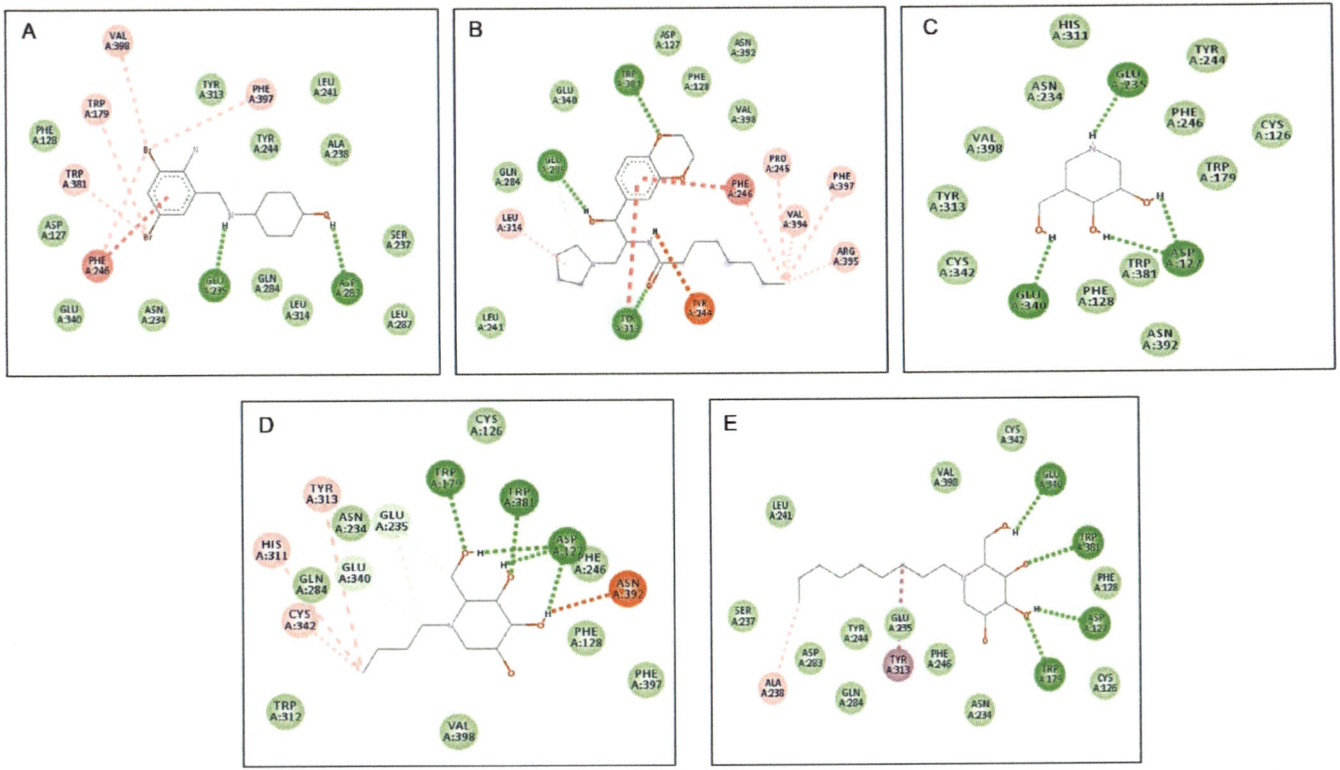

Fig. 3 Visualization of binding amino acids of the protein with N370S mutation to the chaperones. (A) Ambroxol, (B) Eliglustat, (C) Isofagomine, (D) Miglustat, and (E) NN-DNJ.

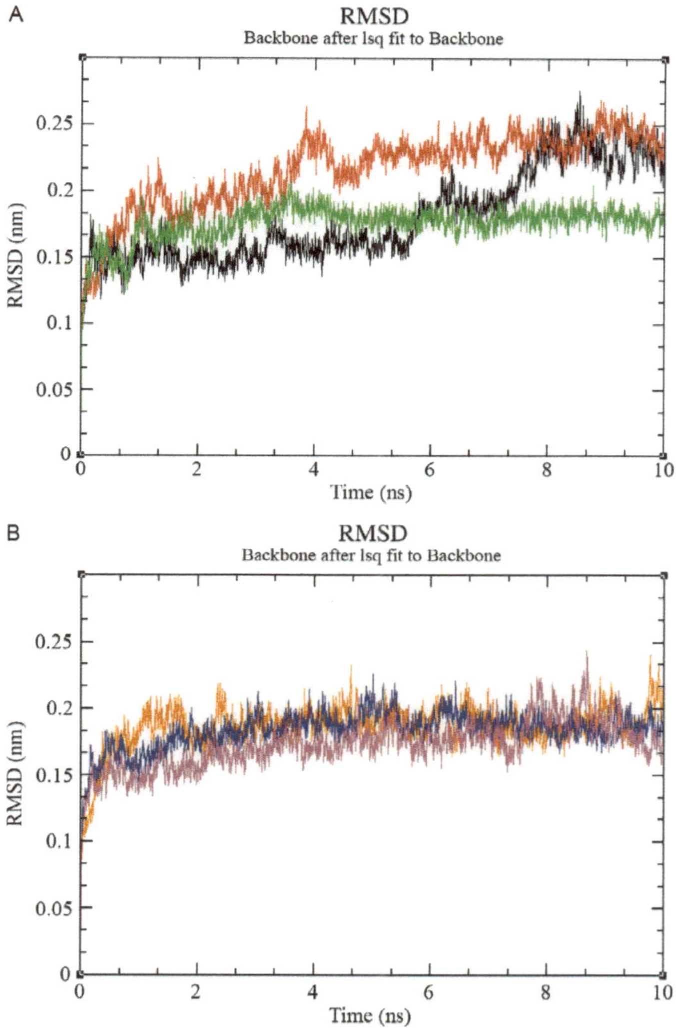

Fig. 4 Root Mean Square Deviation of the protein complexed with chaperones. (A) With NN-DNJ chaperone. Color Scheme: Black—Native protein and NN-DNJ, Green—Protein with L444P mutation and NN-DNJ, Red—Protein with N370S mutation and NN-DNJ. (B) With Ambroxol chaperone. Color Scheme: Orange—Native protein and Ambroxol, Blue—Protein with N370S mutation and Ambroxol, Magenta—Protein with L444P mutation and Ambroxol.

hand, the mutant protein molecules with Ambroxol showed similar deviation pattern throughout the simulation period, and all the molecules have attained convergence at the end of the simulation period (Fig. 4B). The obtained convergence provided a satisfactory condition for the further analysis with the trajectories. The number of intermolecular hydrogen bonds between the protein molecules and the chaperones was analyzed from the trajectories. From the hydrogen bond plot, we observed that the N370S-NN-DNJ complex had possessed the maximum number of intermolecular hydrogen bonds (three hydrogen bonds) over the simulation period (Fig. 5A). On the other hand, a protein with L444P mutation and Ambroxol complex had possessed a maximum number of intermolecular

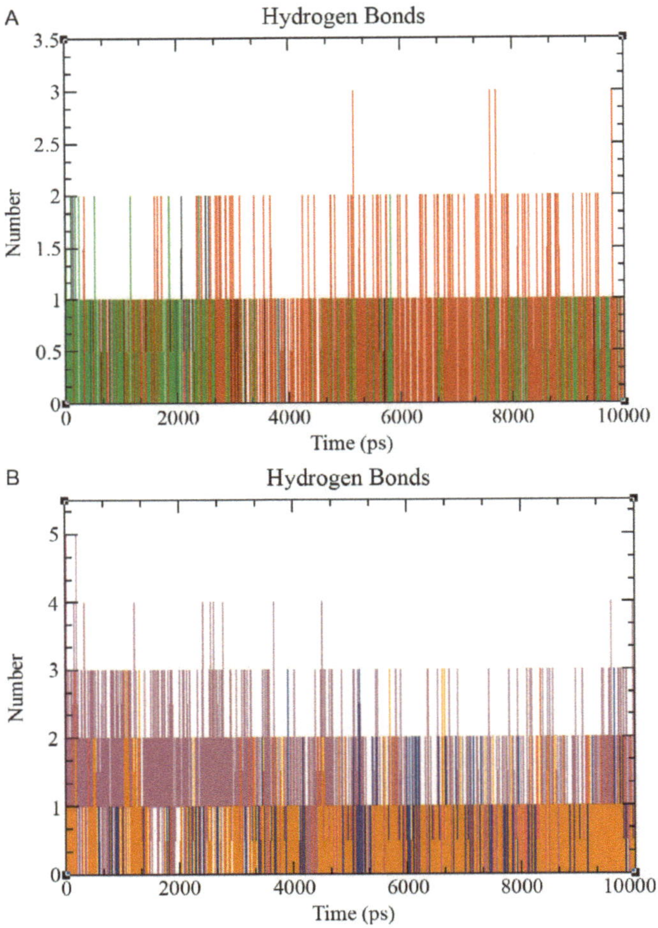

Fig. 5 Change in number of intermolecular in the protein complexed with chaperones. (A) With NN-DNJ chaperone. Color Scheme: Black—Native protein and NN-DNJ, Green—Protein with L444P mutation and NN-DNJ, Red—Protein with N370S mutation and NN-DNJ. (B) With Ambroxol chaperone. Color Scheme: Orange—Native protein and Ambroxol, Blue—Protein with N370S mutation and Ambroxol, Magenta—Protein with L444P mutation and Ambroxol.

hydrogen bonds (five hydrogen bonds) when compared to the other protein molecules (Fig. 5B). Finally, the radius of gyration was analyzed from the trajectories. From the Rg plot, we observed that the native-NN-DNJ have possessed the least Rg and N370S-NN-DNJ have possessed the maximum Rg. The L444P-NN-DNJ have possessed a Rg values in between the other two complexes (native-NN-DNJ and N370S-NN-DNJ) (Fig. 6A). On the

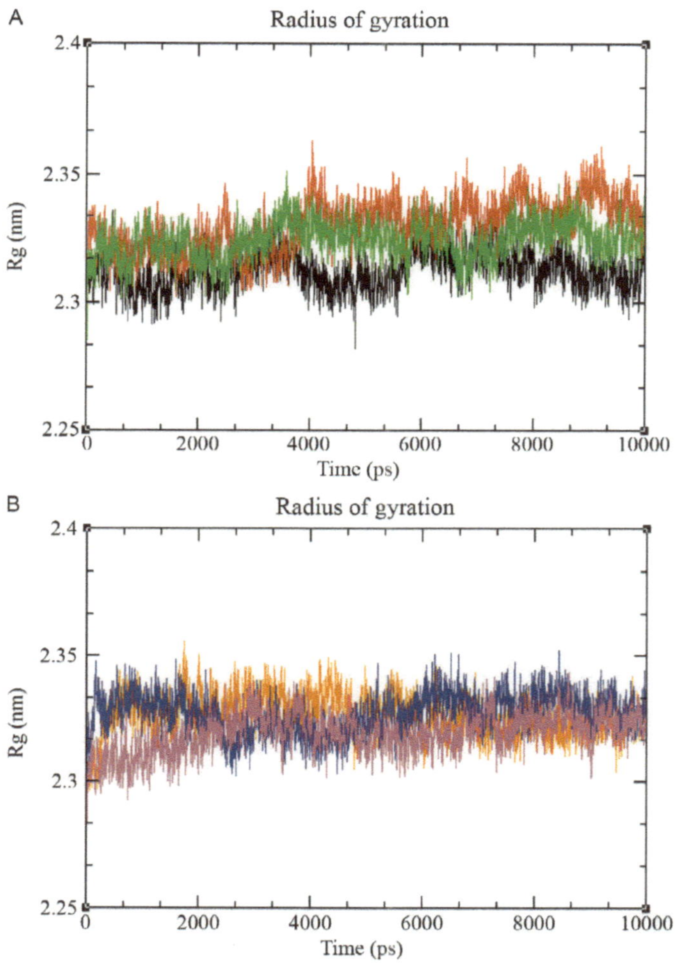

Fig. 6 Change in radius of gyration of the protein complexed with chaperones. (A) With NN-DNJ chaperone. Color Scheme: Black—Native protein and NN-DNJ, Green—Protein with L444P mutation and NN-DNJ, Red—Protein with N370S mutation and NN-DNJ. (B) With Ambroxol chaperone. Color Scheme: Orange—Native protein and Ambroxol, Blue—Protein with N370S mutation and Ambroxol, Magenta—Protein with L444P mutation and Ambroxol.

other hand, all the three protein molecules (Native, a protein with N370S mutation, and protein with L444P mutation) have shown similar Rg pattern throughout the simulation period for Ambroxol complexes.

4. Discussion

Understanding and predicting the molecular pathway of a disease is a very challenging task (Botstein & Risch, 2003). In most genetic diseases mutational heterogeneity plays a significant role with a single amino acid substitutions having multiple effects on the protein. They can render the active site in substrate binding, decreasing the stability leading to protein aggregation and misfolding, decreasing the enzyme leading to accumulation of substrate in a toxic manner and finally changing the whole conformation of the protein non-synonymous Single Nucleotide Polymorphisms (nsSNPs). The occurrence of coding nsSNPs in the genome is relatively less, but they significantly affect the structure and function of the protein. These amino acid variations can disturb the conformation of the protein as nsSNPs can disrupt active sites and bring about deleterious changes in the size, charge, solvent accessibility, and hydrogen bonds. These changes cause loss of thermodynamic stability as well as aberrant folding and aggregation of proteins. nsSNPs can render proteins non-functional and are the causes of several proteins misfolding diseases. Protein misfolding is a significant factor in lysosomal storage disorders; especially Gaucher's disease which has been on a gradual rise in the number of affected cases over the past years. nsSNPs are long known for its association with possible changes in drug response and can also elicit an adverse reaction to the prescribed drugs. This can be due to the presence of nsSNPs in substrate binding pockets, active sites or drug binding pockets. On the other hand, the presence of a few nsSNPs proves to be beneficial to the patients, as they result in an improved outcome for patients with the disease. More and more research material has demonstrated that drug response depends on various factors like genetics, ethnicity, and occurrence of nsSNPs (Roden, Wilke, Kroemer, & Stein, 2011). Therefore more detailed research and information in these fields would help enhance drug delivery, drug response and reduce or eliminate the drug-mediated adverse events.

Due to the above reasons, it is of utmost importance to know which mutation would respond to PCT. Few mutations only render the protein unable to fold correctly, while still maintaining the full catalytic activity. If the protein has lost its catalytic activity or its ability to bind to the substrate due to the mutation, then a chaperone will not be able to rescue such a protein (Valenzano et al., 2011). Therefore it is crucial to document the mutations occurring in various populations, which sure is a daunting task. Then these mutations should be screened, so as to deduce the mutations that would respond to PCT. The biggest challenge scientists face is to understand how population genetics, inheritance patterns, variations in the human genome, and ethnicity feature in drug response. The current world is fast-paced; the number of diseases crippling the human race is on the rise. This has caused a never before seen urgency for results, for drugs, to combat the various diseases that plague the human race. To face this demand and to accelerate disease treatment, the use of a combination of bioinformatic tools could be the immediate solution to enhance this process.

We recently used combinatorial computational tools to recommend certain mutations as potential drug targets (Agrahari, Kumar, Siva, Zayed, George Priya Doss, 2018; Sneha, Thirumal, George Priya Doss, Siva, & Zayed, 2017; Sneha et al., 2018; Thirumal Kumar, Jerushah Emerald, et al., 2018) and developed computational models to evaluate the potential impact of nsSNPs on drug response (George Priya Doss, Chiranjib, Hailong, & Luonan, 2014; George Priya Doss, Rajith, Chiranjib, & Hailong, 2014; GeorgePriya Doss, Rajith, Chiranjib, Balaji, et al., 2014; Nagasundaram et al., 2016, 2015; Thirumal Kumar et al., 2016). In this context, molecular docking and molecular dynamics simulation studies have played a major role in providing molecular insights in the protein-drug interaction studies (Cite PMID: 29978341). These studies shed light on the nature of interaction between ligands and their target (Kumar and Doss, 2016; Thirumal Kumar and George Priya Doss, 2017), which might impact the quality of drug design. The possible treatments for the GD are the ERT and the PCT (Narita et al., 2016). Among these two treatments, the chaperone-based treatment helps in ERT and has become very common as it is quite cost efficient and less laborious (Porto et al., 2009). N370S and L444P mutations of GBA protein were found to be the most significant mutations that involve in the disruption of protein misfolding (Lesage et al., 2011). This was computationally analyzed in our previous work (Thirumal Kumar, Eldous, et al., 2018). The N370S mutation is the most frequently reported mutation in Ashkenazi Jews associated with Type 1 GD (Dimitriou, Moraitou, Troungos, Schulpis, & Michelakakis, 2010; Mata et al., 2008; Sidransky & Lopez, 2012; Walley et al., 1993; Zimran et al., 1989). There is some generalization that the N370S mutation is considered to be neuroprotective as it does not result in neurological symptoms but some forms of Parkinson's showed the presence of this mutation (Goker-Alpan et al., 2005). The L444P is found in all three types of GD and reported to cause type II and type III as homozygotes, but there were cases reported to cause disease as a compound heterozygote and noted to be a mutational hotspot (Saranjam et al., 2013). It is more prevalent in non-Jewish populations (Beutler, 2006). Common PCTs like Isofagomine, Miglustat, Eliglustat, NN-DNJ, and Ambroxol were known to treat the Gaucher's disease. Drug likeliness of the chaperones was analyzed using the SwissADME server (Daina et al., 2017). Similarly, this study uses powerful approaches which identifies the relationship between genetics, pharmacological chaperones, and protein on a structural and functional level in GD. We have used molecular docking studies to understand the interaction of this chaperone with the native and mutant N370S and L444P proteins. The molecular docking analysis was carried out using the AutoDock standalone software. The binding pattern is expressed in terms of binding energy. For a better binding affinity, the binding energy should be lower (Morris et al., 2009; Thirumal Kumar et al., 2017). From the overall interaction analysis, we observed that the Ambroxol exhibited better binding affinity and interaction with the protein molecules when compared to the other chaperones (Isofagomine, Miglustat, Eliglustat, & NN-DNJ) (Figs. 1–3).

Molecular dynamics simulations were performed using the GROMACS simulation package (Abraham et al., 2015). Essential parameters like RMSD, Intermolecular hydrogen bonds, and Radius of Gyration (Rg) were analyzed from the obtained 10 ns trajectories to study the behavior of the protein-chaperone complexes over the period of 10 ns. RMSD was studied to understand the deviation of the mutation-chaperone complexes from the native-chaperone complex. Native protein with NN-DNJ showed a bit higher deviation with mutant proteins with NN-DNJ complexes (Fig. 4A). On the hand, in the case of Ambroxol interaction with native and mutant GBA proteins, all the three molecules showed a similar deviation pattern (Fig. 4B). However, all the complexes have shown convergence at the end of the simulation period. The obtained convergence provided a satisfactory condition for the further analysis with the trajectories. Hydrogen bonds play a significant role in maintaining the interaction between the protein and ligand (Pace et al., 2014; Wade and Goodford, 1989). The number of intermolecular hydrogen bonds between the protein molecules and the chaperones was analyzed from the trajectories. In the case of protein with N370S mutation, it was found to possess a higher number of intermolecular hydrogen bonds with NN-DNJ (Fig. 5A). On the other hand, a protein with L444P mutation was found to possess a higher number of intermolecular hydrogen bonds with Ambroxol (Fig. 5B). The compactness of the protein changes with the interacting molecules. The compactness of the protein increases with the decrease in radius of gyration and vice versa (Lobanov, Bogatyreva, & Galzitskaia, 2008; Thirumal Kumar, Jerushah Emerald, et al., 2018). Chaperone NN-DNJ complexed to a protein with an N370S mutation possesses better compactness when compared to the native and protein with L444P mutation (Fig. 6A). On the other hand, the chaperone Ambroxol has shown similar compactness with the native and mutant proteins elucidating the same activity.

5. Conclusion

In this study, we used the two pathogenic GBA mutant proteins in N370S and L444P to dissect the interaction patterns computationally, with the currently used chaperone molecules (Isofagomine, Miglustat, Eliglustat, NN-DNJ, and Ambroxol) for treatment of GD, compared to the native protein. The chaperone Ambroxol was found to have better binding interaction with the native and protein with N370S and L444P mutations. Molecular dynamics showed the NN-DNJ complex possessed better action toward the protein with N370S mutant, whereas Ambroxol showed better activity against both the mutations. This computational study provides a platform to identify the relationship between pharmacological chaperones and the GBA protein on a structural and functional level. Therefore, more detailed research and information in these fields would help enhance the drug delivery, drug response and decrease or cease the adverse reactions caused by the drug.

Acknowledgments

The authors acknowledge the management of Vellore Institute of Technology, Vellore, India, and BRAF @ CDAC for providing the facilities required to perform this work.

Disclosure statement

No potential conflict of interest exists.

References

Abraham, M. J., Murtola, T., Schulz, R., Páll, S., Smith, J. C., Hess, B., et al. (2015). GROMACS: High-performance molecular simulations through multi-level parallelism from laptops to supercomputers. *SoftwareX*, *1–2*, 19–25.

Agrahari, A. K., Kumar, A., Siva, R., Zayed, H., & George Priya Doss, C. (2018). Substitution impact of highly conserved arginine residue at position 75 in GJB1 gene in association with X-linked Charcot–Marie-tooth disease: A computational study. *Journal of Theoretical Biology*, *437*, 305–317.

Bernier, V., Morello, J. P., Zarruk, A., Debrand, N., Salahpour, A., Lonergan, M., et al. (2006). Pharmacologic chaperones as a potential treatment for X-linked nephrogenic diabetes insipidus. *Journal of the American Society of Nephrology*, *17*(1), 232–243. https://doi.org/10.1681/ASN.2005080854.

Beutler, E. (2006). Gaucher disease: Multiple lessons from a single gene disorder. *Acta Paediatrica*, *95*(451), 103–109. https://doi.org/10.1080/08035320600619039.

Bohra, V., & Nair, V. (2011). Gaucher's disease. *Indian Journal of Endocrinology and Metabolism*, *15*(3), 182–186. https://doi.org/10.4103/2230-8210.83402.

Botstein, D., & Risch, N. (2003). Discovering genotypes underlying human phenotypes: Past successes for mendelian disease, future approaches for complex disease. *Nature Genetics*, *33*, 228–237.

Boyd, R. E., Lee, G., Rybczynski, P., Benjamin, E. R., Khanna, R., Wustman, B. A., et al. (2013). Pharmacological chaperones as therapeutics for lysosomal storage diseases. *Journal of Medicinal Chemistry*, *56*(7), 2705–2725. https://doi.org/10.1021/jm301557k.

Brady, R. O., Kanfer, J., & Shapiro, D. (1965). The metabolism of glucocerebrosides. I. Purification and properties of a glucocerebroside-cleaving enzyme from spleen tissue. *The Journal of Biological Chemistry*, *240*, 39–43.

Castilla, J., Rísquez, R., Cruz, D., Higaki, K., Nanba, E., Ohno, K., et al. (2012). Conformationally-locked N-glycosides with selective β-glucosidase inhibitory activity: Identification of a new non-iminosugar-type pharmacological chaperone for Gaucher disease. *Journal of Medicinal Chemistry*, *55*(15), 6857–6865. https://doi.org/10.1021/jm3006178.

Compain, P., Martin, O. R., Boucheron, C., Godin, G., Yu, L., Ikeda, K., et al. (2006). Design and synthesis of highly potent and selective pharmacological chaperones for the treatment of Gaucher's disease. *Chembiochem*, 7(9), 1356–1359. https://doi.org/10.1002/cbic.200600217.

Coutinho, M. F., Santos, J. I., & Alves, S. (2016). Less is more: Substrate reduction therapy for lysosomal storage disorders. *International Journal of Molecular Sciences*, *17*(7), 1065. https://doi.org/10.3390/ijms17071065.

Cox, T. M. (2010). Gaucher disease: Clinical profile and therapeutic developments. *Biologics: Targets & Therapy*, *4*, 299–313. https://doi.org/10.2147/BTT.S7582.

Daina, A., Michielin, O., & Zoete, V. (2017). SwissADME: A free web tool to evaluate pharmacokinetics, drug-likeness and medicinal chemistry friendliness of small molecules. *Scientific Reports*, 7, 42717.

Dimitriou, E., Moraitou, M., Troungos, C., Schulpis, K., & Michelakakis, H. (2010). Gaucher disease: Frequency of the N370S mutation in the Greek population. *Clinical Genetics*, *78*, 195–196.

Emre, S., Gurakan, F., Yuce, A., Rolf, A., Scott, R., & Ozen, H. (2008). Molecular analysis of Turkish Gaucher disease patients: Identification of novel mutations in glucocerebrosidase (GBA) gene. *European Journal of Medical Genetics*, *51*, 315–321.

George Priya Doss, C., Chiranjib, C., Hailong, Z., & Luonan, C. (2014). Integrating in silico prediction methods, molecular docking, and molecular dynamics simulation to predict the impact of ALK missense mutations in structural perspective. *BioMed Research International*, *2014*, 895831.

George Priya Doss, C., Rajith, B., Chiranjib, C., & Hailong, Z. (2014). Structural signature of the G719S-T790M double mutation in the EGFR kinase domain and its response to inhibitors. *Scientific Reports*, *4*, 5868.

GeorgePriya Doss, C., Rajith, B., Chiranjib, C., Balaji, V., Maitreya, D., Sneha, M., et al. (2014). In silico profiling and structural insights of missense mutations in RET protein kinase domain by molecular dynamics and docking approach. *Molecular BioSystems*, *10*, 421.

Goker-Alpan, O., Hruska, K. S., Orvisky, E., Kishnani, P. S., Stubblefield, B. K., Schiffmann, R., et al. (2005). Divergent phenotypes in Gaucher disease implicate the role of modifiers. *Journal of Medical Genetics*, *42*(6), e37.

Guex, N., & Peitsch, M. C. (1997). SWISS-MODEL and the Swiss-Pdb Viewer: An environment for comparative protein modeling. *Electrophoresis*, *18*(15), 2714–2723. https://doi.org/10.1002/elps.1150181505.

Horowitz, M., Wilder, S., Horowitz, Z., Reiner, O., Gelbart, T., & Beutler, E. (1989). The human glucocerebrosidase gene and pseudogene: Structure and evolution. *Genomics*, *4*, 87–96.

Hruska, K. S., LaMarca, M. E., Scott, C. R., & Sidransky, E. (2008). Gaucher disease: Mutation and polymorphism spectrum in the glucocerebrosidase gene (GBA). *Human Mutation*, *29*, 567–583.

Ishii, S. (2012). Pharmacological chaperone therapy for Fabry disease. Suzuki K., ed. *Proceedings of the Japan Academy. Series B, Physical and Biological Sciences*, *88*(1), 18–30. https://doi.org/10.2183/pjab.88.18.

Kumar, D. T., & Doss, C. G. P. (2016). Investigating the inhibitory effect of Wortmannin in the hotspot mutation at codon 1047 of PIK3CA kinase domain: A molecular docking and molecular dynamics approach. *Advances in Protein Chemistry and Structural Biology*, *102*, 267–297.

Lesage, S., Anheim, M., Condroyer, C., Pollak, P., Durif, F., Dupuits, C., et al. (2011). Large-scale screening of the Gaucher's disease-related glucocerebrosidase gene in Europeans with Parkinson's disease. *Human Molecular Genetics*, *20*(1), 202–210.

Lieberman, R. L., D'aquino, J. A., Ringe, D., & Petsko, G. A. (2009). Effects of pH and iminosugar pharmacological chaperones on lysosomal glycosidase structure and stability. *Biochemistry*, *48*, 4816–4827. https://doi.org/10.1021/bi9002265.

Lobanov, M. I., Bogatyreva, N. S., & Galzitskaia, O. V. (2008). Radius of gyration is indicator of compactness of protein structure. *Molekuliarnaia Biologiia (Mosk)*, *42*(4), 701–706.

Malde, A. K., Zuo, L., Breeze, M., Stroet, M., Poger, D., Nair, P. C., et al. (2011). An automated force field topology builder (ATB) and repository: Version 1.0. *Journal of Chemical Theory and Computation*, 7(12), 4026–4037.

Mata, I. F., Samii, A., Schneer, S. H., Roberts, J. W., Griffith, A., Leis, B. C., et al. (2008). Glucocerebrosidase gene mutations: A risk factor for Lewy body disorders. *Archives of Neurology*, *65*, 379–382.

Mehta, A., Beck, M., Eyskens, F., Feliciani, C., Kantola, I., Ramaswami, U., et al. (2010). Fabry disease: A review of current management strategies. *QJM*, *103*(9), 641–659.

Mistry, P. K., Lukina, E., Ben Turkia, H., et al. (2015). Effect of oral eliglustat on splenomegaly in patients with Gaucher disease type 1The ENGAGE randomized clinical trial. *Journal of the American Medical Association*, *313*(7), 695–706. https://doi.org/10.1001/jama.2015.459.

Montfort, M., Chabas, A., Vilageliu, L., & Grinberg, D. (2004). Functional analysis of 13 GBA mutant alleles identified in Gaucher disease patients: pathogenic changes and 'modifier' polymorphisms. *Human Mutation*, *23*, 567–575. Note: Erratum: Human Mutation 26: 276 only, 2005.

Morris, G. M., Huey, R., Lindstrom, W., Sanner, M. F., Belew, R. K., Goodsell, D. S., et al. (2009). AutoDock4 and AutoDockTools4: Automated docking with selective receptor flexibility. *Journal of Computational Chemistry*, *30*(16), 2785–2791.

Nagasundaram, N., George Priya Doss, C., Chiranjib, C., Karthick, V., Thirumal Kumar, D., Balaji, V., et al. (2016). Mechanism of artemisinin resistance for malaria PfATP6 L263 mutations and discovering potential antimalarials: An integrated computational approach. *Scientific Reports*, *6*, 30106.

Nagasundaram, N., Hailong, Z., Jiming, L., Karthick, V., George Priya Doss, C., Chiranjib, C., et al. (2015). Analyzing the effect of mutation on protein function and discovering potential inhibitors of CDK4: Molecular modeling and dynamics studies. *PLoS One*, *10*(8), e0133969.

Narita, A., Shirai, K., Itamura, S., Matsuda, A., Ishihara, A., Matsushita, K., et al. (2016). Ambroxol chaperone therapy for neuronopathic Gaucher disease: A pilot study. *Annals of Clinical Translational Neurology*, *3*(3), 200–215. https://doi.org/10.1002/acn3.292. eCollection 2016 Mar.

Neudorfer, O., Giladi, N., Elstein, D., Abrahamov, A., Turezkite, T., Aghai, E., et al. (1996). Occurrence of Parkinson's syndrome in type 1 Gaucher disease. *QJM*, *89*(9), 691–694.

Oostenbrink, C., Villa, A., Mark, A. E., & van Gunsteren, W. F. (2004). A biomolecular force field based on the free enthalpy of hydration and solvation: The GROMOS force-field parameter sets 53A5 and 53A6. *Journal of Computational Chemistry*, *25*(13), 1656–1676.

Pace, C. N., Fu, H., Lee Fryar, K., Landua, J., Trevino, S. R., Schell, D., et al. (2014). Contribution of hydrogen bonds to protein stability. *Protein Science*, *23*(5), 652–661.

Porto, C., Cardone, M., Fontana, F., Rossi, B., Tuzzi, M. R., Tarallo, A., et al. (2009). The pharmacological chaperone N-butyldeoxynojirimycin enhances enzyme replacement therapy in Pompe disease fibroblasts. *Molecular Therapy*, *17*(6), 964–971.

Ries, M., Schaefer, E., Lührs, T., Mani, L., Kuhn, J., Vanier, M. T., et al. (2006). Critical assessment of chitotriosidase analysis in the rational laboratory diagnosis of children with Gaucher disease and Niemann–Pick disease type A/B and C. *Journal of Inherited Metabolic Disease*, *29*(5), 647–652. https://doi.org/10.1007/s10545-006-0363-3.

Roden, D. M., Wilke, R. A., Kroemer, H. K., & Stein, C. M. (2011). Pharmacogenomics: The genetics of variable drug responses. *Circulation*, *123*(15), 1661–1670. https://doi.org/10.1161/CIRCULATIONAHA.109.914820.

Saibil, H. (2013). Chaperone machines for protein folding, unfolding and disaggregation. *Nature Reviews: Molecular Cell Biology*, *14*(10), 630–642. https://doi.org/10.1038/nrm3658.

Saranjam, H., Chopra, S. S., Levy, H., Stubblefield, B. K., Maniwang, E., Cohen, I. J., et al. (2013). A germline or de novo mutation in two families with Gaucher disease: Implications for recessive disorders. *European Journal of Human Genetics*, *21*, 115–117.

Sawkar, A. R., Adamski-Werner, S. L., Cheng, W. C., Wong, C. H., Beutler, E., Zimmer, K. P., et al. (2005). Gaucher disease-associated glucocerebrosidases show mutation-dependent chemical chaperoning profiles. *Chemistry & Biology*, *12*, 1235–1244.

Sidransky, E. (1997). New perspectives in type 2 Gaucher disease. *Advances in Pediatrics*, *44*, 73–107.

Sidransky, E., Bottler, A., Stubblefield, B., & Ginns, E. I. (1994). DNA mutational analysis of type 1 and type 3 Gaucher patients: How well do mutations predict phenotype? *Human Mutation*, *3*, 25–28.

Sidransky, E., & Lopez, G. (2012). The link between the GBA gene and parkinsonism. *Lancet Neurology*, *11*, 986–998.

Sneha, P., Thirumal, K. D., George Priya Doss, C., Siva, R., & Zayed, H. (2017). Determining the role of missense mutations in the POU domain of HNF1A that reduce the DNA-binding affinity: A computational approach. *PLoS One*, *12*(4), 1–24.

Sneha, P., Zenith, T. U., Abu Habib, U. S., Evangeline, J., Thirumal Kumar, D., George Priya Doss, C., et al. (2018). Impact of missense mutations in survival motor neuron protein (SMN1) leading to spinal muscular atrophy (SMA): A computational approach. *Metabolic Brain Disease*. https://doi.org/10.1007/s11011-018-0285-4.

Sun, Y., Liou, B., Xu, Y.-H., et al. (2012). Ex vivo and in vivo effects of isofagomine on acid β-glucosidase variants and substrate levels in Gaucher disease. *The Journal of Biological Chemistry*, *287*(6), 4275–4287. https://doi.org/10.1074/jbc.M111.280016.

Thirumal Kumar, D., Eldous, H. G., Mahgoub, Z. A., George Priya Doss, C., & Zayed, H. (2018). Computational modelling approaches as a potential platform to understand the molecular genetics association between Parkinson's and Gaucher diseases. *Metabolic Brain Disease*. https://doi.org/10.1007/s11011-018-0286-3.

Thirumal Kumar, D., & George Priya Doss, C. (2017). Role of E542 and E545 missense mutations of PIK3CA in breast cancer: A comparative computational approach. *Journal of Biomolecular Structure & Dynamics*, *35*(12), 2745–2757.

Thirumal Kumar, D., George Priya Doss, C., Sneha, P., Tayubi, I. A., Siva, R., Chiranjib, C., et al. (2016). Influence of V54 M mutation in Giant muscle protein titin: A computational screening and molecular dynamics approach. *Journal of Biomolecular Structure & Dynamics, 35*(5), 917–928 (IF 2.919).

Thirumal Kumar, D., Jerushah Emerald, L., George Priya Doss, C., Sneha, P., Siva, R., Charles Emmanuel Jebaraj, W., et al. (2018). Computational approach to unravel the impact of missense mutations of proteins (D2HGDH and IDH2) causing D-2-hydroxyglutaric aciduria 2. *Metabolic Brain Disease*. https://doi.org/10.1007/s11011-018-0278-3.

Thirumal Kumar, D., Lavanya, P., George Priya Doss, C., Tayubi, I. A., Naveen Kumar, D. R., Francis Yesurajan, I., et al. (2017). A molecular docking and dynamics approach to screen potent inhibitors against Fosfomycin resistant enzyme in clinical Klebsiella pneumoniae. *Journal of Cellular Biochemistry, 118*, 4088–4094. https://doi.org/10.1002/jcb.26064.

Valenzano, K. J., Khanna, R., PoweJr, A. C., Boyd, R., Lee, G., Flanagan, J. J., et al. (2011). Identification and characterization of pharmacological chaperones to correct enzyme deficiencies in lysosomal storage disorders. *Assay and Drug Development Technologies, 9*(3), 213–235.

Velayati, A., Yu, W. H., & Sidransky, E. (2010). The role of glucocerebrosidase mutations in Parkinson disease and Lewy body disorders. *Current Neurology and Neuroscience Reports, 10*(3), 190–198.

Wade, R. C., & Goodford, P. J. (1989). The role of hydrogen-bonds in drug binding. *Progress in Clinical and Biological Research, 289*, 433–444.

Walley, A. J., Barth, M. L., Ellis, I., Fensom, A. H., & Harris, A. (1993). Gaucher's disease in the United Kingdom: Screening non-Jewish patients for the two common mutations. *Journal of Medical Genetics, 30*, 280–283.

Zimran, A., Sorge, J., Gross, E., Kubitz, M., West, C., & Beutler, E. (1989). Prediction of severity of Gaucher's disease by identification of mutations at DNA level. *Lancet, 334*, 349–352.

Further reading

Bendl, J., Stourac, J., Salanda, O., Pavelka, A., Wieben, E. D., Zendulka, J., et al. (2014). PredictSNP: Robust and accurate consensus classifier for prediction of disease-related mutations. *PLoS Computational Biology, 10*, e1003440.

Chen, C. W., Lin, J., & Chu, Y. W. (2013). iStable: Off-the-shelf predictor integration for predicting protein stability changes. *BMC Bioinformatics, 14*(Suppl. 2), S5.

De Baets, G., Van Durme, J., Reumers, J., Maurer-Stroh, S., Vanhee, P., Dopazo, J., et al. (2012). SNPeffect 4.0: On-line prediction of molecular and structural effects of protein-coding variants. *Nucleic Acids Research, 40*(Database issue), D935–D939. https://doi.org/10.1093/nar/gkr996.

Gaucher, P. C. (2003). *De l'epithelioma primitif de la rate, hypertrophie idiopathique de la rate sans leucemie*. Paris: BIUM.

Glaser, F., Pupko, T., Paz, I., Bell, R. E., Bechor-Shental, D., Martz, E., et al. (2003). ConSurf: Identification of functional regions in proteins by surface-mapping of phylogenetic information. *Bioinformatics, 19*, 163–164.

Grabowski, G. A. (2008). Phenotype, diagnosis, and treatment of Gaucher's disease. *Lancet, 372*, 1263–1271.

CHAPTER NINE

Computational and modeling approaches to understand the impact of the Fabry's disease causing mutation (D92Y) on the interaction with pharmacological chaperone 1-deoxygalactonojirimycin (DGJ)

D. Thirumal Kumar[a], E. Judith[a], J. Priyadharshini Christy[a], R. Siva[a], Iftikhar Aslam Tayubi[b], Chiranjib Chakraborty[c], C. George Priya Doss[a,*], Hatem Zayed[d,*]

[a]Department of Integrative Biology, School of Bio Sciences and Technology, VIT, Vellore, Tamil Nadu, India
[b]Faculty of Computing and Information Technology, King Abdulaziz University, Rabigh, Saudi Arabia
[c]Department of Biotechnology, Adamas University, Kolkata, West Bengal, India
[d]Department of Biomedical Sciences, College of Health and Sciences, Qatar University, Doha, Qatar
*Corresponding authors: e-mail address: georgepriyadoss@vit.ac.in; hatem.zayed@qu.edu.qa

Contents

Advances in Protein Chemistry and Structural Biology, Volume 114
ISSN 1876-1623
https://doi.org/10.1016/bs.apcsb.2018.10.009

Abstract

Fabry's disease (FD) is the second most commonly occurring lysosomal storage disorders (LSDs). The mutations in α-galactosidase A (GLA) protein were widely found to be causative for the Fabry's disease. These mutations result in alternate splicing methods that affect the stability and function of the protein. The mutations near the active site of the protein results in protein misfolding. In this study, we have retrieved the missense mutation data from the three public databases (NCBI, UniProt, and HGMD). We used multiple *in silico* tools to predict the pathogenicity and stability of these mutations. Mutations in the active sites (D92Y, C142Y, D170V, and D266N) of the protein were screened for the phenotyping analysis using SNPeffect 4.0. Mutant D92Y was predicted to increase the amyloid propensity as well as severely reduce the protein stability and the remaining mutations showed no significant results by SNPeffect 4.0. Protein dynamics simulations (PDS) were performed to understand the behavior of the proteins due to the mutations. The simulation results showed that the D92Y mutant was more severe (higher deviation, loss of intramolecular hydrogen bonds, and lower compactness) than the other protein mutants (C142Y, D170V, and D266N). Further, the action of pharmacological chaperone 1-deoxygalactonojirimycin (DGJ) over the severe mutation was studied using the molecular docking analysis. Chaperone DGJ, an iminosugar plays a convincing role in repairing the misfolded protein and helps the protein to achieve its normal function. From the molecular docking analysis, we observed that both the native protein and protein with D92Y mutation followed similar interaction patterns. Further, the docked complexes (native-DGJ and mutant-DGJ) were subjected to PDS analysis. From the simulation analysis, we observed that DGJ had shown the better effect on the protein with the D92Y mutation. This elucidates that DGJ can still be used as a promising chaperone to treat the FD caused by mutations of GLA protein.

1. Introduction

Fabry's disease (FD; OMIM: 301500) is an X-linked rare lysosomal genetic disease with a deficiency of an enzyme called alpha-galactosidase A/GLA (AGAL, E.C. 3.2.1.22). AGAL enzyme helps in catalyzing lipids with terminal α-galactosyl residues. The absence of this enzyme causes glycosphingolipids specifically, globotriaosyl ceramide (Gb3) and related α-galactosyl terminating glycosphingolipids to accumulate in the lysosomes of vascular endothelial cells. The disease affects many parts of the body including eyes, skin, kidney, brain, heart, and gastrointestinal and nervous systems (Desnick & Brady, 2004; Desnick et al., 2003; Germain, 2010; Liguori et al., 2017; Zarate & Hopkin, 2008). AGAL enzyme is encoded by a gene, *GLA*, present on the q arm of the X-chromosome (Xq22) spanning a genomic DNA of 12 kb with seven exons (Bishop, Kornreich, & Desnick, 1988; Kornreich, Bishop, & Desnick, 1989). The GLA cDNA of 1290 bp encodes for a precursor protein of ~50 kDa (429 amino acids), which includes a single peptide of 31 amino acids. The mature protein consists of homodimer structure of 101 kDa, with each monomer consist of two domains namely (alpha/beta)$_8$ N-terminal domain (32–328 amino acids) and a C-terminal domain (329–341 amino acids) (Bishop et al., 1986; Garman & Garboczi, 2004). The N-terminal domain consists of active sites side-chain residues namely W47, D92, Y134, C142, K168, D170, C172, L206, Y207, R227, D231, D266, and M267 (Germain, 2010).

FD incidence rate is estimated as 1 in 40,000 in males. However, the incidence varies depending on the ethnicity and demography. Screening studies on newborn have provided a range of 1 in 40,000 to 1 in 70,000 (Inoue et al., 2013). Although FD is an X-linked inheritance, cases of women manifesting Fabry's have been reported (Wang, Lelis, Mirocha, & Wilcox, 2007). However, the disease manifestations range from severe, classical and non-classical phenotypes. Classical phenotypes are marked by the diminished functionality of GLA and high accumulation of GL-3 while the non-classical type develops symptoms at the later stage in life and also known as the late-onset type. The non-classic phenotype is usually manifested with no severe pain or and no angiokeratoma unlike the cases of classical FDs. The terminal effect of both these types leads to cardiac and renal failures. Symptoms can occur as early as childhood or adolescence in case of classic FD. Cardiac symptoms usually include left ventricular hypertrophy, arrhythmia, angina, and dyspnea. Renal symptoms typically include renal lesions, renal impairment generally due to microalbuminuria and proteinuria and azotemia. Based on the onset and prevalence of the symptoms, this disease can also be classified into cardiac and renal types (Citro et al., 2016; Desnick et al., 2003; Germain, 2010; Liguori et al., 2017; Lukas et al., 2015; Thurberg et al., 2002).

According to the HGMD database (http://www.hgmd.cf.ac.uk/ac/all.php), 720 mutations (missense/nonsense, splicing, insertions, and deletions) were recorded in connection to FD of which 441 were missense mutations. Mutations in the *GLA* gene can affect the alternate splicing methods thereby affecting the stability and functionality of the protein. Several studies have reported the role of missense mutations in changing the functionality of the native protein structure especially the mutations in or nearby the protein active sites resulting in protein misfolding (Garman & Garboczi, 2004). Few residues that are located on or close to the dimer interface may alter proper dimerization of the glycoprotein. These types of structural alterations lead to loss of enzyme activity or stability resulting in the classical phenotype. Also, the normal substitutions occurring in the other regions of the protein (i.e., T41I, M187T, G271V, G271S, P293T, Q312H, and A348P) also express the similar type of structural alterations resulting in loss of enzyme function with a classical phenotype (Shabbeer, Yasuda, Benson, & Desnick, 2006). To date, efforts in establishing the genotype-phenotype correlation in FD remain a daunting task to biologists.

Management of FD comes in the form of enzyme replacement therapy (ERT) or Pharmacological chaperones (PCT) where treatment is a lifelong process. PCT has recently proved to be a possible therapeutic alternative for FD. These chemical molecules bind to the mutant proteins and facilitate correct folding maturation of the protein. FDA-approved Migalastat (1-deoxygalactonojirimycin) (DGJ), an iminosugar, plays a role in revising the misfolded variant of the enzyme to achieve a semblance of its functionality. DGJ is an active site-specific chaperone, and it binds competitively to various mutants. This binding of DGJ to the misfolded protein in the endoplasmic reticulum allows its transport out of the ER to the lysosomes where it carries out degradation of Gb3. This binding leads to subsequent stabilization of the limited mutants which were termed as "amenable" or "responsive." So far, 78 mutations have shown to be amenable to DGJ. However, the action of pharmacological chaperones is applicable only for the misfolded proteins (Ishii, 2012). Treatment with migalastat increases AGAL levels and decreases the accumulation of Gb3. However, concentration studies showed that DGJ has an inhibitory effect on alpha-Gal A at nanomolecular levels (Benjamin et al., 2009). DGJ helps to correct the conformation of the specific misfolded protein and overcomes the ERT associated side effect of antibody formation against the synthetic enzyme (Germain, 2010; Sirrs & Alfadhel, 2011). These treatments can be used together to enhance its functionality and can be used intravenously or orally. This study is the first to systematically validate the impact of the FD mutations and their binding efficiency to the existing chaperone DGJ using molecular docking and protein dynamics analysis. Initially, we screened all the publically available missense mutations associated with FD using *in silico* prediction methods for pathogenicity and protein stability analysis.

2. Methods and materials

2.1 Data compilation

Missense mutations recorded for GLA protein were collected from various public databases like NCBI dbSNP (https://www.ncbi.nlm.nih.gov), HGMD (http://www.hgmd.cf.ac.uk/ac/all.php) and UniProt (https://www.uniprot.org). Comprehensive information regarding various mutants was obtained using OMIM (https://www.omim.org/) and through various literature recorded in databases like PubMed and Google Scholar.

2.2 Prediction of pathogenicity and stability using *in silico* methods

The pathogenicity and stability were assessed using PredictSNP and iStable servers (http://predictor.nchu.edu.tw/istable/indexSeq.php and https://loschmidt.chemi.muni.cz/predictsnp1) respectively (Bendl et al., 2014; Chen, Lin, & Chu, 2013). PredictSNP embeds different algorithms in the form of PANTHER, SIFT, MAPP, PolyPhen 1, PolyPhen 2, SNAP, and nsSNPanalyzer which analyzes and classifies the mutations into deleterious or neutral based on various parameters like functional and structural changes due to mutations, evolutionary evidence, and linkage disequilibrium. iStable predicts the changes in the stability in association with the changes in energy. It embeds different algorithms in the form of I-Mutant and MUpro. The final screening of missense mutations was done based on the mutations present in the active sites. Active sites were elucidated from the online server (http://projects.biotec.tu-dresden.de/metapocket) (Huang, 2009) and compared with literature available online.

2.3 Phenotyping analysis

Phenotyping analysis was performed using online SNPeffect 4.0 server (http://snpeffect.switchlab.org) (De Baets et al., 2012). The server incorporates other tools such as TANGO, WALTZ, LIMBO, and FoldX to predict aggregation-prone regions, amylogenic regions, hsp70 chaperone binding sites, and the effect on structural stability.

2.4 Mutagenesis and molecular docking

The native structure of human GLA was obtained from Protein Data Bank (PDB ID: 1R46) and *in silico* mutagenesis was carried out for the identified mutants (D92Y, C142Y, D170V, and D266N) using Swiss PDB Viewer (Guex & Peitsch, 1997). DGJ structure was obtained from PubChem database with CID 377339 (https://pubchem.ncbi.nlm.nih.gov). The DGJ was docked to the protein structures with shortlisted mutations using AutoDock4.2. A final docking study was conducted using 15 residues of GLA protein as an active site for DGJ (ligand) binding. The Protein (GLA) and ligand (DGJ) were converted to pdbqt format after the first addition of hydrogen and other charges. The ligand was given a flexible torsion while the protein was fixed at a rigid point to coincide with the rotational parameters for protein-ligand interactions. Grid box was set to 60 at *x*, *y*, and z-axis with 0.375 Å around the protein active site. The box was set using the autogrid parameter. The Lamarckian Genetic Algorithm (LGA) under the autodock4 was used for docking the ligand to the protein and to get the best conformations. The protein-ligand complex with the best conformation was then subsequently used for protein dynamics simulation.

2.5 Protein dynamics simulations

Protein dynamics simulations (PDS) were performed in two procedures: (i) native and mutant protein dynamics without DGJ and (ii) native and mutant protein dynamics with the DGJ. PDS were performed using the GROMACS molecular dynamics package (Abraham et al., 2015). GROMOS 53a5 force field was used to obtain the parameters for the protein and parameters for the ligand was obtained from the ATB Server (Malde et al., 2011; Oostenbrink, Villa, Mark, & Van Gunsteren, 2004). The systems were neutralized by adding counter ions (sodium or chloride). The molecules were subjected to energy minimization using the steepest descent method (Petrova & Solov'ev, 1997). The whole energy minimized structures were subjected equilibration through NVT and NPT ensembles for 50,000 steps each. Finally, the dynamics simulations and output trajectories were obtained. The trajectories were analyzed using GROMACS packages. The graphs were analyzed using Grace Software.

3. Results

3.1 Prediction of pathogenicity and stability using *in silico* methods

The deleteriousness and destabilizing properties of the 698 missense mutations were analyzed using PredictSNP and iStable servers. PredictSNP, MAPP, PhD-SNP, PolyPhen1, PolyPhen2, SIFT, SNAP, nsSNPanalyzer, and PANTHER servers predicted 475, 395, 541, 437, 458, 513, 434, 214, and 521 missense mutations to be pathogenic respectively. Further, I-Mutant, MUpro, and iStable predicted 604, 557, and 555 missense mutations to be destabilizing respectively (Tables 1 and 2). The common active sites of GLA protein from the literature source (Germain, 2010) as well as from the MetaPocket prediction was considered as the active site. Further analysis was carried out on the active site mutations (C142Y, D92Y, D170V, and D266N) predicted to be highly deleteriousness and destabilizing.

Table 1 Classification of GLA mutations as deleterious and neutral using PredictSNP server

Mutations	PredictSNP	MAPP	PhD-SNP	Poly Phen-1	Poly Phen-2	SIFT	SNAP	nsSNP Analyzer	PANTHER
M1V	N	U	N	N	N	D	N	U	U
M1I	N	U	N	N	N	D	N	U	U
M1R	D	U	N	D	D	D	D	U	U
M1K	D	U	N	D	D	D	D	U	U
M1L	N	U	N	N	N	D	N	U	U
M1T	D	U	N	D	N	D	D	U	U
Q2P	D	D	N	D	D	D	N	U	U
L3V	N	D	N	N	D	D	N	U	U
L3P	D	D	N	D	D	D	N	U	U
N5D	N	N	N	N	N	N	N	U	U
G11A	N	N	N	N	N	N	N	U	U
G11D	D	D	N	D	D	N	D	U	U
A13V	N	N	N	N	N	N	N	U	N
A13P	D	D	N	D	D	N	D	U	D
A13T	N	N	N	N	N	N	N	U	N
L14P	D	D	D	D	D	D	D	U	D
A15E	N	D	N	N	N	N	N	U	N
A15G	N	N	N	N	N	N	N	U	D
A15P	N	D	D	N	N	N	D	U	D
A15T	N	N	N	N	N	N	N	U	N
L16V	N	N	N	N	N	N	N	U	N
L16R	D	D	D	D	D	D	D	U	N
L16P	D	D	D	D	N	D	N	U	D
L16H	D	N	N	D	D	D	N	U	N
F18S	N	N	N	N	N	N	N	U	N
F18C	N	N	N	D	D	N	N	U	D
L19P	D	N	D	D	N	D	D	U	D
L19Q	N	N	D	D	N	D	N	U	D
A20P	D	D	D	D	D	N	N	U	D

Continued

Table 1 Classification of GLA mutations as deleterious and neutral using PredictSNP server—cont'd

Mutations	PredictSNP	MAPP	PhD-SNP	Poly Phen-1	Poly Phen-2	SIFT	SNAP	nsSNP Analyzer	PANTHER
A20D	D	D	D	D	N	N	N	U	D
A20V	N	D	N	N	N	N	N	U	N
L21F	N	N	N	N	N	N	N	U	D
L21P	D	N	D	D	N	D	D	U	D
L21R	D	D	D	D	D	D	D	U	D
V22G	N	U	D	N	N	D	N	U	N
W24G	N	U	D	N	N	N	D	U	N
W24R	N	U	N	N	N	N	D	U	N
W24C	N	U	N	N	N	N	N	U	N
D25N	N	U	N	N	N	N	N	U	N
I26F	N	U	N	N	N	N	N	U	N
A29T	N	U	N	N	N	N	N	U	N
R30T	N	U	N	N	N	N	N	U	N
A31V	D	U	D	N	D	D	D	U	D
L32P	D	U	D	D	D	D	D	N	D
D33G	D	N	D	N	N	D	D	U	N
D33Y	D	N	D	D	D	D	D	U	D
N34S	D	N	D	N	D	D	D	D	D
N34K	D	N	D	N	D	D	D	N	D
G35E	D	D	D	D	D	D	D	D	D
G35R	D	D	D	D	D	D	D	D	D
L36S	D	D	D	D	D	D	D	N	D
L36W	D	D	D	D	D	D	D	N	D
L36F	D	N	D	D	D	D	D	N	D
A37V	D	D	D	N	D	N	N	N	D
A37T	D	D	D	D	D	D	D	N	D
R38G	N	N	D	N	D	N	N	N	D
R38K	N	N	N	N	N	N	N	N	N
T39M	D	D	D	D	D	D	D	N	D

Table 1 Classification of GLA mutations as deleterious and neutral using PredictSNP server—cont'd

Mutations	PredictSNP	MAPP	PhD-SNP	Poly Phen-1	Poly Phen-2	SIFT	SNAP	nsSNP Analyzer	PANTHER
P40S	D	D	D	D	D	D	D	N	D
P40H	D	D	D	D	D	D	D	N	D
P40A	D	D	D	D	D	D	D	N	D
P40R	D	D	D	D	D	D	D	N	D
P40L	D	D	D	D	D	D	D	N	D
T41S	N	N	D	N	N	D	N	U	N
T41I	N	N	D	N	N	D	N	U	N
M42V	D	D	D	D	D	D	D	N	D
M42R	D	D	D	D	D	D	D	D	D
M42L	D	N	D	N	D	D	D	N	N
M42T	D	D	D	D	D	D	D	D	D
M42K	D	D	D	D	D	D	D	U	D
G43V	D	D	D	D	D	D	D	D	D
G43S	D	D	D	D	D	D	D	D	D
G43R	D	D	D	D	D	D	D	D	D
G43D	D	D	D	D	D	D	D	D	D
W44C	D	D	D	D	D	D	D	D	D
L45P	D	D	D	D	D	D	D	D	D
L45R	D	D	D	D	D	D	D	U	D
H46R	D	D	D	D	N	N	D	U	N
H46Q	N	N	D	N	N	N	D	U	N
H46P	N	D	D	N	N	N	N	U	D
H46Y	D	D	D	N	N	D	D	U	N
H46L	D	D	D	N	N	N	D	U	N
W47G	D	D	D	D	D	D	D	D	D
W47L	D	D	D	D	D	D	D	D	D
W47R	D	D	D	D	D	D	D	U	D
W47C	D	D	D	D	D	D	D	U	D
E48K	D	D	D	D	D	D	D	N	D

Continued

Table 1 Classification of GLA mutations as deleterious and neutral using PredictSNP server—cont'd

Mutations	PredictSNP	MAPP	PhD-SNP	Poly Phen-1	Poly Phen-2	SIFT	SNAP	nsSNP Analyzer	PANTHER
E48Q	N	N	D	N	D	N	N	D	D
E48D	D	D	D	N	N	D	D	N	N
R49S	N	N	D	N	N	D	D	D	N
R49G	D	N	D	D	N	D	D	D	D
R49C	D	N	D	D	D	D	D	D	D
R49P	D	D	D	D	D	D	D	D	D
R49L	D	N	D	D	N	D	D	D	D
F50C	D	D	D	D	D	D	D	D	D
M51I	N	D	N	N	N	D	N	U	N
M51K	N	N	D	N	N	N	N	U	N
C52S	D	N	D	D	D	D	D	D	D
C52W	D	D	D	D	D	D	D	D	D
C52G	D	N	D	D	D	D	D	D	D
C52R	D	D	D	D	D	D	D	D	D
C52Y	D	D	D	D	D	D	D	D	D
N53K	N	U	D	N	D	N	N	N	N
N53D	N	U	N	N	N	N	N	N	N
N53L	D	U	D	D	D	N	D	U	D
L54P	N	U	D	N	N	D	N	U	D
L54F	N	U	N	N	N	N	N	U	N
D55V	D	U	D	D	D	D	D	U	D
C56G	D	U	D	D	D	D	D	D	U
C56F	D	U	D	D	D	D	D	D	U
C56Y	D	U	D	D	D	D	D	D	U
C56S	D	U	D	D	D	D	D	U	U
Q57R	N	U	N	N	N	N	N	U	U
Q57L	N	U	N	N	N	N	N	U	U
E59K	N	U	N	N	N	N	D	U	U
P60L	D	U	D	D	D	D	D	N	U

Table 1 Classification of GLA mutations as deleterious and neutral using PredictSNP server—cont'd

Mutations	PredictSNP	MAPP	PhD-SNP	Poly Phen-1	Poly Phen-2	SIFT	SNAP	nsSNP Analyzer	PANTHER
P60T	D	U	D	D	D	D	D	N	U
P60S	D	U	D	D	D	D	D	U	U
C63Y	D	N	D	D	D	D	D	D	U
C63S	D	N	D	D	D	D	D	U	U
C63R	D	N	D	D	D	D	D	U	U
I64F	D	N	D	D	D	D	N	N	D
S65I	D	D	D	D	D	D	D	D	D
S65T	N	N	N	N	D	D	N	N	D
E66G	D	D	D	D	D	D	D	D	D
E66K	D	D	D	D	D	D	D	N	D
E66Q	D	N	D	N	D	D	D	D	D
L68F	D	N	N	D	D	D	N	D	D
E71G	N	N	N	N	N	N	N	N	D
M72V	N	N	N	N	N	N	N	D	N
M72I	N	N	N	N	N	N	N	D	D
M72R	D	D	D	D	N	D	D	D	D
A73V	D	D	D	N	D	D	N	N	D
A73E	D	D	D	D	D	D	D	N	D
E74D	N	N	N	N	N	N	N	U	N
M76T	D	D	D	D	D	D	D	U	D
M76R	D	D	D	D	D	D	D	U	D
G80D	D	D	D	D	D	D	D	D	D
W81S	D	D	D	D	D	D	D	D	D
W81R	D	D	D	D	D	D	D	D	D
W81C	D	D	D	D	D	D	D	D	D
D83N	N	N	D	N	N	N	N	U	D
G85S	D	D	D	D	D	D	D	D	D
G85D	D	D	D	D	D	D	D	D	D
G85M	D	D	D	D	D	D	D	U	D

Continued

Table 1 Classification of GLA mutations as deleterious and neutral using PredictSNP server—cont'd

Mutations	PredictSNP	MAPP	PhD-SNP	Poly Phen-1	Poly Phen-2	SIFT	SNAP	nsSNP Analyzer	PANTHER
Y86H	D	D	D	D	D	D	D	D	D
Y86C	D	D	D	D	D	D	D	D	D
Y86D	D	D	D	D	D	D	D	D	D
E87K	N	N	D	N	N	N	N	U	N
Y88D	D	D	D	D	D	D	D	D	D
L89H	D	D	D	D	D	D	D	U	D
L89R	D	D	D	D	D	D	D	U	D
L89P	D	D	D	D	D	D	D	U	D
L89F	D	D	D	N	N	D	D	U	D
C90R	D	D	D	D	D	N	D	U	D
C90Y	D	N	D	D	D	D	N	U	D
I91N	D	D	D	D	D	D	D	D	D
I91T	D	D	D	D	D	D	D	D	D
D92V	D	D	D	D	D	D	D	D	D
D92H	D	D	D	D	D	D	D	N	D
D92G	D	D	D	D	D	D	D	D	D
D92N	D	D	D	D	D	D	D	N	D
D92Y	D	D	D	D	D	D	D	D	D
D93V	D	D	D	D	D	D	D	D	D
D93E	D	D	D	D	N	D	D	N	D
D93G	D	D	D	D	D	D	D	D	D
D93N	D	N	D	D	D	D	D	N	D
D93Y	D	D	D	D	D	D	D	D	D
C94S	D	N	D	D	N	D	D	D	D
C94Y	D	D	D	D	D	D	D	D	D
W95S	D	D	D	D	D	D	D	D	D
W95C	D	D	D	D	D	D	D	D	D
W95L	D	D	D	D	D	D	D	D	D
M96I	N	N	N	N	N	N	N	U	N

Table 1 Classification of GLA mutations as deleterious and neutral using PredictSNP server—cont'd

Mutations	PredictSNP	MAPP	PhD-SNP	Poly Phen-1	Poly Phen-2	SIFT	SNAP	nsSNP Analyzer	PANTHER
A97V	N	D	N	N	N	N	N	N	N
A97P	N	D	D	D	N	N	N	N	D
R100K	D	N	D	D	D	D	D	D	D
R100T	D	D	D	D	D	D	D	D	D
S102L	N	N	D	N	N	N	N	N	D
E103Q	N	N	N	N	N	N	N	N	N
G104V	D	D	D	D	D	D	D	D	D
G104S	D	N	D	D	D	D	D	D	D
G104D	D	D	D	D	N	D	D	D	D
R105G	N	N	D	N	N	D	D	N	D
L106R	D	D	D	D	D	D	D	D	D
Q107R	N	N	D	N	N	N	N	U	N
Q107L	N	N	N	N	N	N	N	U	N
A108T	N	N	D	N	N	D	N	U	D
D109H	D	D	D	N	N	D	D	N	D
D109G	D	D	D	D	D	D	D	N	D
R112S	D	D	D	D	D	D	D	D	D
R112H	D	N	D	D	D	D	D	D	D
R112C	D	D	D	D	D	D	D	D	D
R112L	D	D	D	D	D	D	D	D	D
R112G	D	D	D	D	D	D	D	U	D
F113S	D	D	D	D	D	D	D	D	D
F113I	D	D	D	D	D	D	D	D	D
F113L	D	D	D	D	D	D	D	D	D
I117S	D	D	D	D	D	D	D	N	D
R118H	N	N	D	N	N	N	N	U	D
R118C	D	D	D	D	D	D	N	U	D
L120V	N	D	D	N	N	D	N	N	N
L120S	D	D	D	D	D	D	D	U	D

Continued

Table 1 Classification of GLA mutations as deleterious and neutral using PredictSNP server—cont'd

Mutations	PredictSNP	MAPP	PhD-SNP	Poly Phen-1	Poly Phen-2	SIFT	SNAP	nsSNP Analyzer	PANTHER
L120P	D	D	D	D	D	D	D	U	D
A121P	D	D	D	D	D	D	D	N	D
A121T	D	D	D	N	D	D	D	N	D
N122D	N	N	N	N	N	N	N	U	D
Y123C	D	N	D	D	D	D	D	U	D
V124D	D	D	D	D	D	D	D	N	D
H125P	D	D	D	D	D	D	D	N	D
H125Y	D	D	D	D	D	D	D	N	D
H125L	D	D	D	D	D	D	D	U	D
S126I	D	D	D	D	D	D	N	U	D
S126G	N	N	D	N	N	N	N	U	D
G128E	D	D	D	D	D	D	D	D	D
L129P	D	D	D	D	D	D	D	D	D
K130R	D	N	D	N	D	D	D	U	D
L131P	D	D	D	D	D	D	D	D	D
G132E	D	D	D	D	D	D	D	D	D
G132R	D	D	D	D	D	D	D	D	D
G132A	D	D	D	D	D	D	D	U	D
I133M	D	D	D	N	D	D	N	N	D
Y134S	D	D	D	D	D	D	D	D	D
Y134D	D	D	D	D	D	D	D	D	D
Y134H	D	D	D	D	D	D	D	U	D
A135V	N	D	N	N	D	N	N	U	N
D136E	D	D	D	D	N	D	D	N	D
D136H	D	N	D	D	D	D	D	D	D
D136Y	D	D	D	D	D	D	D	D	D
G138E	D	D	D	D	D	D	D	D	D
G138R	D	D	D	D	D	D	D	D	D
N139S	N	N	N	N	N	N	N	U	N

Table 1 Classification of GLA mutations as deleterious and neutral using PredictSNP server—cont'd

Mutations	PredictSNP	MAPP	PhD-SNP	Poly Phen-1	Poly Phen-2	SIFT	SNAP	nsSNP Analyzer	PANTHER
N139T	N	N	N	N	N	N	N	U	N
K140T	N	N	N	N	N	N	N	U	N
T141I	D	D	D	D	D	D	D	D	D
T141N	D	N	D	D	D	D	D	U	D
C142W	D	D	D	D	D	D	D	D	D
C142R	D	D	D	D	D	D	D	D	D
C142Y	D	D	D	D	D	D	D	D	D
A143P	D	D	D	D	D	D	N	U	D
A143T	D	D	D	N	D	D	N	U	D
G144V	D	D	D	D	D	D	D	D	D
G144D	D	N	D	D	N	D	D	D	D
F145S	N	N	D	N	N	D	D	U	D
P146S	D	N	D	N	D	D	N	D	D
G147R	D	D	D	D	D	D	D	D	D
G147E	D	D	D	D	D	D	D	U	D
S148R	D	D	D	D	D	D	D	D	D
S148N	D	D	D	D	D	N	D	D	D
Y152C	D	D	D	D	D	N	N	U	D
D153N	N	N	D	N	N	N	N	U	N
I154T	N	N	N	N	N	N	N	U	N
D155H	D	D	D	D	D	D	D	D	D
A156V	D	D	D	N	D	D	N	N	D
A156D	D	D	D	D	D	D	D	N	D
A156T	D	D	D	D	D	D	N	N	D
W162G	D	D	D	D	D	D	D	D	D
W162R	D	D	D	D	D	D	D	D	D
W162C	D	D	D	D	D	D	D	D	D
W162L	D	D	D	D	D	D	D	D	D
G163V	D	D	D	D	D	D	D	D	D

Continued

Table 1 Classification of GLA mutations as deleterious and neutral using PredictSNP server—cont'd

Mutations	PredictSNP	MAPP	PhD-SNP	Poly Phen-1	Poly Phen-2	SIFT	SNAP	nsSNP Analyzer	PANTHER
V164G	D	D	D	D	D	D	D	N	D
V164L	D	D	D	N	N	D	D	N	D
D165V	D	D	D	D	D	D	D	N	D
D165H	D	D	D	D	D	D	D	N	D
D165Y	D	D	D	D	D	D	D	N	D
D165G	D	D	D	D	D	D	D	U	D
L166V	N	D	N	N	N	D	N	U	D
L166G	D	D	D	D	D	D	D	U	D
L167Q	D	D	D	D	D	D	D	D	D
L167P	D	D	D	D	D	D	D	D	D
K168R	D	D	D	D	D	D	D	N	D
K168N	D	D	D	D	D	D	D	D	D
F169S	D	D	N	D	D	D	D	U	D
F169I	N	N	N	N	N	N	N	U	N
D170V	D	D	D	D	D	D	D	D	D
D170H	D	D	D	D	D	D	D	N	D
D170G	D	D	D	D	D	D	D	D	D
D170N	D	D	D	D	D	D	D	N	D
G171S	N	N	N	D	N	D	N	D	D
G171R	D	D	D	D	D	D	D	D	D
G171C	D	N	D	D	D	D	D	D	D
G171D	D	D	D	D	D	D	D	D	D
C172W	D	D	D	D	D	D	D	D	D
C172G	D	D	D	D	D	D	D	D	D
C172R	D	D	D	D	D	D	D	D	D
C172F	D	D	D	D	D	D	D	D	D
C172Y	D	D	D	D	D	D	D	D	D
C172S	D	D	D	D	D	D	D	U	D
C174G	N	N	D	N	N	N	N	U	N

Table 1 Classification of GLA mutations as deleterious and neutral using PredictSNP server—cont'd

Mutations	PredictSNP	MAPP	PhD-SNP	Poly Phen-1	Poly Phen-2	SIFT	SNAP	nsSNP Analyzer	PANTHER
C174R	N	D	D	N	N	N	N	U	N
D175E	N	N	N	N	N	N	N	N	N
D175N	N	N	N	N	N	N	N	N	N
L180F	N	N	N	N	N	D	N	U	N
G183V	D	D	D	D	D	D	D	D	D
G183S	N	N	N	N	N	D	D	D	D
G183A	N	N	N	N	N	D	N	D	D
G183R	N	N	N	N	D	D	D	D	N
G183D	N	N	D	N	N	D	N	D	D
Y184N	D	D	D	D	D	D	D	D	D
Y184C	D	D	D	D	D	D	D	U	D
K185E	N	N	N	N	N	N	N	U	N
H186R	N	N	N	N	N	N	N	U	N
H186P	N	D	D	N	N	N	N	U	N
M187V	D	D	D	N	D	D	D	D	D
M187I	D	D	D	N	N	D	D	D	D
M187R	D	D	D	D	D	D	D	D	D
M187T	D	D	D	N	D	D	D	D	D
L189F	N	N	N	N	N	N	N	U	D
L191Q	D	D	D	D	D	D	D	D	D
L191P	D	D	D	D	D	D	D	D	D
R193S	N	N	D	N	N	N	N	U	N
T194I	D	N	D	D	D	D	D	N	D
G195V	D	D	D	D	D	D	D	D	D
R196S	D	D	D	D	D	D	D	N	D
S197N	N	N	N	N	N	D	N	U	N
I198T	D	D	D	D	D	D	D	N	D
V199A	N	D	D	N	N	N	D	D	N
V199M	N	N	N	D	D	D	N	N	D

Continued

Table 1 Classification of GLA mutations as deleterious and neutral using PredictSNP server—cont'd

Mutations	PredictSNP	MAPP	PhD-SNP	Poly Phen-1	Poly Phen-2	SIFT	SNAP	nsSNP Analyzer	PANTHER
V199G	D	D	D	D	N	N	D	U	D
Y200C	D	D	D	D	D	D	D	D	D
S201A	D	D	D	N	D	D	N	N	D
S201F	D	D	D	D	D	D	D	D	D
S201Y	D	D	D	D	D	D	D	D	D
C202W	D	D	D	D	D	D	D	D	D
C202Y	D	D	D	D	D	D	D	D	D
C202R	D	D	D	D	D	D	D	U	D
E203K	D	N	D	D	D	D	D	U	D
E203V	D	D	D	D	D	D	D	U	D
E203D	N	N	D	N	N	D	N	U	D
W204R	D	D	D	D	D	D	D	D	D
W204C	D	D	D	D	D	D	D	D	D
P205R	D	D	D	D	D	D	D	D	D
P205L	D	D	D	D	D	D	D	D	D
P205T	D	N	D	D	D	D	D	D	D
P205S	D	N	D	N	D	D	D	U	D
L206V	N	N	N	N	N	N	N	U	N
L206P	N	D	D	N	N	N	N	U	D
Y207S	D	N	D	N	D	D	D	D	D
Y207H	D	N	D	N	D	D	D	D	D
Y207C	D	N	D	N	N	D	D	D	D
P210S	N	N	N	N	N	N	N	U	N
P210L	N	N	N	N	D	N	N	U	N
F211L	N	U	N	N	N	N	N	U	N
K213R	N	U	N	N	N	N	N	N	N
K213M	N	U	N	D	N	N	N	N	N
P214S	N	U	N	N	D	D	N	U	N
P214L	N	U	N	D	D	N	N	U	D

Table 1 Classification of GLA mutations as deleterious and neutral using PredictSNP server—cont'd

Mutations	PredictSNP	MAPP	PhD-SNP	Poly Phen-1	Poly Phen-2	SIFT	SNAP	nsSNP Analyzer	PANTHER
N215S	N	N	D	N	N	D	N	N	D
N215H	N	N	D	N	N	D	N	N	D
N215D	N	D	N	N	N	N	N	U	N
Y216C	D	N	D	D	D	D	D	D	D
Y216D	D	D	D	D	D	D	D	D	D
I219M	N	N	N	N	N	D	N	U	D
I219T	D	N	D	N	D	D	N	U	D
I219N	D	N	D	D	D	D	N	U	D
R220Q	N	N	N	N	N	N	N	U	N
R220P	N	N	D	N	N	N	D	U	N
Q221H	N	N	N	D	N	D	N	U	D
Y222D	D	D	D	N	N	N	D	U	D
C223G	D	N	D	D	N	D	D	D	N
C223R	D	D	D	D	D	D	D	D	D
C223Y	D	D	D	D	D	D	D	D	D
N224S	D	N	D	N	D	D	D	N	D
N224D	D	D	D	N	N	D	D	N	D
H225R	D	D	D	D	N	N	D	U	D
H225D	D	D	D	N	N	D	D	U	D
W226R	D	D	D	D	D	D	D	D	D
W226C	D	D	D	D	D	D	D	D	D
R227Q	D	D	D	D	D	D	D	D	D
R227P	D	D	D	D	D	D	D	D	D
N228S	N	N	D	N	N	D	N	N	D
F229L	N	D	D	N	N	D	N	U	N
A230T	N	N	D	N	N	N	N	U	N
D231V	D	D	D	D	D	D	D	N	D
D231G	D	D	D	D	D	D	D	N	D
D231N	D	D	D	D	D	D	D	N	D

Continued

Table 1 Classification of GLA mutations as deleterious and neutral using PredictSNP server—cont'd

Mutations	PredictSNP	MAPP	PhD-SNP	Poly Phen-1	Poly Phen-2	SIFT	SNAP	nsSNP Analyzer	PANTHER
I232S	D	D	D	D	D	D	D	N	D
I232T	D	D	D	D	D	D	D	N	D
D234E	N	N	D	N	N	D	D	N	D
D234N	N	N	D	N	D	D	N	N	D
D234Y	D	D	D	D	D	D	D	D	D
S235C	D	N	D	N	D	D	N	N	D
S235F	D	N	D	D	D	D	D	N	D
S235Y	D	N	D	D	D	D	D	U	D
W236R	D	D	D	D	D	D	D	D	D
W236C	D	D	D	D	D	D	D	D	D
W236L	D	D	D	D	D	N	D	D	D
S238N	N	N	D	N	N	D	N	D	D
S238R	D	N	D	D	D	D	N	U	D
I239T	N	N	D	N	N	D	D	U	D
S241G	N	N	N	N	N	N	N	N	N
I242V	N	N	N	N	N	N	N	N	D
I242F	D	N	D	D	D	D	D	N	D
I242N	D	N	D	D	D	D	D	N	D
L243W	D	U	N	D	D	D	D	U	D
L243F	N	U	N	N	D	D	N	U	D
D244H	D	U	D	D	D	D	D	N	D
D244N	N	U	N	N	N	N	N	N	N
W245L	N	D	D	N	N	N	N	D	N
W245G	N	N	D	N	N	D	D	U	D
T246S	N	N	N	N	N	N	N	U	N
S247C	D	N	D	D	D	N	D	U	D
S247P	D	D	D	D	D	N	D	U	D
F248L	N	N	N	N	N	N	N	U	N
N249K	N	N	N	D	N	N	D	U	N

Table 1 Classification of GLA mutations as deleterious and neutral using PredictSNP server—cont'd

Mutations	PredictSNP	MAPP	PhD-SNP	Poly Phen-1	Poly Phen-2	SIFT	SNAP	nsSNP Analyzer	PANTHER
Q250P	D	D	D	N	D	D	D	N	D
R252T	N	N	N	N	N	N	N	U	N
I253S	N	N	D	N	N	D	N	U	N
I253T	N	D	D	N	N	D	N	U	D
V254A	N	N	N	N	N	N	N	U	N
A257P	D	D	D	D	D	D	D	N	D
A257G	N	N	D	N	N	D	N	U	D
A257D	D	D	D	D	D	D	D	U	D
G258V	D	D	D	D	D	D	D	D	D
G258R	D	D	D	D	D	D	D	D	N
P259R	D	D	D	D	D	D	D	N	D
P259L	D	D	D	D	D	D	D	N	D
G260E	D	D	D	D	D	D	D	D	D
G260A	D	D	D	D	D	D	D	D	D
G261V	D	D	D	D	D	N	N	D	D
G261S	N	N	D	N	N	N	N	D	N
G261D	D	D	D	D	N	N	D	D	D
W262R	D	D	D	D	D	D	D	D	D
W262C	D	D	D	D	D	D	D	D	D
W262L	D	D	D	D	D	D	D	D	D
N263S	D	N	D	D	D	D	D	D	D
D264V	D	D	D	D	D	D	D	D	D
D264Y	D	D	D	D	D	D	D	D	D
D264A	D	D	D	D	D	D	D	U	D
P265S	D	N	D	D	D	D	D	D	D
P265R	D	D	D	D	D	D	D	D	D
P265L	D	N	D	D	D	D	N	D	N
D266V	D	D	D	D	D	D	D	D	D
D266E	D	D	D	D	D	D	D	N	D

Continued

Table 1 Classification of GLA mutations as deleterious and neutral using PredictSNP server—cont'd

Mutations	PredictSNP	MAPP	PhD-SNP	Poly Phen-1	Poly Phen-2	SIFT	SNAP	nsSNP Analyzer	PANTHER
D266H	D	D	D	D	D	D	D	N	D
D266A	D	D	D	D	D	D	D	D	D
D266N	D	D	D	D	D	D	D	N	D
D266Y	D	D	D	D	D	D	D	D	D
M267I	D	D	D	D	D	D	D	D	D
M267R	D	D	D	D	D	D	D	D	D
M267T	D	D	D	D	D	D	D	D	D
L268S	D	D	D	D	D	D	D	D	D
V269E	N	N	D	N	N	D	N	U	N
V269G	D	D	D	D	N	D	D	U	D
V269A	D	D	D	N	N	D	N	U	D
V269L	N	N	D	N	N	N	N	U	N
V269M	N	N	D	N	D	D	N	U	D
I270M	D	D	D	D	D	D	D	N	D
I270T	D	D	D	N	D	D	N	D	D
G271V	D	D	D	D	D	D	D	D	D
G271S	D	D	D	D	D	D	D	D	D
G271A	D	D	D	D	D	D	D	D	D
G271C	D	D	D	D	D	D	D	D	D
G271D	D	D	D	D	D	D	D	U	D
N272S	D	N	D	N	N	D	D	N	D
N272K	D	N	D	D	D	D	D	N	D
F273S	N	N	D	N	N	D	N	D	N
F273L	N	N	D	N	N	D	N	D	D
G274S	D	N	D	N	D	D	D	D	D
G274V	D	D	D	D	D	D	D	U	D
L275H	D	D	D	D	D	D	D	D	D
L275F	D	N	N	N	D	D	D	D	D
S276G	D	N	D	N	D	D	D	N	D

Table 1 Classification of GLA mutations as deleterious and neutral using PredictSNP server—cont'd

Mutations	PredictSNP	MAPP	PhD-SNP	Poly Phen-1	Poly Phen-2	SIFT	SNAP	nsSNP Analyzer	PANTHER
S276N	N	N	D	N	N	D	N	N	D
W277G	N	N	D	N	N	D	N	U	N
W277C	N	N	D	N	N	D	N	U	D
Q279E	N	N	N	N	N	D	N	N	N
Q279R	D	D	D	D	D	D	D	N	D
Q279K	D	D	D	D	D	D	D	N	D
Q279H	D	D	D	D	D	D	D	U	D
Q280H	D	N	N	D	D	D	N	U	D
Q280L	D	N	D	D	N	D	N	U	D
Q280K	N	N	D	D	N	D	N	U	N
T282I	D	D	D	D	D	D	D	U	D
T282A	N	D	N	N	N	N	N	U	N
T282N	D	D	D	D	D	D	D	U	D
Q283P	D	D	D	D	D	D	D	D	D
Q283R	D	D	D	D	D	D	D	U	D
M284T	D	D	D	D	D	D	D	D	D
M284V	D	D	D	N	N	D	N	U	D
A285P	D	D	D	D	D	D	D	N	D
A285D	D	D	D	D	D	D	D	N	D
W287C	D	D	D	D	D	D	D	D	D
W287G	D	D	D	D	D	D	D	U	D
A288P	D	D	D	D	D	D	D	N	D
A288D	D	D	D	D	D	D	D	N	D
I289V	N	N	N	N	N	N	N	N	N
I289S	D	D	D	D	D	D	D	N	D
I289F	D	N	D	D	D	D	D	N	D
M290I	N	N	D	N	N	D	N	N	D
M290L	N	N	N	N	N	N	N	N	N
A291T	D	D	D	D	D	D	D	N	D

Continued

Table 1 Classification of GLA mutations as deleterious and neutral using PredictSNP server—cont'd

Mutations	PredictSNP	MAPP	PhD-SNP	Poly Phen-1	Poly Phen-2	SIFT	SNAP	nsSNP Analyzer	PANTHER
A292V	D	D	D	D	D	D	D	N	D
A292G	D	D	D	D	D	D	D	N	D
A292P	D	D	D	D	D	D	D	N	D
A292T	D	D	D	D	D	D	N	N	D
P293S	D	D	D	D	D	D	D	D	D
P293A	D	D	D	D	D	D	D	D	D
P293R	D	D	D	D	D	D	D	D	D
P293L	D	D	D	D	D	D	D	D	D
P293T	D	D	D	D	D	D	D	D	D
P293H	D	D	D	D	D	D	D	U	D
L294S	D	D	D	D	D	D	D	D	D
F295C	D	D	D	D	D	D	D	U	D
M296V	D	D	D	N	N	D	N	N	D
M296I	N	N	D	N	N	D	N	N	N
M296L	N	N	D	N	N	D	N	N	N
M296T	D	D	D	N	N	D	D	D	D
S297C	D	D	D	D	D	D	D	N	D
S297F	D	D	D	D	D	D	D	D	D
N298S	N	N	D	N	N	N	N	U	D
N298H	D	N	D	D	D	D	D	U	D
N298K	D	D	D	D	D	N	D	U	D
D299E	D	D	D	D	D	D	D	N	D
D299G	D	D	D	D	D	D	D	D	D
L300H	D	D	D	D	D	D	D	D	D
L300F	D	N	D	D	D	D	D	D	D
L300P	D	D	D	D	D	D	D	D	D
R301G	D	N	D	N	D	D	D	D	D
R301Q	D	N	D	D	D	D	D	D	D
R301P	D	D	D	N	D	D	D	D	D

Table 1 Classification of GLA mutations as deleterious and neutral using PredictSNP server—cont'd

Mutations	PredictSNP	MAPP	PhD-SNP	Poly Phen-1	Poly Phen-2	SIFT	SNAP	nsSNP Analyzer	PANTHER
R301L	D	N	D	D	D	D	D	D	D
I303F	D	N	D	D	N	D	D	N	D
I303N	D	D	D	D	D	D	D	N	D
S304N	N	N	D	N	N	N	N	N	D
S304T	N	N	N	N	N	N	N	U	N
A307V	N	N	N	N	N	N	N	N	D
A307T	N	N	N	N	N	N	N	U	N
K308N	D	N	N	D	D	D	N	N	D
A309V	N	D	N	N	N	N	N	U	D
A309P	D	D	D	D	D	N	N	U	D
L310R	D	D	D	D	D	D	D	U	D
L310F	D	N	D	D	D	D	D	U	D
L311V	D	D	D	N	D	D	N	N	D
L311R	D	D	D	D	D	D	D	D	D
L311F	D	N	D	D	D	D	N	D	D
L311P	D	D	D	D	D	D	D	U	D
Q312H	N	N	N	D	D	D	N	N	D
Q312R	N	N	N	N	N	N	N	D	D
D313G	D	D	D	N	N	D	D	U	D
D313N	N	N	N	D	N	N	N	U	D
D313Y	D	D	D	D	D	D	D	U	D
D315G	N	N	D	N	N	N	N	U	D
D315N	N	N	N	N	N	N	N	U	N
V316E	D	D	D	D	D	D	D	N	D
V316I	N	N	N	N	N	N	N	N	N
V316A	N	D	D	N	N	N	N	N	D
V316G	D	D	D	D	D	D	D	U	D
I317S	D	D	D	D	D	D	D	N	D
I317N	D	D	D	D	D	D	D	N	D

Continued

Table 1 Classification of GLA mutations as deleterious and neutral using PredictSNP server—cont'd

Mutations	PredictSNP	MAPP	PhD-SNP	Poly Phen-1	Poly Phen-2	SIFT	SNAP	nsSNP Analyzer	PANTHER
I317T	D	D	D	D	D	D	D	N	D
I319F	D	D	D	N	D	D	D	N	D
I319T	D	D	D	D	D	D	D	N	D
N320I	D	D	D	D	D	D	D	D	D
N320K	D	D	D	D	D	D	D	D	D
N320Y	D	D	D	D	D	D	D	D	D
Q321E	D	D	D	D	D	D	D	N	D
Q321H	D	D	D	D	D	D	D	N	D
Q321R	D	D	D	D	D	D	D	N	D
Q321L	D	D	D	D	D	D	D	D	D
D322E	D	D	D	D	D	D	D	N	D
D322N	D	D	D	D	D	D	D	U	D
P323R	N	D	D	N	N	D	N	N	D
P323T	N	N	D	N	N	D	D	N	D
G325S	D	N	D	D	D	D	D	U	D
G325D	D	D	D	D	D	D	D	U	D
G325R	D	D	D	D	D	D	D	U	D
Q327E	D	D	D	D	D	D	D	U	D
Q327R	D	D	D	D	D	D	D	U	D
Q327K	D	D	D	D	D	D	D	U	D
Q327L	D	N	D	D	D	D	D	U	D
G328V	D	N	D	D	D	D	D	U	D
G328E	D	D	D	D	D	D	D	U	D
G328A	N	N	D	N	D	D	N	U	N
G328R	D	D	D	D	D	D	D	U	D
G328W	D	D	D	D	D	D	D	U	D
Q330R	N	N	N	N	N	N	N	U	D
L331F	N	N	N	N	N	N	N	U	D
R332G	N	N	N	N	N	N	D	U	D
R332K	N	N	N	N	N	N	N	U	N

Table 1 Classification of GLA mutations as deleterious and neutral using PredictSNP server—cont'd

Mutations	PredictSNP	MAPP	PhD-SNP	Poly Phen-1	Poly Phen-2	SIFT	SNAP	nsSNP Analyzer	PANTHER
Q333R	N	N	N	N	N	D	N	U	N
G334E	N	N	N	N	N	N	N	U	D
E338K	D	N	D	D	D	D	N	U	D
V339A	D	D	D	N	D	D	D	U	D
V339E	D	D	D	D	D	D	D	U	D
V339G	D	D	D	D	D	D	D	U	D
W340R	D	D	D	D	D	D	D	U	D
E341K	N	D	D	N	N	N	N	U	N
E341L	N	N	D	N	N	N	N	U	N
E341D	N	D	N	N	N	N	N	U	D
R342G	N	N	N	N	D	D	D	U	N
R342Q	D	N	D	D	D	D	D	U	D
R342P	D	D	D	D	D	D	D	U	D
R342L	D	D	D	D	D	D	D	U	D
P343S	D	N	D	N	D	D	N	U	D
P343L	D	D	D	N	D	D	D	U	D
L344P	D	D	D	D	D	D	D	U	D
S345P	D	D	D	D	N	D	N	U	D
A348P	D	D	D	D	D	D	N	U	D
W349S	N	N	N	N	N	N	D	U	N
W349R	N	D	N	N	N	D	N	U	N
A350P	D	D	D	D	D	D	D	U	D
A352G	N	N	D	N	N	N	N	U	D
A352D	D	D	D	D	D	D	D	U	D
A352V	N	N	D	N	D	N	N	U	N
A352P	D	D	D	D	D	D	N	U	D
I354V	N	N	N	N	N	N	N	U	N
I354K	D	D	D	D	N	D	N	U	D
N355K	D	D	D	D	D	D	D	U	D
R356W	D	N	D	D	D	D	D	U	D

Continued

Table 1 Classification of GLA mutations as deleterious and neutral using PredictSNP server—cont'd

Mutations	PredictSNP	MAPP	PhD-SNP	Poly Phen-1	Poly Phen-2	SIFT	SNAP	nsSNP Analyzer	PANTHER
R356Q	N	N	D	N	D	N	N	U	D
R356P	D	D	D	D	D	D	D	U	D
R356G	D	N	D	D	D	D	D	U	D
E358G	D	U	D	D	D	N	N	U	D
E358A	N	U	N	N	D	N	N	U	N
E358K	D	U	D	D	D	N	N	U	N
E358D	N	U	N	N	N	N	N	U	N
I359T	N	U	N	N	N	N	N	U	N
G360S	N	N	D	N	D	D	N	U	N
G360A	N	N	N	N	D	D	N	U	N
G360C	D	N	D	D	D	D	N	U	D
G360D	N	N	D	N	N	N	N	U	D
G361A	N	N	N	N	N	N	N	U	U
G361R	N	N	N	N	D	N	N	U	U
G361E	N	N	D	N	D	N	N	U	U
P362L	D	D	D	D	D	D	N	U	D
P362T	D	N	D	N	D	D	N	U	D
R363H	N	N	D	N	N	N	N	U	D
R363C	N	N	D	N	D	N	N	U	N
R363P	D	D	D	D	D	N	N	U	D
Y365N	D	D	N	D	D	N	N	U	D
T366N	N	N	D	N	N	N	N	U	N
A368P	N	N	D	N	N	N	N	U	N
A368T	N	N	N	N	N	N	N	U	N
L372R	D	D	D	D	D	D	D	U	D
L372Q	D	D	D	D	D	D	N	U	D
L372P	D	D	D	D	D	D	N	U	D
G373S	D	U	D	N	D	D	N	U	D
G373R	D	U	D	D	D	D	N	U	D
G373D	D	U	D	N	D	D	D	U	D

Table 1 Classification of GLA mutations as deleterious and neutral using PredictSNP server—cont'd

Mutations	PredictSNP	MAPP	PhD-SNP	Poly Phen-1	Poly Phen-2	SIFT	SNAP	nsSNP Analyzer	PANTHER
K374T	N	U	N	N	N	N	N	U	U
G375A	N	N	N	N	N	N	N	U	U
A377D	N	N	D	N	N	N	N	U	N
C378R	D	N	D	D	N	N	D	U	N
C378Y	N	N	D	D	N	N	N	U	D
A381P	N	U	D	N	N	N	N	U	D
C382W	N	U	D	D	N	N	N	U	D
C382Y	N	U	D	D	N	N	N	U	D
I384N	N	D	D	D	N	N	N	U	N
T385A	N	D	N	N	N	N	N	U	N
T385P	N	D	N	N	N	N	N	U	D
Q386P	D	D	D	D	D	D	N	U	D
L387P	D	D	D	D	D	N	N	U	D
P389R	D	U	N	D	D	N	D	U	D
P389L	N	U	N	N	N	N	N	U	N
P389A	N	U	N	N	N	N	N	U	N
V390M	N	N	N	D	N	N	N	U	D
K391T	N	N	N	N	N	N	N	U	N
R392S	N	N	N	N	N	N	N	U	N
R392T	N	N	N	N	N	N	N	U	N
L394P	N	D	D	N	N	N	N	U	D
G395A	N	N	N	N	D	D	N	U	D
G395E	D	D	D	D	D	D	D	U	D
F396Y	N	N	N	N	N	N	N	U	N
E398A	N	N	N	N	N	N	N	U	N
E398K	N	N	N	N	N	N	N	U	N
W399S	N	N	N	N	N	N	N	U	N
L403S	D	N	D	D	D	D	D	U	D
S405R	D	D	N	D	N	D	D	U	N
H406R	N	N	N	N	N	N	N	U	N

Continued

Table 1 Classification of GLA mutations as deleterious and neutral using PredictSNP server—cont'd

Mutations	PredictSNP	MAPP	PhD-SNP	Poly Phen-1	Poly Phen-2	SIFT	SNAP	nsSNP Analyzer	PANTHER
I407K	D	D	D	D	D	D	D	U	D
I407V	N	N	N	N	N	N	N	U	N
I407R	D	D	D	D	D	D	D	U	D
N408Y	D	D	D	D	D	D	D	U	D
P409S	D	N	D	D	D	D	N	U	D
P409H	D	N	D	D	D	D	D	U	D
P409A	D	N	D	N	D	D	D	U	N
P409T	D	N	D	D	D	D	D	U	D
T410I	D	D	N	D	D	D	N	U	D
T410A	N	D	N	N	N	D	N	U	N
T410K	D	D	D	D	N	D	N	U	N
T410P	D	D	N	D	D	D	N	U	D
G411D	D	D	D	D	D	D	D	U	D
T412N	D	D	N	D	D	D	D	U	D
V413I	N	N	N	N	N	D	N	U	D
L414S	D	D	D	D	D	D	D	U	D
L415F	N	N	N	N	N	N	N	U	N
L415P	D	D	D	D	D	D	D	U	D
Q416P	N	D	N	N	N	D	N	U	D
L417P	D	D	D	D	D	D	D	U	D
L417R	D	D	D	D	D	D	D	U	D
E418G	N	N	N	N	N	N	D	U	N
N419I	N	N	N	D	N	N	N	U	N
N419D	N	N	N	N	N	N	N	U	N
M421V	N	N	N	N	N	N	N	U	N
L425F	D	U	N	D	D	N	D	U	U
L429P	D	U	N	D	D	D	N	U	U

N, neutral; D, deleterious; U, unknown.

Table 2 Classification of GLA mutations based on increase and decrease in the stability of the protein.

Variant	Position	I-Mutant	DDG	MUpro	MUpro	iStable	iStable
M1I	1	D	−0.56	D	−0.18764	D	0.650499
M1K	1	D	−1.02	D	−0.42392	D	0.673508
M1L	1	D	−0.75	D	0.287415	I	0.512499
M1R	1	I	−0.16	I	0	I	0.606355
M1T	1	D	−0.64	D	0	D	0.60843
M1V	1	D	−0.5	I	0.101842	I	0.526831
Q2P	2	D	−0.61	D	−0.33773	D	0.730712
L3P	3	D	−0.37	D	−1	D	0.837869
L3V	3	D	−1.36	D	−1	D	0.84709
N5D	5	D	−0.26	D	−0.2672	D	0.830353
G11A	11	I	−0.38	D	−0.72355	D	0.510916
G11D	11	D	−0.97	D	−0.44704	D	0.661912
A13P	13	I	−0.05	D	−0.38043	I	0.53829
A13T	13	D	−0.6	D	−0.45188	D	0.695274
A13V	13	I	0.37	I	0.20582	I	0.781071
L14P	14	D	−1.43	D	−1	D	0.845055
A15E	15	D	−0.16	D	−0.25666	D	0.86752
A15G	15	N	−0.93	D	−0.97523	D	0.810052
A15P	15	I	0.08	D	−0.67182	D	0.651135
A15T	15	D	−0.05	D	−0.94712	D	0.876172
L16H	16	D	−2.12	D	−1	D	0.86191
L16P	16	D	−1.36	D	−1	D	0.842799
L16R	16	D	−1.41	D	−1	D	0.838455
L16V	16	D	−1.18	D	−1	D	0.851033
F18C	18	D	−1.61	D	−0.69033	D	0.764198
F18S	18	D	−1.62	D	−1	D	0.78147
L19P	19	D	−1.57	D	−0.48192	D	0.777525
L19Q	19	D	−1.99	D	−0.38258	D	0.78458

Continued

Table 2 Classification of GLA mutations based on increase and decrease in the stability of the protein.—cont'd

Variant	Position	I-Mutant	DDG	MUpro	MUpro	iStable	iStable
A20D	20	D	−0.6	D	−0.57128	D	0.707634
A20P	20	I	−0.04	D	−1	I	0.53269
A20V	20	I	0.17	D	−0.19054	I	0.519153
L21F	21	D	−1.1	D	−0.98085	D	0.83842
L21P	21	D	−1.7	D	−1	D	0.836226
L21R	21	D	−1.76	D	−1	D	0.81947
V22G	22	D	−2.34	D	−1	D	0.888814
W24C	24	D	−1.69	D	−0.80692	D	0.640834
W24G	24	D	−2.18	D	−1	D	0.731553
W24R	24	D	−1.25	D	−0.42329	D	0.627686
D25N	25	D	−1.08	D	−0.68783	D	0.762537
I26F	26	D	−1.99	D	−1	D	0.792296
A29T	29	D	−0.84	D	−0.88692	D	0.797467
R30T	30	D	−0.86	D	−0.33711	D	0.723594
A31V	31	D	0.28	I	0.059979	D	0.571389
L32P	32	D	−1.86	D	−1	D	0.826206
D33G	33	D	−0.83	D	−1	D	0.607117
D33Y	33	I	0.11	D	−0.83035	I	0.688045
N34K	34	D	−0.63	D	−0.12604	D	0.781495
N34S	34	D	−0.34	D	−0.47554	D	0.791212
G35E	35	D		D	−0.233	D	0.731943
G35R	35	D	−1.24	D	−0.18081	D	0.713139
L36F	36	D	−1.2	D	−1	D	0.689328
L36S	36	D	−2.69	D	−1	D	0.724003
L36W	36	D	−1.52	D	−1	D	0.649433
A37T	37	D	−0.88	D	−0.62842	D	0.873369
A37V	37	D	0.19	I	0.571138	D	0.593353
R38G	38	D	−0.93	D	−1	D	0.814581

Table 2 Classification of GLA mutations based on increase and decrease in the stability of the protein.—cont'd

Variant	Position	I-Mutant	DDG	MUpro	MUpro	iStable	iStable
R38K	38	D	−0.52	D	−1	D	0.778149
T39M	39	D	−0.3	D	−0.07067	D	0.727281
P40A	40	D	−1.22	D	−1	D	0.874001
P40H	40	D	−1.33	D	−0.64308	D	0.86549
P40L	40	D	−0.46	I	0.322076	D	0.55455
P40R	40	D	−0.72	D	−0.52458	D	0.84038
P40S	40	D	−1.03	D	−0.86846	D	0.873739
T41I	41	D	−0.61	D	−0.81342	D	0.610754
T41S	41	D	−0.75	D	−0.7821	D	0.711603
M42K	42	D	−1.39	D	−0.3274	D	0.834061
M42L	42	D	−0.55	I	0.621783	D	0.52961
M42R	42	D	−0.51	I	0.176374	I	0.511703
M42T	42	D	−0.92	D	−0.48074	D	0.850459
M42V	42	D	−0.12	I	0.008988	D	0.520846
G43D	43	D	−1.09	D	−0.29052	D	0.763114
G43R	43	D	−0.42	D	−0.16415	D	0.746657
G43S	43	D	−1.22	D	−1	D	0.798367
G43V	43	D	−0.23	I	0.031282	I	0.538033
W44C	44	D	−1.57	D	−1	D	0.714696
L45P	45	D	−1.76	D	−0.69018	D	0.849339
L45R	45	D	−1.53	D	−0.66563	D	0.840544
H46L	46	I	0.82	I	0.608234	I	0.795446
H46P	46	I	−0.57	I	0.13969	I	0.801057
H46Q	46	D	−0.29	D	−0.03691	D	0.738718
H46R	46	D	−0.15	I	0.568985	I	0.643048
H46Y	46	D	0.43	I	0.546501	I	0.630856
W47C	47	D	−1.55	D	−0.74439	D	0.796946
W47G	47	D	−2.32	D	−1	D	0.843933

Continued

Table 2 Classification of GLA mutations based on increase and decrease in the stability of the protein.—cont'd

Variant	Position	I-Mutant	DDG	MUpro	MUpro	iStable	iStable
W47L	47	D	−0.77	I	0.660506	I	0.52288
W47R	47	D	−0.87	I	0.233081	I	0.571501
E48D	48	D	−0.24	I	0.479549	D	0.525217
E48K	48	D	−0.44	D	−0.07279	D	0.843613
E48Q	48	D	−0.17	D	−0.05031	D	0.843077
R49C	49	D	−0.62	D	−0.7607	D	0.735066
R49G	49	D	−1.02	D	−1	D	0.791299
R49L	49	D	−0.47	I	0.265016	I	0.617817
R49P	49	D	−1.25	D	−0.6464	D	0.727192
R49S	49	D	−1.71	D	−1	D	0.753861
F50C	50	D	−1.53	D	−1	D	0.864173
M51I	51	D	−0.79	D	−0.5544	D	0.595073
M51K	51	D	−1.2	D	−0.93403	D	0.693939
C52G	52	D	−1.18	D	−0.61894	D	0.8513
C52R	52	D	−0.12	I	0.601539	I	0.56975
C52S	52	D	−0.77	D	−0.45098	D	0.831302
C52W	52	D	−0.03	I	0.346474	I	0.575722
C52Y	52	I	−0.15	I	0.466375	I	0.774403
N53D	53	D	−0.37	D	−0.66382	D	0.79653
N53K	53	D	−0.66	D	−0.91913	D	0.812213
N53L	53	I	0.46	D	−0.3199	D	0.598709
L54F	54	D	−0.57	D	−1	D	0.875937
L54P	54	D	−1.57	D	−1	D	0.866539
D55V	55	I	0.07	D	−0.43648	D	0.529586
C56F	56	D	−0.17	I	0.071509	I	0.526663
C56G	56	D	−1.36	D	−0.87432	D	0.843808
C56S	56	D	−0.92	D	−0.95248	D	0.809072
C56Y	56	D	0.13	I	0.063033	I	0.572614

Table 2 Classification of GLA mutations based on increase and decrease in the stability of the protein.—cont'd

Variant	Position	I-Mutant	DDG	MUpro	MUpro	iStable	iStable
Q57L	57	I	0.26	I	0.948615	I	0.819326
Q57R	57	I	0.24	I	0.320278	I	0.829502
E59K	59	D	−0.8	D	−0.5697	D	0.867272
P60L	60	D	−0.28	D	−0.70667	D	0.821296
P60S	60	D	−1.33	D	−1	D	0.830302
P60T	60	D	−1.18	D	−1	D	0.840247
C63R	63	I	−0.09	D	−0.26489	D	0.588252
C63S	63	D	−0.82	D	−0.98843	D	0.851963
C63Y	63	N	−0.29	D	−0.29481	D	0.71573
I64F	64	D	−1.11	D	−1	D	0.802026
S65I	65	I	0.36	D	−0.40465	D	0.600471
S65T	65	D	−0.91	D	−0.26134	D	0.843639
E66G	66	D	−0.99	D	−0.96983	D	0.841359
E66K	66	D	−0.46	D	−0.21626	D	0.794715
E66Q	66	D	0.29	D	−0.28767	D	0.810709
L68F	68	D	−0.8	D	−1	D	0.915416
E71G	71	D	−1.13	D	−1	D	0.921803
M72I	72	D	0	D	−0.42138	D	0.522159
M72R	72	D	−0.66	D	−0.17501	D	0.587662
M72V	72	D	−1.12	I	0.162802	I	0.744719
A73E	73	D	−1.07	D	−0.64114	D	0.837038
A73V	73	D	−0.23	I	0.076742	I	0.506246
E74D	74	D	−0.2	I	0.119716	I	0.536916
M76R	76	D	−0.53	D	−0.66203	D	0.824549
M76T	76	D	−0.76	D	−1	D	0.85514
G80D	80	D	−1.45	D	−0.73572	D	0.845638
W81C	81	D	−1.59	D	−0.89565	D	0.811441
W81R	81	D	−1.02	D	−0.70241	D	0.778986

Continued

Table 2 Classification of GLA mutations based on increase and decrease in the stability of the protein.—cont'd

Variant	Position	I-Mutant	DDG	MUpro	MUpro	iStable	iStable
W81S	81	D	−1.8	D	−1	D	0.818915
D83N	83	D	−0.6	I	0.030421	I	0.686539
G85D	85	D	−1.89	D	−0.45408	D	0.802423
G85M	85	D	−0.28	I	0.031852	I	0.581776
G85S	85	D	−1.44	D	−1	D	0.836653
Y86C	86	D	0.03	D	−0.62646	D	0.798206
Y86D	86	D	−1.27	I	0.007248	I	0.599204
Y86H	86	D	−1.25	D	−0.282	D	0.802179
E87K	87	D	−0.61	I	0.174502	I	0.572432
Y88D	88	D	−1.33	D	−1	D	0.792822
L89F	89	D	−0.8	D	−1	D	0.863211
L89H	89	D	−1.81	D	−1	D	0.877414
L89P	89	D	−1.54	D	−0.78075	D	0.858917
L89R	89	D	−1.02	D	−1	D	0.849298
C90R	90	I	0.06	I	0.672157	I	0.733011
C90Y	90	I	0.13	I	0.320336	I	0.739686
I91N	91	D	−1.61	D	−1	D	0.810328
I91T	91	D	−3.98	D	−1	D	0.834016
D92G	92	I	−0.41	D	−1	D	0.554741
D92H	92	D	−0.42	D	−1	D	0.751321
D92N	92	I	−0.01	D	−1	I	0.524033
D92V	92	I	0.3	D	−1	D	0.524287
D92Y	92	I	0.12	D	−0.79906	I	0.517331
D93E	93	I	0.34	D	−1	I	0.514149
D93G	93	D	−0.58	D	−1	D	0.763231
D93N	93	D	−0.55	D	−1	D	0.701104
D93V	93	I	0.23	D	−0.67139	I	0.5
D93Y	93	I	0.49	D	−0.50103	I	0.526883

Table 2 Classification of GLA mutations based on increase and decrease in the stability of the protein.—cont'd

Variant	Position	I-Mutant	DDG	MUpro	MUpro	iStable	iStable
C94S	94	D	−1.28	D	−1	D	0.777652
C94Y	94	D	0.16	D	−0.28333	D	0.748589
W95C	95	D	−1.53	D	−1	D	0.814147
W95L	95	D	−0.77	D	−0.47567	D	0.821558
W95S	95	D	−2.27	D	−1	D	0.834221
M96I	96	D	−0.79	D	−0.70951	D	0.707847
A97P	97	I	−0.36	D	−0.41217	D	0.529191
A97V	97	D	−0.72	I	0.085474	I	0.553949
R100K	100	D	−1.45	D	−1	D	0.754342
R100T	100	D	−1.22	D	−1	D	0.773384
S102L	102	I	0.58	I	0.52487	I	0.732759
E103Q	103	D	−0.44	D	−0.08517	D	0.664164
G104D	104	D	−0.86	I	0.184601	D	0.516849
G104S	104	D	−0.27	D	−0.70526	D	0.876595
G104V	104	D	−0.59	I	0.448363	D	0.572495
R105G	105	D	−1.09	D	−0.97901	D	0.813717
L106R	106	D	−1.4	D	−0.98457	D	0.836727
Q107L	107	I	0.2	D	−0.47726	I	0.508566
Q107R	107	I	0.24	I	0.015292	I	0.878463
A108T	108	D	−0.58	D	−0.88192	D	0.741192
D109G	109	D	−1.38	D	−0.90406	D	0.826317
D109H	109	D	−0.81	D	−0.60192	D	0.813515
R112C	112	D	0.32	D	−1	D	0.695296
R112G	112	D	−0.97	D	−1	D	0.742521
R112H	112	D	−0.42	D	−1	D	0.70727
R112L	112	D	−0.09	D	−0.20044	D	0.6798
R112S	112	D	−1.26	D	−1	D	0.72452
F113I	113	D	−0.39	D	−0.10658	D	0.85972

Continued

Table 2 Classification of GLA mutations based on increase and decrease in the stability of the protein.—cont'd

Variant	Position	I-Mutant	DDG	MUpro	MUpro	iStable	iStable
F113L	113	D	−0.59	I	0.204683	D	0.631288
F113S	113	D	−1.39	D	−0.88565	D	0.904558
I117S	117	D	−3.52	D	−1	D	0.860957
R118C	118	D	−1.19	I	0.571172	I	0.659295
R118H	118	D	−1.97	I	0.81083	I	0.640721
L120P	120	D	−1.84	D	−0.99282	D	0.777236
L120S	120	D	−2.77	D	−1	D	0.801031
L120V	120	D	−0.58	D	−1	D	0.787554
A121P	121	I	−0.32	I	0.402823	I	0.8013
A121T	121	D	−0.59	D	−0.25451	D	0.808908
N122D	122	I	−0.22	D	−1	D	0.610942
Y123C	123	D	−0.97	D	−0.76944	D	0.823853
V124D	124	D	−1.41	D	−1	D	0.805668
H125L	125	I	0.06	I	0.216014	I	0.743684
H125P	125	I	0.05	D	−0.20037	D	0.630091
H125Y	125	I	−0.01	D	−0.11037	D	0.591
S126G	126	D	−1.49	D	−0.80737	D	0.849146
S126I	126	I	0.47	I	0.29281	I	0.774792
G128E	128	I	0.35	D	−0.48985	D	0.563934
L129P	129	D	−1.83	D	−1	D	0.8651
K130R	130	D	−0.54	I	0.107485	I	0.610186
L131P	131	D	−2.22	D	−1	D	0.894294
G132A	132	D	−1.45	D	−0.15235	D	0.820532
G132E	132	D	−1.17	I	0.099449	D	0.561403
G132R	132	D	−0.94	I	0.416118	I	0.562935
I133M	133	D	−1.53	D	−1	D	0.774111
Y134D	134	D	−0.91	I	0.106054	I	0.636268
Y134H	134	D	−1.09	D	−0.00923	D	0.787744

Table 2 Classification of GLA mutations based on increase and decrease in the stability of the protein.—cont'd

Variant	Position	I-Mutant	DDG	MUpro	MUpro	iStable	iStable
Y134S	134	D	−1.23	D	−0.45505	D	0.803303
A135V	135	D	−1.05	I	0.03915	I	0.5
D136E	136	D	−0.39	I	0.338921	I	0.589366
D136H	136	D	−0.7	D	−1	D	0.789731
D136Y	136	D	−0.21	D	−0.65058	D	0.767706
G138E	138	D	−0.96	D	−0.55207	D	0.825775
G138R	138	D	−1.92	D	−0.17668	D	0.819171
N139S	139	D	−0.17	D	−0.94843	D	0.744154
N139T	139	N	0.01	D	−0.91708	D	0.633022
K140T	140	D	−1.4	I	0.028802	I	0.633487
T141I	141	I	−0.46	D	−0.86659	I	0.559845
T141N	141	D	−0.71	D	−0.82769	D	0.726583
C142R	142	D	−1.66	D	−0.19878	D	0.775306
C142W	142	D	−0.81	D	−0.48443	D	0.777564
C142Y	142	D	−0.59	D	−0.39163	D	0.782694
A143P	143	I	−1.13	I	0.521445	I	0.85217
A143T	143	D	−1.12	I	0.13365	I	0.692661
G144D	144	D	−1.14	D	−0.211	D	0.752801
G144V	144	D	−1.25	D	−0.55664	D	0.747772
F145S	145	D	−1.93	D	−1	D	0.825403
P146S	146	D	−2.42	D	−0.33747	D	0.727029
G147E	147	D	−0.76	I	0.597509	I	0.535489
G147R	147	D	−0.58	I	0.387424	I	0.564393
S148N	148	D	0.37	I	0.300728	I	0.709691
S148R	148	I	−0.56	I	0.285782	I	0.810119
Y152C	152	D	−0.98	D	−0.26821	D	0.803903
D153N	153	D	−0.93	I	0.139645	I	0.582392
I154T	154	D	−2.38	D	−1	D	0.831293

Continued

Table 2 Classification of GLA mutations based on increase and decrease in the stability of the protein.—cont'd

Variant	Position	I-Mutant	DDG	MUpro	MUpro	iStable	iStable
D155H	155	D	−0.96	D	−0.81794	D	0.716983
A156D	156	D	−0.63	D	−0.37937	D	0.677105
A156T	156	D	−1.78	D	−1	D	0.708071
A156V	156	D	−1.53	D	−0.61655	D	0.730112
W162C	162	D	−2.06	D	−0.73042	D	0.817386
W162G	162	D	−2.08	D	−0.92991	D	0.858454
W162L	162	D	−1.12	I	0.068047	I	0.520098
W162R	162	D	−2.05	D	−0.14068	D	0.793971
G163V	163	D	−2.51	D	−0.55613	D	0.883396
V164G	164	D	−2.34	D	−1	D	0.795999
V164L	164	D	−1.14	D	−0.79564	D	0.766795
D165G	165	D	−0.73	D	−0.89583	D	0.82234
D165H	165	D	−0.38	D	−0.47768	D	0.789895
D165V	165	D	−0.32	I	0.302325	I	0.574637
D165Y	165	I	0.44	D	−0.03938	I	0.524077
L166G	166	D	−3.26	D	−1	D	0.869905
L166V	166	D	−1.82	D	−1	D	0.832756
L167P	167	D	−1.65	D	−1	D	0.757316
L167Q	167	D	−2.02	D	−1	D	0.770708
K168N	168	I	0.01	D	−0.73155	I	0.5
K168R	168	D	0.03	D	−0.02284	D	0.719928
F169I	169	D	−0.57	D	−1	D	0.708194
F169S	169	D	−1.7	D	−1	D	0.811083
D170G	170	D	−1.13	D	−0.56011	D	0.76604
D170H	170	D	−0.74	D	−0.7522	D	0.748259
D170N	170	D	−0.84	D	−0.51444	D	0.697513
D170V	170	D	−1.08	D	−0.30012	D	0.721606
G171C	171	D	−1.39	D	−0.79884	D	0.739882

Table 2 Classification of GLA mutations based on increase and decrease in the stability of the protein.—cont'd

Variant	Position	I-Mutant	DDG	MUpro	MUpro	iStable	iStable
G171D	171	D	−1.83	D	−0.47434	D	0.741632
G171R	171	D	−0.48	D	−0.36537	D	0.733938
G171S	171	D	−1.36	D	−0.93545	D	0.774623
C172F	172	D	−0.2	D	−1	D	0.73589
C172G	172	D	−1.55	D	−1	D	0.813376
C172R	172	D	−0.57	D	−0.93113	D	0.721138
C172S	172	D	−1.02	D	−1	D	0.782445
C172W	172	D	−0.24	D	−1	D	0.730507
C172Y	172	D	0.24	D	−1	D	0.744095
C174G	174	D	−0.24	D	−1	D	0.872285
C174R	174	D	−0.26	D	−0.6715	D	0.832314
D175E	175	I	−0.05	I	0.370673	I	0.73926
D175N	175	D		I	0.286612	I	0.559274
L180F	180	D	−1	D	−1	D	0.844842
G183A	183	D	−0.77	D	−0.97554	D	0.727689
G183D	183	D	−1.55	D	−0.75614	D	0.734364
G183R	183	D	−0.57	D	−0.06722	D	0.733641
G183S	183	D	−1.51	D	−1	D	0.777358
G183V	183	D	−0.62	D	−1	D	0.758078
Y184C	184	D	−0.79	D	−1	D	0.750425
Y184N	184	D	−1.12	D	−1	D	0.728051
K185E	185	D	−0.9	D	0.231976	I	0.541142
H186P	186	I	0.4	D	−0.84003	D	0.691612
H186R	186	D	−0.31	D	−0.36159	D	0.855531
M187I	187	D	−1.03	D	−0.72977	D	0.823121
M187R	187	D	−0.83	D	−0.53972	D	0.864296
M187T	187	D	−0.81	D	−1	D	0.863211
M187V	187	D	0.05	D	−0.28152	D	0.881337

Continued

Table 2 Classification of GLA mutations based on increase and decrease in the stability of the protein.—cont'd

Variant	Position	I-Mutant	DDG	MUpro	MUpro	iStable	iStable
L189F	189	D	−1.32	D	−0.67149	D	0.816265
L191P	191	D	−1.51	D	−1	D	0.781835
L191Q	191	D	−2.81	D	−1	D	0.780277
R193S	193	D	1.86	D	−0.8949	D	0.744956
T194I	194	D	−0.93	D	−0.28309	D	0.658293
G195V	195	D	−0.76	D	−0.01603	D	0.887737
R196S	196	D	−1.79	D	−0.74601	D	0.714684
S197N	197	D	−1.01	I	0.746583	I	0.518253
I198T	198	D	−3.02	D	−0.59665	D	0.889606
V199A	199	D	−1.99	D	−1	D	0.847507
V199G	199	D	−3.2	D	−1	D	0.86549
V199M	199	D	−1.42	D	−0.64698	D	0.810404
Y200C	200	D	−1.34	D	−1	D	0.834828
S201A	201	D	−0.67	D	−0.75781	D	0.669435
S201F	201	I	0.45	I	0.433728	I	0.748278
S201Y	201	I	0	I	0.768291	I	0.801808
C202R	202	D	−0.15	I	0.265784	I	0.603665
C202W	202	D	−0.69	I	0.121514	I	0.614096
C202Y	202	D	−0.14	I	0.162963	I	0.586037
E203D	203	D	−0.31	D	0.551687	D	0.722737
E203K	203	D	−0.69	D	−1	D	0.741489
E203V	203	D	0.27	D	−0.04291	D	0.754498
W204C	204	D	−1.74	D	−1	D	0.841872
W204R	204	D	−0.92	D	−0.93387	D	0.829496
P205L	205	D	−0.13	D	−0.16917	D	0.88475
P205R	205	D	−0.62	D	−0.36888	D	0.869006
P205S	205	D	−1.71	D	−0.73341	D	0.892832
P205T	205	D	−1.69	D	−0.79066	D	0.882297

Table 2 Classification of GLA mutations based on increase and decrease in the stability of the protein.—cont'd

Variant	Position	I-Mutant	DDG	MUpro	MUpro	iStable	iStable
L206P	206	D	−1.38	D	−1	D	0.83251
L206V	206	D	−1.44	D	−0.97782	D	0.843936
Y207C	207	D	−0.96	D	−1	D	0.808226
Y207H	207	D	−1.2	D	−1	D	0.826583
Y207S	207	D	−1.52	D	−1	D	0.825951
P210L	210	D	−0.02	I	0.699975	I	0.641571
P210S	210	D	−1.71	D	−1	D	0.706285
F211L	211	D	−1.03	D	−0.56814	D	0.803642
K213M	213	D	−0.16	I	0.06459	I	0.58386
K213R	213	D	0	I	−0.23452	D	0.790682
P214L	214	D	−0.1	I	0.495143	I	0.53574
P214S	214	D	−1.91	D	−0.43796	D	0.805375
N215D	215	D	−0.33	D	−1	D	0.861557
N215H	215	D	−0.71	D	−1	D	0.878285
N215S	215	D	−1.27	D	−1	D	0.882505
Y216C	216	D	−0.96	D	−0.31073	D	0.77544
Y216D	216	D	−1.28	D	−0.08385	D	0.756962
I219M	219	D	−1.45	D	−1	D	0.858291
I219N	219	D	−0.84	D	−1	D	0.86418
I219T	219	D	−1.53	D	−1	D	0.774111
R220P	220	D	−0.91	D	−0.13996	D	0.71026
R220Q	220	D	−0.78	D	−0.43728	D	0.692872
Q221H	221	D	−0.77	D	−1	D	0.822978
Y222D	222	D	−1.56	D	−1	D	0.700381
C223G	223	D	−1.44	D	−0.16765	D	0.840601
C223R	223	D	−0.31	I	0.900485	I	0.520851
C223Y	223	D	−0.39	I	0.614662	I	0.5
N224D	224	D	−0.53	D	0	D	0.647804

Continued

Table 2 Classification of GLA mutations based on increase and decrease in the stability of the protein.—cont'd

Variant	Position	I-Mutant	DDG	MUpro	MUpro	iStable	iStable
N224S	224	D	−0.61	D	0	D	0.68006
H225D	225	D	−0.78	I	0.474637	I	0.642604
H225R	225	D	−0.25	I	0.34369	I	0.626128
W226C	226	D	−1.75	D	−0.72463	D	0.829388
W226R	226	D	−1.74	D	−0.44566	D	0.824973
R227P	227	D	−1.07	I	0.052525	I	0.587345
R227Q	227	D	−1.79	D	−0.35752	D	0.770654
N228S	228	D	−0.5	D	−0.89513	D	0.811997
F229L	229	D	−1.39	I	0.026867	I	0.587571
A230T	230	D	−1.17	D	−1	D	0.748273
D231G	231	D	−0.97	D	−0.69104	D	3,679,057
D231N	231	D	−2.98	D	−0.77319	D	0.852577
D231V	231	D	−0.46	I	0.240901	I	0.5
I232S	232	D	−2.18	D	−1	D	0.740349
I232T	232	D	−2.1	D	−1	D	0.751394
D234E	234	D	−1.64	D	−0.76597	D	0.547113
D234N	234	D	−1.42	D	−0.93227	D	0.52368
D234Y	234	I	−0.51	D	−0.16373	I	0.718986
S235C	235	D	−1.84	D	−1	D	0.817854
S235F	235	I	0	D	−0.72996	D	0.634038
S235Y	235	I	−0.45	D	−0.51209	D	0.589974
W236C	236	D	−1.91	D	−1	D	0.79838
W236L	236	D	−2.05	D	−0.43675	D	0.809535
W236R	236	D	−0.3	D	−0.91229	D	0.784798
S238N	238	D	−0.79	D	−0.58246	D	0.868731
S238R	238	I	−0.22	I	0.072459	I	0.668219
I239T	239	D	−2.13	D	−1	D	0.890665
S241G	241	D	−1.16	D	−0.71497	D	0.86706

Table 2 Classification of GLA mutations based on increase and decrease in the stability of the protein.—cont'd

Variant	Position	I-Mutant	DDG	MUpro	MUpro	iStable	iStable
I242F	242	D	−1.76	D	−0.27923	D	0.858965
I242N	242	D	−2.57	D	−0.61261	D	0.831441
I242V	242	D	−1.15	D	−0.54082	D	0.864881
L243F	243	D	−0.85	D	−1	D	0.802181
L243W	243	D	0.51	D	−1	D	0.773013
D244H	244	D	−0.96	D	−1	D	0.830733
D244N	244	D	−3.42	D	−1	D	0.80294
W245G	245	D	0	D	−1	D	0.811199
W245L	245	D	−1.01	I	0.058645	I	0.563586
T246S	246	D	−0.84	D	−0.1946	D	0.78135
S247C	247	D	−0.87	D	−0.34819	D	0.712582
S247P	247	I	−0.66	I	0.254341	I	0.804215
F248L	248	D	−1.46	I	0.314234	I	0.561435
N249K	249	D	−0.49	I	0.108487	I	0.702124
Q250P	250	D	−0.78	I	0.516201	I	0.632155
R252T	252	D	−1.42	D	−0.94261	D	0.90575
I253S	253	D	−2.34	D	−1	D	0.904351
I253T	253	D	−2.35	D	−1	D	0.908082
V254A	254	D	−1.4	D	−1	D	0.833113
A257D	257	D	−0.5	I	0.192483	I	0.646337
A257G	257	D	−1.15	D	−0.51954	D	0.807059
A257P	257	D	−0.66	D	−0.16524	D	0.771937
G258R	258	D	−1.19	D	−0.15258	D	0.766976
G258V	258	N	−0.95	D	−0.5854	D	0.693634
P259L	259	D	0.63	I	0.149896	I	0.591966
P259R	259	D	−0.75	D	−0.01984	D	0.703994
G260A	260	D	−1.3	I	0.719792	I	0.522936
G260E	260	D	−0.79	I	0.651404	I	0.507583

Continued

Table 2 Classification of GLA mutations based on increase and decrease in the stability of the protein.—cont'd

Variant	Position	I-Mutant	DDG	MUpro	MUpro	iStable	iStable
G261D	261	D	−1.44	D	−0.79419	D	0.587799
G261S	261	D	−1.67	D	−1	D	0.62755
G261V	261	D	−0.992	D	−0.87839	D	0.597948
W262C	262	D	−1.67	D	−0.3984	D	0.696777
W262L	262	D	−1.13	I	0.284734	I	0.63067
W262R	262	D	−1.14	I	0.131505	I	0.670181
N263S	263	D	−0.98	D	−0.6564	D	0.817513
D264A	264	D	−1.19	D	−0.67288	D	0.737436
D264V	264	D	−1.63	D	−0.65024	D	0.756801
D264Y	264	I	−1.4	D	−0.61875	I	0.517702
P265L	265	D	−0.59	D	0.964264	D	0.773635
P265R	265	D	−0.95	D	−1	D	0.756496
P265S	265	D	−1.56	D	−1	D	0.801906
D266A	266	D	−1.11	D	−0.26831	D	0.871203
D266E	266	I	−0.43	D	−0.82771	D	0.664087
D266H	266	I	−0.44	D	−0.86175	D	0.68536
D266N	266	D	−1.17	D	−1	D	0.867713
D266V	266	I	−1.07	D	−0.33497	D	0.67978
D266Y	266	I	−0.58	D	−0.39794	D	0.652261
M267I	267	D	−1.12	D	−1	D	0.714106
M267R	267	D	−0.56	D	−1	D	0.75553
M267T	267	D	−0.76	D	−1	D	0.781466
L268S	268	D	−2.2	D	−1	D	0.854008
V269A	269	D	−1.37	D	−1	D	0.855304
V269E	269	D	−1.38	D	−0.96124	D	0.842325
V269G	269	D	−2.35	D	−1	D	0.86969
V269L	269	D	−1.45	D	−0.71716	D	0.86294
V269M	269	D	−1.02	D	−0.3127	D	0.820875

Table 2 Classification of GLA mutations based on increase and decrease in the stability of the protein.—cont'd

Variant	Position	I-Mutant	DDG	MUpro	MUpro	iStable	iStable
I270M	270	N	−1.53	D	−1	D	0.731924
I270T	270	D	−2.34	D	−1	D	0.865757
G271A	271	D	−0.81	D	−0.45699	D	0.730318
G271C	271	D	−1.36	D	−0.00961	D	0.742986
G271D	271	D	−1.03	I	0.411573	I	0.665078
G271S	271	D	−1.6	D	−0.23074	D	0.776771
G271V	271	D	−0.79	I	0.191559	I	0.639166
N272K	272	D	−1.55	D	−1	D	0.66214
N272S	272	D	−1.61	D	−1	D	0.706129
F273L	273	D	−1.63	I	1	I	0.648234
F273S	273	D	−2.16	D	−0.22966	D	0.768993
G274S	274	D	−1.59	D	−0.86694	D	0.749307
G274V	274	D	−0.6	D	−0.37851	D	0.719207
L275F	275	D	−0.69	D	−1	D	0.859995
L275H	275	D	−2.07	D	−1	D	0.876809
S276G	276	D	−0.8	D	−0.22567	D	0.703656
S276N	276	I	0.35	D	−0.13894	I	0.607596
W277C	277	D	−1.62	D	−0.67595	D	0.71047
W277G	277	D	−2.1	D	−1	D	0.77869
Q279E	279	I	0.52	D	−0.31655	D	0.538316
Q279H	279	D	−0.5	D	−0.68394	D	0.805435
Q279K	279	D	−0.29	D	−0.68547	D	0.791425
Q279R	279	D	0.14	D	−0.50449	D	0.774624
Q280H	280	D	−0.5	D	−1	D	0.761287
Q280K	280	D	−0.29	D	−0.82984	D	0.731252
Q280L	280	I	0.08	D	−0.41125	I	0.532669
T282A	282	D	−0.63	D	−0.58759	D	0.845664
T282I	282	I	−0.07	I	0.600932	I	0.76269

Continued

Table 2 Classification of GLA mutations based on increase and decrease in the stability of the protein.—cont'd

Variant	Position	I-Mutant	DDG	MUpro	MUpro	iStable	iStable
T282N	282	I	−0.26	I	0.030138	I	0.733847
Q283P	283	D	−0.47	I	0.108527	I	0.720206
Q283R	283	I	0.19	D	−0.00297	I	0.527538
M284T	284	D	−0.27	D	−0.80982	D	0.771861
M284V	284	N	−0.53	D	−0.1729	D	0.670532
A285D	285	D	−0.75	I	0.609932	I	0.52768
A285P	285	D	−1.65	I	0.140349	I	0.5
W287C	287	D	−0.32	I	0.114801	I	0.57561
W287G	287	D	−2.25	D	−0.32385	D	0.844901
A288D	288	D	−1.11	I	1	D	0.528353
A288P	288	D	−0.46	I	0.938073	D	0.546679
I289F	289	D	−2.06	D	−1	D	0.750892
I289S	289	D	−2.08	D	−1	D	0.774617
I289V	289	D	−0.79	D	−1	D	0.764451
M290I	290	D	−0.66	D	−0.25481	D	0.767032
M290L	290	D	0.18	I	0.466415	I	0.526021
A291T	291	D	−0.98	D	−0.88667	D	0.7358
A292G	292	D	−1.43	D	−1	D	0.778941
A292P	292	D	−0.72	D	−0.93777	D	0.719198
A292T	292	D	−1.9	D	−1	D	0.740177
A292V	292	D	−0.27	I	0.347818	I	0.657849
P293A	293	D	−0.96	D	−0.98352	D	0.832601
P293H	293	D	−1.37	D	−0.9677	D	0.847667
P293L	293	I	−0.07	I	0.106515	I	0.760406
P293R	293	D	−0.64	D	−0.36671	D	0.820827
P293S	293	D	−1.45	D	−1	D	0.846441
P293T	293	D	−1.07	D	−1	D	0.835119
L294S	294	D	−2.49	D	−0.6576	D	0.865783

Table 2 Classification of GLA mutations based on increase and decrease in the stability of the protein.—cont'd

Variant	Position	I-Mutant	DDG	MUpro	MUpro	iStable	iStable
F295C	295	D	−1.62	D	−0.81635	D	0.778763
M296I	296	D	−0.29	D	−0.39569	D	0.606691
M296L	296	D	−0.92	I	0.158605	I	0.656107
M296T	296	D	−1.02	D	−0.73305	D	0.723059
M296V	296	D	−1.07	D	0	D	0.531512
S297C	297	D	−0.41	D	−0.45996	D	0.686679
S297F	297	I	0.25	I	0.057024	I	0.797818
N298H	298	D	−1.02	D	0	D	0.508254
N298K	298	D	−1.54	D	−0.34231	D	0.64251
N298S	298	D	−0.72	D	−0.56221	D	0.685065
D299E	299	I	0	D	−0.44616	D	0.57249
D299G	299	D	−0.92	D	−1	D	0.864967
L300F	300	D	−0.35	D	−1	D	0.868875
L300H	300	D	−2.35	D	−1	D	0.874473
L300P	300	0	0		0	0	0
R301G	301	D	−1.24	D	−1	D	0.840113
R301L	301	D	−0.5	D	−1	D	0.821722
R301P	301	D	−0.77	D	−1	D	0.816928
R301Q	301	D	−0.77	D	−1	D	0.819828
I303F	303	D	−1.48	D	−0.67376	D	0.82055
I303N	303	D	−1.5	D	−0.79048	D	0.778642
S304N	304	I	0	D	−0.33511	D	0.508456
S304T	304	I	−0.15	D	0.446898	D	0.553927
A307T	307	D	−0.67	D	−1	D	0.860309
A307V	307	D	−0.36	D	0.636171	D	0.853756
K308N	308	I	−0.17	D	−0.22304	D	0.509666
A309P	309	D	−0.34	D	−0.82963	D	0.866543
A309V	309	D	−0.25	I	0.158654	D	0.581901

Continued

Table 2 Classification of GLA mutations based on increase and decrease in the stability of the protein.—cont'd

Variant	Position	I-Mutant	DDG	MUpro	MUpro	iStable	iStable
L310F	310	D	−0.73	D	−1	D	0.804066
L310R	310	D	−1.63	D	−1	D	0.78042
L311F	311	D	−0.76	D	−0.19118	D	0.780822
L311P	311	D	−1.67	D	−0.50148	D	0.779361
L311R	311	D	−1.6	I	0.232712	I	0.612917
L311V	311	D	−1.76	I	0.313748	I	0.511974
Q312H	312	D	−0.79	D	−1	D	0.805274
Q312R	312	D	0.01	D	−0.62793	D	0.765421
D313G	313	D	−0.55	D	−0.89791	D	0.792614
D313N	313	D	−0.56	D	−0.55603	D	0.710334
D313Y	313	D	−0.46	D	−0.30939	D	0.716995
D315G	315	D	0.24	D	−1	D	0.879394
D315N	315	D	−0.39	D	−0.33357	D	0.8034
V316A	316	D	−0.74	D	−1	D	0.831878
V316E	316	D	−0.53	D	−0.88411	D	0.851438
V316G	316	D	−1.76	D	−1	D	0.887293
V316I	316	I	−0.33	D	−0.86098	D	0.563348
I317N	317	D	−1.54	D	−1	D	0.867592
I317S	317	D	−2.32	D	−1	D	0.902263
I317T	317	D	−2.02	D	−1	D	0.899873
I319F	319	D	−1.55	D	−1	D	0.792283
I319T	319	D	−2.17	D	−1	D	0.808881
N320I	320	I	1.47	I	0.235502	I	0.828012
N320K	320	D	−1.02	D	−0.09539	D	0.809937
N320Y	320	D	−0.35	I	0.014311	I	0.59458
Q321E	321	D	0.1	D	−0.87826	D	0.660292
Q321H	321	D	−0.61	D	−1	D	0.688579
Q321L	321	D	−0.31	D	−0.06685	D	0.680587

Table 2 Classification of GLA mutations based on increase and decrease in the stability of the protein.—cont'd

Variant	Position	I-Mutant	DDG	MUpro	MUpro	iStable	iStable
Q321R	321	D	0.09	D	−0.92404	D	0.640268
D322E	322	I	−0.19	D	−0.20735	D	0.552891
D322N	322	D	−0.93	D	−0.64799	D	0.766999
P323R	323	D	−0.77	D	−0.63409	D	0.821401
P323T	323	D	−0.76	D	−1	D	0.833545
G325D	325	D	−0.89	D	−0.64164	D	0.803305
G325R	325	D	−0.49	D	−0.4828	D	0.807279
G325S	325	D	−1.58	D	−1	D	0.841188
Q327E	327	I	0.09	I	0.783599	I	0.810581
Q327K	327	D	−0.25	I	1	I	0.615769
Q327L	327	D	−0.14	I	1	I	0.601618
Q327R	327	I	0.14	I	1	I	0.81571
G328A	328	D	−1.08	D	−1	D	0.797679
G328E	328	D	−0.39	D	−0.40562	D	0.809275
G328R	328	D	−0.67	D	−0.4446	D	0.797293
G328V	328	D	−0.83	D	−1	D	0.809334
G328W	328	I	−0.18	D	−0.63809	D	0.649772
Q330R	330	I	0.15	I	0.743799	I	0.840921
L331F	331	D	−0.75	D	−1	D	0.875881
R332G	332	D	−1	D	−0.80891	D	0.809533
R332K	332	D	−0.74	D	−0.51274	D	0.774501
Q333R	333	I	0.21	I	0.338082	I	0.774365
G334E	334	I	−0.49	D	−0.77047	I	0.530944
E338K	338	D	−0.99	D	−1	D	0.836227
V339A	339	D	−1.2	D	−1	D	0.821935
V339E	339	D	−1.1	D	−0.53449	D	0.820606
V339G	339	D	−2.44	D	−1	D	0.866957
W340R	340	D	−1.66	D	−0.70029	D	0.775024

Continued

Table 2 Classification of GLA mutations based on increase and decrease in the stability of the protein.—cont'd

Variant	Position	I-Mutant	DDG	MUpro	MUpro	iStable	iStable
E341D	341	D	−0.45	D	−0.10913	D	0.843422
E341K	341	D	−1.12	D	−0.39012	D	0.848423
E341L	341	D	0.1	I	0.968671	D	0.564684
R342G	342	D	−1.05	D	−1	D	0.838481
R342L	342	D	−0.42	D	−0.07536	D	0.779827
R342P	342	D	−0.73	D	−0.61545	D	0.765553
R342Q	342	D	−1.06	D	−0.93511	D	0.778359
P343S	343	D	−1.78	D	−0.9862	D	0.875052
P343L	343	D	−0.42	D	−0.02634	D	0.869836
L344P	344	D	−1.96	D	−0.27298	D	0.801824
S345P	345	I	−0.18	I	0.289889	I	0.833745
A348P	348	D	−0.46	D	−0.3761	D	0.730256
W349R	349	D	−1.01	D	−0.62156	D	0.73886
W349S	349	D	−1.81	D	−1	D	0.78954
A350P	350	D	−0.61	I	0.410037	D	0.524304
A352D	352	D	−0.64	I	0.14045	I	0.5
A352G	352	D	−1.27	D	−1	D	0.858762
A352P	352	D	−0.51	D	−0.57639	D	0.826771
A352V	352	D	−0.03	D	−0.1784	D	0.840618
I354K	354	D	−1.99	D	−0.4982	D	0.783311
I354V	354	D	−0.89	D	−0.65911	D	0.778076
N355K	355	D	−0.67	D	−0.62984	D	0.851502
R356G	356	D	−1.3	D	−0.66374	D	0.770573
R356P	356	D	−0.81	I	0.275039	I	0.701351
R356Q	356	D	−0.87	D	−0.37487	D	0.729307
R356W	356	D	−0.81	I	0.131433	I	0.719353
E358A	358	D	−0.7	D	−1	D	0.814306
E358D	358	D	−0.17	D	−0.79676	D	0.81403

Table 2 Classification of GLA mutations based on increase and decrease in the stability of the protein.—cont'd

Variant	Position	I-Mutant	DDG	MUpro	MUpro	iStable	iStable
E358G	358	D	−1.18	D	−1	D	0.84948
E358K	358	D	−0.98	D	−1	D	0.831177
I359T	359	D	−2.38	D	−1	D	0.85937
G360A	360	D	−1	D	−1	D	0.808625
G360C	360	D	−0.14	D	−1	D	0.822
G360D	360	D	−1.25	D	−0.99313	D	0.82235
G360S	360	D	−1.72	D	−1	D	0.841888
G361A	361	D		D	−0.68124	D	0.722134
G361E	361	D	−0.99	D	−0.7806	D	0.753707
G361R	361	D	−0.86	D	−0.43015	D	0.718042
P362L	362	D	−0.29	I	0.378817	D	0.543971
P362T	362	D	−1.51	D	−0.66281	D	0.848242
R363C	363	D	−1.04	I	0.939126	I	0.539854
R363H	363	D	−1.07	I	0.849376	I	0.514594
R363P	363	D	−0.85	I	0.922037	I	0.542754
Y365N	365	D	−1.42	D	−0.45528	D	0.71225
T366N	366	D	−1.06	D	−0.60718	D	0.774511
A368P	368	I	−0.14	I	−0.21296	D	0.635505
A368T	368	D	−0.81	D	−0.02046	D	0.872657
L372P	372	D	−1.69	D	−0.49957	D	0.821952
L372Q	372	D	−2.06	D	−1	D	0.823127
L372R	372	D	−2.01	D	−0.37328	D	0.79559
G373D	373	D	−1.83	D	−0.34052	D	0.766895
G373R	373	D	−1	D	−0.26361	D	0.781618
G373S	373	D	−1.65	D	−1	D	0.805316
K374T	374	D	−0.57	D	−0.29177	D	0.714099
G375A	375	D	−1.39	I	0.831844	I	0.563192
A377D	377	D	−0.43	I	0.376061	I	0.578387

Continued

Table 2 Classification of GLA mutations based on increase and decrease in the stability of the protein.—cont'd

Variant	Position	I-Mutant	DDG	MUpro	MUpro	iStable	iStable
C378R	378	D	−0.38	D	−0.16461	D	0.777519
C378Y	378	D	−0.64	D	−0.69687	D	0.786267
A381P	381	I	−0.02	D	−0.03286	D	0.632283
C382W	382	D	−0.03	D	−0.29695	D	0.847602
C382Y	382	D	−0.08	D	−0.33068	D	0.855648
I384N	384	D	−1.49	D	−1	D	0.83898
T385A	385	D	−1.16	D	−0.03877	D	0.645135
T385P	385	D	−0.91	I	0.026107	I	0.69963
Q386P	386	D	−0.7	I	0.279363	I	0.629326
L387P	387	D	−1.63	D	−1	D	0.843327
P389A	389	D	−1.53	D	−1	D	0.785968
P389L	389	D	−0.48	D	−0.7346	D	0.830136
P389R	389	D	−0.91	D	−0.39086	D	0.813918
V390M	390	D	−1.21	D	−0.33326	D	0.874752
K391T	391	D	−0.26	D	−0.65851	D	0.810194
R392S	392	D	−1.36	D	−1	D	0.832855
R392T	392	D	−0.79	D	−1	D	0.828646
L394P	394	D	−1.61	D	−0.89557	D	0.812946
G395A	395	D	−0.63	D	−0.12784	D	0.787826
G395E	395	I	−0.53	I	0.0184	I	0.742985
F396Y	396	D	−0.7	D	−0.4666	D	0.789283
E398A	398	D	−0.79	D	−0.8325	D	0.843749
E398K	398	D	−1.05	D	−0.85642	D	0.839556
W399S	399	D	−2.06	D	−0.61366	D	0.818602
L403S	403	D	−2.83	D	−1	D	0.874415
S405R	405	I	0.04	I	0.007865	I	0.688172
H406R	406	0	0		0	0	0
I407K	407	D	−2.19	D	−1	D	0.738993

Table 2 Classification of GLA mutations based on increase and decrease in the stability of the protein.—cont'd

Variant	Position	I-Mutant	DDG	MUpro	MUpro	iStable	iStable
I407R	407	D	0	D	−0.28287	D	0.711831
I407V	407	D	−0.69	D	−0.63334	D	0.740544
N408Y	408	I	0.18	D	−0.44147	D	0.592149
P409A	409	D	−1.51	D	−1	D	0.790784
P409H	409	D	−1.64	D	−1	D	0.837184
P409S	409	D	−1.82	D	−1	D	0.842818
P409T	409	D	−1.47	D	−1	D	0.825812
T410A	410	D	−0.92	D	−0.24744	D	0.821924
T410I	410	D	−0.71	D	−0.37676	D	0.748646
T410K	410	D	−1.16	D	−0.18216	D	0.813588
T410P	410	D	−0.98	D	−0.14968	D	0.812854
G411D	411	D	−1.08	D	−0.22401	D	0.879206
T412N	412	D	−0.76	D	−1	D	0.727321
V413I	413	D	0.17	D	−0.22252	D	0.816484
L414S	414	D	−2.68	D	−1	D	0.806286
L415F	415	N	−0.65	D	−0.54743	D	0.735651
L415P	415	D	−0.99	D	−0.59356	D	0.815527
Q416P	416	D	−0.45	D	−0.39151	D	0.713425
L417P	417	D	−1.41	D	−1	D	0.84044
L417R	417	D	−1.34	D	−1	D	0.834465
E418G	418	D	−1.07	D	−1	D	0.786457
N419D	419	I	0.15	I	0.002671	I	0.788345
N419I	419	I	1.41	I	0.066649	I	0.834726
M421V	421	D	0.02	I	0	D	0.766887
L425F	425	D	−0.36	D	−1	D	0.85774
L429P	429	D	−1.45	D	−1	D	0.697375

D, decrease; I, increase.

3.2 Phenotyping analysis

Phenotyping analysis was carried out using the SNPeffect 4.0 server. From the analysis we observed that the mutation D92Y alone was predicted to increase the amyloid propensity as well as severely reduces the protein stability (Table 3).

3.3 Protein-dynamics without DGJ

Protein-dynamics without ligand were performed for 30 ns using GROMACS package. The behavior of proteins with mutations was compared against the native protein. Root Mean Square Deviation (RMSD) of the protein was analyzed using the GROMACS package to understand the convergence of the molecules over 30 ns. From the RMSD plot, we observed that all the protein molecules with mutations and the native protein molecule attained the convergence over the period of 30 ns. We observed the least deviation pattern in the native molecule, and the highest deviation was observed in the protein molecule with D92Y mutation. Protein molecules with mutations D170V, D266N, and C142Y showed deviations in between the native protein and protein molecule with D92Y mutation (Fig. 1). However, the convergence at the end of 30 ns provided confidence for the further analysis. We further analyzed the radius of gyration to understand the change in the compactness the different mutant proteins. Similar to the RMSD plot, least deviation pattern was observed in the native protein, and the highest deviation was observed in the protein

Table 3 Phenotyping analysis deleterious and destabilizing active site mutations of GLA protein.

Molecules	TANGO	WALTZ	LIMBO	FoldX
C142Y	Does not affect the aggregation tendency	Does not affect the amyloid propensity	Does not affect the chaperone binding tendency	Reduces the protein stability
D92Y	Does not affect the aggregation tendency	Increases the amyloid propensity	Does not affect the chaperone binding tendency	Severely reduces the protein stability
D170V	Does not affect the aggregation tendency	Does not affect the amyloid propensity	Does not affect the chaperone binding tendency	Reduces the protein stability
D266N	Does not affect the aggregation tendency	Does not affect the amyloid propensity	Does not affect the chaperone binding tendency	Reduces the protein stability

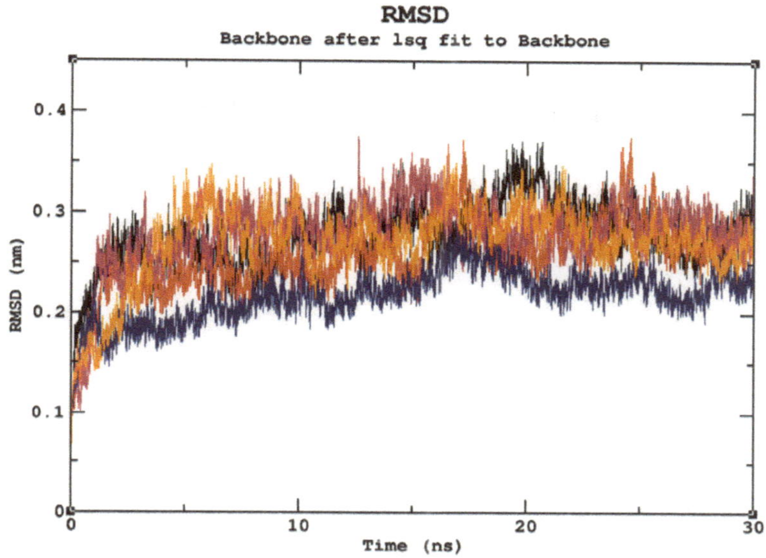

Fig. 1 The root mean square deviation plot of the native and mutants for 30 ns. *Color code:* Native: Blue; D92Y: Red; C142Y: Black; D170V: Orange; D266N: Magenta.

molecule with the D92Y mutation. Protein molecules with the mutations D170V, D266N, and C142Y showed deviations in between the native protein and protein molecule with D92Y mutation (Fig. 2). Finally, change in a number of intramolecular hydrogen bonds due to the mutation was analyzed. From the plot, we observed drastic loss of intramolecular hydrogen in the protein structure with the mutation D92Y, followed by the protein structure with the mutation C142Y. The native protein structure retained a higher number of intramolecular hydrogen bonds throughout the simulation period. The proteins with D170V and D266N showed a number of intramolecular molecular similar to that of the native protein molecule except at ~9000 ps, the protein structure with D170V mutation showed drastic fall in a number of intramolecular hydrogen bonds (Fig. 3).

3.4 Protein-DGJ interaction study

AutoDock standalone package was used to study the interaction between the protein molecules (native and mutant) and the DGJ. The binding energy between DGJ and native protein was found to be −7.73 kcal/mol. The binding energy of the protein with C142Y, D92Y, D170V, and D266N mutations were found to possess binding energies of −6.94, −5.16, −6.62, and − 6.54 kcal/mol respectively.

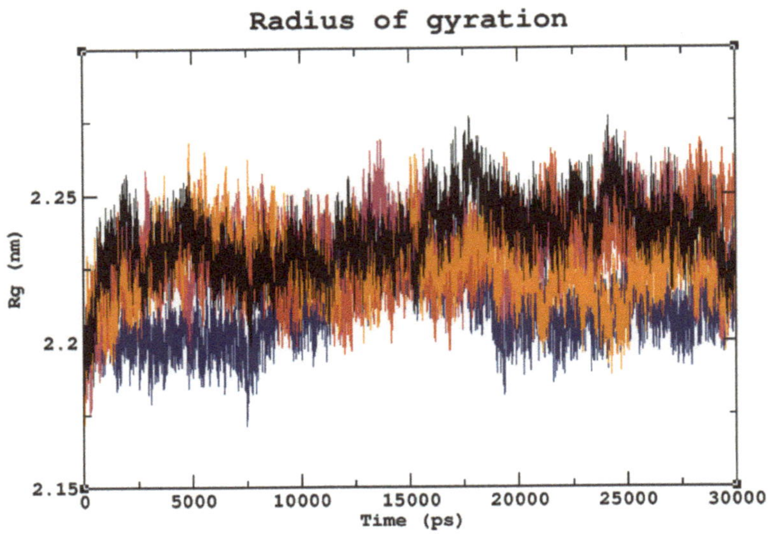

Fig. 2 The radius of gyration plot of the native and mutants for 30 ns. *Color code*: Native: Blue; D92Y: Red; C142Y: Black; D170V: Orange; D266N: Magenta.

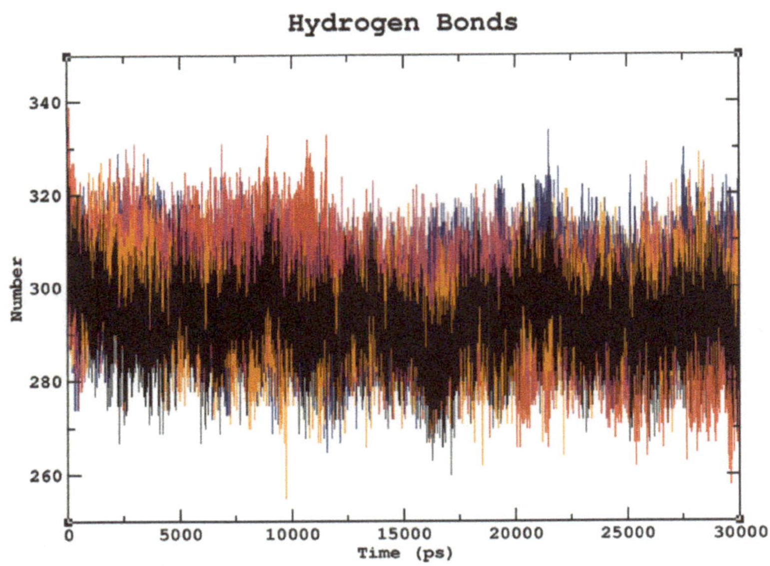

Fig. 3 The hydrogen bond analysis of the native and mutants for 30 ns. *Color code*: Native: Blue; D92Y: Red; C142Y: Black; D170V: Orange; D266N: Magenta.

3.5 Protein-dynamics with DGJ

Protein-dynamics with DGJ was performed for 50 ns to study the behavior of the DGJ with the native protein and protein with the D92Y mutation. RMSD was used to analyze the convergence of the protein complexes over the period of 50 ns. From the RMSD plot, we observed that the protein-DGJ complex with D92Y mutation showed a minor deviation than the native protein-DGJ complex (Fig. 4). However both the complexes showed convergence at the end of the simulation period and provided confidence for the further analysis. Root Mean Square Fluctuation (RMSF) was analyzed to understand the movement of each residue due to the binding of DGJ. From the RMSF plot, we observed that the overall fluctuation of the residues to be minimal in the protein-DGJ complex with D92Y mutation when compared to the native protein-DGJ complex (Fig. 5). Finally, change in compactness of the protein due to the DGJ binding was analyzed using the radius of gyration parameter. The radius of gyration was found to be lesser in the protein-DGJ complex with D92Y mutation when compared to the native protein-DGJ complex (Fig. 6).

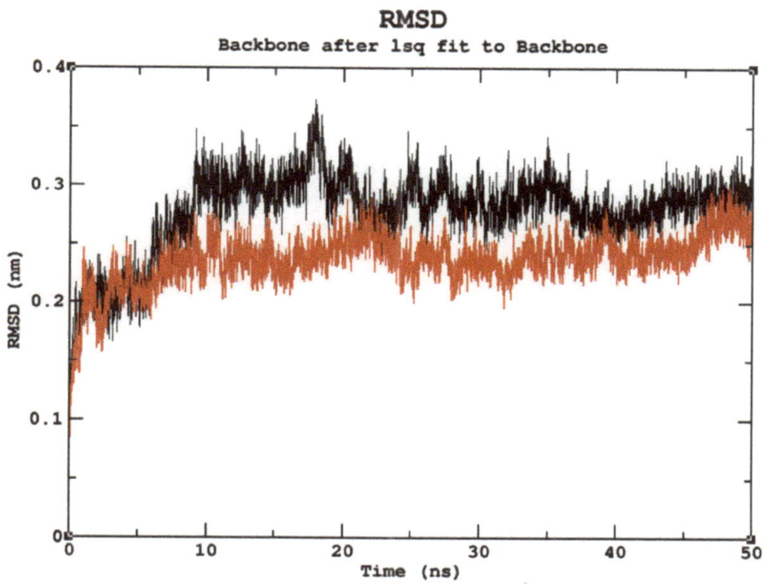

Fig. 4 The root mean square deviation plot for the native (black) and mutant D92Y (red) proteins bound with DGJ.

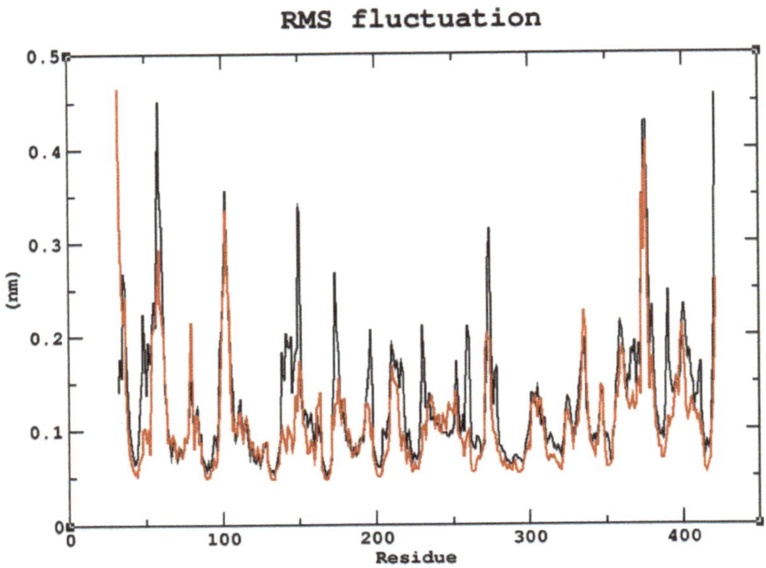

Fig. 5 The root mean square fluctuation plot for the native (black) and mutant D92Y (red) proteins bound with DGJ.

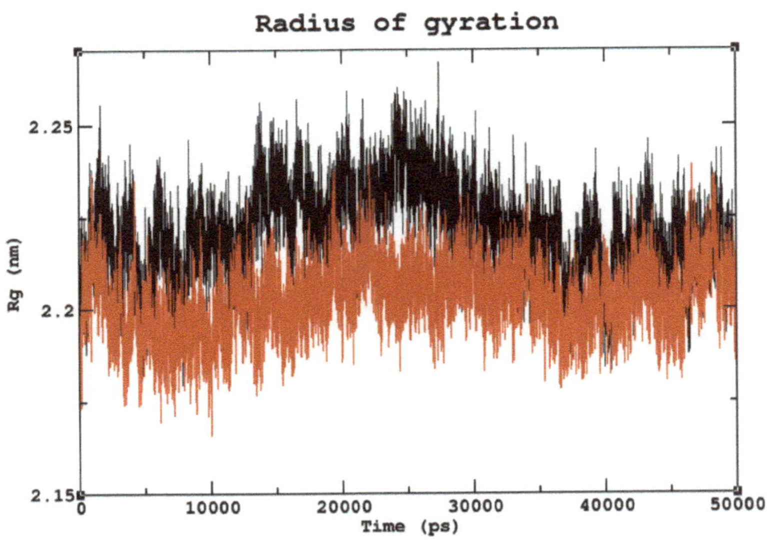

Fig. 6 Analysis of the radius of gyration for the native (black) and mutant D92Y (red) proteins bound with DGJ.

4. Discussion

In this work, to the best of our knowledge, we present the first study that uses sophisticated computational tools to investigate the pathogenicity of causative mutations for FD and their impact in the efficiency of binding to DGJ. We collected 698 missense mutations in GLA protein from NCBI, UniProt, and HGMD databases. PredictSNP server was used to classify the mutations as disease and neutral whereas iStable server was used to identify the destabilizing mutations (Bendl et al., 2014; Chen et al., 2013). PredictSNP, MAPP, PhD-SNP, PolyPhen1, PolyPhen2, SIFT, SNAP, nsSNPanalyzer, and PANTHER servers predicted 475, 395, 541, 437, 458, 513, 434, 214, and 521 mutations to be deleterious respectively (Table 1). Further, I-Mutant, MUpro, and iStable predicted 604, 557, and 555 mutations to be destabilizing respectively (Table 2). The use of chaperones is a potential treatment option for LSDs, and well-proven method for treatment of many genetic diseases that result in protein misfolding; however, their use is limited to certain diseases. Chaperones play an essential role in the transport of Gb3 to the lysosome for its breakdown. Therefore any alteration in the GLA due to mutations will lead to protein misfolding. The active sites of the GLA protein play an essential role in DGJ binding and stabilizing the protein (Ishii, 2012). A huge number of mutations are deposited in the public databases every day, and therefore there is a current demand to interpret and classify them as neutral or deleterious. In recent years, in silico prediction tools have been used successfully to classify hundreds of mutations in different disease models, which proved to be a good alternative to laboratory experiments, saving time, efforts, and resources (Agrahari, Muskan, George Priya Doss, Siva, & Zayed, 2018; Agrahari, Sneha, George Priya Doss, Siva, & Zayed, 2018; George Priya Doss, Chakraborty, Narayan, & Thirumal Kumar, 2014; Sneha, Ebrahimi, et al., 2018; Sneha, Thirumal Kumar, George Priya Doss, Siva, & Zayed, 2017; Sneha, Thirumal Kumar, Tanwar, et al., 2017; Sneha, Zenith, et al., 2018; Thirumal Kumar, Eldous, Mahgoub, George Priya Doss, & Zayed, 2018; Thirumal Kumar et al., 2016; Thirumal Kumar, Jerushah Emerald, et al., 2018). The use of more than one algorithm to predict the effect of missense mutations increases the confidence of the prediction (Adzhubei, Jordan, & Sunyaev, 2013; Agrahari, Sneha, et al., 2018; Doss & Sethumadhavan, 2009; George Priya Doss & Zayed, 2017; Rajasekaran, Sudandiradoss, Doss, & Sethumadhavan, 2007; Sneha, Ebrahimi, et al., 2018; Sneha, Thirumal Kumar, George Priya Doss, et al., 2017; Sneha, Thirumal Kumar, Tanwar, et al., 2017; Zaki et al., 2017).

The MetaPocket active site prediction identified TRP47, ASP92, TYR134, CYS142, LYS168, ASP170, CYS172, LEU206, TYR207, ARG227, ASP231, ASP266, and MET267 as the active sites in the GLA protein and matches with the previous literature (Germain, 2010; Huang, 2009). Out of 698 missense mutations, 54 were found to be positioned in the active site region of the protein. Out of which four missense mutations (D92Y, C142Y, D170V, and D266N) were predicted to be deleterious and destabilizing by all the *in silico* prediction tools. Lastly, these molecules were subjected to phenotyping analysis, and except D92Y all the mutations showed no significant output. The mutation D92Y alone was predicted to increase the amyloid propensity, as well as was found to be involved in severely reducing the protein stability (Table 3). All the mutations were subjected to PDS to understand the change in behavior of the protein due to the mutation. Protein dynamics was performed using the GROMACS simulation package (Abraham et al., 2015). RMSD measures the convergence of the proteins during the simulation period (Thirumal Kumar et al., 2017; Unissa et al., 2017). The simulations were initially performed for 30 ns, and the molecules were found to be converged at the end of 30 ns. From the RMSD plot, we observed the highest deviation in the protein molecule with D92Y mutation whereas the highest deviation was observed in the native protein. However, other protein molecules with D170V, D266N, and C142Y mutations showed deviations in between the native and D92Y mutant (Fig. 1). This elucidates that the protein molecule with D92Y mutation exhibited higher structural changes when compared to the native protein and other mutant proteins. The radius of gyration is used to measure the change in the compactness of the protein due to the mutations. The decrease in radius of gyration favors higher compactness of the protein (Lobanov, Bogatyreva, & Galzitskaia, 2008; Thirumal Kumar et al., 2017). From the plot, it was observed that the protein with D92Y mutation exhibited the highest radius of gyration which elucidates that it possessed the minimal compactness. On the other hand, the native protein molecule exhibited the lowest radius of gyration favoring the highest compactness. Similar to the RMSD, proteins with other mutations (D170V, D266N, and C142Y) possessed compactness in between the native protein and mutant D92Y protein (Fig. 2). Hydrogen bonds play a significant role in maintaining the stability and compactness of the protein (Gerlt, Kreevoy, Cleland, & Frey, 1997; Sudhakar et al., 2015). To understand this phenomenon, the change in a number of intramolecular hydrogen bonds was analyzed. From the plot, it was observed that the native protein molecule had retrained a maximum number of hydrogen bonds whereas the protein molecule with D92Y mutation retained the least number of hydrogen bonds. The proteins with D170V and D266N showed a number of intramolecular molecular similar to that of the native protein molecule except with drastic fall in a number of hydrogen bonds in protein with the D170V mutation (Fig. 3). DGJ also known as Migalastat is an orally active pharmacological chaperone. A synthetic derivative of Nojirimycin, DGJ, is known to possess potent galactosidase inhibition activity (Karjalainen, Passiniemi, & Koskinen, 2010). Pharmacological chaperone therapy has been studied to increase enzyme activity in mutations with residual enzyme activity (Fan & Ishii, 2007; Ishii, 2012). Due to decreased specificity or functionality, the proteins tend to misfold and accumulate in the form of aggregates. These aggregates can be either in amorphous or fibrillar (amyloid). Amyloid deposit formations have been associated with many neurodegenerative disorders like Parkinson's, Huntington's and many more (Castello et al., 2017). DGJ applies to only misfolded proteins and is not suitable for mutations that give rise to non-functional proteins. Various studies have shown that enzyme substrates and their analogs have been associated with enzyme stability. Studies have revealed that DGJ activity has known to reduce the substrate activity in case of misfolded proteins. Since DGJ has a similarity to alpha-galactose, which binds to the active site of alpha-Gal A, various models have been studied to unravel the binding site of DGJ. Studies revealed that it binds reversibly to different active site residues in a favorable manner leading to alteration in the folding process of the misfolded enzyme (Fan & Ishii, 2007). We have used a molecular docking approach using AutoDock to understand the interaction patterns of the DGJ with the native protein and protein with mutations. Higher the binding energy between the ligand and the protein elucidates that the binding is weak and has a weaker affinity and *vice versa* (Kumar & George Priya Doss, 2016; Morris et al., 2009). In this study we found that the lowest binding energy (−7.73 kcal/mol) was observed between the native protein and DGJ whereas highest binding energy (−5.16 kcal/mol) was observed between the protein molecule with D92Y mutation and DGJ molecule. Proteins with other mutations exhibited binding energy between these two protein complexes. This interaction study supports that the mutation D92Y could be more deleterious than the other mutations and can hinder the drug activity. Finally, the behavior of DGJ with the protein with the D92Y mutation was studied using protein dynamics simulations of 50 ns. In RMSD analysis we observed that mutant-DGJ complex exhibited a lower deviation than the native-DGJ complex. However, both, native-DGJ complex and mutant-DGJ complex exhibited convergence at the end of 50 ns (Fig. 4). These well-converged trajectories provided confidence for further analysis like RMSF and radius of gyration (Rg). Similar to the RMSD, RMSF has also shown minimal fluctuation in the mutant-DGJ complex when compared to the native-DGJ complex (Fig. 5). Finally, the radius of gyration was performed to understand the change in the compactness of the proteins due to the interaction of the DGJ. The radius of gyration was found to be higher in the native-DGJ complex when compared to the mutant-DGJ complex elucidating that the mutant-DGJ complex has attained protein compactness with the DGJ interaction (Fig. 6).

5. Conclusion

In conclusion, we used several computational tools to analyze the impact of GLA mutations. Out of the 54 missense mutations located in the active-site region of the GLA protein, 4 mutations (C142Y, D92Y, D266N, and D170V) were predicted to be highly deleterious and destabilizing by all the algorithms that are incorporated in the PredictSNP and iStable servers. From the phenotypic analysis, we observed that the protein with mutation D92Y could increase the amyloid propensity as well as severely reduce the protein stability. This mutation D92Y showed more structural alterations with the highest deviation, lowest compactness with least number of intramolecular hydrogen bonds using protein dynamic simulations. Molecular docking confirmed that this mutant D92Y might interfere with the binding with the DGJ chaperone. Finally, protein dynamics simulations were performed to understand the behavior of the protein with DGJ. From the analysis, DGJ has profoundly helped in maintaining the structural stability which is observed using RMSD, RMSF, and Rg. Our study confirms the benefits of the use of chaperones for correcting the misfolding of protein mutations and opens the way for designing chaperones for treatment of diseases that result from misfolding of protein due to missense mutations.

Acknowledgments

The authors acknowledge the management of Vellore Institute of Technology, Vellore, India, and BRAF @ CDAC for providing the facilities required to perform this work.

Disclosure statement

No potential conflict of interest exists.

References

Abraham, M. J., Murtola, T., Schulz, R., Páll, S., Smith, J. C., Hess, B., et al. (2015). GROMACS: High performance molecular simulations through multi-level parallelism from laptops to supercomputers. *SoftwareX, 1–2*, 19–25.

Adzhubei, I., Jordan, D. M., & Sunyaev, S. R. (2013). Predicting functional effect of human missense mutations using PolyPhen-2. *Current Protocols in Human Genetics*, *76*(1), 7.20.1–7.20.41.

Agrahari, A. K., Muskan, M., George Priya Doss, C., Siva, R., & Zayed, H. (2018). Computational insights of K1444N substitution in GAP-related domain of NF1 gene associated with neurofibromatosis type 1 disease: A molecular modeling and dynamics approach. *Metabolic Brain Disease*, *33*(5), 1443–1457.

Agrahari, A. K., Sneha, P., George Priya Doss, C., Siva, R., & Zayed, H. (2018). A profound computational study to prioritize the disease-causing mutations in PRPS1 gene. *Metabolic Brain Disease*, *33*(2), 589–600.

Bendl, J., Stourac, J., Salanda, O., Pavelka, A., Wieben, E. D., Zendulka, J., et al. (2014). PredictSNP: Robust and accurate consensus classifier for prediction of disease-related mutations. *PLoS Computational Biology*, *10*(1), e1003440.

Benjamin, E. R., Flanagan, J. J., Schilling, A., Chang, H. H., Agarwal, L., Katz, E., et al. (2009). The pharmacological chaperone 1-deoxygalactonojirimycin increases α-galactosidase A levels in Fabry patient cell lines. *Journal of Inherited Metabolic Disease*, *32*(3), 424–440.

Bishop, D. F., Calhoun, D. H., Bernstein, H. S., Hantzopoulos, P., Quinn, M., & Desnick, R. J. (1986). Human alpha-galactosidase A: Nucleotide sequence of a cDNA clone encoding the mature enzyme. *Proceedings of the National Academy of Sciences of the United States of America*, *83*(13), 4859–4863.

Bishop, D. F., Kornreich, R., & Desnick, R. J. (1988). Structural organization of the human alpha-galactosidase a gene: Further evidence for the absence of a 3′ untranslated region. *Proceedings of the National Academy of Sciences of the United States of America*, *85*(11), 3903–3907.

Castello, F., Paredes, J. M., Ruedas-Rama, M. J., Martin, M., Roldan, M., Casares, S., et al. (2017). Two-step amyloid aggregation: Sequential lag phase intermediates. *Scientific Reports*, 7(1), 40065.

Chen, C.-W., Lin, J., & Chu, Y.-W. (2013). iStable: Off-the-shelf predictor integration for predicting protein stability changes. *BMC Bioinformatics*, *14*(Suppl. 2), S5.

Citro, V., Cammisa, M., Liguori, L., Cimmaruta, C., Lukas, J., Cubellis, M., et al. (2016). The large phenotypic spectrum of Fabry disease requires graduated diagnosis and personalized therapy: A meta-analysis can help to differentiate missense mutations. *International Journal of Molecular Sciences*, *17*(12), 2010.

De Baets, G., Van Durme, J., Reumers, J., Maurer-Stroh, S., Vanhee, P., Dopazo, J., et al. (2012). SNPeffect 4.0: On-line prediction of molecular and structural effects of protein-coding variants. *Nucleic Acids Research*, *40*(Database issue), D935–D939.

Desnick, R. J., & Brady, R. O. (2004). Fabry disease in childhood. *The Journal of Pediatrics*, *144*(5), S20–S26.

Desnick, R. J., Brady, R., Barranger, J., Collins, A. J., Germain, D. P., Goldman, M., et al. (2003). Fabry disease, an under-recognized multisystemic disorder: Expert recommendations for diagnosis, management, and enzyme replacement therapy. *Annals of Internal Medicine*, *138*(4), 338.

Doss, C. G., & Sethumadhavan, R. (2009). Investigation on the role of nsSNPs in HNPCC genes—A bioinformatics approach. *Journal of Biomedical Science*, *16*, 42.

Fan, J.-Q., & Ishii, S. (2007). Active-site-specific chaperone therapy for Fabry disease. *FEBS Journal*, *274*(19), 4962–4971.

Garman, S. C., & Garboczi, D. N. (2004). The molecular defect leading to Fabry disease: Structure of human α-galactosidase. *Journal of Molecular Biology*, *337*(2), 319–335.

George Priya Doss, C., Chakraborty, C., Narayan, V., & Thirumal Kumar, D. (2014). Computational approaches and resources in single amino acid substitutions analysis toward clinical research. *Advances in Protein Chemistry and Structural Biology*, *94*, 365–423.

George Priya Doss, C., & Zayed, H. (2017). Comparative computational assessment of the pathogenicity of mutations in the aspartoacylase enzyme. *Metabolic Brain Disease*, *32*(6), 2105–2118. https://doi.org/10.1007/s11011-017-0090-5. Epub 2017 Sep 6.

Gerlt, J. A., Kreevoy, M. M., Cleland, W. W., & Frey, P. A. (1997). Understanding enzymic catalysis: The importance of short, strong hydrogen bonds. *Chemistry & Biology*, *4*(4), 259–267.

Germain, D. P. (2010). Fabry disease. *Orphanet Journal of Rare Diseases*, *5*(1), 30.

Guex, N., & Peitsch, M. C. (1997). SWISS-MODEL and the Swiss-Pdb viewer: An environment for comparative protein modeling. *Electrophoresis*, *18*(15), 2714–2723.

Huang, B. (2009). MetaPocket: A meta approach to improve protein ligand binding site prediction. *OMICS: A Journal of Integrative Biology*, *13*(4), 325–330.

Inoue, T., Hattori, K., Ihara, K., Ishii, A., Nakamura, K., & Hirose, S. (2013). Newborn screening for Fabry disease in Japan: Prevalence and genotypes of Fabry disease in a pilot study. *Journal of Human Genetics*, *58*(8), 548–552.

Ishii, S. (2012). Pharmacological chaperone therapy for Fabry disease. *Proceedings of the Japan Academy, Series B*, *88*(1), 18–30.

Karjalainen, O. K., Passiniemi, M., & Koskinen, A. M. P. (2010). Short and straightforward synthesis of (−)-1-deoxygalactonojirimycin. *Organic Letters*, *12*(6), 1145–1147.

Kornreich, R., Bishop, D. F., & Desnick, R. J. (1989). The gene encoding alpha-galactosidase A and gene rearrangements causing Fabry disease. *Transactions of the Association of American Physicians*, *102*, 30–43.

Kumar, D. T., & George Priya Doss, C. (2016). Investigating the inhibitory effect of Wortmannin in the hotspot mutation at codon 1047 of PIK3CA kinase domain: A molecular docking and molecular dynamics approach. *Advances in Protein Chemistry and Structural Biology*, *102*, 267–297.

Liguori, R., Incensi, A., de Pasqua, S., Mignani, R., Fileccia, E., Santostefano, M., et al. (2017). Skin globotriaosylceramide 3 deposits are specific to Fabry disease with classical mutations and associated with small fibre neuropathy. *PLoS One*, *12*(7), e0180581.

Lobanov, M. I., Bogatyreva, N. S., & Galzitskaia, O. V. (2008). Radius of gyration is indicator of compactness of protein structure. *Molekuliarnaia biologiia*, *42*(4), 701–706.

Lukas, J., Scalia, S., Eichler, S., Pockrandt, A.-M., Dehn, N., Cozma, C., et al. (2015). Functional and clinical consequences of novel α-galactosidase a mutations in Fabry disease. *Human Mutation*, *37*(1), 43–51.

Malde, A. K., Zuo, L., Breeze, M., Stroet, M., Poger, D., Nair, P. C., et al. (2011). An automated force field topology builder (ATB) and repository: Version 1.0. *Journal of Chemical Theory and Computation*, 7(12), 4026–4037.

Morris, G. M., Huey, R., Lindstrom, W., Sanner, M. F., Belew, R. K., Goodsell, D. S., et al. (2009). AutoDock4 and AutoDockTools4: Automated docking with selective receptor flexibility. *Journal of Computational Chemistry*, *30*(16), 2785–2791.

Oostenbrink, C., Villa, A., Mark, A. E., & Van Gunsteren, W. F. (2004). A biomolecular force field based on the free enthalpy of hydration and solvation: The GROMOS force-field parameter sets 53A5 and 53A6. *Journal of Computational Chemistry*, *25*(13), 1656–1676.

Petrova, S. S., & Solov'ev, A. D. (1997). The origin of the method of steepest descent. *Historia Mathematica*, *24*(4), 361–375.

Rajasekaran, R., Sudandiradoss, C., Doss, C. G., & Sethumadhavan, R. (2007). Identification and in silico analysis of functional SNPs of the BRCA1 gene. *Genomics*, *90*(4), 447–452.

Shabbeer, J., Yasuda, M., Benson, S. D., & Desnick, R. J. (2006). Fabry disease: Identification of 50 novel α-galactosidase a mutations causing the classic phenotype and three-dimensional structural analysis of 29 missense mutations. *Human Genomics*, *2*(5), 297–309.

Sirrs, S., & Alfadhel, M. (2011). Enzyme replacement therapy for Fabry disease: Some answers but more questions. *Therapeutics and Clinical Risk Management*, 7, 69.

Sneha, P., Ebrahimi, E. A., Ghazala, S. A., Thirumal Kumaran, D., Siva, R., Priya Doss, C. G., et al. (2018). Structural analysis of missense mutations in galactokinase 1 (GALK1) leading to galactosemia type-2. *Journal of Cellular Biochemistry, 119*(9), 7585–7598. https://doi.org/10.1002/jcb.27097.

Sneha, P., Thirumal Kumar, D., George Priya Doss, C., Siva, R., & Zayed, H. (2017). Determining the role of missense mutations in the POU domain of HNF1A that reduce the DNA-binding affinity: A computational approach. *PLoS One, 12*(4), e0174953.

Sneha, P., Thirumal Kumar, D., Tanwar, H., Siva, R., George Priya Doss, C., & Zayed, H. (2017). Structural analysis of G1691S variant in the human filamin B gene responsible for Larsen syndrome: A comparative computational approach. *Journal of Cellular Biochemistry, 118*(7), 1900–1910.

Sneha, P., Zenith, T., Abu Habib, U., Evangeline, J., Thirumal Kumar, D., George Priya Doss, C., et al. (2018). Impact of missense mutations in survival motor neuron protein (SMN1) leading to spinal muscular atrophy (SMA): A computational approach. *Metabolic Brain Disease, 33*, 1823–1834. https://doi.org/10.1007/s11011-018-0285-4. Epub ahead of print.

Sudhakar, N., George Priya Doss, C., Thirumal Kumar, D., Chakraborty, C., Anand, K., & Suresh, M. (2015). Deciphering the impact of somatic mutations in exon 20 and exon 9 of PIK3CA gene in breast tumors among Indian women through molecular dynamics approach. *Journal of Biomolecular Structure and Dynamics, 34*(1), 29–41.

Thirumal Kumar, D., Eldous, H. G., Mahgoub, Z. A., George Priya Doss, C., & Zayed, H. (2018). Computational modelling approaches as a potential platform to understand the molecular genetics association between Parkinson's and Gaucher diseases. *Metabolic Brain Disease, 33*, 1835–1847. https://doi.org/10.1007/s11011-018-0286-3. Epub ahead of print.

Thirumal Kumar, D., George Priya Doss, C., Sneha, P., Tayubi, I. A., Siva, R., Chakraborty, C., et al. (2016). Influence of V54M mutation in giant muscle protein titin: A computational screening and molecular dynamics approach. *Journal of Biomolecular Structure & Dynamics, 35*(5), 917–928.

Thirumal Kumar, D., Jerushah Emerald, L., George Priya Doss, C., Sneha, P., Siva, R., Charles Emmanuel Jebaraj, W., et al. (2018). Computational approach to unravel the impact of missense mutations of proteins (D2HGDH and IDH2) causing D-2-hydroxyglutaric aciduria 2. *Metabolic Brain Disease, 33*, 1699–1710. https://doi.org/10.1007/s11011-018-0278-3. Epub ahead of print.

Thirumal Kumar, D., Lavanya, P., George Priya Doss, C., Tayubi, I. A., Naveen Kumar, D. R., Yesurajan, F., et al. (2017). A molecular docking and dynamics approach to screen potent inhibitors against fosfomycin resistant enzyme in clinical *Klebsiella pneumoniae*. *Journal of Cellular Biochemistry, 118*(11), 4088–4094.

Thurberg, B. L., Rennke, H., Colvin, R. B., Dikman, S., Gordon, R. E., Collins, A. B., et al. (2002). Globotriaosylceramide accumulation in the Fabry kidney is cleared from multiple cell types after enzyme replacement therapy. *Kidney International, 62*(6), 1933–1946.

Unissa, A. N., Doss, C., P, G., Kumar, T., Swathi, S., Lakshmi, A. R., et al. (2017). Analysis of interactions of clinical mutants of catalase-peroxidase (KatG) responsible for isoniazid resistance in Mycobacterium tuberculosis with derivatives of isoniazid. *Journal of Global Antimicrobial Resistance, 11*, 57–67.

Wang, R. Y., Lelis, A., Mirocha, J., & Wilcox, W. R. (2007). Heterozygous Fabry women are not just carriers, but have a significant burden of disease and impaired quality of life. *Genetics in Medicine, 9*, 34–45.

Zaki, O. K., Priya Doss, C. G., Ali, S. A., Murad, G. G., Elashi, S. A., Ebnou, M. S. A., et al. (2017). Genotype-phenotype correlation in patients with isovaleric acidaemia: Comparative structural modelling and computational analysis of novel variants. *Human Molecular Genetics, 26*(16), 3105–3115. https://doi.org/10.1093/hmg/ddx195.

Zarate, Y. A., & Hopkin, R. J. (2008). Fabry's disease. *The Lancet, 372*(9647), 1427–1435.

FSC
www.fsc.org
MIX
Papier | Fördert gute Waldnutzung
FSC® C083411

Kolibri 360 GmbH
Ferdinand-Jühlke-Straße 7
99095 Erfurt, Deutschland
produktsicherheit@kolibri360.de